S0-AND-654

LANDOLT-BÖRNSTEIN

Numerical Data and Functional Relationships in Science and Technology

New Series

Editors in Chief: K.-H. Hellwege · O. Madelung

Group V: Geophysics and Space Research

Volume 3

Oceanography

H. G. Gierloff-Emden · G. Koslowski
L. Magaard · E. Mittelstaedt · L. A. Mysak
D. Olbers · W. Rosenthal · W. Zahel

Editor: J. Sündermann

Subvolume c

Springer-Verlag Berlin Heidelberg New York
London Paris Tokyo

LANDOLT-BÖRNSTEIN

Zahlenwerte und Funktionen aus Naturwissenschaften und Technik

Neue Serie

Gesamtherausgabe: K.-H. Hellwege · O. Madelung

Gruppe V: Geophysik und Weltraumforschung

Band 3

Ozeanographie

H. G. Gierloff-Emden · G. Koslowski
L. Magaard · E. Mittelstaedt · L. A. Mysak
D. Olbers · W. Rosenthal · W. Zahel

Herausgeber: J. Sündermann

Teilband c

Springer-Verlag Berlin Heidelberg New York
London Paris Tokyo

ISBN 3-540-15955-X Springer-Verlag Berlin Heidelberg New York
ISBN 0-387-15955-X Springer-Verlag New York Heidelberg Berlin

CIP-Kurztitelaufnahme der Deutschen Bibliothek
Zahlenwerte und Funktionen aus Naturwissenschaften und Technik/Landolt-Börnstein. – Berlin; Heidelberg; London; Paris; New York; Tokyo: Springer Teilw. mit d. Erscheinungsorten Berlin, Heidelberg, New York. – Parallelt.: Numerical data and functional relationships in science and technology
NE: Landolt, Hans [Begr.]; PT Landolt-Börnstein, ... N.S. Gesamthrsg.: K.-H. Hellwege; O. Madelung. Gruppe 5, Geophysik und Weltraumforschung. Bd. 3. Ozeanographie. Teilbd. c. H. G. Gierloff-Emden ... Hrsg.: J. Sündermann. – 1986.
ISBN 3-540-15955-X (Berlin ...)
ISBN 0-387-15955-X (New York ...)
NE: Hellwege, Karl-Heinz [Hrsg.]; Gierloff-Emden, Hans-Günter [Mitverf.]; Sündermann, Jürgen [Hrsg.]

Printed in Germany

Typesetting: Brühlsche Universitätsdruckerei, Giessen
Printing: Druckhaus Langenscheidt KG, Berlin
Bookbinding: Lüderitz & Bauer-GmbH, Berlin

2163/3020–543210

Editor

J. Sündermann
Institut für Meereskunde der Universität Hamburg,
Heimhuder Str. 71, D-2000 Hamburg 13

Contributors

H. G. Gierloff-Emden
Institut für Geographie der Universität München,
Luisenstr. 37, D-8000 München 2

G. Koslowski
Deutsches Hydrographisches Institut,
Bernhard-Nocht-Str. 78, D-2000 Hamburg 4

L. Magaard
Department of Oceanography, University of Hawaii at Manoa,
1000 Pope Road, Honolulu, Hawaii 96822, USA

E. Mittelstaedt
Deutsches Hydrographisches Institut,
Bernhard-Nocht-Str. 78, D-2000 Hamburg 4

L. A. Mysak
Department of Oceanography, The University of British Columbia,
Vancouver BC V6T 1W5, Canada

D. Olbers
Alfred-Wegener-Institut für Polar- und Meeresforschung, Columbus-Straße,
D-2850 Bremerhaven

W. Rosenthal
Institut für Physik, GKSS-Forschungszentrum Geesthacht GmbH,
Max-Planck-Straße, D-2054 Geesthacht

W. Zahel
Institut für Meereskunde der Universität Hamburg,
Heimhuder Str. 71, D-2000 Hamburg 13

Vorwort

In der 6. Auflage des Landolt-Börnstein war das Sachgebiet Ozeanographie in den Band Astronomie und Geophysik (1952) integriert und umfaßte 116 Seiten. Hiermit wird jetzt im Rahmen der Neuen Serie des Landolt-Börnstein ein eigener Band (V/3) mit dem Titel "Oceanography" vorgelegt, der einen Umfang von ca. 1100 Seiten aufweist. Dieser Band wird in drei Teilbänden (V/3a, b, c) erscheinen. Schon daran lassen sich die stürmische Zunahme unserer Kenntnisse und die wachsende Bedeutung der Ozeanographie in einem Zeitraum von 34 Jahren ermessen. Ich bin dem Springer-Verlag dankbar, daß er diese Entwicklung erkannt und der Ozeanographie entsprechenden Raum in der neuen Serie bereitgestellt hat.

Es kann dennoch nicht behauptet werden, daß die Ozeanographie bislang den Reifegrad und die Abgeschlossenheit der klassischen Physik erreicht hätte. Viele, auch grundlegende, Fragen befinden sich noch in der Bearbeitung oder sind gar noch unerforscht. Entsprechend schwer ist es gewesen, in der Darstellung der Ozeanographie den traditionellen Stil des Landolt-Börnstein, der ja Figuren und Tabellen liefern soll, zu treffen. Ich habe mich sehr bemüht, im Zusammenwirken mit den Autoren hier eine gute Mischung aus grundlegender Information und quantitativen Angaben zu finden. Es hat sich indessen nicht vermeiden lassen, daß auch mehr erläuternder Text und mehr Formeln in den Band eingegangen sind als in diejenigen Bände, die den besser erforschten Teilen der Physik gewidmet sind. Ich danke hier vor allem dem Gesamtherausgeber für sein Verständnis. Ich glaube, daß die schließlich gefundene Form einen guten Kompromiß darstellt und für die ozeanographische Fachwelt interessant und nützlich ist.

An dieser Stelle möchte ich den Begriff Ozeanographie für dieses Werk präzisieren, ohne einen Anspruch auf Allgemeingültigkeit zu stellen. Gemeint ist *physikalische* Ozeanographie, ein Teilgebiet der Geophysik und damit der Physik. Die Kapitel "Topography" und "Coastal Oceanography" stellen unter diesem Gesichtspunkt Randbereiche dar, da sie mehr geographisch orientiert sind. Andererseits sind sie für die Physik des Ozeans von großer Bedeutung.

Damit bin ich bei der generellen Gliederung des Bandes. Obwohl diese Gliederung zweifellos eine größere Systematik und Vollständigkeit aufweist als die Gliederung in der 6. Auflage des Landolt-Börnstein, ist sie sicher immer noch etwas subjektiv und lückenhaft. Hier mußte ich mich nach dem aktuellen Bedarf der Ozeanographie richten, aber auch nach den Spezialgebieten und den Interessen der beitragenden Fachkollegen. Es ist daher nicht auszuschließen, daß – vom Standpunkt einer lehrbuchartigen Gesamtdarstellung – manches Kapitel zu speziell erscheint, während vielleicht ein anderes zu allgemein ist. Es kann auch sein, daß ein wünschenswertes Kapitel ganz fehlt. Ich habe mich aber nach Kräften bemüht, in Zusammenarbeit mit den Autoren das Gesamtgebiet der physikalischen Ozeanographie einigermaßen zu überdecken, Lücken zu schließen und Beiträge aufeinander abzustimmen. Mehr erscheint bei einem solchen Gemeinschaftswerk, das ja die speziellen Kenntnisse eines jeden Autors nutzen soll, nicht geboten.

Ich bin den Autoren, die sich alle sehr schnell und engagiert bereiterklärt haben mitzuarbeiten, für ihren Enthusiasmus, ihre Kompromißbereitschaft und ihre Geduld zu größtem Dank verpflichtet. Es gehört einiger Idealismus dazu, in einer Zeit, in der kurze Originalveröffentlichungen mehr gefragt sind als längere systematische Beiträge, dieses Arbeitspensum aufzubringen. Es bedarf auch eines gewissen Mutes, auf einem sich schnell entwickelnden Fachgebiet Wissen zu einem bestimmten Zeitpunkt festzuschreiben. Dies kann nur unter der Prämisse geschehen, daß es sich eben um den gegenwärtigen Erkenntnisstand handelt, der künftig gewiß Änderungen unterworfen sein wird. Es gehört schließlich einige Einsicht dazu, im Interesse einer Homogenisierung des Gesamtwerkes und im Interesse seiner Anpassung an den Stil des Landolt-Börnstein, die Einwände und Korrekturwünsche von Herausgeber, Gesamtherausgeber und Redaktionsstab (Dr. W. Finger, Frau G. Burfeindt und Frau H. Hämmer) zu akzeptieren. Alles dies ist in weitgehender Harmonie geschehen.

Leider hat es sich bei den unterschiedlichen Belastungen der Autoren, bei ihren vielseitigen anderweitigen Verpflichtungen, nicht vermeiden lassen, daß Verzögerungen bei einzelnen Beiträgen aufgetreten sind, die den Abschluß des Gesamtwerkes doch erheblich beeinträchtigt haben. Um nicht weitere Zeit zu verlieren, ist deswegen mit dem Druck des am weitesten fortgeschrittenen Teilbandes begonnen worden; dies ist Teilband V/3c, der die Kapitel 6 bis 9 umfaßt. Die Teilbände V/3a (Kapitel 1 bis 3) und V/3b (Kapitel 4 und 5) werden – in dieser Reihenfolge – möglichst bald folgen.

Dieser Band wird – wie alle Bände des Landolt-Börnstein – ohne finanzielle Unterstützung durch andere Stellen gedruckt.

Hamburg, Mai 1986 **Der Herausgeber**

Preface

In the 6th edition of Landolt-Börnstein, the field of oceanography was incorporated into the volume on Astronomy and Geophysics (1952) and covered 116 pages. In this new edition a whole volume (V/3) of the New Series of Landolt-Börnstein is devoted to oceanography with a total of, approximately, 1200 pages, to be published in three subvolumes (V/3a, b, c). This demonstrates the tremendous increase in knowledge and the growing significance of oceanography over the past 34 years. I am grateful to Springer-Verlag for recognizing this development and providing the appropriate space in the New Series for this subject.

In spite of such an impressive accumulation of new knowledge, it can hardly be claimed that oceanography has reached the maturity and conclusiveness of classical physics. Many questions, even fundamental issues, are still in early stages of development or have not yet been explored. Accordingly, it has been rather difficult to represent oceanography in terms of the traditional Landolt-Börnstein style by means of figures and tables.

Editor and authors have made a considerable effort to find the proper combination of basic information and data. However, it has not been possible to avoid including more explanatory text and more formulas than have been necessary in the volumes dealing with more developed areas of physical sciences. I am particularly grateful to the editor-in-chief for his understanding. I believe the final form represents a good compromise and will be interesting and helpful to the oceanographic community.

Let me clarify here that under our general topic we are dealing with *physical* oceanography, a part of geophysics and physics. From this point of view the chapters “Topography” and “Coastal Oceanography” represent peripherical areas, since they are more geographically oriented. On the other hand, they are very pertinent to the physics of the ocean.

This brings me to the general organization of this volume. Although it is undoubtedly more systematic and comprehensive than in the previous (6th edition) Landolt-Börnstein volume, it is certainly still somewhat subjective and incomplete. Here it was necessary to consider not only the present needs of oceanography but also the specific fields and interests of the contributors. It is therefore possible that, in comparison to a textbook, some chapters might appear to be too specialized or too general. It is also possible that a desirable chapter is entirely missing. I have tried my best, in cooperation with the authors, to cover more or less completely the entire field of physical oceanography, to close gaps and to balance the individual contributions. The remaining inhomogeneity is surely more than compensated by the advantages of letting the authors demonstrate their individual expertise.

I am much obliged to the authors who all promptly committed their cooperation, for their enthusiasm, patience and willingness to compromise. It requires a certain amount of idealism to take on the extra work involved in writing long and systematic articles in a time when short, original publications are the order of the day. It requires courage to draw the line at a particular point in time and write about factual knowledge in such a rapidly developing field; this must be done under the premise that this transient state of the art will be revised in the future. Subsequently, it requires mutual sense to accept the reservations and editorial modifications of the editor-in-chief, the editor and the editorial staff (Dr. W. Finger, Frau G. Burfeindt, and Frau H. Hämmer) in the interests of a homogeneous whole in the Landolt-Börnstein style. This all proved to be an agreeable harmonious process.

Unfortunately, the heavy work load of some of the authors prevented them from submitting their contributions punctually, which delayed completion of the subvolumes. To avoid further delay, the last subvolume has been printed first. This is the present subvolume V/3c containing Chapters 6 to 9. Subvolumes V/3a (Chapters 1 to 3) and V/3b (Chapters 4 and 5) will follow – in this order – as soon as possible.

Like all other volumes of Landolt-Börnstein, this volume is published without any outside financial support.

Hamburg, May 1986 **The Editor**

Contents

6 Ocean Waves

6.1 Classification and basic features

6.1.0 List of symbols

a	wave amplitude
c	phase speed, $c = L/T$
$\boldsymbol{c}$	phase velocity, $\boldsymbol{c} = (\omega/k^2)\boldsymbol{k}$
$\boldsymbol{c}_g$	group velocity, with components $\partial\omega/\partial k_i$, $i = 1, 2, 3$
D	deep-sea depth far from the coast
d	largest depth of sloping shelf
E	energy density
f	Coriolis parameter, $f = 2\,\Omega \sin\phi$
g	gravitational acceleration, $g = 9.81\ \mathrm{m\,s^{-2}}$
g'	reduced gravity
H	water depth
h_n	equivalent depth for internal waves, $n = 1, 2, \ldots$
$\boldsymbol{k}$	wavenumber vector
k_h	magnitude of the horizontal component of $\boldsymbol{k}$
k_i	$i = 1, 2, 3$: east, north and vertical (upward positive) component of $\boldsymbol{k}$, respectively
L	wavelength, in [m]
m	meridional mode number, $m = 0, 1, 2, \ldots$
n	(vertical) mode number
N	Brunt-Väisälä (angular) frequency
p	pressure
R_e	earth radius, $R_e = 6371\ \mathrm{km}$
R_0	external Rossby radius of deformation, $R_0 = (gH)^{1/2}/f$
R_n	internal Rossby radius of deformation for the nth internal mode; for linear stratification, $R_n = N_0 H/(n\pi f)$
S	phase function
T	wave period
t	time
$\boldsymbol{U}$	mean velocity of moving medium
u	horizontal current associated with interfacial waves
W	width of ocean bassin
$\boldsymbol{x}$	position vector
x	Cartesian coordinate directed eastward; normal distance from shore
y	Cartesian coordinate directed northward
z	vertical coordinate directed upwards
α	slope of beach profile
β	parameter characterizing the change of f with latitude, $\beta = \frac{\mathrm{d}f}{\mathrm{d}y}$
ε	ratio of nonlinear to linear terms in the equation of motion
η	sea level elevation
λ	generic symbol describing the basic state of the ocean
ϱ	fluid density
σ	dispersion relation function, $\omega = \sigma(\boldsymbol{k})$
σ_s	surface tension coefficient
τ	characteristic time of energy transfer between waves
ϕ	geographical latitude
Ω	earth's angular velocity, $\Omega = 7.2921 \cdot 10^{-5}\ \mathrm{s^{-1}}$
ω	angular frequency

Magaard, Mysak

6.1.1 Introduction

A propagating plane wave, as a member of a certain type of waves, is described, in the linear approximation, by its wavenumber vector $\boldsymbol{k}$, its angular frequency ω, and its amplitude $a(\boldsymbol{k}, \omega)$, where $\boldsymbol{k}$ and ω satisfy a dispersion relation, $\omega = \sigma(\boldsymbol{k})$, characteristic of that type of waves. Other important wave parameters are the wavelength $L = 2\pi/k$, where $k = |\boldsymbol{k}|$, the wave period $T = 2\pi/\omega$, the phase speed $c = L/T$, the phase velocity $\boldsymbol{c} = (\omega/k^2)\boldsymbol{k}$, and the group velocity $\boldsymbol{c}_g$, whose components are $\partial\omega/\partial k_i$, where k_1, k_2, k_3 are the eastward, northward, and upward components of $\boldsymbol{k}$, respectively. If, in the dispersion relation, σ is a function of the modulus of $\boldsymbol{k}$ only, the waves are isotropic. For such waves $\boldsymbol{k}$ and $\boldsymbol{c}_g$ are parallel and $|\boldsymbol{c}_g| = c_g = c - L\mathrm{d}c/\mathrm{d}L$. Isotropic waves can exhibit normal dispersion ($\mathrm{d}c/\mathrm{d}L > 0$, $c_g < c$), anomalous dispersion ($\mathrm{d}c/\mathrm{d}L < 0$, $c_g > c$), or no dispersion ($\mathrm{d}c/\mathrm{d}L = 0$, $c_g = c$). The dispersion relation for a wave type is obtained from the linearized hydrodynamic equations and boundary conditions that characterize that type of waves. A detailed discussion of the governing equations for various wave types is given in LeBlond and Mysak [78 L 1]. The dispersion relations of the various ocean waves are presented in the subsequent sections of this article.

The most elementary wave solutions of the linearized hydrodynamic equations are those in which each field variable is described by a constant times $\exp i(\boldsymbol{k} \cdot \boldsymbol{x} - \omega t)$, where $\boldsymbol{x}$ is the position vector and t is the time. These solutions describe propagating plane waves; the loci of constant phase, $\boldsymbol{k} \cdot \boldsymbol{x} - \omega t$, are planes perpendicular to the direction of phase propagation of the waves. Plane waves can only occur if the basic state of the ocean (defined, for example, by its depth and its mean stratification), of which the waves are small perturbations, does not vary in space and time. Therefore, for most oceanic conditions plane waves are a highly idealized description of the actual perturbations. A more realistic approach allows the amplitude, wavenumber, and frequency to vary slowly in space and time, i.e. on length scales large compared to L and on time scales large compared to T. Local values of the wavenumber vector and the frequency are then defined by $\boldsymbol{k}(\boldsymbol{x}, t) = \nabla S$ and $\omega(\boldsymbol{x}, t) = -\partial S/\partial t$, respectively, where $S = S(\boldsymbol{x}, t)$ is the phase function, i.e. field variables are proportional to $\exp iS(\boldsymbol{x}, t)$. Such waves are called nearly-plane waves; their behavior can be described by means of ray theory, first developed in optics, and often referred to as geometrical optics, eikonal method, or WKB approximation. In ray theory it is assumed that the dispersion relation, $\omega = \sigma(\boldsymbol{k})$, is still valid locally: $\omega(\boldsymbol{x}, t) = \sigma[\boldsymbol{k}(\boldsymbol{x}, t), \lambda(\boldsymbol{x}, t)]$ where $\lambda(\boldsymbol{x}, t)$ describes the variability of the basic state of the ocean, like, for instance, its varying depth or stratification. Also, the phase function S satisfies a first-order nonlinear partial differential equation (the eikonal equation). This equation can be obtained by replacing $\boldsymbol{k}$ and ω by ∇S and $-\partial S/\partial t$, respectively, in the locally valid dispersion relation. The corresponding system of ordinary differential equations for the wave characteristics or wave rays is

$$\frac{\mathrm{d}\boldsymbol{x}}{\mathrm{d}t} = \boldsymbol{c}_g, \tag{1}$$

where $\mathrm{d}/\mathrm{d}t = \partial/\partial t + \boldsymbol{c}_g \cdot \nabla$. The system of eqs. (1) describes the rays $\boldsymbol{x}(t)$, along which the wave groups propagate at group velocity $\boldsymbol{c}_g$. The behavior of the wavenumber vector and the frequency of the groups as they travel along are described by the equations

$$\frac{\mathrm{d}\boldsymbol{k}}{\mathrm{d}t} = -\frac{\partial\sigma}{\partial\lambda}\nabla\lambda, \qquad \frac{\mathrm{d}\omega}{\mathrm{d}t} = \frac{\partial\sigma}{\partial\lambda}\frac{\partial\lambda}{\partial t}. \tag{2}$$

The phenomenon that $\boldsymbol{k}$ changes direction as the waves propagate through an inhomogeneous medium is called refraction.

The most significant feature of waves is their ability to transport, at group velocity $\boldsymbol{c}_g$, energy and information over large distances (e.g., several thousand kilometers) without substantial transport of matter. The energy transport of free nearly-plane waves is described by the conservation of wave-action equation

$$\frac{\partial}{\partial t}\left(\frac{E}{\omega}\right) + \nabla \cdot \left(\frac{E}{\omega}\boldsymbol{c}_g\right) = 0, \tag{3}$$

where E is the energy density (in a volume moving with the local group velocity and containing a wave group) and E/ω is called the wave-action density. From classical mechanics terminology, the quantity E/ω is called an adiabatic invariant.

If the waves propagate in a moving medium with slowly varying mean velocity $\boldsymbol{U}(\boldsymbol{x}, t)$, one distinguishes between the frequency ω with respect to a stationary frame of reference (Doppler shifted frequency) and the frequency ω_0 with respect to a frame of reference moving with the mean flow (intrinsic frequency). The relation between the frequencies is

$$\omega = \omega_0 + \boldsymbol{k} \cdot \boldsymbol{U} = \sigma_0[\boldsymbol{k}(\boldsymbol{x}, t), \lambda(\boldsymbol{x}, t)] + \boldsymbol{k} \cdot \boldsymbol{U}.$$

The equation for the wave rays is then

$$\frac{\mathrm{d}\boldsymbol{x}}{\mathrm{d}t}=\boldsymbol{c}_{g0}+\boldsymbol{U}, \tag{4}$$

where $\mathrm{d}/\mathrm{d}t=\partial/\partial t+(\boldsymbol{c}_{g0}+\boldsymbol{U})\cdot\nabla$. For $\boldsymbol{k}$ and ω one has

$$\frac{\mathrm{d}\boldsymbol{k}}{\mathrm{d}t}=-\frac{\partial\sigma_0}{\partial\lambda}\nabla\lambda-(\nabla\boldsymbol{U})\cdot\boldsymbol{k}, \quad \frac{\mathrm{d}\omega}{\mathrm{d}t}=\frac{\partial\sigma_0}{\partial\lambda}\frac{\partial\lambda}{\partial t}+\left(\frac{\partial\boldsymbol{U}}{\partial t}\right)\cdot\boldsymbol{k}. \tag{5}$$

The equation of conservation for the wave action becomes

$$\frac{\partial}{\partial t}\left(\frac{E}{\omega_0}\right)+\nabla\cdot\left[\frac{E}{\omega_0}(\boldsymbol{c}_{g0}+\boldsymbol{U})\right]=0. \tag{6}$$

If the waves are not free but under the influence of a generating or damping mechanism, the corresponding source or sink functions appear on the right hand side of the equation for the wave-action density, eqs. (3) or (6).

In the above discussion it has been tacitly assumed that the propagation of waves of any type is not impeded by boundaries or topographic irregularities. However, lateral and horizontal boundaries as well as bathymetric features in the ocean serve to reflect and scatter wave energy whose incident direction is determined by the group velocity. The propagation of upward and downward wave rays in a laterally unbounded ocean of constant depth results in the formation of vertical normal modes and the quantization of the vertical wavenumber k_3 in the dispersion relation. In a closed ocean basin the horizontal wavenumber components k_1 and k_2 become quantized as well. Thus a discrete set of eigenfrequencies replaces the usual continuous dispersion relation. Examples of vertical normal modes and closed basin eigenfrequencies are given in LeBlond and Mysak [78 L 1]. The presence of isolated topographic features (e.g., islands, seamounts, ridges) and small-scale bottom or coastal irregularities will diffract and scatter incident wave energy into many directions. In addition, other types of waves can be generated by the topographic scattering or diffraction of a prescribed incident wave. For a discussion of various ocean wave scattering and diffraction problems, see LeBlond and Mysak [78 L 1] and Mysak [78 M].

It is convenient to define here the terms barotropic and baroclinic, which are commonly used to characterize waves in a stratified ocean. If for a given wave type the isopycnals (surfaces of constant density) and isobars (surfaces of constant pressure) are not parallel, the waves are said to be baroclinic. Mathematically, this defining criterion is expressed as $\nabla\varrho\times\nabla p\neq 0$, where ϱ and p are the fluid density and pressure, respectively. An example of baroclinic waves are internal gravity waves. Barotropic waves, on the other hand, have the property that $\nabla\varrho\times\nabla p=0$. By this definition all waves in homogeneous media ($\varrho=$const) are barotropic. If the influence of the stratification is negligible for a certain class of waves such waves are also called barotropic. Surface gravity waves fit into this category. The horizontal components of the velocity field associated with free long surface gravity waves (defined in subsect. 6.1.2) in a homogeneous inviscid ocean of constant or slowly varying depth are independent of the vertical coordinate. This feature is sometimes invoked to serve as a definition of barotropic motion. This definition is, of course, not fully equivalent to the general definition given above.

An important concept which has received considerable attention in the oceanographic literature is that of a wave guide. In fact, the presence of a variety of wave guides in the ocean has led to the discovery of several new classes of ocean waves (see subsect. 6.1.3). Wave guides occur in the ocean's interior (e.g., at a sharp pycnocline, along the equator), along coastal boundaries (e.g., the continental shelf), or in the neighborhood of isolated topographic features (e.g., a trench or a seamount). The main characteristics of waves in a wave guide are that their propagation is one-dimensional and their amplitudes decay (on the average) with distance away from the wave guide. Because of the latter property the waves' energy is trapped in the wave guide, and therefore such waves are often called "trapped waves". Clearly, the dispersion relation for trapped waves is a function of one wavenumber only, the other two components being effectively purely imaginary. Because trapped waves are not subject to spatial alteration (e.g., radial spreading), they can travel vast distances in the ocean and thus serve as transmitters of energy and information from one region to another.

The discussion up to this point has focussed on linear wave theory (i.e. waves of infinitesimal amplitude) for which linearized equations of motion apply. However, important new wave phenomena (finite-amplitude effects) arise when nonlinearities are introduced into the equations of motion. Examples of such phenomena are wave steepening, wave-wave interactions, solitary waves, and strong wave-mean flow interactions (e.g., the instability of shear flows). For a discussion of these topics, see LeBlond and Mysak [78 L 1].

In the theory of wave-wave interactions, energy is slowly exchanged between a set of waves with different frequencies and wavenumbers. A particularly efficient energy transfer takes place when the waves form a resonant triad in which the waves' frequencies and wavenumbers satisfy the resonance conditions

$$\omega_1+\omega_2=\omega_3, \tag{7}$$

$$\boldsymbol{k}_1+\boldsymbol{k}_2=\boldsymbol{k}_3, \tag{8}$$

where each ω_j, $\boldsymbol{k}_j$ pair also satisfies the dispersion relation of the wave type in question. The conditions in eqs. (7) and (8) imply that the three waves are phase-locked as they trade energy amongst themselves. If ε is a measure of the ratio (nonlinear terms/linear terms) in the equations of motion and T is a characteristic wave period, then, for $\varepsilon<1$, the transfer of energy between the waves in a resonant triad occurs over the long time scale $\tau_1=T\varepsilon^{-1}$. For an elementary treatment of the theory and a discussion of observations of resonant wave interactions in the ocean, see chapter 6 in LeBlond and Myask [78 L 1].

Solitary waves can exist when there is a balance between weak dispersion and weak nonlinearity. Their motion is often governed by the Korteweg-de Vries (KdV) equation (for reviews of this equation, see Miura [76 M] and Miles [81 M 2])

$$\frac{\partial\eta}{\partial t}+c_0\frac{\partial\eta}{\partial x}+A\eta\frac{\partial\eta}{\partial x}+B\frac{\partial^3\eta}{\partial x^3}=0, \tag{9}$$

where η is, for example, the sea level elevation and c_0, A, and B are constants that depend on the basic state of the fluid. The first two terms in eq. (9) of course describe a nondispersive wave travelling with speed c_0. The third and fourth terms counter this motion with wave steepening and wave dispersion, respectively. When all the terms in eq. (9) are of equal importance, solitary wave solutions can occur. An example of such a solution is given by the travelling "hump"

$$\eta=a\cosh^{-2}[(Aa/12B)^{1/2}(x-ct)], \tag{10}$$

where $c=c_0+aA/3$. Note that the speed of propagation c depends on both the wave amplitude a and the size of the nonlinearity A. Another important fact concerns the time scale for the development of solitary waves in a fluid. It tends, for $\varepsilon<1$, to be very long, of the order $\tau_2=T\varepsilon^{-2}$, as compared to the shorter time scale $\tau_1=T\varepsilon^{-1}$ which characterizes the time scale of energy transfer between waves in a resonant triad. Nevertheless, because of the highly coherent nature of solitary waves, they have been observed in the laboratory and in the ocean. In particular, both surface and internal gravity solitary waves have frequently been observed. The latter have received considerable attention in the recent literature because of their significant influence on oil drilling operations in continental shelf and slope waters.

Waves travelling in the presence of a slowly varying mean current generally experience advection and refraction. During such processes the slow exchange of energy between the waves and current is governed by the wave-action equation (6). However, if the current has large vertical or horizontal shears and is therefore not slowly varying, ray theory breaks down and eq. (6) does not apply. In such cases the waves may, under certain conditions, extract energy from the mean current fairly rapidly, on a time scale comparable to that of the wave period. When this occurs the mean flow is said to be unstable with respect to these amplifying wave perturbations. In the ocean there are three basic types of instability of interest: Kelvin-Helmholtz, baroclinic, and barotropic. The first two types of instability arise because of small-scale and large-scale vertical shears, respectively; the corresponding amplifying waves are internal gravity waves and baroclinic Rossby waves. Barotropic instability arises in the presence of strong lateral shears, and the corresponding amplifying waves are barotropic Rossby waves. The initial growth of the wave perturbations in any of these instabilities is exponential with time and can be found in detail by solving the appropriate linearized boundary value problem which contains the mean flow as a prescribed energy source. However, after one or two wave periods, the waves have become so large that the linear approximation breaks down. Then the fully nonlinear equations must be used to describe the energy exchanges between the mean current and perturbation field. The significance of shear flow instabilities is that they provide a mechanism for the onset of various scales of turbulent motions in the ocean. Examples of various mean flow instability problems are given in chapter 7 of LeBlond and Mysak [78 L 1].

To get an overview of the great variety of oceanic wave motions it is adequate to seek classifications of these motions. In what follows a few of the many possible classifications will be presented, and characteristic features of different types of waves will be described. For a more detailed discussion on ocean waves we refer to LeBlond and Mysak [78 L 1] and to Lighthill [78 L 2]. Useful references on the more mathematical aspects of waves are the books by Whitham [74 W] and Pedlosky [79 P].

6.1.2 Classification of waves with respect to restoring forces

The five basic restoring forces in the ocean are elasticity, surface tension, gravity, Coriolis force, and a restoring force that results from spatial changes of the equilibrium potential vorticity. For the case of barotropic motions the latter quantity is given by f/H, where $f=2\,\Omega\sin\phi$ is the Coriolis parameter, Ω is the magnitude of the earth's angular velocity $\boldsymbol{\Omega}$ and has the numerical value $\Omega=7.2921\cdot10^{-5}\,\mathrm{s}^{-1}$, ϕ is the geographical latitude, and H is the depth of the water. The corresponding five basic types of oceanic waves are acoustic waves, capillary waves, gravity waves, inertial waves, and Rossby waves, respectively. In this order the waves have increasing wave periods.

Since all restoring forces act simultaneously in the ocean, mixed types of waves can occur. However, here we shall mostly discuss the separate wave types that arise because of each restoring force.

6.1.2.1 Elasticity of the sea water

The compressibility or elasticity of the sea water can be neglected for the study of most oceanic dynamical processes. However, it is the restoring force for acoustic waves (also called sound waves). These waves are isotropic, nondispersive, and longitudinal, i.e. velocity and wavenumber vectors are parallel. The phase speed c (the speed of sound) is of the order of $1.5\cdot10^{3}\,\mathrm{m\,s^{-1}}$. The exact value depends on the temperature, salinity, and pressure of the sea water (see sect. 3.2). The dispersion relation

$$\omega=ck \tag{11}$$

is displayed in Fig. 1. The ocean can act as a wave guide for horizontally propagating acoustic wave modes. Their dispersion relation for homogeneous water and constant depth,

$$\omega=c\left(k_{\mathrm{h}}^{2}+\frac{n^{2}\pi^{2}}{H^{2}}\right)^{1/2}, \tag{12}$$

where $k_{\mathrm{h}}^{2}=k_{1}^{2}+k_{2}^{2}$ and n is the vertical mode number, is also displayed in Fig. 1.

Acoustic waves are used as a tool in oceanography to measure the water depth (echo sounding) and the water density (acoustic tomography, see Munk and Wunsch [79 M 1]). In oceanographic applications acoustic waves with cyclic frequencies ($\omega/2\pi$) lying between 1 Hz and 100 kHz are used. The corresponding periods cover the range from 10^{-5} s to 1 s, and the corresponding wavelengths cover the range from about one centimeter to a few kilometers.

6.1.2.2 Surface tension

When the sea surface is not flat, the surface tension tends to restore it to a planar configuration. This tension is the restoring force for the capillary waves of the sea surface. Their dispersion relation,

$$\omega^{2}=\frac{\sigma_{\mathrm{s}}}{\varrho}k_{\mathrm{h}}^{3} \tag{13}$$

is displayed in Fig. 1. Here σ_{s} is the surface tension coefficient and ϱ is the density of the water. σ_{s} depends on the temperature and the salinity of the water (see sect. 3.1). A typical value is $\sigma_{\mathrm{s}}=7.4\cdot10^{-2}$ N/m (for air/fresh-water interface at 10 °C).

Capillary waves are isotropic and show anomalous dispersion; their group speed,

$$c_{\mathrm{g}}=\frac{3}{2}\left(\frac{\sigma_{\mathrm{s}}}{\varrho}k_{\mathrm{h}}\right)^{1/2},$$

is larger than their phase speed,

$$c=\left(\frac{\sigma_{\mathrm{s}}}{\varrho}k_{\mathrm{h}}\right)^{1/2};$$

shorter waves travel faster than longer ones. The wavelength of pure capillary waves is small compared to 1.73 cm, while the period is small compared to $7.5\cdot10^{-2}$ s, see the following subsection. These waves are generated by the wind.

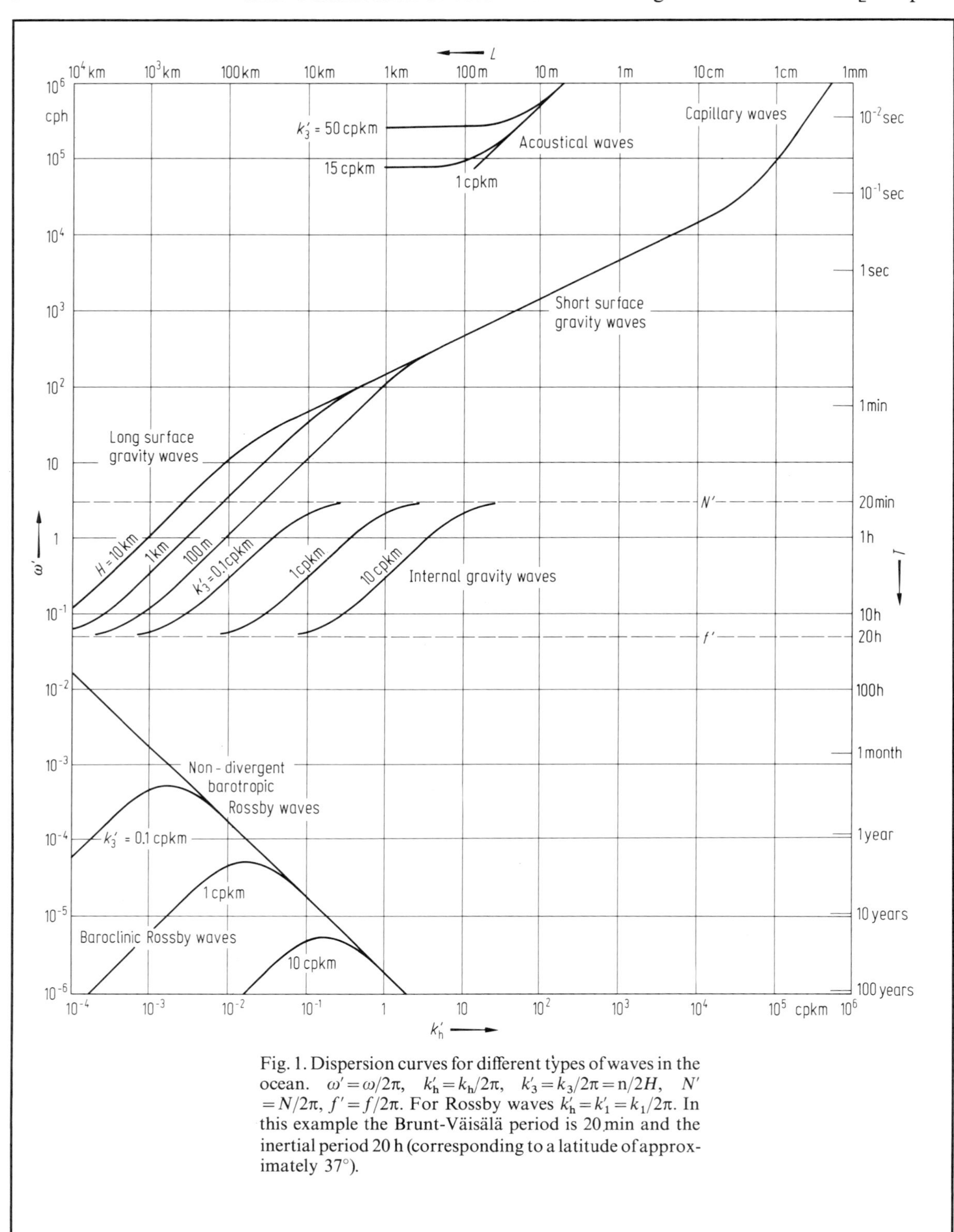

Fig. 1. Dispersion curves for different types of waves in the ocean. $\omega' = \omega/2\pi$, $k'_h = k_h/2\pi$, $k'_3 = k_3/2\pi = n/2H$, $N' = N/2\pi$, $f' = f/2\pi$. For Rossby waves $k'_h = k'_1 = k_1/2\pi$. In this example the Brunt-Väisälä period is 20 min and the inertial period 20 h (corresponding to a latitude of approximately 37°).

6.1.2.3 Gravity

There are many different types of gravity waves that will be mentioned in various parts of ch. 6. Here we consider gravity waves at the sea surface and internal gravity waves in an unbounded ocean and in a laterally unbounded ocean with a flat bottom.

The dispersion relation of surface gravity waves is

$$\omega^2 = g k_h \tanh k_h H, \tag{14}$$

where g is the gravitational acceleration. These waves are isotropic and show normal dispersion. Their group speed,

$$c_g = \frac{c}{2}\left(1 + \frac{2k_h H}{\sinh 2k_h H}\right),$$

is smaller than their phase speed c. Also, longer waves travel faster than shorter ones. In the limiting case $k_h H \to \infty$ one has $c \to (g/k_h)^{1/2}$ and $c_g \to c/2$. In case of $k_h H \to 0$ it follows that $c \to (gH)^{1/2}$ and $c_g \to c$ (no dispersion). Usually it is assumed that, for $L < 2H$ (short waves, deep water waves), $c = (g/k_h)^{1/2}$ and $c_g = c/2$ are appropriate approximations. Similarly, for $L \gg 2H$ (long waves, shallow water waves) one uses $c = (gH)^{1/2}$ and $c_g = c$.

In surface gravity waves the particle trajectories are ellipses in a vertical plane containing the wavenumber vector. For short waves the particle orbits become circles whose radii decay with depth. The influence of these waves is limited to a surface layer of approximate thickness $L/2$. For long waves the horizontal velocity is independent of the vertical coordinate; these waves influence the whole water column in a uniform manner.

If one considers gravity and surface tension simultaneously, a mixed type of surface wave (capillary-gravity wave) results. In the case of deep water its phase speed is

$$c = \left(\frac{gL}{2\pi} + \frac{2\pi\sigma_s}{\varrho L}\right)^{1/2}.$$

For $g = 9.81\ \mathrm{m\,s^{-2}}$, $\varrho = 1024\ \mathrm{kg\,m^{-3}}$, and $\sigma_s = 7.6 \cdot 10^{-2}\ \mathrm{N\,m^{-1}}$, c has a minimum $c_m = 0.23\ \mathrm{m\,s^{-1}}$ at $L_m = 1.73$ cm, corresponding to $T_m = 7.5 \cdot 10^{-2}$ s.

Examples of short surface gravity waves are wind waves and swell (sect. 6.2); tsunamis are an example of long gravity waves (subsect. 6.1.4.3).

In the preceding discussion of gravity waves it was assumed that the influence of the earth's rotation can be neglected. This assumption is acceptable if ω^2 is large compared to f^2, or if, for waves in a channel, the width of the channel is small compared to the (external) Rossby radius of deformation, $R_0 = (gH)^{1/2}/f$. The dispersion relation for long gravity waves under the influence of the earth's rotation (Sverdrup waves) is

$$\omega^2 - f^2 = gHk_h^2. \tag{15}$$

Note that these waves have frequencies larger than f, i.e. wave periods smaller than the inertial period $\pi/\Omega \sin\phi$. The velocity vector of these waves is no longer confined to a vertical plane containing $\boldsymbol{k}$. Its horizontal component describes an ellipse (current ellipse). The ratio of its minor to its major axis is f/ω. When Sverdrup waves are reflected from a vertical wall Poincaré waves result. These waves no longer have crests that form straight lines (infinite crests). They have sinusoidal crests (finite crests). Gravity waves under the influence of the earth's rotation and their relation to the tides (sect. 6.4) are described by Platzman [71 P]. The dispersion relation of surface gravity waves (long waves, short waves, and mixed gravity – capillary waves) is displayed in Fig. 1.

Internal gravity waves in an unbounded ocean are baroclinic and anisotropic. Their existence depends on the stratification of the water, which is usually described by a stratification parameter N, the Brunt-Väisälä frequency, defined by

$$N^2 = -\frac{g}{\varrho_0}\frac{\mathrm{d}\bar{\varrho}}{\mathrm{d}z},$$

where ϱ_0 is a constant reference density and $\bar{\varrho}(z)$ the mean potential density as function of the vertical coordinate z, directed upwards. See subsect. 4.1.5 for definition of potential density. For the special case of a linear stratification, i.e. constant $N = N_0$, their dispersion relation is

$$(N_0^2 - \omega^2)k_h^2 - (\omega^2 - f^2)k_3^2 = 0. \tag{16}$$

Magaard, Mysak

For real values of $\boldsymbol{k}$ the frequency range of these waves is limited by the inertial frequency, f, and the stability frequency (Brunt-Väisälä frequency), $N: f<\omega<N$. (For $N<f$, see subsect. 6.1.2.4.) For a given frequency, the locus of possible wavenumber vectors is a double cone whose angle is determined by the ratio $(k_h/k_3)^2=(\omega^2-f^2)/(N_0^2-\omega^2)$. The phase velocity vector,

$$\boldsymbol{c}=\frac{\omega}{k^2}\boldsymbol{k},$$

and the group velocity vector,

$$\boldsymbol{c}_g=\frac{N_0^2-f^2}{k^4}\frac{k_3}{\omega}\begin{pmatrix}k_3k_1\\ k_3k_2\\ -k_h^2\end{pmatrix},$$

are orthogonal and form a vertical plane containing $\boldsymbol{k}$: when the phase velocity is directed downward the group velocity is directed upward and vice versa.

Internal gravity waves in a linearly stratified ocean with a flat bottom at $z=-H$ propagate horizontally and are isotropic. Their dispersion relation is

$$(N_0^2-\omega^2)k_h^2-(\omega^2-f^2)\frac{\mathrm{n}^2\pi^2}{H^2}=0, \tag{17}$$

where n is the vertical mode number. Eq. (17) is displayed in Fig. 1. For $f^2 \ll \omega^2 \ll N_0^2$ the dispersion relation becomes

$$\frac{\omega}{k_h}=\frac{N_0 H}{\mathrm{n}\pi},$$

i.e. the waves are nondispersive. The depth dependence of the horizontal velocity components of these waves is described by $\cos(\mathrm{n}\pi z/H)$, that of the vertical velocity component by $\sin(\mathrm{n}\pi z/H)$.

A more thorough discussion of internal gravity waves will be given in sect. 6.3.

6.1.2.4 Coriolis force

When a water particle is displaced horizontally on the rotating earth, the Coriolis force tends to drive it back to its original position. This force is the restoring force of inertial waves. In an unbounded ocean their dispersion relation is the same as for internal gravity waves, eq. (16). However, the frequency range of the inertial waves is $N<\omega<f$ implying that a condition for these waves to exist is $N<f$. In the ocean this latter condition is rarely satisfied. Therefore the role of these waves is limited in the ocean. However, the limiting case, $\omega \to f$, is of considerable oceanographic interest. The corresponding phenomenon is called inertial oscillation and means a circular horizontal motion of the water at the inertial period. At frequencies just above f there is usually high energy in the oceanic motions. Inertial oscillations as well as waves near the inertial frequency, i.e. mixed inertial-gravitational waves with $\omega \gtrsim f$ have been studied and observed extensively in the ocean (see Webster [68 W] and Fu [81 F]).

6.1.2.5 Spatial changes of the equilibrium potential vorticity

The Coriolis parameter f changes with latitude. This change leads to the restoring force of Rossby waves. Other names are planetary waves, quasigeostrophic waves, vorticity waves, oscillations of the second class (as opposed to oscillations of the first class, another name of gravity waves). In his book [79 P], pp. 102···104, Pedlosky describes how the change of f results in a restoring force driving a water particle displaced to the north or to the south back to its original position. A spatial change of H has similar consequences. It leads to the restoring force of topographic Rossby waves. The dispersion relation of mid-latitude barotropic (surface) Rossby waves in case of a flat ocean floor is

$$\omega=\frac{-\beta k_1}{k_h^2+R_0^{-2}}, \tag{18a}$$

where $\beta=\mathrm{d}f/\mathrm{d}y$, y is a Cartesian coordinate directed northward, and k_1 is the east component of $\boldsymbol{k}$. $R_0=(gH)^{1/2}/f$ is the external Rossby radius of deformation. Eq. (18a) shows that these anisotropic waves can only propagate in a direction which has a positive westward component. Whenever $R_0^{-2} \ll k_h^2$, i.e. if the wavelength is small compared to $2\pi R_0$, eq. (18a) reduces to

$$\omega = -\beta k_1/k_h^2. \tag{18b}$$

Eq. (18b) is the exact dispersion relation for the case that the wave associated sea level elevation is neglected (rigid lid approximation), and is displayed in Fig. 1.

Waves satisfying eq. (18b) are called nondivergent barotropic Rossby waves while waves satisfying eq. (18a) are called divergent. At a latitude of 37°, which was used in Fig. 1, the dispersion curves of nondivergent and divergent barotropic Rossby waves can not be distinguished in Fig. 1 for $H \geqq 5000$ m. Dispersion relation (18a) can also be described by a so-called slowness curve in the wavenumber plane. Such a curve describes the possible locus of the wavenumber vector at constant frequency (Fig. 2). This figure also displays the direction of the group velocity vector.

The dispersion relation of mid-latitude baroclinic (internal) Rossby waves in case of a flat ocean floor is

$$\omega = \frac{-\beta k_1}{k_h^2 + R_n^{-2}}, \tag{19}$$

where R_n is the internal Rossby radius of deformation for the nth internal mode. In case of linear stratification, $R_n = N_0 H/(n\pi f)$. Eq. (19) is displayed in Fig. 1.

Rossby waves can only exist for wave periods longer than a cut-off period T_c. This period is $T_c = 4\pi/\beta R_0$ for barotropic Rossby waves and $T_c = 4\pi/\beta R_n$ for baroclinic Rossby waves. Typical values for T_c are 1 to 5 days for barotropic and 3 to 12 months for baroclinic Rossby waves.

The horizontal component of the Rossby wave velocity vector is perpendicular to the wavenumber vector. Baroclinic Rossby waves manifest themselves most notably in internal isotherm displacements.

To date, mid-latitude baroclinic Rossby waves were only observed in the North Pacific Ocean (e.g. Kang and Magaard [80 K]). The observed waves are all of the first mode (n = 1) and travel predominantly northwestward. Their periods range from the cut-off period to about 10 years, their wavelengths from 300 to about 1500 km.

For bottom slopes larger than 10^{-3} in the deep ocean, the restoring force due to the variable Coriolis parameter is dominated by the restoring force due to the variable topography and the above described Rossby waves are changed into topographic Rossby waves. For very long waves (wavelength about 1000 km) the influence of stratification on topographic Rossby waves is negligible and the wave associated currents are barotropic. Also, the dispersion relation is very similar in form to that for barotropic Rossby waves [eq. (18b)]. For a uniformly sloping bottom which shoals to the north [i.e. $H(y) = H_0 - \alpha y$], the dispersion relation (under the rigid lid approximation) is given by

$$\omega = -\frac{\alpha f}{H_0}\frac{k_1}{k_h^2}.$$

Note that in the northern hemisphere the phase travels westward in a manner which has the shallower water on the right. Comparison of the above dispersion relation with eq. (18b) reveals that an equivalent topographic β is given by $\alpha f/H_0$.

For short topographic Rossby waves (wavelength about 100 km or less) the motions are exponentially trapped near the bottom and the dispersion relation takes the form $\omega = -\alpha N_0 k_1/k_h$ in an ocean of linear stratification. Such bottom intensified motions, with a period of about one week, have frequently been observed on the New England continental slope (e.g. Thompson and Luyton [76 T]).

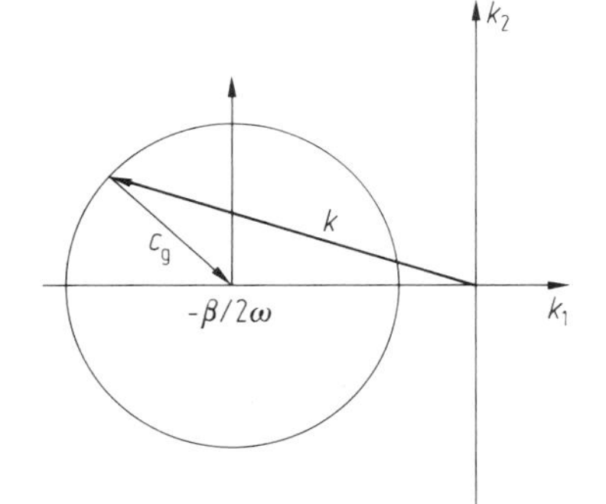

Fig. 2. Slowness curve for Rossby waves,

$$\left(k_1 + \frac{\beta}{2\omega}\right)^2 + k_2^2 = \frac{\beta^2}{4\omega^2} - \frac{1}{R_n^2},$$

according to eq. (18a) (n = 0) and eq. (19) (n ≧ 1).

6.1.3 Classification of waves with respect to wave guides

Many of the wave types described in the previous section become trapped when they propagate near or along special oceanic features such as a thermocline, an equatorial zone, a coastal region, a trench, an island, or a seamount. As a consequence it is sometimes convenient to classify oceanic waves according to the different types of wave guides that exist in the ocean. In this section we shall discuss the following types of trapped waves:

1) thermocline-trapped internal gravity waves,
2) equatorially trapped waves,
3) Kelvin waves trapped against a vertical wall,
4) surface gravity waves trapped along a sloping beach (edge waves),
5) topographic Rossby waves trapped along a continental shelf (shelf waves) or a trench (trench waves), and
6) waves trapped around islands and seamounts.

6.1.3.1 Internal gravity wave trapping in a thermocline

In a linearly stratified ocean ($N=N_0=\text{const}$), the dispersion relation (16) implies that internal gravity waves can propagate (vertically) with any frequency lying in the range $f<\omega<N_0$. A similar frequency passband holds for vertical propagation in an ocean with a variable $N(z)$: $f<\omega<N(z)$. In particular, if $N(z)$ has the form shown in Fig. 3, with N having a large maximum in the thermocline, then an internal wave with the relatively high frequency ω shown in Fig. 3 cannot propagate upward, above the depth $z=z_a$, or downward, below the depth $z=z_b<z_a$. This is because above and below these depths, the z-dependence of the vertical velocity, $v_z(z)$, of each normal mode has a decaying behavior. Inside the depth range $z_b<z<z_a$, however, $v_z(z)$ has an oscillatory behavior (see Fig. 3). Thus the waves are confined to travelling horizontally in the thermocline wave guide $z_b<z<z_a$.

In the extreme limit where $N(z)$ becomes proportional to $N_0\delta(z-z_0)$, corresponding to a two-layer fluid with a density jump at $z=z_0$, thermocline-trapped waves become interfacial waves which propagate along the surface formed by the density discontinuity. They are analogous to surface gravity waves which travel along the surface separating air and water. For interfacial waves that are long compared to the depth of each fluid layer, the dispersion relation is given by

$$\omega=k(g'H_1H_2/H)^{1/2}, \tag{20}$$

where $g'=g(\varrho_2-\varrho_1)/\varrho_2$ (reduced gravity), $H=H_1+H_2$, and ϱ_i, $H_i (i=1,2)$ are the density and depth in layer i. For such waves the currents in each layer are horizontal and depth-independent, with the lower layer current u_2 being 180° out of phase with the upper layer current u_1, viz.,

$$u_2=-H_1u_1/H_2. \tag{21}$$

Since eq. (21) implies that $u_1H_1+u_2H_2=0$ there is no net horizontal transport over the water column associated with long interfacial wave motions.

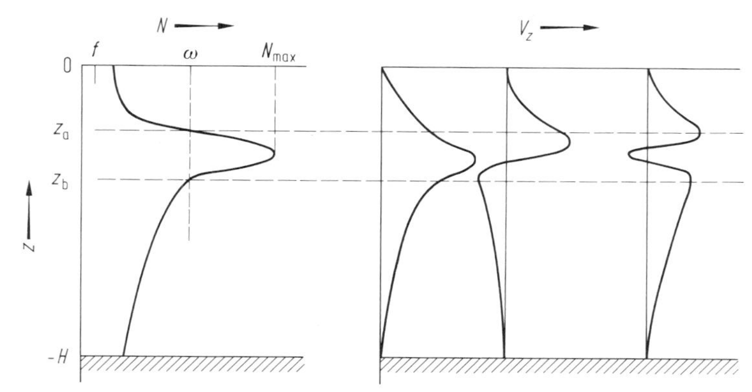

Fig. 3. Internal wave trapping in a thermocline. The relative values of f, ω and $N(z)$ are shown on the left. The vertical dependence of the vertical velocity $v_z(z)$ is shown on the right for the first three modes (from LeBlond and Mysak [78 L 1]).

6.1.3.2 Equatorially trapped waves

A narrow band of latitudes centered symmetrically about the equator can act as a wave guide for westward propagating Rossby waves and westward or eastward propagating long gravity waves. The reason for this is that near the equator the Coriolis parameter f is a rapidly varying function of distance y from the equator, viz., $f=(2\Omega/R_e)y\equiv\beta y$, where R_e is the earth's radius. Thus the usual mid-latitude two-dimensional dispersion relations for horizontally propagating Rossby waves [eq. (18)] or gravity waves [eq. (17)] are not valid near the equator since in these relations f varies rapidly with y. As a consequence only zonally (eastward or westward) propagating waves can exist whose amplitudes are oscillatory in a latitude band (typically, a few degrees) symmetric about the equator and decaying outside this latitude band. The dispersion relation for equatorially trapped waves is given by

$$\frac{\omega^2}{gH}-k^2-\frac{\beta k}{\omega}=(2\mathrm{m}+1)\beta(gH)^{-1/2}, \tag{22}$$

where $\mathrm{m}=0,1,2,\ldots$ is the north-south mode number. Since eq. (22) is a cubic polynomial in ω, for a given k and $\mathrm{m}\geqq 1$, there are three roots for ω, with one corresponding to a westward travelling Rossby wave and two corresponding to eastward or westward travelling gravity waves. The dispersion curves for these waves are shown in Fig. 4. If $\mathrm{m}=0$, the cubic polynomial (22) can be factored, giving one inadmissible root, $\omega=-(gH)^{1/2}k$ (the dashed line in Fig. 4), and one admissible root, the latter being called a mixed Rossby-gravity wave or Yanai wave. For large positive wavenumber k it behaves like a high frequency eastward travelling gravity wave, whereas for small negative k it behaves like a low-frequency westward travelling Rossby wave.

In addition to the above waves, there is another eastward-travelling trapped wave known as an equatorial Kelvin wave for which $\omega=(gH)^{1/2}k$ (see Fig. 4) and the north-south velocity $v\equiv 0$. Several oceanographers have recently established that this wave plays an important role in the occurrence of the El Niño, an oceanic phenomenon in which abnormally warm water appears every few years off the coast of Peru.

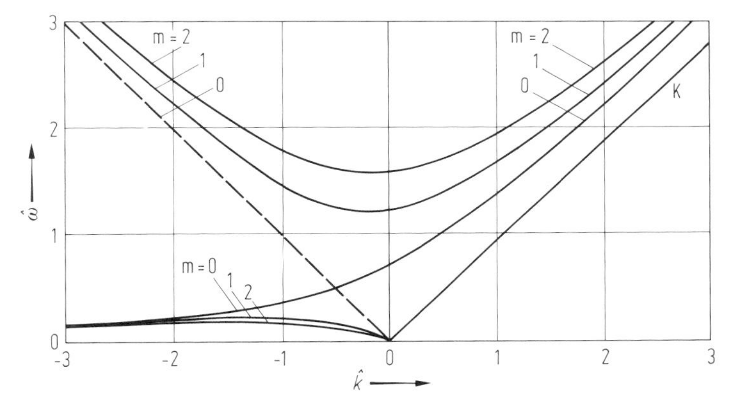

Fig. 4. Dispersion curves for barotropic equatorial wave modes $\mathrm{m}=0,1,2$, according to eq. (22).

$$\hat{\omega}=(gH)^{-1/4}(2\beta)^{-1/2}\omega, \quad \hat{k}=(gH)^{1/4}(2\beta)^{-1/2}k.$$

The solid lines are, from top to bottom, the gravity waves ($\mathrm{m}=2$, $\mathrm{m}=1$), the mixed gravity – Rossby wave (Yanai wave, $\mathrm{m}=0$), and the Rossby waves ($\mathrm{m}=1$, $\mathrm{m}=2$). The dashed line represents an inadmissible solution for $\mathrm{m}=0$, while the line labelled K is the Kelvin wave for which the meridonal velocity vanishes. Note that in each of the gravity wave and Rossby wave curves there is a $\hat{k}$ at which the tangent has zero slope, corresponding to zero group velocity (from LeBlond and Mysak [78 L 1]).

6.1.3.3 Kelvin waves

Kelvin waves are exponentially trapped against a vertical wall which acts as a side boundary to an ocean of constant depth. They are maintained by an exact balance between the Coriolis force and a pressure gradient normal to the wall. In the northern (southern) hemisphere they propagate their phase with the wall on the right (left). The phase speed for the barotropic (surface) mode is simply

$$c=(gH)^{1/2}, \tag{23}$$

which is identical to that for long gravity waves in a nonrotating ocean of depth H. The e-folding (decay) distance away from the wall is $c/f=R_0$, the external Rossby radius of deformation. Another interesting feature about this wave, discovered over a century ago by Lord Kelvin, is that the associated current fluctuations are entirely parallel to the wall.

Kelvin waves have often been used to model the propagation of the tides along continental margins and in channel regions.

Internal Kelvin waves can also exist in the ocean and, like the barotropic mode, have currents entirely parallel to the coast. However, they propagate with a much slower speed. For the nth vertical mode, the phase speed is given by

$$c_n=(gh_n)^{1/2}, \tag{24}$$

where h_n is the equivalent depth. In terms of the internal Rossby radius R_n it can be written as $h_n=f^2R_n^2/g$. For typical oceanic stratification, $h_1 \lesssim 1$ m, $h_2 \approx 0.2$ m, $h_3 \approx 0.1$ m, etc. Since $H \approx 4 \cdot 10^3$ m, comparison of eqs. (23) and (24) shows that $c_n \ll c$. Further, the offshore e-folding distance of the internal Kelvin wave is $c_n/f=R_n$, the nth mode internal Rossby radius of deformation. Since $R_n \ll R_0$, internal Kelvin waves are trapped relatively close to the coast and thus play an important role in coastal upwelling.

6.1.3.4 Edge waves

Edge waves are long gravity waves which slowly travel parallel to the shore, along a sloping beach. They are trapped against the coast by refraction, a mechanism that can be easily understood in terms of ray theory (see LeBlond and Mysak [78 L 1], ch. 4). The first mathematical solution for the gravest mode edge wave was obtained by Stokes well over a century ago, and for decades this wave solution was regarded as a hydrodynamical curiosity [45 L]. However, recently it has been established that edge waves are an essential ingredient to our understanding of such phenomena as rip currents, beach cusps, crescentic bars, and nearshore sediment transport.

For a uniformly sloping beach with depth profile

$$H(x)=\alpha x, \qquad x \geqq 0 \tag{25}$$

the dispersion relation for the nth mode edge wave is given by

$$\omega_n^2=gk\alpha(2n+1), \qquad n=0,1,2,\ldots. \tag{26}$$

For $n=0$, the wave amplitude decays exponentially away from the coast like e^{-kx}. For the nth mode, there are n nodal lines parallel to the coast; this oscillatory behavior, however, is damped by the same exponential decay e^{-kx} as for the gravest $(n=0)$ mode.

From eq. (26) it follows that the phase speed of the nth mode edge wave is given by

$$c_n=[g\alpha(2n+1)/k]^{1/2}. \tag{27}$$

Since typically $\alpha=0(10^{-3})$ for most beaches, eq. (27) implies that the first few mode edge waves travel much slower than deep water gravity waves, whose phase speed is simply $c=(g/k)^{1/2}$.

6.1.3.5 Trapped topographic Rossby waves

Reid [58 R] showed that in the presence of uniform rotation, a low-frequency trapped wave is possible on the sloping beach profile, eq. (25), in addition to the classical edge waves discussed above. This wave has always subinertial frequencies $(\omega<f)$ and was therefore called a "quasigeostrophic wave" by Reid. Furthermore, its direction of phase propagation is consistent with that of a topographic Rossby wave: in the northern hemisphere the shallower water (or coast) lies to the right when following the wave phase. However, because its amplitude

decays exponentially away from the coast, its energy is also coastally trapped, like that of a Kelvin wave or an edge wave. The dispersion curves for the first three edge wave modes and the first three topographic wave modes are shown as dashed lines in Figs. 5a and 5b, respectively. Note that in the presence of rotation, the edge wave frequencies are split at $k=0$: the waves with $\omega/f>0$, which travel with the coast on the right (in the northern hemisphere), have zero frequency at $k=0$, whereas the edge waves with $\omega/f<0$, which travel, in the northern hemisphere, with the coast on the left, have $\omega=-f$ at $k=0$. Note also that the gravest mode (n = 0) topographic wave is simply the inertial oscillation $\omega=f$ and is not trapped against the coast.

If a more realistic topography is used to model the continental shelf and slope region, with an abyssal plane region at the base of the continental slope (see Fig. 6), the dispersion curves for the edge waves and topographic waves change considerably. Also, it can be shown that the barotropic Kelvin wave discussed above can be regarded as a very long edge wave that is trapped against the continental slope (see Fig. 5). The dispersion curves for all three wave types that exist on the discontinuous topographic model

$$\begin{aligned} H(x) &= \mathrm{d}x/L, \quad 0<x<L \\ &= D, \quad L<x<\infty, \end{aligned} \tag{28}$$

where $d/D \ll 1$, are shown as solid lines in Fig. 5. We note that at low wavenumbers the curves diverge considerably from the sloping beach curves. Also, at sufficiently low wavenumbers the edge waves are no longer trapped: this is because the refraction mechanism necessary for trapping breaks down.

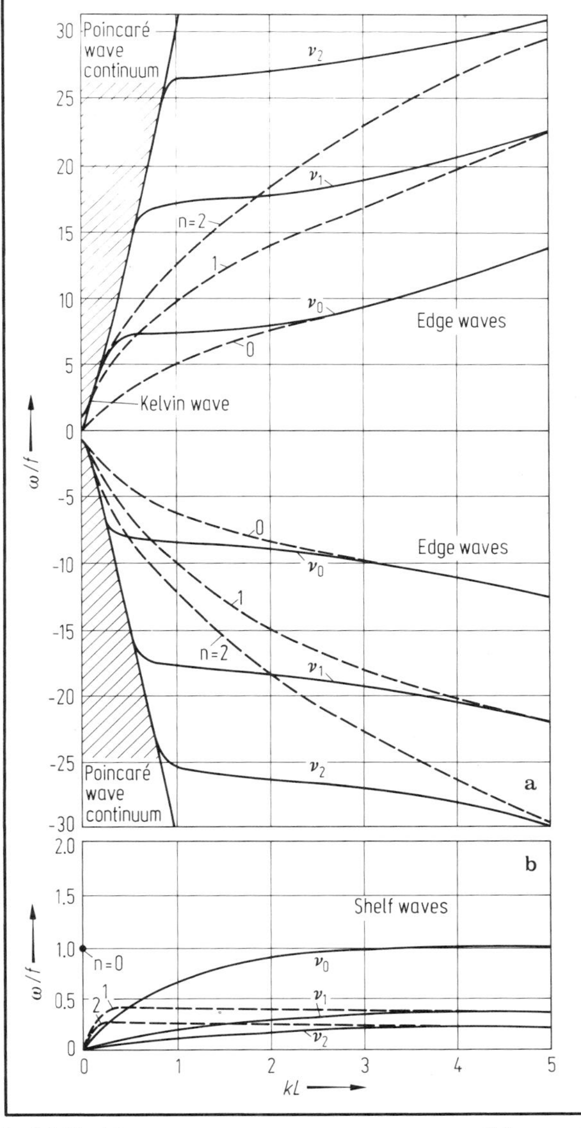

Fig. 5. (a) Edge wave, and (b) shelf wave dispersion curves for a sloping shelf of finite width as described in eq. (28) (solid lines) and semi-infinite width as described in eq. (25) (dashed lines), with $f=7.3\cdot10^{-5}\,\mathrm{s}^{-1}$, $d=200\,\mathrm{m}$, $L=10^5\,\mathrm{m}$, i.e. $\alpha=2\cdot10^{-3}$, (adapted from Mysak [68 M]). The different modes ν_n and n are indicated on the curves; the ν_n's are the roots of $L_\nu(2kL)=0$, with L_ν the Laguerre functions, and a cubic relation between ν and ω. The shaded region corresponds to the continuous spectrum of topographically modified Poincaré waves and is bounded by the dispersion relation of Sverdrup waves, $\omega=\pm(f^2+gDk^2)^{1/2}$ where $D=5\cdot10^3\,\mathrm{m}$ (deep-sea depth). The gravest mode (n = 0) shelf wave coincides with the inertial oscillation $\omega=f$ and is not trapped against the coast. (From Mysak [80 M 1].)

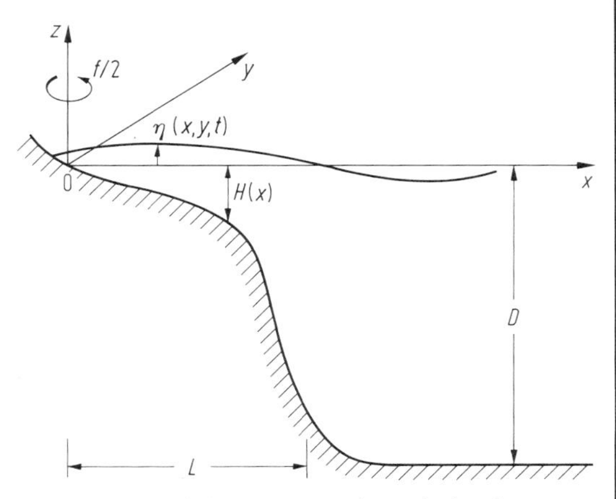

Fig. 6. Typical shelf/slope topography. L is the characteristic shelf-slope width and D is the deep-sea depth far from the coast. $\eta(x, y, t)$ is the sea level elevation.

Magaard, Mysak

Because the topographic waves are effectively trapped on the shelf, they are frequently called "continental shelf waves" or "shelf waves" for short. The very low frequency shelf waves ($\omega \ll f$), which are nondispersive, were first discovered by Robinson [64 R]. During the last two decades they have attracted the attention of many theoreticians and observationalists for it is now believed that these waves play an important role in Gulf Stream meanders, coastal upwelling, shelf circulation, storm surges, and maritime weather and climate. A recent review on this wave type has been given by Mysak [80 M 1].

When the depth of the ocean does *not* increase monotonically away from the coast because of the presence of a trench, for example, another low frequency trapped Rossby wave can exist. This wave has its energy trapped over the trench region and propagates its phase in the opposite direction to that of the shelf wave. Such "trench waves" have recently been described by Mysak et al. [79 M 2], who also found observational evidence of these waves in sea level records from the Japan-Kuril trench and the Peru-Chile trench. Finally, if one eliminates the coastal shelf region from a shelf/trench region, one is left with an escarpment, which can also serve as a wave guide for the trapping of low-frequency topographic wave energy in the open ocean. Such waves, sometimes called "double Kelvin waves", were first discussed in the literature by Longuet-Higgins [68 L] and Rhines [69 R]. However, to date there has been no observational evidence for these waves.

6.1.3.6 Waves trapped around islands and seamounts

The trapping of long waves by islands and seamounts has been intensively studied during the past decade. Generally speaking, the three topographic trapped wave types discussed above (Kelvin waves, edge waves, and topographic Rossby waves) can also exist around islands and seamounts. However, there is one new feature of these waves: because the waves are circularly travelling, the longshore wavenumber is now quantized, there being an integral number of wavelengths around the island or seamount. As a consequence there is a discrete spectrum for both the frequency and wavenumber of island or seamount trapped waves. In principle, this should make it easy to identify such waves from observed current or sea level spectra. However, in practice there has been limited success in this direction. For references on these trapped waves, see LeBlond and Mysak [78 L 1], ch. 4, and Mysak [80 M 2].

6.1.4 Classification with respect to generating forces or mechanisms

In the governing equations for ocean waves there are three important forces: Restoring forces, the inertia of the moving water particles, and generating forces (including negative generating forces produced by damping mechanisms). A study of the various restoring and generating forces leads to convenient classifications of ocean waves. In subsect. 6.1.2 a classification with respect to restoring forces was given. In this subsection a classification with respect to generating forces is presented.

6.1.4.1 Astronomical forces

The orbits of the earth around the sun and of the moon around the earth are defined by an equilibrium of gravitational and centrifugal forces. Only at the centers of gravity of the celestial bodies is this equilibrium perfect. At all other points, and hence at the surface of the earth, there are residual forces. The horizontal components of these residual forces are the tide generating forces that excite the astronomical tides in the ocean. These tides will be discussed in sect. 6.4.

Other astronomical wave generating agents are the oscillations of the earth's poles. These cause periodic oscillations of the sea level, the pole tides. The period of a pole tide is the same as the period of the corresponding pole oscillation (e.g. the Chandler period of about 14 months). The pole tide due to the Chandler wobble has been studied by Haubrich and Munk [59 H] (see also Munk and MacDonald [60 M]).

6.1.4.2 Meteorological forces

The fluctuating fields of atmospheric pressure and wind stress can excite virtually all types of ocean waves except acoustic waves. The best known meteorologically generated waves are the high frequency (wave periods smaller than about 30 s) capillary and short surface gravity waves, usually referred to as wind waves. The extensive observational and theoretical studies of these waves and their generation are presented in sect. 6.2.

Studies of wind waves go back to the middle of the last century. Only in the last two decades it became increasingly clear that atmospheric disturbances can also generate Kelvin waves, edge waves, shelf waves (LeBlond and Mysak [78 L 1]), internal gravity waves (sect. 6.3), and surface and internal Rossby waves (Müller and Frankignoul [81 M 1]).

Meteorological forces manifest themselves as momentum and buoyancy fluxes at the free sea surface. These forces lead to boundary sources of wave energy where the boundary is the free ocean-atmosphere interface. The solid boundaries of the ocean can also be sources of wave energy.

6.1.4.3 Boundary sources

Sudden deformations of the sea floor by seaquakes or volcanic eruptions can act as generating forces for long surface gravity waves called tsunamis. These waves propagate radially from the center of generation (epicenter) approximately along great circles with speed $c=(gH)^{1/2}$ where H is the mean depth of water. The periods of these waves range from 10 min to 2 h. Smaller periods occur more often than larger ones. Typical values are 15···30 min. Typical phase speeds are about 200 m s^{-1}. Thus, typical wavelengths are 200 to 300 km. In the open ocean tsunamis have small amplitudes but as the depth decreases their amplitudes can become very large (effective crests as high as 30 m). Tsunamis occur mostly in the Pacific Ocean because of its relatively more frequent volcanic and seismic activity. The coast of Japan is most frequently hit by tsunamis.

Topographic irregularities like sea mounts, variations in depth, or continental slopes can become boundary sources of waves: Steady flows over such irregularities produce stationary waves called lee waves; periodic currents due to wave motion produce scattered waves emanating from the irregularities.

Reflection of an oceanic disturbance from a coastal boundary can lead to the generation of waves of a nature different from that of the incident disturbance. Bryan and Ripa [78 B], and White and Saur [81 W] have studied cases where the eastern coastline of the North Pacific becomes a line source of Rossby waves.

6.1.4.4 Moving bodies

A moving ship generates short surface gravity waves (called wakes). If a ship moves through a thin boundary layer of lighter water overlying heavier water it can moreover excite internal interfacial waves at the interface of the two water masses. Similarly, a submerged submarine generates internal waves as it moves through a stratified fluid (Keller and Munk [70 K]).

6.1.4.5 Mechanisms to enhance existing waves

Existing waves can gain energy from mean shear flow and from other waves. Gravity waves at the interface between two layers with different mean flow can be enhanced by means of Kelvin-Helmholtz instability. Barotropic instability of a current with horizontal shear and baroclinic instability of a current with vertical shear in a stratified ocean can lead to Rossby wave growth.

Nonlinear interactions between various components of a wave field lead to energy exchanges between such components. Such interactions are a fundamental process in the formation of the peak in wind wave spectra (sect. 6.2).

6.1.4.6 Free waves

When a wave is no longer under the direct influence of its generating force it is called a free wave. When wind generated waves are no longer under the influence of the generating wind field (because they have travelled out of the area of the wind field or because the wind has decayed) they are called swell.

In enclosed basins long free gravity waves can superimpose to form a system of standing waves, the so-called seiches. For a rectangular basin with length L, width W, and depth H the period T_n of the nth surface seiche mode is

$$T_n = \frac{2L}{n(gH)^{1/2}}. \tag{29}$$

This equation, called Merian's formula, is only applicable if the influence of the earth's rotation on the waves is negligible, i.e. if the Rossby radius of deformation, $R_0=(gH)^{1/2}/f$, is large compared to W.

6.1.4.7 Damping mechanisms

The viscosity of the water has a damping influence on all ocean waves. For surface waves this influence is small except for extremely short gravity and capillary waves (length scales of centimeters). Such short waves are damped even more effectively by thin layers of contaminants (slicks) at the sea surface. The longer wave components of the wind wave and swell spectrum travel over thousands of kilometers with only moderate attenuation. When entering shallower water the waves become subject to bottom friction that influences all long waves.

The most effective dissipation mechanism is breaking by which wave energy is converted to turbulent energy. It occurs at sea (whitecapping) and in onshore surf zones. Internal gravity waves can also become unstable and break. But little is known about the dissipation rate due to breaking.

Waves can loose energy to turbulence (by breaking), to other waves (by nonlinear interaction), and to mean currents. Most of the processes by which waves loose energy to mean currents are not well understood yet. One such process, the critical layer absorption of internal gravity waves, will be described in sect. 6.3.

6.1.5 References for 6.1

45 L Lamb, H.: Hydrodynamics, Dover: New York, 6th edition **1945**, 738 pp.
58 R Reid, R.O.: J. Mar. Res. **16** (1958) 109.
59 H Haubrich, R., Munk, W.: J. Geophys. Res. **64** (1959) 2373.
60 M Munk, W.H., MacDonald, G.: The Rotation of the Earth, London: Cambridge University Press **1960**, 323 pp.
64 R Robinson, A.R.: J. Geophys. Res. **69** (1964) 367.
68 L Longuett-Higgins, M.S.: J. Fluid Mech. **34** (1968) 49.
68 M Mysak, L.A.: J. Mar. Res. **26** (1968) 24.
68 W Webster, F.: Rev. Geophys. **6** (1968) 473.
69 R Rhines, P.B.: J. Fluid Mech. **37** (1969) 161.
70 K Keller, J.B., Munk, W.H.: Phys. Fluids **13** (1970) 1425.
71 P Platzman, G.W.: Ocean tides and related waves, in: Reid, W.H. (ed.): Mathematical Problems in the Geophysical Sciences, Providence (Rhode Island): Am. Math. Soc. **14** (1971) 239.
74 W Whitham, G.B.: Linear and Nonlinear Waves, New York: Wiley **1974**, 636 pp.
76 M Miura, R.M.: SIAM Soc. Ind. Appl. Math. Rev. **18** (1976) 412.
76 T Thompson, R.O.R.Y., Luyten, J.R.: Deep-Sea Res. **23** (1976) 629.
78 B Bryan, K., Ripa, P.: J. Geophys. Res. **83** (1978) 2419.
78 L 1 LeBlond, P.H., Mysak, L.A.: Waves in the Ocean, Amsterdam, Oxford, New York: Elsevier Scientific Publ. Comp. **1978**, 602 pp.
78 L 2 Lighthill, J.: Waves in Fluids. Cambridge, London, New York, Melbourne: Cambridge University Press **1978**, 504 pp.
78 M Mysak, L.A.: Rev. Geophys. Space Phys. **16** (1978) 233.
79 M 1 Munk, W.H., Wunsch, C.: Deep-Sea Res. **26** (1979) 123.
79 M 2 Mysak, L.A., LeBlond, P.H., Emery, W.J.: J. Phys. Oceanogr. **9** (1979) 1001.
79 P Pedlosky, J.: Geophysical Fluid Dynamics, New York, Heidelberg, Berlin: Springer **1979**, 624 pp.
80 K Kang, Y.Q., Magaard, L.: J. Phys. Oceanogr. **10** (1980) 1159.
80 M 1 Mysak, L.A.: Rev. Geophys. Space Phys. **18** (1980) 211.
80 M 2 Mysak, L.A.: Ann. Rev. Fluid Mech. **12** (1980) 45.
81 F Fu, L.-L.: Rev. Geophys. Space Phys. **19** (1981) 141.
81 M 1 Müller, P., Frankignoul, C.: J. Phys. Oceanogr. **11** (1981) 287.
81 M 2 Miles, J.W.: J. Fluid Mech. **106** (1981) 131.
81 W White, W.B., Saur, J.F.T.: J. Phys. Oceanogr. **11** (1981) 1452.

6.2 Wind waves and swell

6.2.0 List of symbols

$a(\boldsymbol{k}, \omega)$	(complex) Fourier transform of the space and time dependence of surface displacements, [m^2]
a_n, b_n	(real) Fourier coefficients of directional distribution $D(f, \theta)$; n = 0, 1, 2, in [rad^{-1}]
$b(\omega)$	(complex) Fourier transform of the time dependence of surface displacements, in [mHz$^{-1/2}$]
C	phase speed
C_{10}	drag coefficient, $C_{10} = U_*^2/U_{10}^2$
c_f	friction coefficient
D	depth of shallow water, in [m]
d	depth at which waves break
$D(f, \theta)$	directional distribution of wave energy, in [rad^{-1}]
E	energy density (energy per surface unit), in practical unit [m^2]
$E(f)$	one-dimensional energy spectrum, in practical unit [m^2 Hz^{-1}]
$E(f, \theta)$	two-dimensional energy spectrum, in practical unit [m^2 Hz^{-1} rad^{-1}]
F	fetch (normal distance from shore), in [m]
$F(k)$	energy distribution in wavenumber space, in practical unit [m^4]
f	frequency, in [Hz]
f_m	peak frequency of spectrum
$\tilde{f}_m$	nondimensional peak frequency, $\tilde{f}_m = f_m U_{10}/g$
f_{PM}	Pierson-Moskowitz frequency, $f_{PM} = 0.13\, g/U_{10}$
g	gravitational acceleration, $g = 9.806$ m s^{-2}
H	average wave height
H_b	breaker height
H_s	significant wave height, $H_s = 4 m_0^{1/2}$, in [m]
h	wave height: $h(x, y)$ = local depth
I	energy flux, in practical unit [m^3 s^{-1}]
$\boldsymbol{k}$	(horizontal) wavenumber vector, Cartesian components k_i, i = 1, 2
L	wavelength
$\boldsymbol{M}$	momentum density, in practical unit [ms]
m_n	nth spectral moment, n = 0, 1, 2, ..., in practical units [m^2 Hz^n]
S	source function for energy generation and dissipation, eq. (23)
s	(frequency-dependent) exponent characterizing $D(f, \theta)$, eq. (33)
T	time interval; wave period, in [s]
t	time, in [s]
U	local wind speed
U_b	current velocity at sea bottom
U_{10}	local wind speed, 10 m above sea surface, in [ms^{-1}]
U_*	local friction velocity, $U_* = (\tau/\varrho_{air})^{1/2}$
$\boldsymbol{u}$	particle velocity
v_g	group velocity
$\boldsymbol{x}$	(horizontal) position vector
x	normal distance from shore (fetch)
$\tilde{x}$	nondimensional fetch, $\tilde{x} = gx/U_{10}^2$
z	vertical coordinate
$\alpha(\boldsymbol{k}, \omega)$, $\beta(\boldsymbol{k}, \omega)$	(real) Fourier coefficients of surface displacement, in [m^2]
Δ	directional spread of $D(f, \theta)$
ε	nondimensional energy, $\varepsilon = Eg^2/U_{10}^4$; spectral width, eq. (44)
θ	azimuth of $\boldsymbol{k}$
λ	wavelength
ζ	sea-air surface displacement, in [m]
ϱ	sea water density, $\varrho = 1.030 \cdot 10^3$ kg m^{-3}
σ	surface tension, in [m^3 s^{-2}]
τ	momentum transfer, in [Pa]

Rosenthal

Φ velocity potential, in [$m^2 s^{-1}$]
Ω oceanic surface area, in [m^2]
ω angular frequency, in [$rad\, s^{-1}$]

Abbreviations

CERC Coastal Engineering Research Center
JONSWAP Joint North Sea Wave Project
WMO World Meteorological Organization

6.2.1 Introduction

This section treats ocean surface waves with periods less than 30 s. We restrict ourselves to the linearized theory if possible. Nonlinearities are only taken into account if they are necessary for the basic understanding. The deterministic description of monochromatic surface waves is shortly reviewed in subsect. 6.2.2. For a more detailed description, see [77 P]. For practical purposes most important are parameters describing the statistical behavior of surface waves under given oceanographical and meteorological conditions. The emphasis in sect. 6.2 will be on the statistical aspect of surface waves.

6.2.2 The deterministic description of a linear plane wave, its energy and momentum density

The dispersion relation for surface gravity waves as given in sect. 6.1 is

$$\omega^2 = gk\left(1 + \frac{\sigma}{g}k^2\right)\tanh kh\,. \tag{1}$$

In eq. (1) and the remainder of this section we omit the suffix h, for horizontal, of the wavenumber vector, introduced in sect. 6.1, because for surface waves $\boldsymbol{k}_h$ and $\boldsymbol{k}$ are identical. ω denotes the angular frequency, k the modulus of $\boldsymbol{k}_h$, g the gravitational acceleration, and σ the surface tension constant. h is the sea bottom depth. A convenient description of surface waves is by the surface displacement ζ as a function of the horizontal position vector $\boldsymbol{x}$ with components x_1, x_2. In our notation the vertical coordinate z has its origin at the averaged surface. Then a linear sinusoidal wave with amplitude a generates a surface displacement

$$\zeta = \zeta(\boldsymbol{x}, t) = a\cos(\boldsymbol{k}\cdot\boldsymbol{x} - \omega t + \varphi)\,. \tag{2}$$

φ is a constant phase angle that depends on the chosen coordinate system.

The particle velocity $\boldsymbol{u}$ connected with such a plane wave is given by

$$\boldsymbol{u} = \nabla\Phi$$

$$\Phi = \frac{\omega a \cosh k(z+h)}{k \sinh kh}\sin(\boldsymbol{k}\boldsymbol{x} - \omega t + \varphi)\,. \tag{3}$$

Φ is called the velocity potential.

The amplitude a determines the crest height of the wave, which is one half of the wave height h. h is defined as the difference between crest and trough of a wave.

The energy density E (energy per surface unit) for the plane wave can be calculated by:

$$E = \frac{\varrho g}{\Omega}\int_\Omega \zeta^2(\boldsymbol{x}, t)\,d\boldsymbol{x} = \frac{1}{2}\varrho g a^2 = \varrho g\overline{\zeta^2}\,. \tag{4}$$

ϱ is the density of sea water and g is the gravitational acceleration. The bar in $\overline{\zeta^2}$ means the spatial average of $\zeta^2(\boldsymbol{x}, t)$ over the surface patch Ω. It is convenient to use the average of the squared surface displacement $\overline{\zeta^2}$ itself as a measure of energy density instead of eq. (4). ζ is normally given in [m]. Therefore, in this practical unit, energy density is given in [m^2]. The conversion to SI units depends on water density and gravity constant:

$$1.01\cdot 10^4\,J\,m^{-2} \mathrel{\hat=} 1\,m^2$$

for

$$g = 9.806\,\mathrm{m\,s^{-2}}$$
$$\varrho = 1.030 \cdot 10^3\,\mathrm{kg\,m^{-3}}. \tag{5}$$

The momentum density $\boldsymbol{M}$, in this case equal to the mass transport of the wave, is connected with E by the phase velocity c defined in sect. 6.1:

$$\frac{E}{c} = M. \tag{6}$$

The direction of $\boldsymbol{M}$ is parallel to $\boldsymbol{k}$.

6.2.3 The energy density spectrum

The surface of the ocean will be considered in the following as a linear superposition of monochromatic waves characterized by their wave vector, amplitude, and phase as described in subsect. 6.2.2.

If the sea surface elevation $\zeta(\boldsymbol{x}, t)$ together with its time derivative $\dot{\zeta}(\boldsymbol{x}, t)$ is known over an oceanic surface patch Ω the coefficients $a(\boldsymbol{k}, \omega)$ of the connected Fourier series

$$\zeta(\boldsymbol{x}, t) = \frac{1}{\sqrt{\Omega}} \sum_{\boldsymbol{k}, \omega} a(\boldsymbol{k}, \omega) \exp i(\boldsymbol{k}\boldsymbol{x} - \omega t), \tag{7}$$

where $\boldsymbol{x}$ lies within Ω, can be calculated by standard mathematical methods [62 C].

The sum in eq. (7) goes over all $\boldsymbol{k}$ vectors and ω takes the two values $+\omega(\boldsymbol{k})$ and $-\omega(\boldsymbol{k})$, where $\omega(\boldsymbol{k})$ is defined in eq. (1). Ω is chosen to be much larger in extension than the longest wavelength under consideration. For convenience we choose for Ω a square with side length L. By imposing periodic boundary conditions

$$\zeta(x_1, x_2, t) = \zeta\left(x_1 + \frac{2\pi}{L}, x_2, t\right) = \zeta\left(x_1, x_2 + \frac{2\pi}{L}, t\right) \tag{8}$$

we quantizie the wave vector $\boldsymbol{k}$. It is restricted to the discrete values

$$k_{1,2} = \mathrm{n} \cdot \frac{2\pi}{L}, \qquad \mathrm{n} = \pm 1, \pm 2, \pm 3 \ldots. \tag{9}$$

Ω is always taken large enough that the differences in $a(\boldsymbol{k}, \omega)$ for adjacent $\boldsymbol{k}$ are small enough to replace sums over $\boldsymbol{k}$ (as in eq. (7)) by integrals:

$$\sum_{k_1 k_2} \to \frac{\Omega}{4\pi^2} \int \mathrm{d}k_1 \mathrm{d}k_2, \tag{10}$$

where k_1, k_2 are considered to be continuous variables. Because $\zeta(\boldsymbol{x}, t)$ is always real, the coefficients $a(\boldsymbol{k}, \omega)$ obey the conditions:

$$a(\boldsymbol{k}, \omega) = a^*(-\boldsymbol{k}, -\omega), \tag{11}$$

where the asterisk denotes the complex conjugate.

A series equivalent to eq. (7) is:

$$\zeta(\boldsymbol{x}, t) = \sqrt{\frac{2}{\Omega}} \sum_{\boldsymbol{k}} (\alpha(\boldsymbol{k}, \omega) \cos(\boldsymbol{k}\boldsymbol{x} - \omega t) + \beta(\boldsymbol{k}, \omega) \sin(\boldsymbol{k}\boldsymbol{x} - \omega t)). \tag{12}$$

Here the summation is over all $\boldsymbol{k}$ values, whereas $\omega = \omega(k)$, eq. (1), is restricted to positive values only.

The connection for monochromatic waves between surface elevation and velocity potential, eqs. (2) and (3), is applicable to each component of eq. (7). Therefore the velocity potential connected with $\zeta(\boldsymbol{x}, t)$ of eq. (7) is

$$\Phi = \sqrt{\frac{1}{\Omega}} \sum_{\boldsymbol{k}, \omega(\boldsymbol{k})} \frac{\omega \cosh k(z+h)}{ik \sinh kh} a(\boldsymbol{k}, \omega) \exp i(\boldsymbol{k}\boldsymbol{x} - \omega t). \tag{13}$$

Expansion (12) for $\zeta(x, t)$ leads to a similar expansion for the velocity potential:

$$\Phi = \sqrt{\frac{2}{\Omega}} \sum_{\boldsymbol{k}} \frac{\omega \cosh k(z+h)}{k \sinh kh} (\alpha(\boldsymbol{k}, \omega) \sin(\boldsymbol{k}\boldsymbol{x} - \omega t) - \beta(\boldsymbol{k}, \omega) \cos(\boldsymbol{k}\boldsymbol{x} - \omega t)). \tag{14}$$

The spatial energy density E is given as in the monochromatic case by averaging $\zeta^2(\boldsymbol{x}, t)$ over a sufficiently large patch of the ocean, for which we may choose our normalization area Ω. In practical units:

$$E = \frac{1}{\Omega} \int \zeta^2(\boldsymbol{x}, t)\mathrm{d}\boldsymbol{x} = \frac{1}{\Omega} \sum_{\boldsymbol{k}, \omega(\boldsymbol{k})} a(\boldsymbol{k}, \omega) a^*(k, \omega) = \frac{1}{\Omega} \sum_{\boldsymbol{k}} (\alpha^2(\boldsymbol{k}, \omega) + \beta^2(\boldsymbol{k}, \omega)). \tag{15}$$

With the aid of eq. (10) we can switch to continuous variables. In that limiting case, the result is independent of the normalization area Ω:

$$E = \int \mathrm{d}\boldsymbol{k} F(\boldsymbol{k})$$
$$F(\boldsymbol{k}) = \frac{1}{2\pi^2} |a(\boldsymbol{k}, \omega)|^2 . \tag{16}$$

In eq. (16) the summation over the two possible values of $\omega(k)$ has been taken into account.

$F(\boldsymbol{k})$ is the energy per surface unit per wavenumber space unit and is usually measured in [m^4]. For fixed meteorological and oceanographical conditions $F(\boldsymbol{k})$ does not show the same value when measured at independent occasions. It is distributed as a χ_2^2 distribution ([52 L], [68 J]). For most applications the expectation value of $F(\boldsymbol{k})$ is the relevant quantity and in the remainder of this section we will associate this mean statistical value with $F(\boldsymbol{k})$. Instead of k_1, k_2 it has become customary to use frequency $f(k) = \omega(k)/2\pi$ and angle θ to determine a fixed $\boldsymbol{k}$ value. We will count θ clockwise from north. This leads to an energy density $E(f, \theta)$ in f, θ space. $E(f, \theta)$ and $F(\boldsymbol{k})$ are connected by

$$F(\boldsymbol{k}) k \frac{\mathrm{d}k}{\mathrm{d}f} = E(f, \theta) . \tag{17}$$

The data obtained from time series $\zeta(x_0, t)$ of the sea surface elevation at a fixed location x_0 (e.g. by wave rider buoys or by wave staffs) allow a determination of $E(f, \theta)$ integrated over the angle θ, a quantity known as the one-dimensional wave spectrum:

$$E(f) = \int E(f, \theta)\mathrm{d}\theta . \tag{18}$$

$E(f)$ is connected with the temporal Fourier transform of $\zeta(\boldsymbol{x}_0, t)$ which is given by:

$$\zeta(\boldsymbol{x}_0, t) = \sqrt{\frac{1}{2T}} \sum b(\omega) \mathrm{e}^{-\mathrm{i}\omega t}, \qquad 0 < t < T$$
$$\omega = 2\pi f = \mathrm{n} \cdot \frac{2\pi}{T}, \qquad \mathrm{n} = \pm 1, \pm 2, \pm 3, \ldots . \tag{19}$$

The discrete ω values result from imposing periodic boundary conditions for the time interval T. The one-dimensional spectrum is then given by

$$E(f) = |b(\omega)|^2, \qquad \omega = 2\pi f . \tag{20}$$

The energy density E defined in eq. (16) is then given by

$$E = \int E(f)\mathrm{d}f . \tag{21}$$

The energy or energy density are quantities that can depend on space and time. In that case we have to add the variables $\boldsymbol{x}$ and t as arguments in $F(\boldsymbol{k})$. $F(\boldsymbol{k}, \boldsymbol{x}, t)$ fulfils the transport equation:

$$\frac{\mathrm{d}}{\mathrm{d}t} F(\boldsymbol{k}, \boldsymbol{x}, t) = \left(\frac{\partial}{\partial t} + \dot{\boldsymbol{x}}(\boldsymbol{k}, \boldsymbol{x}) \cdot \frac{\partial}{\partial \boldsymbol{x}} + \dot{\boldsymbol{k}}(\boldsymbol{k}, \boldsymbol{x}) \cdot \frac{\partial}{\partial \boldsymbol{k}} \right) F(\boldsymbol{k}, \boldsymbol{x}, t) = S$$
$$\dot{\boldsymbol{x}}(\boldsymbol{k}, \boldsymbol{x}) = \frac{\partial \omega}{\partial \boldsymbol{k}} \tag{22}$$
$$\dot{\boldsymbol{k}}(\boldsymbol{k}, \boldsymbol{x}) = -\frac{\partial \omega}{\partial \boldsymbol{x}} .$$

For generation and dissipation processes the source function S has to be introduced in eq. (22). If there is no wave generation or dissipation involved, S vanishes.

6.2.4 Surface waves in deep water

With "deep water" we designate situations for which the dispersion relation (1) does not depend on the water depth h, or, equivalently, for which the relation

$$\tan k \cdot h \approx 1$$

holds.

This situation occurs if the wavelength $\lambda = 2\pi/k$ is small compared to the water depth h. Since in that case the dispersion relation (1) is independent of $\boldsymbol{x}$, the last term on the left hand side of eq. (22) for $F(\boldsymbol{k}, \boldsymbol{x}, t)$ vanishes. The transport equation is then usually written in f, θ coordinates:

$$\frac{\mathrm{d}}{\mathrm{d}t} E(f, \theta) = \left(\frac{\partial}{\partial t} + \frac{g}{2\omega} \frac{\boldsymbol{k}}{k} \cdot \frac{\partial}{\partial \boldsymbol{x}}\right) E(f, \theta) = S\,. \tag{23}$$

Here contributions due to surface tension have been neglected.

In this subsection we consider deep water waves only, for which case there is much more empirical knowledge available than for shallow water situations.

6.2.4.1 The definition of swell and sea (wind waves)

Empirically, swell is defined as waves originating from a distant area and being no longer under the influence of wind forcing. Contrary, waves under active generation by local wind are called sea or wind waves.

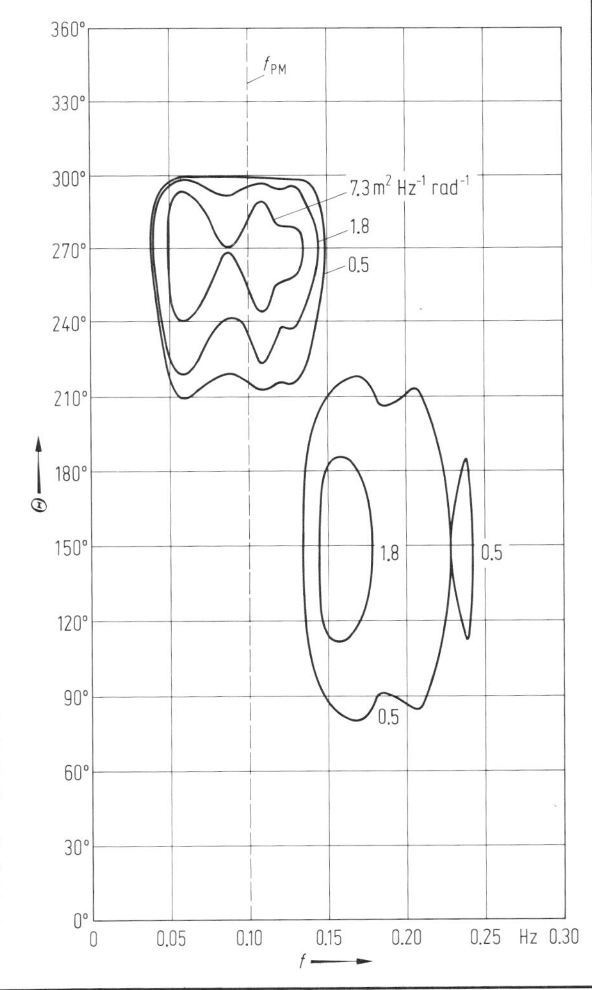

Fig. 1. An example of a two-dimensional energy density spectrum taken from a data set described by Crabb [80 C]. Solid lines are isolines of $E(f, \theta)$ with numbers giving $E(f, \theta)$ in [$\mathrm{m^2\,Hz^{-1}\,rad^{-1}}$]. The significant wave height for this spectrum is 4.22 m, the local wind speed and direction 25 knots and 150°, respectively (the wind wave part shows the same mean direction). The dotted line depicts the Pierson-Moskowitz frequency for the local wind speed. The main energy is concentrated in the swell regime at low frequencies.

Normal oceanic conditions show both sea and swell. Swell is generally of longer wavelength, and propagation direction can differ substantially from the local wind direction. Sea is more short crested, contains different wavelengths, and in general propagates along the wind direction. There are ideal cases where these visual properties of the sea surface can be used to characterize each wave component f, θ space as belonging to the swell or sea regime (Figs. 1, 2).

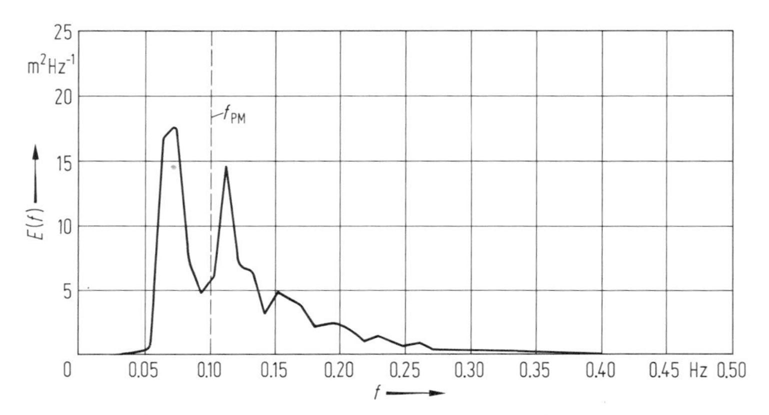

Fig. 2. The one-dimensional energy density spectrum related to the two-dimensional spectrum of Fig. 1. The dotted line depicts the Pierson-Moskowitz frequency for the local wind speed.

In cases where swell and sea is not separated by a spectral gap, the following definition is frequently used:

For a given one-dimensional spectrum and a local wind speed $U_{19.5}$ measured in 19.5 m height, the energy at frequencies below the so called Pierson-Moskowitz frequency f_{PM}, determined by

$$\frac{U_{19.5} \cdot f_{PM}}{g} = 0.14, \tag{24}$$

is called swell, energy at higher frequencies is called sea. For a wind speed U_{10} measured at 10 m height, assuming a neutral atmospheric boundary layer eq. (24) can be expressed by:

$$\frac{U_{10} \cdot f_{PM}}{g} = 0.13. \tag{25}$$

6.2.4.2 The general behavior of wind waves

Yet the exact physical mechanism by which wind generates waves is still not completely understood. However, the statistical behavior of the spectrum for a nonvanishing atmospheric momentum and energy input in the absence of swell is well known and documented. In that case the spectrum has a characteristic uniform shape that is described by a few parameters of which peak frequency and total energy density are the most important. Under a constant atmospheric energy and momentum input, the peak frequency of the spectrum shifts towards lower values, while the total energy increases with increasing distance from shore (Fig. 3). The spectral shape, however, is retained. The spectrum stabilizes as soon as the phase velocity of the spectral peak reaches the Pierson-Moskowitz frequency f_{PM} defined in eq. (24). The averaged phase velocity approximately equals in that case the local wind velocity measured in a height of the order of about 10 m. This final state of development is called fully developed sea.

The described behavior originates from the properties of the source function S in eq. (23). S is usually subdivided into three contributions

$$S = S_{in} + S_{dis} + S_{nl}.$$

S_{in} is the input of energy from the atmosphere. Detailed measurements of this part have been done by Snyder et al. [81 S] in the "Bight of Abaco" experiment. The dissipation processes are lumped together in S_{dis}. This part of S is the least well known. Hasselmann [64 H] developed a formalism to parametrize S_{dis} if the main mechanism is wave breaking.

S_{nl} denotes the nonlinear interaction processes between the monochromatic wave components and redistributes energy in the spectral domain. Hasselmann et al. [73 H, 76 H] discussed this term and its interplay with S_{in} and S_{dis} in detail. The conclusion is that the universal shape of the wind sea spectrum and the migration of the peak frequency towards lower values with increasing distance from shore is mainly caused by S_{nl}. A parametrization of S for actual wave forecasts can be found in Günther et al. [79 G 1, 79 G 2]. Modifications of S for varying wind fields may be found in [81 G].

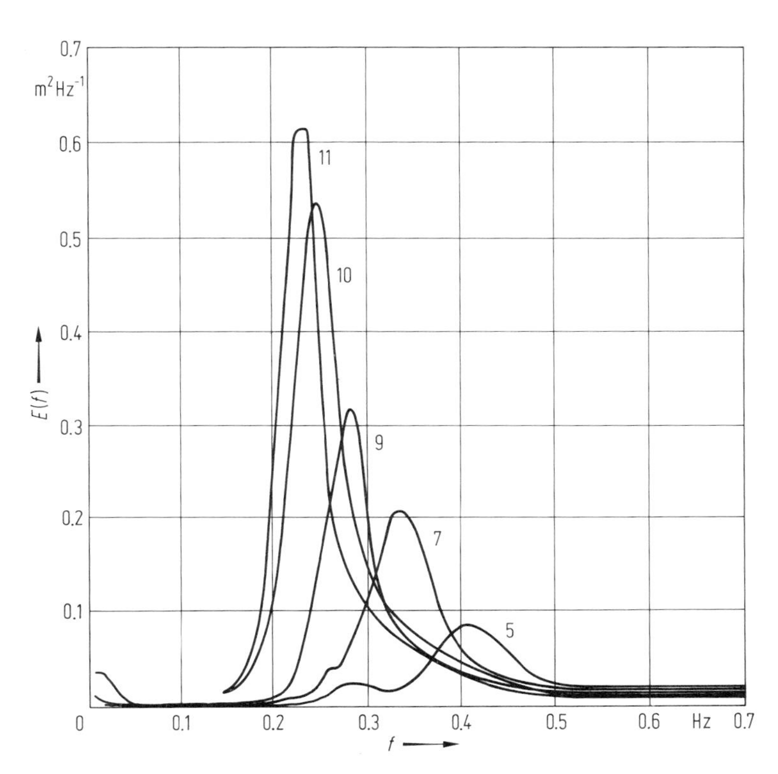

Fig. 3. Evolution of a wave spectrum with increasing fetch for offshore winds. Increasing numbers refer to stations with increasing distance from shore [73 H].

6.2.4.3 The spectral shape for wind waves

Fig. 4 shows the one-dimensional spectrum characterizing the fully developed sea, the so called Pierson-Moskowitz spectrum. The analytical formula was given by Pierson and Moskowitz (1964):

$$E_{PM}(f)=\alpha g^2(2\pi)^{-4}f^{-5}\exp\{-\tfrac{5}{4}(f/f_m)^{-4}\} \tag{26}$$

where $g=9.806\,\mathrm{m\,s^{-2}}$.

f_m denotes the peak frequency. For the fully developed seastate, α has the value 0.008 and f_m is given by the Pierson-Moskowitz frequency f_{PM} defined by eqs. (24) and (25) and therefore determined by the local wind speed. It has become practice to use the name Pierson-Moskowitz spectrum also for $f_m \neq f_{PM}$ and $\alpha \neq 0.008$.

Fig. 3 shows measured spectra for growing wind sea [73 H]. There are several analytical expressions proposed to describe the shape of a growing wind sea. We want to cite three of them.

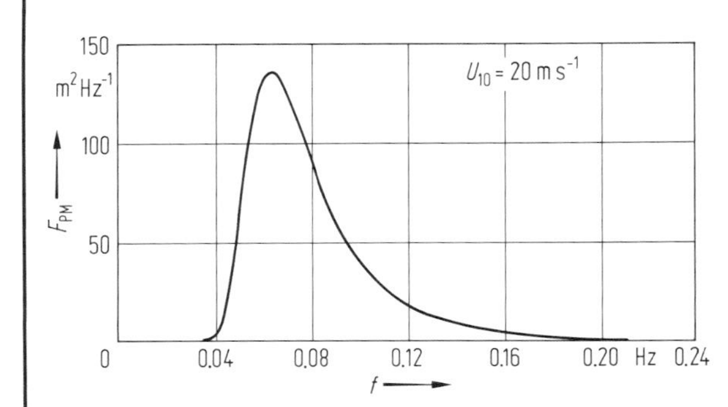

Fig. 4. The Pierson-Moskowitz spectrum, calculated for a wind speed, U_{10} of $20\,\mathrm{m\,s^{-1}}$ [85 S]. For the corresponding Toba spectrum, see Fig. 7.

6.2.4.3.1 The JONSWAP spectrum

The JONSWAP spectrum (Fig. 5) modifies the Pierson-Moskowitz spectrum, eq. (26), to account for the sharp maximum in a growing sea:

$$E_J(f)=E_{PM}(f)\exp\left\{\ln\gamma\left[\exp\frac{-(f-f_m)^2}{2\sigma^2 f_m^2}\right]\right\},$$

with

$$\begin{aligned}\sigma&=\sigma_a=0.07 \quad \text{for} \quad f\leqq f_m,\\ \sigma&=\sigma_b=0.09 \quad \text{for} \quad f\geqq f_m,\end{aligned} \tag{27}$$

and f_m the peak frequency. The average value of γ is 3.3.

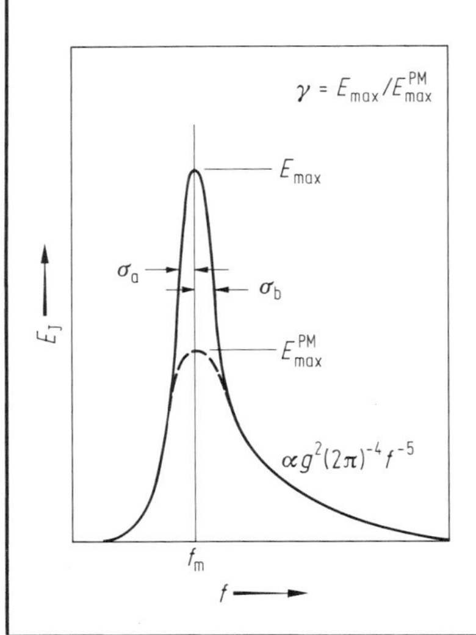

Fig. 5. Illustration for the definition of the five JONSWAP parameters [73 H] σ_a, σ_b, E_{max}, E_{max}^{PM}, and γ.

6.2.4.3.2 The Kruseman spectrum

The Kruseman spectrum (Fig. 6) has a simpler analytical shape [76 K]

$$\begin{aligned} E_K(f) &= af^{-5} && \text{for} \quad f_m \leqq f, \\ E_K(f) &= \frac{a}{1-b} f_m^{-6}(f - bf_m) && \text{for} \quad bf_m \leqq f \leqq f_m, \\ E_K(f) &= 0 && \text{for} \quad 0 \leqq f < bf_m. \end{aligned} \tag{28}$$

For a and b having the values

$$\begin{aligned} a &= 0.80\alpha g^2 (2\pi)^{-4} \\ b &= 0.72, \end{aligned} \tag{29}$$

the Kruseman and JONSWAP spectrum with the same peak frequency f_m contain the same total energy (Fig. 6).

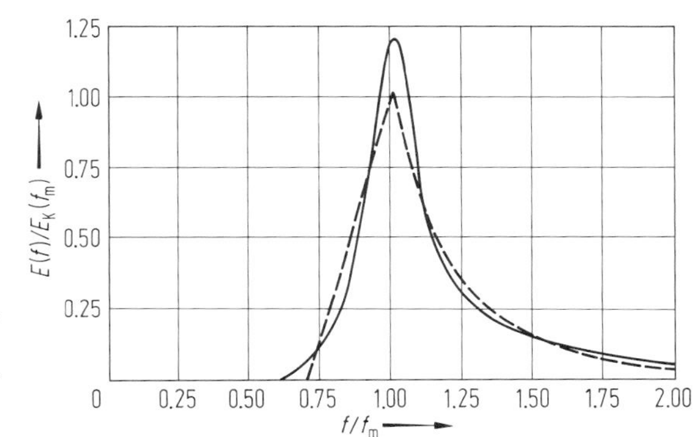

Fig. 6. A comparison between the model spectrum proposed by Kruseman (dashed line) and the JONSWAP spectrum (solid line). Both spectra contain the same total energy [76 W].

6.2.4.3.3 The Toba spectrum

Toba [78 T] proposed a spectral shape for a growing sea by including the local friction velocity U_* which is defined by the momentum transfer τ through the air-sea interface,

$$\varrho_{air} U_*^2 = \tau \tag{30}$$

with ϱ_{air} the air density. The ratio between U_*^2 and U_{10}^2, where U_{10} is the wind speed measured at a height of 10 m, is given by the drag coefficient C_{10}:

$$C_{10} U_{10}^2 = U_*^2. \tag{31}$$

Toba used in his work $C_{10} = 1.2 \cdot 10^{-3}$. The spectral shape (Fig. 7) is given by:

$$\begin{aligned} E_T(f) &= (2\pi)^{-3} \alpha_s g U_* f^{-4} && \text{for} \quad f \geqq f_m \\ E_T(f) &= (2\pi)^{-3} \alpha_s g U_* f_m^{-8} f^4 && \text{for} \quad f \leqq f_m \end{aligned} \tag{32}$$

with f_m the peak frequency and α_s a constant.

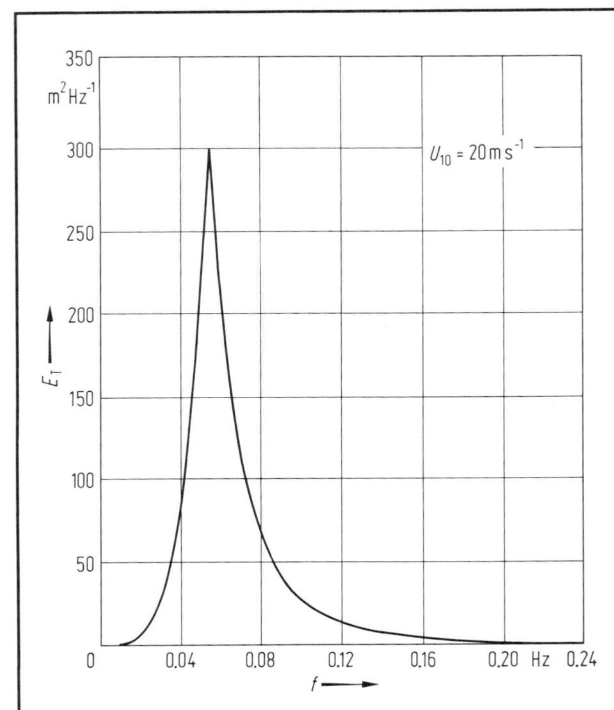

Fig. 7. The model spectrum proposed by Toba, calculated for a wind speed U_{10} of $20\,\mathrm{m\,s^{-1}}$ as stationary state is reached. For the corresponding Pierson-Moskowitz spectrum, see Fig. 4 [85 S].

6.2.4.4 The directional distribution of wave energy

If the two-dimensional spectrum is written in the form

$$E(f,\theta)=E(f)\cdot D(f,\theta),$$

where $E(f)$ is the one-dimensional spectrum, $D(f,\theta)$ is called the directional distribution.

$D(f,\theta)$ has been proposed by Mitsuyasu et al. [75 M] to be

$$D(f,\theta)=\frac{\Gamma(s+1)}{2\sqrt{\pi}\,\Gamma\left(s+\frac{1}{2}\right)}\left(\cos^2\frac{\theta}{2}\right)^s, \tag{33}$$

where Γ denotes the gamma function and s is frequency-dependent. $\theta=0$ corresponds to the direction of wind. $D(f,\theta)$ is normalized such that

$$\int D(f,\theta)\mathrm{d}\theta=1. \tag{34}$$

Mitsuyasu et al. [75 M] found for s the frequency dependence:

$$\begin{aligned}
\frac{s}{s_m}&=\left(\frac{f}{f_m}\right)^{-2.5} \quad \text{for} \quad f\geqq f_m\\
\frac{s}{s_m}&=\left(\frac{f}{f_m}\right)^{5} \quad \text{for} \quad f<f_m\\
s_m&=0.116\tilde{f}_m^{-2.5}\\
\tilde{f}_m&=\frac{U_{10}\cdot f_m}{g}.
\end{aligned} \tag{35}$$

This means that the directional spread for wind waves is narrowest at the peak frequency f_m. Hasselmann et al. [80 H] got similar relations between s and f from measurements in the North Sea, but reported for s_m the values:

$$\begin{aligned}
s_m&=9.77 \quad \text{for} \quad f\geqq f_m\\
s_m&=6.97 \quad \text{for} \quad f<f_m.
\end{aligned} \tag{36}$$

6.2.4.5 The statistical properties of the spectral parameters for wind waves

As explained in subsect. 6.2.4.2, wind wave spectra show a self similar shape. Therefore it is possible to use parametric model spectra as given in subsect. 6.2.4.3. For the parameters defining the model spectra, several empirical statistical relationships have been established for simple, well defined situations.

As an example we consider a straight coast line with perpendicular off-shore blowing wind. This is called an ideal fetch limited geometry and several measurements for seastate development in that case are known [69 M, 73 H]. The results are given in nondimensional quantities ($g=9.806\ \mathrm{m\,s^{-2}}$):

Nondimensional energy:

$$\varepsilon = E\frac{g^2}{U_{10}^4}. \tag{37}$$

Nondimensional fetch:

$$\tilde{x} = \frac{gx}{U_{10}^2}, \quad x = \text{normal distance from shore} \tag{38}$$

The JONSWAP relation [73 H] between energy and fetch is

$$\varepsilon = 1.6\cdot 10^{-7}\,\tilde{x} \tag{39}$$

Since it belongs to a special location and to special meteorological conditions, this result may be biased. To show the uncertainty for relation (39), Fig. 8 shows a number of ε vs. $\tilde{x}$ relations used by numerical wave prediction models [85 S]. As an average relation for the North Sea

$$\varepsilon = 2.012\cdot 10^{-7}\,\tilde{x} \tag{40}$$

achieved good results in hindcast studies.

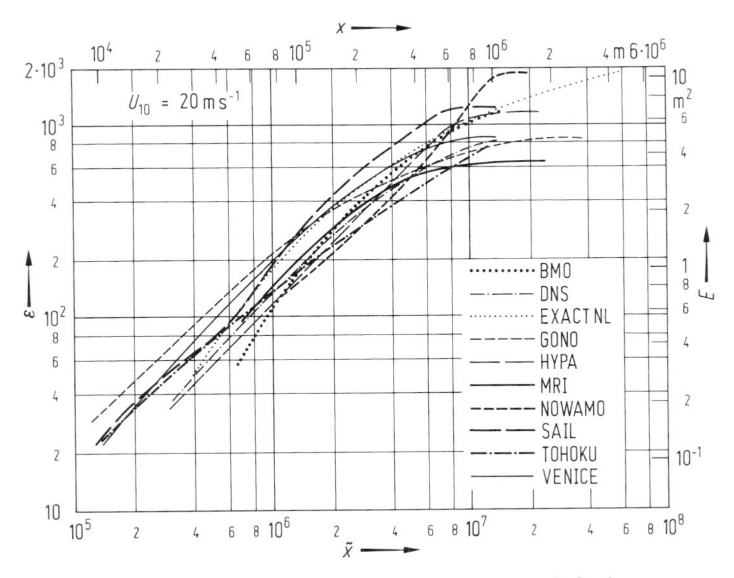

Fig. 8. The relation between energy E and fetch x calculated by ten different wave prediction models [85 S] with $U_* = 0.85\ \mathrm{m\,s^{-1}}$ and $U_{10} = 20\ \mathrm{m\,s^{-1}}$. ε and $\tilde{x}$ denote nondimensional energy and fetch, respectively.

Measurements for the spectral peak versus fetch are reported in the same publications as for the ε vs. $\tilde{x}$ relations above. The JONSWAP relation [73 H] is

$$\tilde{f}_m = 3.5\,\tilde{x}^{-0.33} \tag{41}$$

with the nondimensional peak frequency $\tilde{f}_m$ defined in eq. (35).

To give an estimate of the uncertainty for this relation, Fig. 9 shows eq. (41) together with similar relations used by various numerical wave prediction models [85 S].

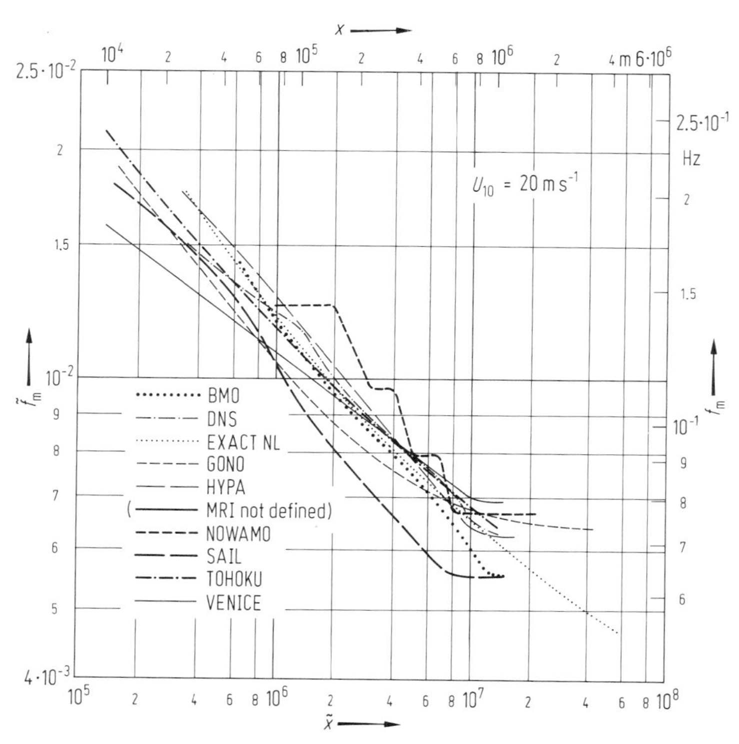

Fig. 9. The relation between peak frequency f_m and fetch x calculated by ten different wave prediction models [85 S] with $U_* = 0.85\,\mathrm{m\,s^{-1}}$ and $U_{10} = 20\,\mathrm{m\,s^{-1}}$. $\tilde{f}_m$ and x denote nondimensional peak frequency and fetch, respectively.

Phillips parameter versus peak frequency:

The parameter α of the JONSWAP spectrum, eqs. (26) and (27), has been proposed originally by Phillips [58 P] to be a constant. The JONSWAP experiment [73 H] showed α to be dependend on fetch, and Hasselmann et al. [76 H] showed that there is a general relation between f_m and α for a growing sea which is not restricted to fetch limited conditions. This relation is:

$$\alpha = 0.033\,\tilde{f}_m^{2/3}. \tag{42}$$

From the statistical behavior of energy and peak frequency in a fetch limited case, the development for the duration limited situation can be calculated. This situation means the development of a wind wave spectrum with time in the case of negligible influence of an upwind coastline (the ideal is an infinite ocean, starting with no waves at time $t=0$, and exhibiting a wind field constant in time and space). Fig. 10 shows a diagram to estimate energy and peak frequency of a spectrum for duration and time limited situation (from [76 G]).

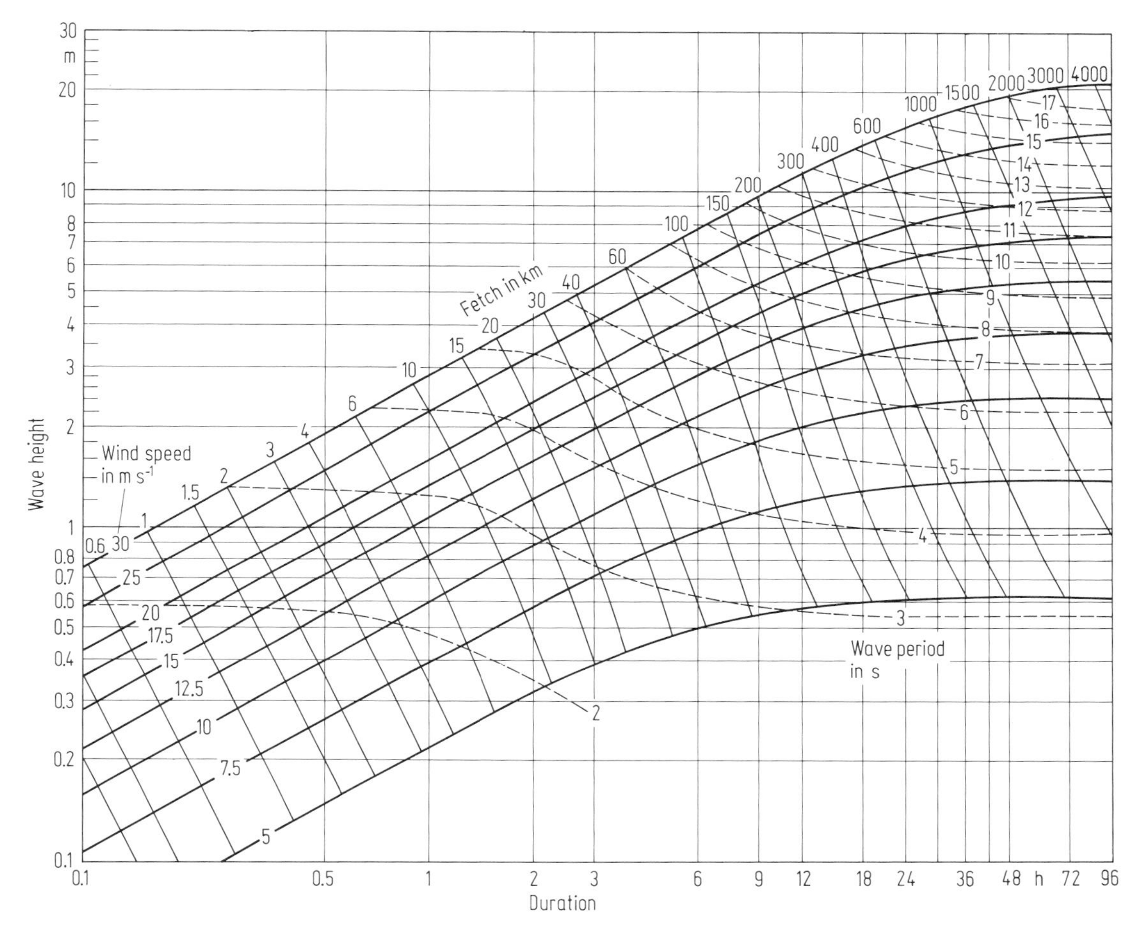

Fig. 10. Diagram to determine fetch limited and duration limited wave height and period. Example for fetch limitation: for a wind speed of 15 m s^{-1} and a fetch of 200 km the wave height is 4 m for the final stationary sea state and the period is $T=7.4$ s. Example for duration limitation: For a wind speed of 15 m s^{-1} after 6 hours the wave height is 2.9 m and the period $T=5.8$ s [76 G].

6.2.4.6 Frequently used spectral characteristics to classify sea state

Especially if a wave spectrum does not belong to any of the classes of model spectra described in subsect. 6.2.4.3, for instance in case of a swell or inhomogeneous wind fields, the spectral moments

$$m_n = \int f^n E(f) \mathrm{d}f, \quad n=0,1,2\ldots \tag{43}$$

can be used to describe the spectral behavior. m_0 denotes the total energy. Together with m_1 it provides information about the average frequency of the spectrum and can be related to the peak frequency f_m if the shape of the spectrum is known.

A parameter derived from m_0 is the significant wave height:

$$H_s = 4\sqrt{m_0}\,.$$

The following parameters are used to describe a wave spectrum [76 W]:

$$\begin{aligned}
&f_m \quad \text{peak frequency (modal frequency)}\\
&T_m = f_m^{-1} \quad \text{peak period (modal period)}\\
&f_l \quad \text{lower limit of energy containing wave frequencies}\\
&T_l = f_l^{-1}\\
&f_u \quad \text{upper limit of energy containing wave frequencies}\\
&T_u = f_u^{-1}\\
&T_{02} = \sqrt{\frac{m_0}{m_2}} \quad \text{wave period derived from the second moment}\\
&\varepsilon^2 = \frac{m_0 m_4 - m_2^2}{m_0 m_4} \quad \text{width of the spectrum}\,.
\end{aligned} \tag{44}$$

For the directional properties the averaged wavenumber vector for each frequency

$$\begin{aligned}
\bar{k}_1(f) &= k(f) \int D(f,\theta) \sin\theta \,\mathrm{d}\theta\\
\bar{k}_2(f) &= k(f) \int D(f,\theta) \cos\theta \,\mathrm{d}\theta
\end{aligned} \tag{45}$$

is used to derive the mean direction θ_0,

$$\theta_0(f) = \arctan \frac{\bar{k}_1(f)}{\bar{k}_2(f)}, \tag{46}$$

and the directional spread Δ,

$$\Delta^2 = 2\left(1 - \frac{(\bar{k}_1^2 + \bar{k}_2^2)^{1/2}}{k(f)}\right). \tag{47}$$

$k(f)$ denotes the wavenumber corresponding to frequency f, $2\pi f = \omega(k)$, with $\omega(k)$ the dispersion relation. Δ goes to zero for a narrow angular distribution. The parameter $s(f)$, determining the distribution (33) can be expressed by the corresponding Δ:

$$s = \frac{2}{\Delta^2} - 1\,. \tag{48}$$

The Fourier coefficients a_n, b_n for the directional distribution

$$D(f,\theta) = a_0 + \sum_{n=1}^{\infty} (a_n \cos n\theta + b_n \sin n\theta) \tag{49}$$

are also in use, because for lower numbers n they can be obtained from the cross spectra of directional sensitive buoys [63 L].

6.2.4.7 Parameters to classify sea state from time series

Before spectra of time series of wave records became a major tool of monitoring sea state, the registered time series itself has been used to extract characteristic parameters. We subsequently list some of them. As far as possible, the definitions are taken from WMO publication 446 [76 W]:

$H = m_0^{1/2}$ average wave height

H_{max} maximum wave height occuring in a record

$\bar{T}$ average wave period; the time obtained by dividing the record length by the number of downcrossings in the record

$H_{1/n}$ the averaged height of the 1/n highest waves (i.e.: All relative wave maxima from the record are ordered in descending magnitude. If N is the total number of waves in the record, the N/n highest waves are taken and $H_{1/n}$ is then computed as the average height of these waves).

Frequently used values for n are:

n=3: $H_{1/3}$ is approximately equal to H_s defined by the spectrum

n=10: $H_{1/10}$ average height of the ten per cent highest waves

n=100: $H_{1/100}$ average height of the one per cent highest waves.

Usually wave records of 20···30 min are used to derive the above parameters. $H_{1/3}$ resembles H_s derived from the spectrum very closely. T_{02} from eq. (44) is often used. If $\bar{T}$ has to be expressed by the spectral quantities of eq. (44), the relation

$$\bar{T} = T_{02} \tag{50}$$

is in use. For a narrow spectrum, of course,

$$\bar{T} \approx f_m^{-1} \approx T_{02}. \tag{51}$$

If it is necessary to construct a model wind sea spectrum from the two parameters $\bar{T}$ and $H_{1/3}$, the connected Pierson-Moskowitz spectrum may be taken:

$$E_{PM}(f) = 0.25 H_{1/3}^2 f^{-5} \bar{T}^{-4} \exp[-(f\bar{T})^{-4}] \tag{52}$$

(Brettschneider spectrum) [76 B].

6.2.4.8 Visual observations

Most of todays wave data come from visual observations on board of ships. They are delivered to the national weather services and used to establish the synoptic weather maps. Especially the official weather ships are supplied with the encoded sea state on these maps. The following parameters are defined for visual observation in the WMO publication [71 W]:

sea — system of waves observed at a point which lies within the wind field producing the waves.

swell — system of waves observed at a point remote from the wind field which produced the waves, or observed when the wind field which generated the waves no longer exists.

breaker — the collapse of a whole wave resulting from its running into very shallow water (of the order of twice the wave height).

surf — the broken water between the shore line and the outermost line of the breakers.

breaking sea — the partial collapse of the crest of a wave caused by:
(a) action of the wind;
(b) steepening of waves due to their encountering a contrary current or tidal stream;
(c) steepening of waves due to their running into shoal water not shallow enough to cause a breaker.

wavelength — the horizontal distance between successive crests or throughs.

wave height — the vertical distance between trough and crest.

wave period — the time between the passage of two successive wave crests past a fixed point.

wave speed — the distance travelled by a wave in a unit of time.

The mentioned WMO publication gives the detailed procedure and the typical mistakes made by estimating the sea state visually. There are several publications showing visual observations against measurements. Fig. 11 shows results from [69 N].

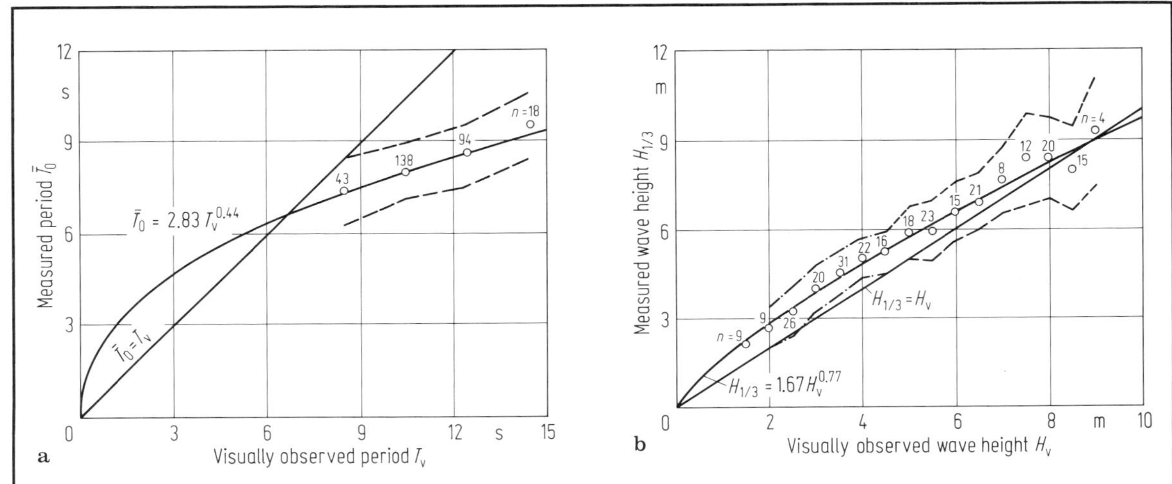

Fig. 11. Comparison between visually observed and measured wave period (a) and wave height (b) [69 N]. Circles give the mean values of the respective measured quantities, dashed lines the standard deviation, n denotes the number of observations. Full lines represent analytic relations between measured and visually observed quantities.

6.2.5 Surface waves in shallow water

The shallow water case will not be treated in its full complexity. The basic assumption will always be that the bottom topography is smoothly varying. Smoothly means that the depth variations over length scales of several wavelengths are negligible in the sense that we can subdivide the ocean under consideration into patches large enough to contain several wavelengths of the longest waves under consideration, and small enough to neglect varying water depth. For such a topography, plane waves are still a meaningful tool to describe surface gravity waves if we allow the k-vector of a wave train to vary while propagating over large distances. (WKB-method mentioned in sect. 6.1.)

A wave mode in that case is defined by the mode parameters k_x, k_y at any location of the ocean, but k_x, k_y change slowly with depth according to the laws defined in subsect. 6.2.5.1, while the frequency stays constant. The phenomenon of changing $\boldsymbol{k}$-vector is called refraction.

The energy density changes if the propagation velocity of the energy is inhomogeneous. This is denoted as shoaling and will be treated in subsect. 6.2.5.2. Bottom induced dissipation processes in shallow water substantially complicate the energy balance. They are discussed in subsect. 6.2.5.3. Subsect. 6.2.5.4 treats the problem of wave breaking induced by decreasing water depth. In subsect. 6.2.5.5 statistical properties of wind waves in shallow water are discussed.

A detailed study of shallow water processes may be found in the shore protection manual from CERC [73 C].

6.2.5.1 Refraction

Continuity of the motion requires that, in the absence of an explicit time dependence, the frequency ω_0 of each wave mode is a constant also for varying water depth. The spatial and temporal change of the wavenumber vector of a propagating wave group can be calculated from

$$\omega_0 = \omega(k, h(x, y)), \tag{53}$$

where $h(x, y)$ denotes the local depth, and from

$$\dot{\boldsymbol{k}} = -\frac{\partial \omega}{\partial \boldsymbol{x}}, \qquad \dot{\boldsymbol{x}} = \frac{\partial \omega}{\partial \boldsymbol{k}}. \tag{54}$$

For the physical meaning of $\dot{\boldsymbol{k}}$, $\dot{\boldsymbol{x}}$ for wave groups, the reader is referred to sect. 6.1.

6.2.5.2 Shoaling of waves

If we consider an ascending bottom with contours of equal depth running parallel to a straight shore line, the energy propagation equation (22) becomes particularly simple. If the coordinate x_1 is chosen normal to shore, we get for stationary conditions, vanishing source function S, and for homogeneous energy density in x_2-direction:

$$\frac{\partial}{\partial x_1}(\dot{x}_1 F(\boldsymbol{k}, x_1)) + \frac{\partial}{\partial k_1}(\dot{k}_1 F(\boldsymbol{k}, x_1)) = 0$$

$$\dot{x}_1 = \frac{\partial \omega}{\partial k_1} \tag{55}$$

$$\dot{k}_1 = -\frac{\partial \omega}{\partial x_1}.$$

Integration of eq. (55) yields for arbitrary x_1:

$$I = \int \dot{x}_1 \cdot F(\boldsymbol{k}, x_1)\mathrm{d}\boldsymbol{k} = a_1$$

$$\dot{x}_1 = \frac{\partial \omega}{\partial k_1}. \tag{56}$$

I is called the energy flux, a_1 is an integration constant.

For a narrow banded energy spectrum centered at wavenumber $\boldsymbol{k}_0$ and with total energy $E(x_1) = \int F(\boldsymbol{k}, x_1)\mathrm{d}\boldsymbol{k}$, $\dot{x}_1$ can be taken outside the integral and eq. (56) becomes

$$\dot{x}_1(\boldsymbol{k}_0) \cdot E(x_1) = a_1. \tag{57}$$

For waves running from deep into shallow water, the group velocity $\dot{x}_1(k_0)$ defined by eqs. (55) and (1) increases and, after reaching a maximum, decreases again. Because of eq. (57), this means that the total energy decreases first and then increases again. To get an impression of the magnitude of variations involved, we assume that waves with energy E_0 and wavelength L_0 in deep water run into a direction perpendicular to shore. Then energy will reach its minimum of $E \approx 0.8\, E_0$ at a depth of $D = 0.15\, L_0$. At a depth of $0.05\, L_0$ the energy E reaches its deep water value E_0 again. Fig. 12 gives the detailed variation of the significant wave height H_s with decreasing water depth.

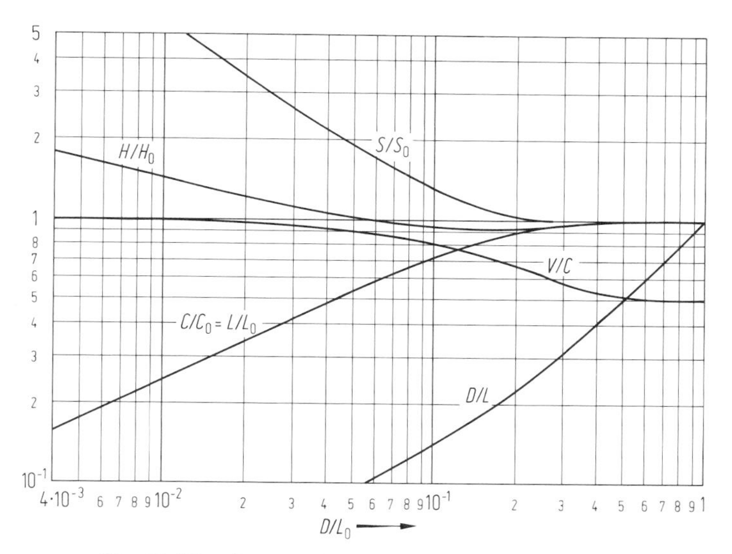

Fig. 12. The diagram shows the variation of various sea state parameters for waves approaching shallow water. D denotes water depth, H wave height, L wavelength, C phase speed, V group velocity, $S = H/L$ steepness. The subscript "0" refers to deep water values.

6.2.5.3 Bottom-induced dissipation

In this and the following subsection we consider changes in the source function S of eq. (22) induced by finite water depth.

There are four mechanisms discussed in the literature as a sink for wave energy:
Bottom friction
Percolation
Sediment motion
Wave reflection by varying bottom topography.

All four are connected with the action of the orbital motion and the varying pressure near the bottom. There is an additional loss of energy caused by increased breaking in shallow water. This process is discussed in subsect. 6.2.5.4.

Bottom friction

This mechanism is treated theoretically by Hasselmann and Collins [68 H]. By assuming a quadratic friction law for currents U_b at the bottom, the momentum loss τ can be written:

$$\tau = -\varrho c_f |\boldsymbol{U}_b| \boldsymbol{U}_b,$$

where τ is the momentum loss of the current per time and surface unit, ϱ is the density of water, and c_f the friction coefficient. Hasselmann and Collins concluded that for special conditions it is possible to express the spatial decrease of the energy flux I for a monochromatic wave by

$$\frac{dI}{d\xi} = -\lambda I$$

$$\xi = \int^{x} \frac{k^2}{\omega^2 v_g \cosh^2 k \cdot h} dx \tag{58}$$

$$v_g = \dot{x} = \text{group velocity}.$$

λ depends on additional parameters like, for instance, the mean current. It has, however, become customary to assume a law like eq. (58) and calibrate λ by measurements.

Percolation

This term describes the energy and momentum lost by a current in a porous medium. If the sediment is sufficiently porous, the orbital wave motion and thereby the total wave energy is efficiently attenuated. Shemdin et al. in [80 S] discussed this process and give references to earlier literature.

Bottom motion

As the sea bed is not rigid but able to move under the varying pressure of the surface waves, energy can be exchanged with the sediment, provided the phase difference of pressure and vertical motion, differs from $\pm 90°$.

An example for the calculation of the energy transfer has been published by Rosenthal [78 R]. A review of this process can be found in [80 S], together with references for further literature.

It seems reasonable that all three effects, bottom friction, percolation, and bottom motion, are important for wave attenuation, depending on the properties of the sediment. As they have similar dependence on orbital wave velocity, a parametrization of the form of eq. (58) seems applicable to all of them.

Wave reflection by varying bottom topography

This effect is treated by Long [78 L]: irregular bottom topography for which the size of the irregularities matches the wavelength of the surface waves can partly reflect the incoming waves. Therefore the incoming wave energy decreases.

6.2.5.4 Wave breaking in shallow water

Although the phenomenon of wave breaking in shallow water is very well known qualitatively, an exact theory from first principles is not yet available. The breakers are phenomenologically described as

spilling breaker continuous loss of energy by generating foam on top of wave crest. It exists on gently sloping beaches.

plunging breaker exists on steep beaches. Top of wave crests "overtakes" the rest of the waves and falls down on the front side. Energy is almost completely lost in a short time (different from the spilling breaker).

surging breaker appear on very steep coast lines. Therefore the wave crest is raised before it can develop a plunging breaker. There is only a thin top foam layer generated and the up and down moving water surface resembles the movement of a standing wave.

Wave breaking, which occurs also in deep water if the waves are sufficiently high, can be defined by the condition that the particle velocity at the wave crests is larger than the phase speed, which is the propagation velocity of the crests. In shallow water the wave crests grow due to shoaling and the phase speed decreases. From experience it is known that waves break at a depth of

$$d = 1.2\,H_b,$$

where H_b is the breaker height. It should be mentioned that in shallow water the linear approximation for waves breaks down and a nonlinear description is necessary [74 W].

For a more detailed discussion of breaking in shallow water we may refer to the Shore Protection Manual [73 C].

6.2.5.5 Statistical behavior of wind waves in shallow water

In this subsection we consider only cases seaward from the surf zone. Although there is additional dissipation and a change in the dispersion relation as compared to deep water, there is an indication of the existence of a uniform spectral shape also for shallow water wave spectra. Such a feature has been earlier postulated by Kitaigorodskii et al. [75 K] for the high frequency portion of the wave spectrum. The statistical growth of wave energy or the development of the spectral parameters with increasing fetch is not as well documented as in the deep water case. Best known are limiting values of wave height in shallow water. Groen and Dorrestein [76 G] published a diagram (Fig. 13) that allows the extration of significant wave height and period.

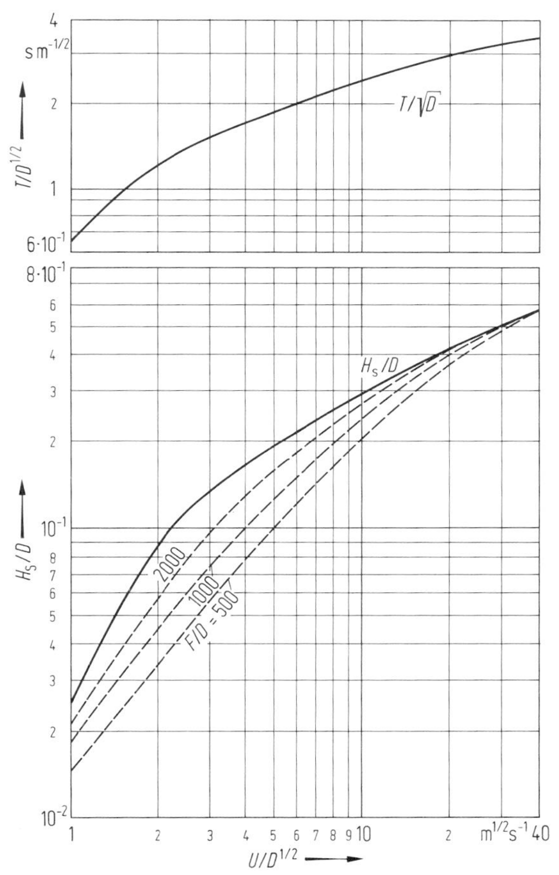

Fig. 13. Estimated behaviour of sea state (wave period T [s] and significant wave height H_s [m]) over constant finite water depth D [m] for wind speed U [m s^{-1}]. Solid lines give the result for the fully developed sea state without fetch or time limitation. Dotted lines show behaviour for limited fetch F [m]. Example: For a wind speed of 12 m s^{-1}, $D = 4$ m, $F = 20$ km, we get from the solid line (because $F/D \gg 2000$): $H_s \approx 4 \cdot 0.22\,\text{m} = 0.88$ m and $T \approx 2 \cdot 2.0\,\text{s} = 4$ s.

Rosenthal

6.2.6 References for 6.2

52 L Longuet-Higgins, M.S.: J. Mar. Res. Vol. **XI**, Nr. 3 (1952) 245.
58 P Phillips, O.M.: J. Fluid Mech. **4** (1958a) 426.
62 C Courant, R., Hilbert, D.: Methods of Mathematical Physics, New York, London: Springer **1962**.
63 L Longuet-Higgins, M.S., Cartwright, D.E., Smith, N.D., in: Ocean Wave Spectra, Englewood Cliffs, N.J.: Prentice-Hall Inc. **1963**, pp. 111–136.
64 H Hasselmann, K.: Boundary-Layer Meteorol. **6** (1964) 107.
68 H Hasselmann, K., Collins, J.I.: J. Mar. Res. **26** (1968) 1.
68 J Jenkins, G.M., Watts, D.G.: Spectral Analysis and its Applications, Holden-Day, **1968**.
69 M Mitsuyasu, H.: On the growth of the spectrum of wind-generated waves. 2. Rep. Res. Inst. Appl. Mech., Kyushu Univ. **17** (1969) 235–243.
69 N Nordenstrøm, N.: Methods for predicting long term distribution of wave loads and probability of failure for ships. Appendix II. Relations between visually estimated and theoretical wave heights and periods. Det Norske Veritas, Research Department, Report no. 69–22-S, **1969**.
71 W World Meteorological Organization: Guide to Meteorological Instrument and Observing Practices, WMO Publication No. 8 TP 3, **1971**.
73 C CERC: Shore Protection Manual, U.S. Army Coastal Engineering Research Center, Corps of Engineers. U.S. Government Printing Office Washington, D.C. 20402, **1973**.
73 H Hasselmann, K., Barnett, T.P., Bouws, E., Carlson, H., Cartwright, D.E., Enke, K., Ewing, J.A., Gienapp, H., Hasselmann, D.E., Kruseman, P., Meerburg, A., Müller, P., Olbers, D.J., Richter, K., Sell, W., Walden, H.: Dtsch. Hydrogr. Z. **A** (**8°**), No. 12 (1973).
74 W Witham, G.B.: Linear and Nonlinear Waves. New York: John Wiley and Sons, **1974**.
75 K Kitaigorodskii, S.A., Krasitskii, V.P., Zaslavskii, M.M.: J. Phys. Oceanogr. **5** (1975) 410.
75 M Mitsuyasu, H., Tasai, F., Suhara, T., Mizuno, S., Okhuso, M., Honda, T., Rikiishi, K.: J. Phys. Oceanogr. **5** (1975) 750.
76 B Brettschneider, C.L., Tamage, E.E.: Hurricane Wind and Wave Forecasting Techniques Proc. 15th Coastal Engng. Conf., Honolulu, ASCE, **1976**, 202.
76 G Groen, P., Dorrestein, R.: Zeegolven, Staatsdrukkerij en Uitgeverijkbedrijkt, 's-Gravenhage, **1976**.
76 H Hasselmann, K., Ross, D.B., Müller, P., Sell, W.: J. Phys. Oceanogr. **6** (1976) 200.
76 K Kruseman, P.: Two practical methods for forecasting wave components with periods between 10 and 25 seconds near Hoek van Holland, Koninklijk Nederlands Meteorologisch Instituut, Wetenschapelijk Rapport 76-1, **1976**.
76 W World Meteorological Organization: Handbook on Wave Analysis and Forecasting, WMO-No. 446, **1976**.
77 P Phillips, O.M.: The Dynamics of the Upper Ocean, Cambridge: University Press, **1977**.
78 L Long, R.B.: J. Geophys. Res. **78** (33) (1973) 7861.
78 R Rosenthal, W.: J. Geophys. Res. **83** (1978) 1980.
78 T Toba, Y.: J. Phys. Oceanogr. **8** (1978) 494.
79 G 1 Günther, H., Rosenthal, W., Weare, T.J., Worthington, B.A., Hasselmann, K., Ewing, J.A.: J. Geophys. Res. **84** (1979) 5727.
79 G 2 Günther, H., Rosenthal, W., Richter, K.: J. Geophys. Res. **84** (1979) 4855
80 C Crabb, J.A., in: Power from sea waves Count, B.M., (ed.) (Based on Conference on "Power from sea waves" organized by Institute of Mathematics and its applications, Edinburgh, 26–28 June, 1979), London: Academic Press, **1980**.
80 H Hasselmann, D.E., Dunckel, M., Ewing, J.A.: J. Phys. Oceanogr. **10** (1980) 1264.
80 S Shemdin, O.H., Hsiao, S.V., Carlson, H.E., Hasselmann, K., Schulze, K.: J. Geophys. Res. **85**, No. C 9 (1980) 5012.
81 G Günther, H., Rosenthal, W., Dunckel, M.: J. Phys. Oceanogr. **11**, No. 5 (1981) 718
81 S Snyder, R.L., Dobsen, F.W., Elliot, J.A., Long, R.B.: J. Fluid Mech. **102** (1981) 1.
85 S SWAMP group (Allender, J.H., Barnett, T.P., Bertotti, L., Bruinsma, J., Cardone, V.J., Cavaleri, L., Ephraums, J., Golding, B., Greenwood, A., Guddal, J., Günther, H., Hasselmann, K., Hasselmann, S., Joseph, P., Kawai, S., Komen, G.J., Lawson, L., Linné, H., Long, R.B., Lybanon, M., Maeland, E., Rosenthal, W., Toba, Y., Uji, T., de Voogt, W.J.P.): An intercomparison study of wind wave prediction models, part 1, in: Ocean Wave Modelling, New York: Plenum Press **1985**.

6.3 Internal gravity waves

6.3.0 List of symbols and indices

Symbols

A	action spectrum
$A^{s}_{\nu\boldsymbol{K}}$	amplitude of the state vector, corresponding to mode ν, wavevector $\boldsymbol{K}$ and branch s
$a(\boldsymbol{k})$	amplitude of wave with wavevector $\boldsymbol{k}$
$a^{\nu}_{\boldsymbol{K}}$	amplitude of wave, corresponding to mode ν and wavevector $\boldsymbol{K}$
B	buoyancy associated with large-scale motion
$B^{s}_{a\nu\boldsymbol{K}}$	decomposition operator, eq. (46)
c	phase speed, $c=\omega/k$
E	energy spectrum
F	source term arising from nonlinearities in the equation of motion
F^{a}	spectrum of atmospheric forcing
F^{g}	surface wave spectrum
F_{i}	source term in velocity equation, $i=1,2,3$, including viscous effects
F^{τ}	wind stress spectrum
f	Coriolis parameter, $f=2\Omega_{e}\sin\phi$
g	gravity
H	mean depth of sea bottom
H_{ab}	operator governing evolution of state vector, a, b = 1, 2, 3
h	bottom topography
$\boldsymbol{K}$	horizontal wavevector, components K_1, K_2
$\boldsymbol{k}$	wavevector, components k_i, $i=1,2,3$
$\mathscr{L}$	horizontal Laplace operator, $\mathscr{L}=\dfrac{\partial^2}{\partial x_\alpha \partial x_\alpha}$
$\mathscr{M}$	integral operator, eq. (38)
$\mathscr{N}$	integral operator, eq. (38)
N	buoyancy (angular) frequency
P	pressure associated with large-scale motion; power spectrum
p	pressure, scaled by constant reference density
Q	source term in buoyancy equation, including diffusive effects
q_{a}	source terms, $a=1,2,3$
Ri	Richardson number, $Ri=N^2/(\partial U/\partial z)^2$
$\boldsymbol{r}$	refraction rate, components r_i, $i=1,2,3$
S_{n}	source term, $n=1,\ldots,7$
T_{a}	transfer function related to atmospheric forcing
t	time
$\mathbf{U}$	shear flow current, components U_i, $i=1,2,3$, associated with large scale motion
$\boldsymbol{\mathscr{U}}$	eigenvector of small-scale wave representation, components $\mathscr{U}_i$, $i=1,2,3$
u_{i}	velocity component, $i=1,2,3$
$\boldsymbol{v}$	intrinsic group velocity, components v_i, $i=1,2,3$
w	vertical velocity, $w=u_3$
$\boldsymbol{X}$	horizontal position vector, components x_1, x_2
$\boldsymbol{x}$	position vector, components x_i, $i=1,2,3$, $x_3=z$ upward position
z	vertical coordinate, upward position
α	modulus of horizontal wavevector
β	modulus of horizontal wavevector
β_{a}	eigenfunction of H_{ab}, $a=1,2,3$
δ_{ij}	Kronecker symbol
ε	wave energy density
ε_{a}	polarization associated with β_a, $a=1,2,3$
ε_{ij}	Levi-Civita symbol in 2 dimensions, $\varepsilon_{ij}=\delta_{i1}\delta_{j2}-\delta_{i2}\delta_{j1}$
ζ	sea surface displacement
η	deviation of sea bottom from mean depth
θ	inclination of $\boldsymbol{k}$

λ	eigenvalue, $\lambda^2 = f^2K^2/(\omega^2 - f^2)$
ν	modal index, $\nu = 0, 1, 2, \ldots$
ξ	wave displacement of fluid
ϱ	density, scaled by constant reference density
ϕ	geographic latitude
$\boldsymbol{\phi}$	energy flux vector
$\phi(z)$	vertical eigenfunction
$\varphi_{\boldsymbol{K}}^{\nu}$	vertical eigenfunction for mode ν and wavevector $\boldsymbol{K}$, $\nu = 0, 1, 2, \ldots$
χ	vertical eigenfunction in Taylor-Goldstein equation
ψ_a	state vector, $a = 1, 2, 3$
Ω	eigenfrequency, $\Omega = f(1 + K^2/\lambda^2)^{1/2}$; local dispersion relation, eq. (53)
Ω_e	angular velocity of earth's rotation
ω	angular frequency
$\omega_{\boldsymbol{K}}^{\nu}$	eigenfrequency for mode ν and wavevector $\boldsymbol{K}$, $\nu = 0, 1, 2, \ldots$

Indices

$i, j = 1, 2, 3$
$\alpha, \beta = 1, 2$
Summation convention is used throughout.

6.3.1 Introduction

Internal gravity waves arise in a stably stratified fluid through the restoring force of gravity on water particles displaced from their equilibrium levels. Interfacial waves occurring between two superposed layers of different density are a familiar phenomenon, in particular at the upper free surface of the ocean in form of surface waves. In the continuously stratified interior of the ocean the restoring force of gravity is much weaker (by a factor $\delta\varrho/\varrho \approx 10^{-3}$), and the periods and wavelengths of internal waves are much larger than those of surface gravity waves. Internal gravity waves have periods between the inertial period ($2\pi/f$, where f is the Coriolis parameter) and the local buoyancy period ($2\pi/N$, where N is the buoyancy frequency). The inertial period is, e.g., 25 h at 30° latitude, and the buoyancy period is typically in the range 10 min···3 h. This interval of about 10 octaves is precisely fixed by kinematical reasons. Spatial scales range from a few meters to a few tens of kilometers. In the spectrum of oceanic motions internal gravity waves are thus embedded between (and partly overlap) small-scale three-dimensional turbulence and the geostrophic turbulence of the oceanic eddy field. Amplitudes of internal gravity waves are remarkably large of the order of 10 m, and current speeds are typically 5 cm s^{-1}. The wave motion is therefore not difficult to observe, in fact it is the dominant signal in many measurements.

The first scientific observations of oceanic internal gravity waves were reported by the Norwegian explorer Fridtjof Nansen in the last decade of the 19th century. During his passage across the Barents Sea he noted that the forward motion of his ship "Fram" was considerably reduced when sailing on a thin layer of fresh water overlying saltier water. This phenomenon, which he called "dead water", was later explained by Ekman as due to the drag by ship-generated interfacial waves. A historical discussion of observations of internal gravity waves in the first half of this century can be found in Defant's [61 D] book. Theoretical investigations of internal gravity waves had preceded the observations by half a century. Interfacial waves were studied by Stokes [1847 S] and the extension to continuous stratification was done by Rayleigh [1883 R]. However, the important role of internal gravity waves in the spectrum of oceanic varibility has been recognized only since a few decades, and we may look back on a period of intense research on this subject. The milestone in this development was the work of Garrett and Munk [72 G, 75 G] on modelling the spectral distribution of wave energy. This work initiated a series of field experiments, such as the IWEX experiment (Briscoe [75 B 2]) and the GATE internal wave experiment (Käse and Siedler[80 K]), and started as well theoretical investigations on the role of internal gravity waves in the energetics of the oceanic circulation. The time has definitely passed when oceanographers merely considered internal gravity waves as a passive contaminant, as unwanted noise (lumped together with any other kind of "turbulence") in the measurements.

For supplement reading we refer to the recent reviews of Garrett and Munk [79 G 2] and Olbers [83 O] on internal gravity waves in the ocean, and Munk [81 M 3] on the interrelation of internal waves and smaller-scale structures. Some of the theoretical framework can be found in the textbooks of LeBlond and Mysak [78 L 1], Phillips [77 P], Lighthill [78 L 2] and Gill [82 G], and in the papers of Müller and Olbers [75 M] and Thorpe [75 T]. A comprehensive bibliography of the internal gravity wave research in the last years has been given and discussed in the IUGG reviews by Briscoe [75 B 3], Gregg and Briscoe [79 G 3], and Levine [83 L 1].

In this section we will abbreviate "internal gravity waves" by "internal waves" since there will be no confusion with other types of internal wave motion.

6.3.2 Observational techniques

The present state of knowledge about the structure and importance of the oceanic internal wave field is strongly based on experimental evidence of the wave motion. Most early measurements of internal waves are records of temperature variations from thermometers at a fixed depth. The measurements of today are more sophisticated in adaption to the formidable task of obtaining a view of the highly complex space-time structure of the internal wave field. A great variety of devices has been designed to measure either the velocity associated with the wave motion or temperature and salinity fluctuations which are due to the vertical displacements of the stratified fluid induced by the waves.

Moored current observations constitute the main data base (see ch. 2). The common current meter measures the speed of the horizontal flow by a rotor and the direction by a vane. Other devices use acoustic or propellor sensors. The technique of mooring these instruments in the deep ocean was mainly developed by members of the Woods Hole Oceanographic Institution (e.g. Fofonoff and Webster [71 F]). Temperature sensors are frequently used in connection to moored current meter to obtain the vertical component of the current velocity through the "mutilated" heat equation $w = -(\partial T/\partial t)/(\partial \bar{T}/\partial z)$ where $\partial \bar{T}/\partial z$ is the mean vertical temperature gradient. Sensors which move through the wave field with speeds much larger than the phase speed of the waves yield a spatial snap-shot of the wave fluctuations. Towed and dropped devices have been designed to collect data of the spatial structure of the motion. There are many kinds of dropped instruments measuring profiles of temperature as, e.g., the rapidly repeated sounding system on the platform FLIP (Pinkel [75 P]). Thermistors arranged in a chain which is lowered from a ship and towed through the upper layers of the ocean have been used, e.g., by Lafond [62 L] to document the evidence of internal wave motion. This technique maps a two-dimensional section of the thermal structure down to about 200 m depth. A towed depth-controlled "fish" was used by Katz [75 K 1] to follow particular isotherms at greater depths (700···800 m) and thereby measuring vertical displacements.

There are many other instruments and measuring techniques which reflect the effort required to collect reliable information on the wave motion. Noteworthy examples are briefly discussed. Fig. 1 shows the time history of displacements of selected isopycnals in the upper ocean extracted by interpolation from a time series of 66 successive CTD casts (Käse and Clarke [78 K]). The associated motion appears to be well correlated over the complete depth range, which is a characteristic feature of internal waves in the upper ocean.

Measurements of the velocity field in the upper ocean using a Doppler sonar have been made by Pinkel [79 P, 81 P]. The sound beam is scattered off drifting organisms in the sea, the velocity parallel to the beam is obtained from the Doppler shift of the returning signal. Fig. 2 displays perturbations of the velocity field dominated by inertial motions.

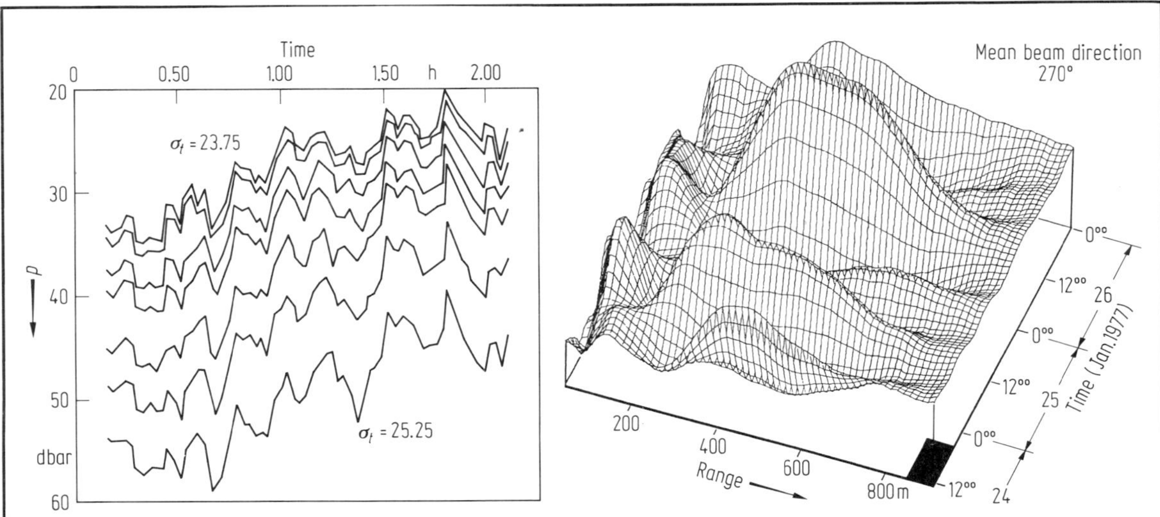

Fig. 1. Depth of selected isopycnals ($\sigma_t = 23.75$–25.25) interpolated from rapid successive CTD casts (Käse and Clarke [78 K]).

Fig. 2. Low-frequency velocity variations taken with a sonar at 50 m depth slanted 7° down from the horizontal. Profiles are low-pass filtered in range and time as indicated by the black square. Greatest range corresponds to approximately 225 m depth (Pinkel [81 P]).

An ingenious instrument measuring vertical displacements of isotherms in the deeper ocean is the yo-yo-capsule of Cairns [75 C]. This capsule freely drifts with the mean current yo-yoing up and down approximately 15 m about a selected isotherm while sensing temperature and pressure. The data segment given in Fig. 3 shows a couple of high-frequency waves riding on a large amplitude wave with a period of about half a day.

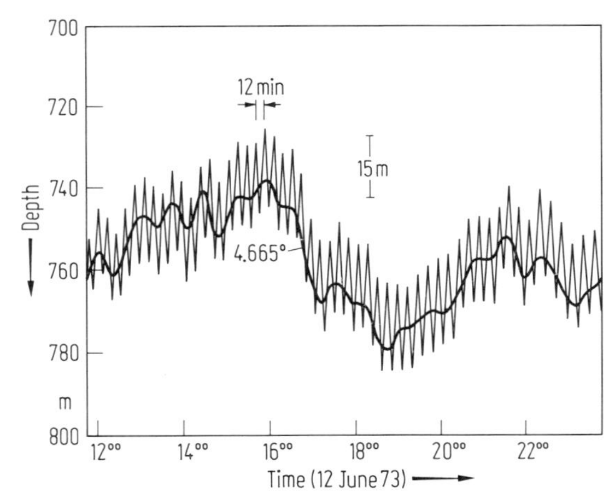

Fig. 3. A segment of yo-yo data. The heavy line shows the depth of the 4.665 °C isotherm, the zig-zag line is the position of the vertically yo-yoing instrument (Cairns [75 C]).

Profiles of relative horizontal velocity have been obtained by the electromagnetic velocity profiler of Sanford [75 S] which measures the electric potential in the sea induced by the motion of the salty sea water in the magnetic field of the earth. Fig. 4 shows two vertical profiles of each of the horizontal current components taken about half an inertial period apart at the same location. This picture is a simple but convincing illustration of the internal wave variability of the oceanic motion at all depths.

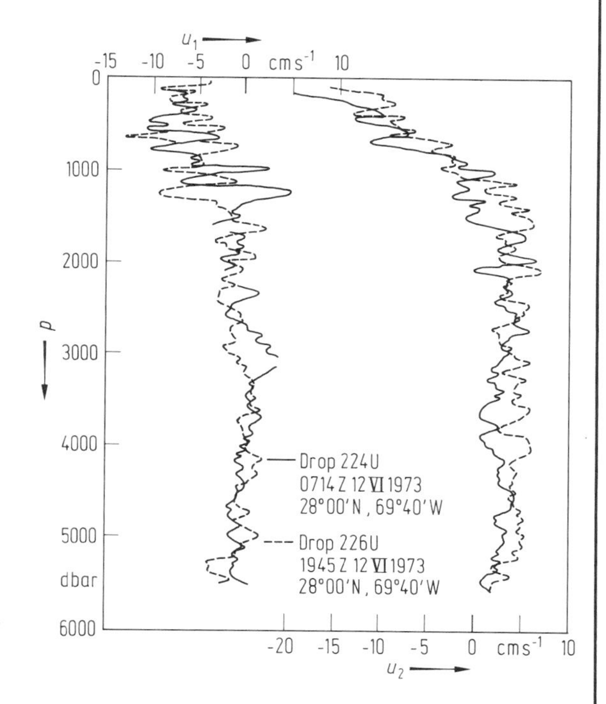

Fig. 4. Vertical profiles of eastward (u_1) and northward (u_2) velocity, taken 12.5 h (half an inertial period) apart at the same location (Sanford [75 S]).

Surface manifestations of internal waves are visible from satellites by different instruments. Fig. 5 displays a synthetic aperture radar (SAR) image of the Sicily Straits (Alpers and Salusti [83 A]). The circular pattern off the coast of Calabria visible on the SAR image is most likely due to internal waves which originate from interaction of the tidal current with the sill topography in the strait.

Internal waves are a three-dimensional phenomenon (the four-dimensional space-time continuum is constrained by the dispersion relation), but any of the commonly used sensors get only one- or two-dimensional cuts of the three-dimensional variability: moored sensors record the time history of the fluctuations at a fixed point, towed and dropped sensors record a mixture of temporal and spatial structure. Single point or section measurements thus get only a limited amount of information, and quantitative interpretation of these data is extremely difficult. Further, most of the more conventional observation techniques may yield a severely disturbed picture of the true fluctuations. Moored measurements suffer from a substantial unknown Doppler shift if the phase speed of the waves is not large compared to the velocity of lower frequency currents. Freely drifting devices such as Cairn's capsule avoid this problem. Moored temperature records may also be contaminated by the vertical migration of layered temperature fine-structure past the sensor induced by the wave motion (e.g. Phillips [71 P]). Decontamination of data from such effects, i.e. filtering out the true wave fluctuations, is only possible if experiments collect separable information about time and space structure and if certain knowledge about the contaminating fields is available. Many recent experiments have indeed proceeded in this direction. The step from a single mooring to complex arrays of moored instruments was taken some years ago. The most ambitious attempts were pursued with the IWEX experiment (Briscoe [75 B 2]) and the H-mooring of the GATE experiment (Käse and Siedler [80 K]). In these experiments up to 20 current meters and temperature sensors were deployed for some weeks in a three-dimensional configuration with separations ranging from 1 m to about 1000 m. To obtain uncontaminated measurements the requirements on the mooring stability is severe: for IWEX the motion of the instruments was less than ± 0.2 m at 600 m depth and ± 3 m at 1500 m depth. Analysis and results of these experiments will be presented at many places in this section.

Fig. 5. Optically processed SAR image from SEASAT orbit 1149 showing Calabria, Sicily, the Strait of Messina and part of the Tyrrhenian Sea. The three rings in the Tyrrhenian Sea are surface manifestations of internal wave solitons generated by tidal bores in the Strait of Messina (Alpers and Salusti [83 A]).

6.3.3 Space-time scales

Measurements yield series of data points. The adequate tool for interpretation is spectral analysis which converts time series or spatial series into auto- and cross-spectra of frequency or wavenumber. This section introduces the basic structure of these one-dimensional spectra and the relevant space and time scales of the wave field in the ocean. The construction of the complete frequency-wavenumber distribution of the wave energy is considered in subsect. 6.3.6.

By kinematical reasons internal waves can only exist in the frequency range between the local Coriolis frequency $f = 2\Omega_e \sin\phi$ (where Ω_e is the rotation frequency of the earth and ϕ the geographical latitude) and the local buoyancy frequency $N(z)$ (see subsect. 6.3.4). Profiles of $N(z)$ (also called Brunt-Väisälä frequency) which are typical of midocean conditions are displayed in Fig. 6. Observations have revealed that spectral levels and shapes show remarkable differences between the upper layer and deeper layers of the ocean. Further it is useful to classify the wave motions with respect to frequency into

(a) near-inertial waves, i.e. waves with frequencies close to the local Coriolis (=inertial) frequency f.
(b) near-buoyancy waves, i.e. waves with frequencies close to the local buoyancy frequency $N(z)$.
(c) intermediate (or continuum) waves, i.e. waves with frequencies ω in the range $f \ll \omega \ll N$.
and
(d) internal tides, i.e. waves with tidal frequencies, in particular the frequency of the semidiurnal M_2-tide.

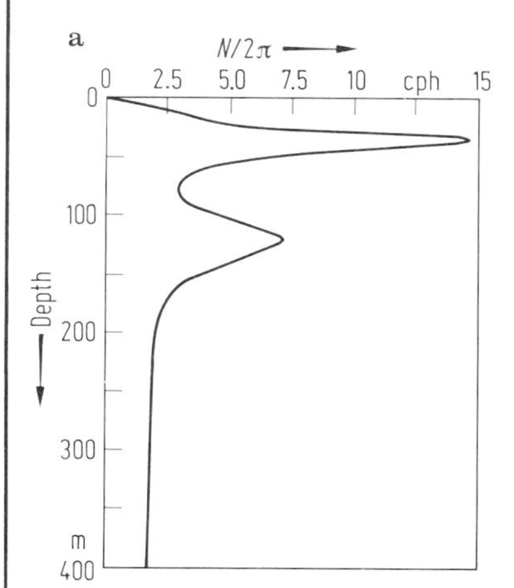

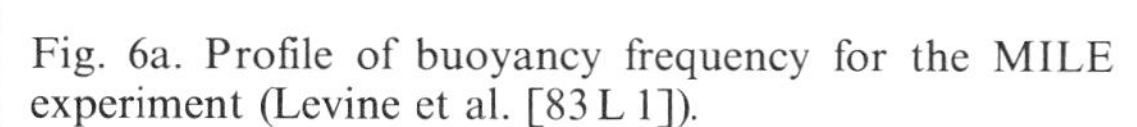
Fig. 6a. Profile of buoyancy frequency for the MILE experiment (Levine et al. [83 L 1]).

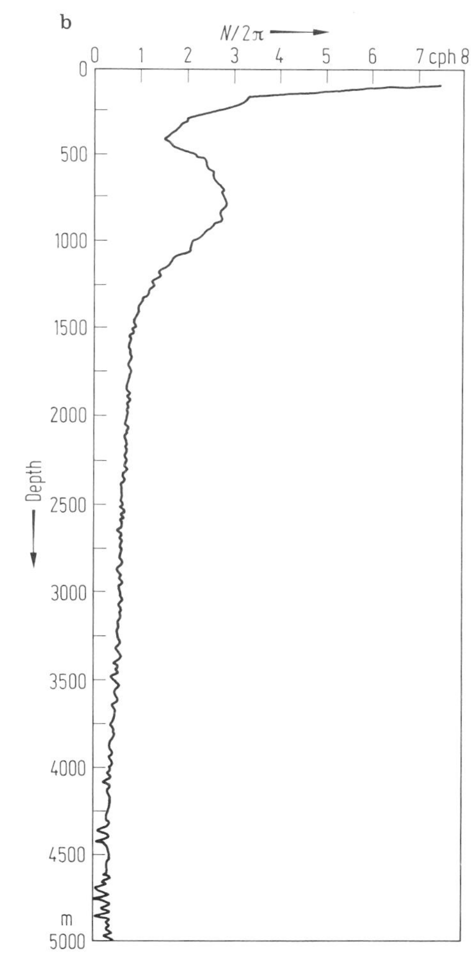

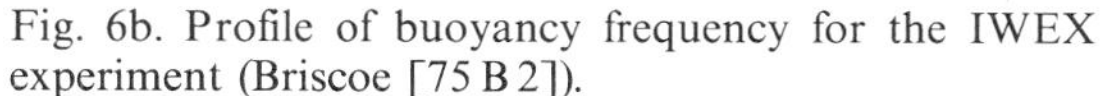
Fig. 6b. Profile of buoyancy frequency for the IWEX experiment (Briscoe [75 B 2]).

Note that waves in the classes (a) and (b) are close to turning points of the wave guide which occur in the horizontal at the latitude where $\omega = \pm f = \pm 2\Omega_e \sin\phi$ and in the vertical at the depth z where $\omega = \pm N(z)$.

The internal tide is an internal wave at tidal frequency. It is usually distinguished from the ambient internal wave by its large amplitude. Unlike the waves in the classes (a), (b), and (c), which are random phenomena, internal tides appear more as a deterministic structure in space and time. They are strongly related to the barotropic tide by their generation processes. Internal tides will not be described here. We refer to the review by Wunsch [75 W].

In this subsection we describe the spectral properties of intermediate-frequency, near-inertial, and near-buoyancy waves in the upper and the deep ocean. The description is illustrated by typical observed spectra and some approximate numbers for the spectral parameters are given. The average spectral properties are represented in subsect. 6.3.6 on spectral modeling of the wave energy.

6.3.3.1 Intermediate waves in the deep ocean

The deep ocean is defined here as the main thermocline (roughly between 500 and 1000 m) and the region below. There is overwhelming observational evidence that the structure and intensity of the internal wave field in this part of the ocean is universal: the shape of observed spectra is similar at different sites and the energy level (when scaled appropriately) does not vary by more than a factor of 3. Illustrative examples of frequency and wavenumber spectra are displayed in the Figs. 7 and 8.

Frequency spectra:

The frequency power spectra ("moored" spectra) (Fig. 7) generally show a significant drop of the variance below f and above the local $N(z)$. This suggests strong relations of the fluctuations with internal wave kinematics. The energy at the inertial period dominates, followed by a steep decrease with a slope of about -2. The total variance of the displayed spectra in the range $f \cdots N$ corresponds to roughly an rms displacement of 5 m and rms speed of 5 cm s^{-1}.

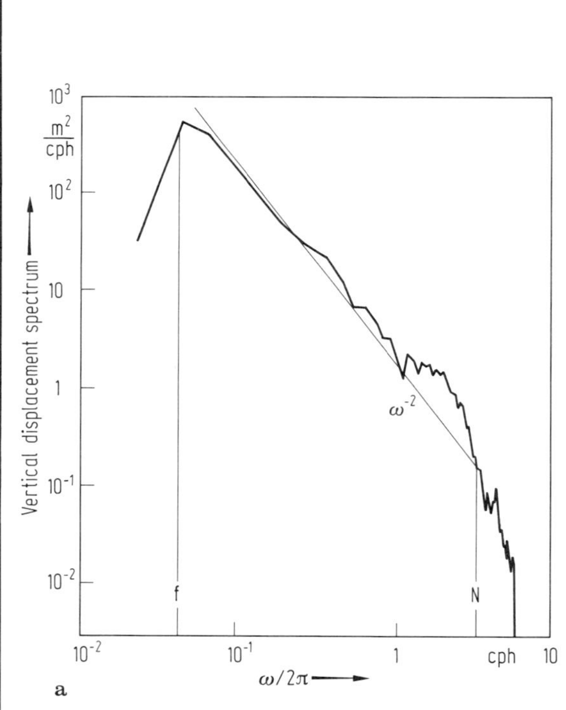

Fig. 7a. Power spectrum of the vertical displacement of an isotherm (Cairns and Williams [76 C]).

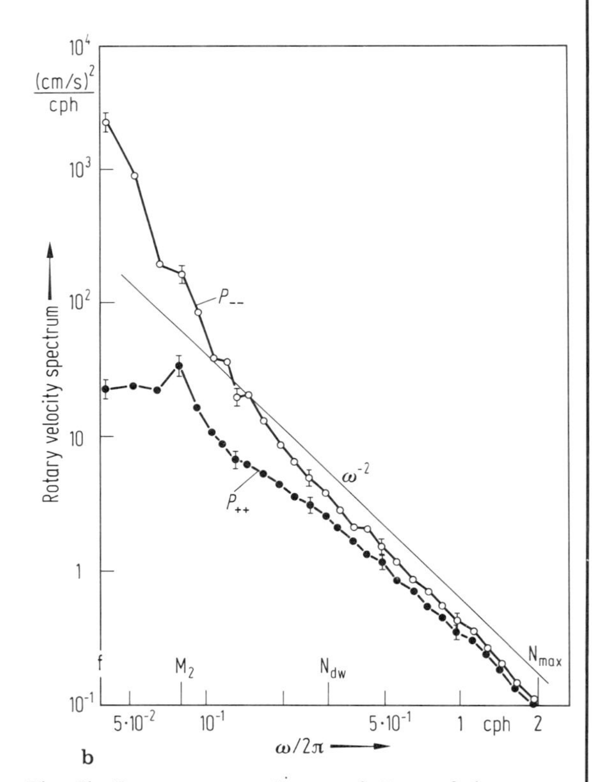

Fig. 7b. Power spectra P_{++} and P_{--} of the rotary components $u_+ = u_1 + iu_2$ and $u_- = u_1 - iu_2$, respectively (Müller et al. [78 M]).

Wavenumber spectra:

A typical spectrum of horizontal wavenumber ("towed" spectrum) as shown in Fig. 8 decreases with slight changes in the spectral slope. Spectra of vertical wavenumbers ("dropped" spectra) reveal more pronounced changes in the slope. Temperature as well as horizontal velocity spectra decay with a slope of about -2 down to 10^{-1} cpm. In the range 10^{-1} to 1 cpm the slope is steeper (-3) and the behavior at wavelengths smaller than 1 m appears more irregular. It is intriguing to ascribe in some way the spectral changes to cutoffs in the wave regime. However, there are no natural bounds in the wavenumbers as in the frequencies.

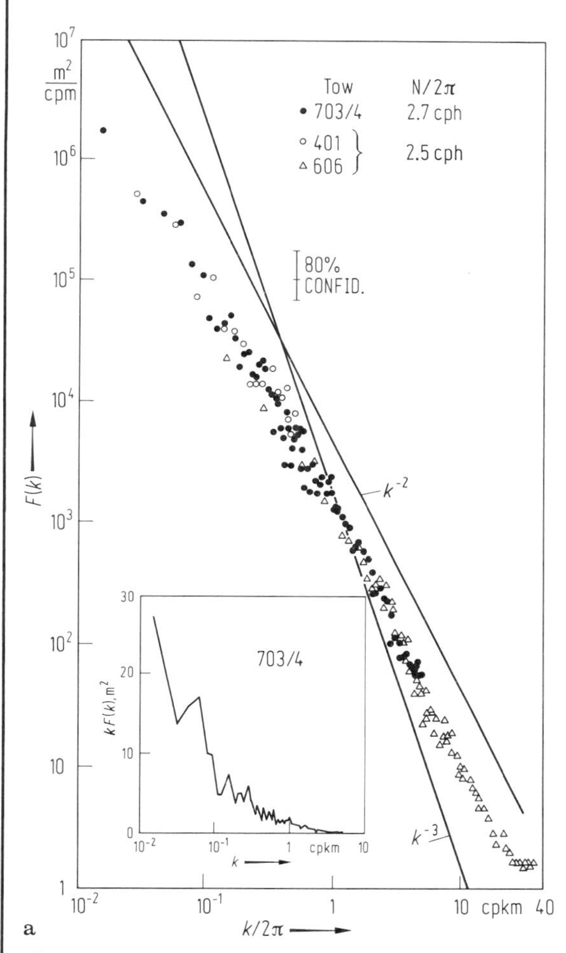

Fig. 8a. Power spectra of the displacement of the 12 °C isotherm from several horizontal tracks (Katz [75 K 1]). A variance preserving presentation is given in the inlet.

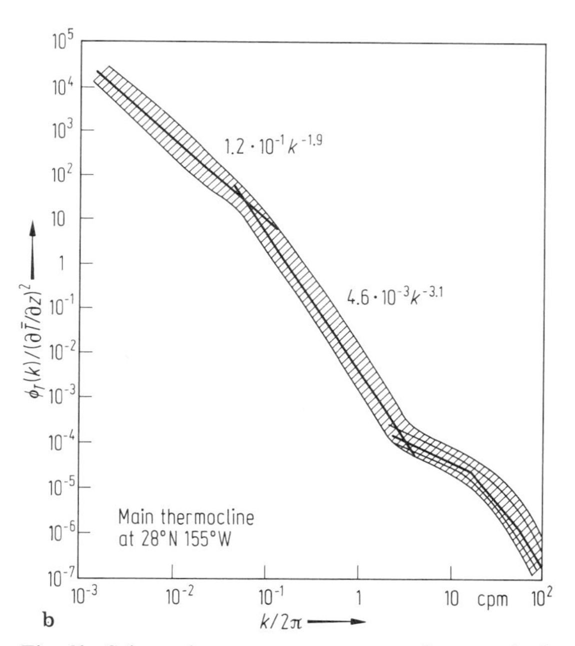

Fig. 8b. Schematic temperature spectra from vertically dropped instruments. These levels represent spectra that have been averaged vertically over 100 m or more. Spectra averaged over shorter vertical intervals exhibit much more variation in the microstructure range. (Gregg [77 G]).

WKB or N-scaling:

According to internal wave theory (see subsects. 6.3.5 and 6.3.6.3) spectra of horizontal currents (or horizontal kinetic energy) should scale by $N^{-1}(z)$, and spectra of vertical velocity (vertical displacement, temperature, and potential energy) by $N(z)$. The same scaling, of course, then applies to the total variances contained in the internal wave range. Note further that the vertical wavenumber scales by $N^{-1}(z)$ while the horizontal is constant (see subsect. 6.3.5.3).

Cross-spectra:

Though frequency spectra appear more useful to identify internal waves they do not allow a unique identification: fluctuations in the frequency range $f<\omega<N$ may have horizontal and vertical wavelengths which are inconsistent with the dispersion relation. Corresponding difficulties apply to purely spatial information. A unique identification of internal waves requires information linking the structure of the fluctuations in the time and the space domain. This kind of information is contained in the cross-spectra (or coherences and phases) relating the time histories of fluctuations at different positions, or spatial snap-shots of fluctuations at different times or different vertical levels. The coherence of the same field variables is a measure of relatedness of the fluctuations, and the phase is a measure of phase propagation. For different field variables the cross-spectra contain information on the consistency of the fluctuations with the internal kinematics (for this problem, see subsect. 6.3.6.2). For illustrative purposes Fig. 9 displays the coherences of vertical displacements at various frequencies as a function of horizontal separation. The decay of the coherence at a fixed frequency with separation is related to the wavenumber bandwidth of frequency-wavenumber spectrum. For details we refer to subsect. 6.3.6.3.

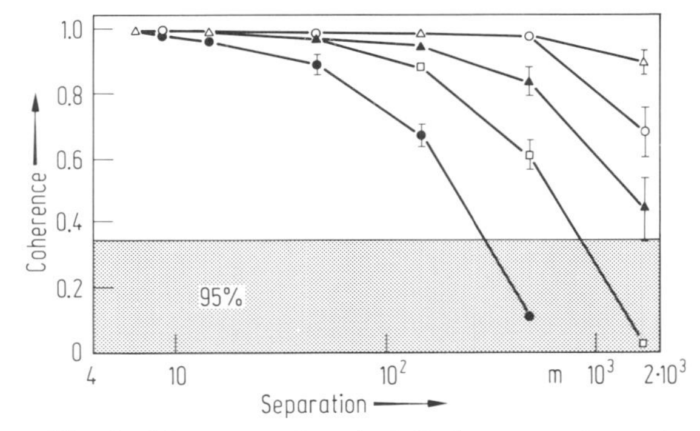

Fig. 9. Coherences of vertical displacements for various periods vs. horizontal separation. Open circle, (inertial); open triangle, (tidal); full triangle, (5.0 h): open square, (2.0 h); full circle, (0.67 h). (Müller et al. [78 M]).

6.3.3.2 Intermediate waves in the upper ocean

The upper ocean is here defined as the region above the main thermocline.

A clear picture of upper ocean wave properties has only begun to emerge in recent years. Very few high quality measurements have been made in the near-surface layers with appropriate space-time resolution. Basic results came from the temperature profiling data of Pinkel [75 P] collected on the stable platform FLIP, from the H-mooring experiment during GATE in the equatorial North Atlantic (Käse and Siedler [80 K]), and from the field experiments MILE in the central Pacific (Mixed Layer Experiment, Levine et al. [83 L 2]) and JASIN in the North Atlantic (Joint Air-Sea Interaction Experiment, Levine et al. [83 L 3]). The similarity of the frequency spectra observed in these experiments is demonstrated in Fig. 10. The variance-conserving plot of the energy spectrum (Fig. 11) shows the dominance of waves in three frequency bands: near-inertial waves, tidal waves, and high-frequency waves near 3 cph. The strong deviation of the spectral shape from the deep-ocean universal form is obvious. The greatest deviations are found at near-inertial frequencies and in a high-frequency band (roughly $2\cdots5$ cph) preceding the spectral roll-off above the local N (roughly at $10\cdots15$ cph).

The spectral level at these high frequencies is well above the (appropriately scaled) deep-ocean level and does not follow at simple power-law extrapolation from intermediate frequencies. These high-frequency waves (visible also in Fig. 1) are more intermittent and anisotropic – the waves occur in distinct groups – and have a simple vertical structure: with higher frequency the first mode becomes increasingly prominent. An explanation of these features includes both kinematical and dynamical reasons. Waves of high-frequency are trapped in the thin seasonal pycnocline in which a standing mode structure can be established much more easily than the diffuse permanent pycnocline. They are in a region of large shear and close to the surface forcing (atmospheric fields and surface waves, see subsect. 6.3.7.3.1). The wave guide and the forcing function have substantial space and time variations. This variability may also cause the invalidity of the N-scaling in the upper ocean.

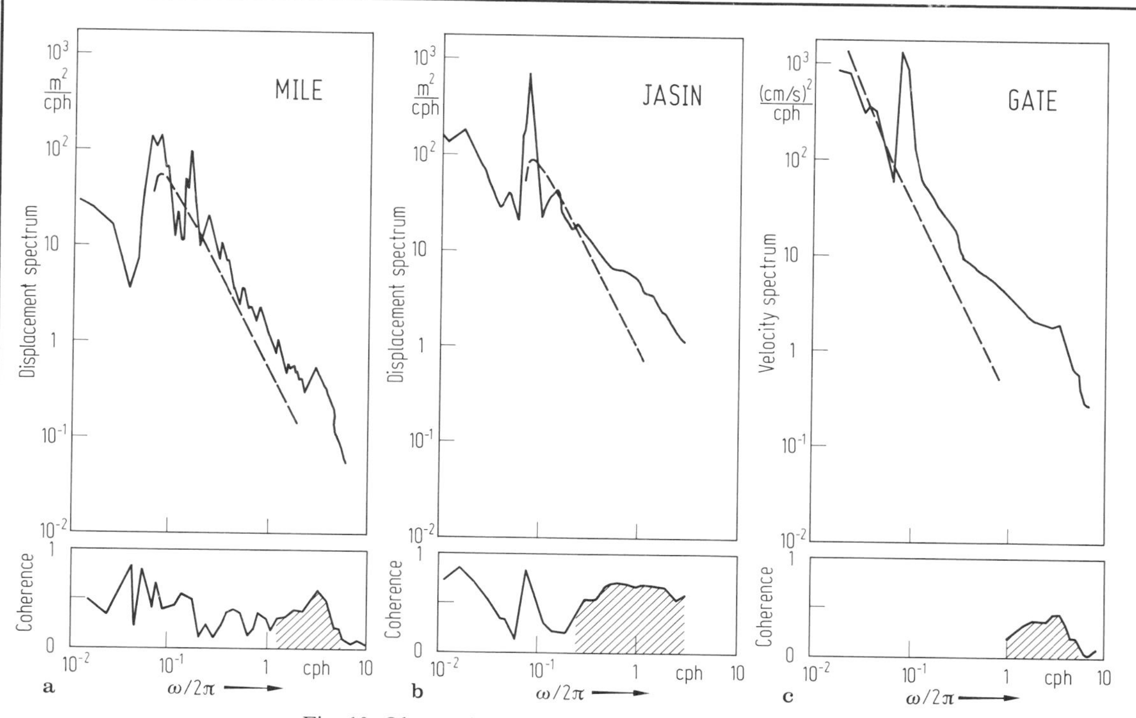

Fig. 10. Observed frequency spectra along with vertical coherences from (a) MILE, (b) JASIN, and (c) GATE. Garrett-Munk spectra based on local average N are shown by the dashed lines. Displacement spectra from JASIN and MILE are from 50 and 41 m, respectively; the clockwise rotary velocity spectrum from GATE is from 56 m. Coherences are from depths between 45 and 60 m (JASIN). 41 and 175 m (MILE), and 7 and 51 m (GATE); coherences that are significantly non-zero at the 95% confidence limit are shaded (Levine et al. [83 L 3]).

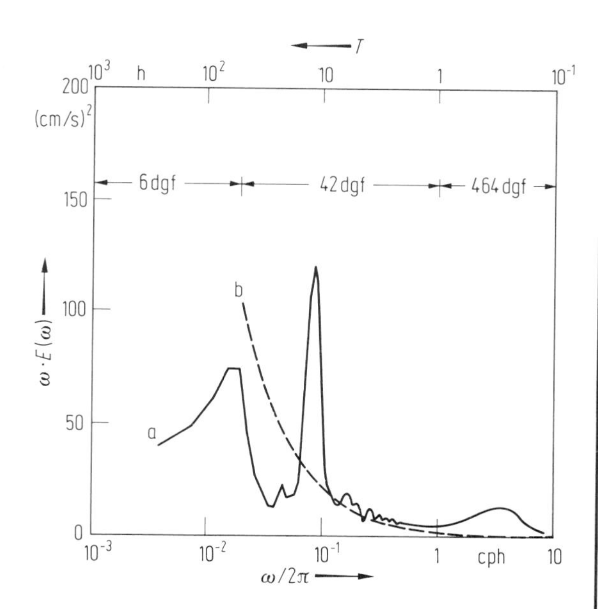

Fig. 11. (a): Variance preserving presentation of total internal wave energy of the GATE spectrum. (b): Theoretical Garrett-Munk spectrum with energy in the frequency band 0.1...1 cph equal to that observed in GATE spectrum (a). (Käse and Siedler [80 K 1]).

6.3.3.3 Near-inertial waves

Since in the ocean N is generally much larger than f inertial waves occur at all depths. They represent a major contribution to the kinetic energy of oceanic motions in the upper and the deep ocean (see, e.g., Fig. 4 and the spectra in the preceding subsections). Considerable experimental effort has been spent to establish their basic structure, namely the clockwise (in the northern hemisphere) rotation of the horizontal current vector with time, upward propagation of phase (and thus downward energy flux, see subsect. 6.3.5), and clockwise turning of the velocity vector with depth. Current speeds in the near-inertial range may be as large as $1\ \mathrm{ms}^{-1}$ in the upper few tens of meters of the water column in extreme conditions. Typical values range between 5 to $10\ \mathrm{cm\,s}^{-1}$.

An extensive description of the spectral properties of inertial waves away from horizontal and vertical boundaries has recently been presented by Fu [81 F], using data from POLYMODE arrays (U.S. Polymode Organizing Committee [76 U]) which provided an excellent data base for the study of inertial waves because of the large regional covering (subtropical to temperature latitudes in the Western North Atlantic) and the long duration of the observations (at least 9 months at each mooring site). In most of the data there is a prominent peak at frequencies slightly exceeding f (about 1···5%), but an universal frequency spectrum close to f does not exist. The parameters characterizing the level and shape of the inertial peak vary substantially with latitude and environmental oceanic conditions. The peak height (measured by the ratio of the peak level to a power-law extension from the higher frequencies in the intermediate internal wave band) is roughly 10 dB. It shows large values close to potential sources of inertial waves (near the Mid-Atlantic ridge, at depths less than 2000 m, in the deep ocean over rough topography and under the Gulf Stream) as compared to regions which are supposed to be dynamically inactive (the deep ocean over smooth topography). The bandwidth of the peak (roughly $0.1\,f$) decreases with increasing latitude, the peak frequency seems to be inversely correlated to the peak height. The horizontal coherence scale is some tens of kilometers and decreases with depth. The vertical coherence scale is in the range 10···100 m. This behavior is caused by the small inclination of the rays of near-inertial waves with resepect to the horizontal (see subsect. 6.3.5): wave groups with small vertical separations may arise from uncorrelated generation events with large horizontal separations.

6.3.3.4 Near-buoyancy waves

A prominent feature of frequency spectra is an increase of the energy level very close to the local buoyancy frequency (see, e.g., Fig. 7a; this must not be confused with the high-frequency bump in upper ocean spectra which is well below the local N). Associated with the spectral bump is an increase in coherence of the fluctuations. This property can be attributed to local phase coupling between the incident and reflected waves near the local turning point at $\omega = N(z)$: here the waves interfere constructively and increase energy and coherence (Desaubies [75 D]).

6.3.4 Equations of motion

This subsection gives various forms of the equations of motion used for the theoretical description of oceanic internal waves. They all follow from equations for small-scale motions on a f-plane (see sects. 4.1 and 5.2). These equations are here discussed in the subsects. 6.3.4.1 and 6.3.4.2, including the appropriate boundary conditions and a larger scale mean flow. The following subsections are devoted to derive projections of the equations and to introduce the concept of decomposition into vertical normal modes and the possible wave branches.

6.3.4.1 Equations of motion for a Boussinesq fluid on a f-plane

The ocean will be treated here as an incompressible, stratified, rotating fluid of infinite horizontal extent. Vertically, the fluid is bounded by a rigid bottom and a freely moving surface. We use local Cartesian coordinates x_1, x_2, $x_3 = z$, with x_3 upward.

Equations for the interior:

In the Boussinesq approximation the equations of motion appropriate for small-scale motions (see sect. 4.1) take the form

$$\frac{\partial u_i}{\partial t}+u_j\frac{\partial u_i}{\partial x_j}-f\varepsilon_{ij}u_j-\delta_{i3}b+\frac{\partial p}{\partial x_i}=F_i\,, \tag{1a}$$

$$\frac{\partial b}{\partial t}+u_j\frac{\partial b}{\partial x_j}+u_3N^2=Q\,, \tag{1b}$$

$$\frac{\partial u_j}{\partial x_j}=0\,, \tag{1c}$$

where $u_i(i=1,2,3)$ denote the velocity components, $b=-g\varrho$ the buoyancy, and p the kinematic pressure. More specifically, density and pressure are scaled by a constant reference density, the total density is given by $\varrho+\bar{\varrho}(x_3)$, and the total pressure by $p+\bar{p}(x_3)$, where $\bar{\varrho}(x_3)$ is the horizontally averaged density field associated with the pressure field $\bar{p}(x_3)$ such that

$$\frac{\mathrm{d}\bar{p}}{\mathrm{d}x_3}=-g\bar{\varrho}(x_3)\,. \tag{2}$$

The buoyancy frequency is then defined as

$$N(x_3)=\left(-g\frac{\mathrm{d}\bar{\varrho}}{\mathrm{d}x_3}\right)^{1/2}. \tag{3}$$

Other names are Brunt-Väisälä or stability frequency. We have neglected the forces induced by the horizontal components of the angular rotation vector of the earth so that only (twice) the vertical component (the Coriolis parameter)

$$f=2\Omega_e\sin\phi \tag{4}$$

appears in the equations. This will be treated as a constant appropriate to the local latitude. This is the *f-plane approximation.*

The terms F_i and Q denote viscous and diffusive effects. We will neglect these forces in this subsection and thus restrict the discussion to inviscid, adiabatic motions. Viscous and diffusive terms will be discussed in subsect. 6.3.4.2.

Boundary conditions:

The appropriate boundary conditions at the sea surface $x_3=\zeta(x_\alpha,t)$ and the bottom $x_3=h(x_\alpha)$ are then given by $(\alpha=1,2)$

$$\frac{\partial\zeta}{\partial t}+u_\alpha\frac{\partial\zeta}{\partial x_\alpha}-u_3=0\,,\quad \text{at}\quad x_3=\zeta(x_\alpha,t)\,, \tag{5a}$$

$$u_\alpha\frac{\partial h}{\partial x_\alpha}-u_3=0\,,\quad \text{at}\quad x_3=h(x_\alpha)\,, \tag{6a}$$

and

$$p=p_a-\bar{p}\,,\quad \text{at}\quad x_3=\zeta(x_\alpha,t)\,, \tag{7a}$$

where p_a is the atmospheric pressure at the sea surface. Equations (5a) and (6a) are kinematic conditions stating that there is no flow through the vertical boundaries. Eq. (7a) is the dynamic boundary condition for an inviscid flow at a free surface stating the continuity of the pressure field across the surface (see sect. 4.1). It is convenient (and appropriate for the small variations of ζ associated with the motions considered here) to expand the boundary condition at the free surface $x_3=\zeta(x_\alpha,t)$ about a mean sea surface $x_3=0$ so that eqs. (5a) and (7a) become

$$\frac{\partial\zeta}{\partial t}+u_\alpha\frac{\partial\zeta}{\partial x_\alpha}-u_3=Q_\zeta\,,\quad \text{at}\quad x_3=0\,, \tag{5b}$$

$$p-g\zeta=Q_p\,,\quad \text{at}\quad x_3=0\,, \tag{7b}$$

where generally only the first few terms of the source terms Q_ζ and Q_p will be retained,

$$Q_\zeta = \left(\zeta \frac{\partial u_3}{\partial x_3} + \cdots \right)_{x_3=0},$$
$$Q_p = \left(-\zeta \frac{\partial p}{\partial x_3} - \frac{1}{2} N^2 \zeta^2 + \cdots \right)_{x_3=0} + (p_a - \bar{p}(0)). \tag{8}$$

Similarly, the bottom boundary condition is expanded about a level $x_3 = -H$ where H is the constant average depth. With $h(x_\alpha) = -H + \eta(x_\alpha)$ one finds

$$u_3 = u_\alpha \frac{\partial \eta}{\partial x_\alpha} - \eta \frac{\partial u_3}{\partial x_3} - \frac{1}{2} \eta^2 \frac{\partial^2 u_3}{\partial x_3^2} + \eta \frac{\partial}{\partial x_3}\left(u_\alpha \frac{\partial \eta}{\partial x_\alpha}\right) + \cdots, \quad \text{at} \quad x_3 = -H. \tag{6b}$$

The above set of equations describes the evolution of the six field variables u_1, u_2, u_3, p, b, and ζ in an implicit way since there are only five prognostic equations [(1a), (1b) and (5a)]. The continuity equation (1c) is a diagnostic constraint and eqs. (6a) and (7a) are the two vertical boundary conditions appropriate for the problem. As outlined in sect. 4.1, and also more elaborated below, the state of the motion may completely be described by a state vector of only three dimensions, in accordinace with the fact that the linearized set of equations contains only three wave branches, two branches describing internal gravity wave motion and one describing stationary geostrophic motion (cf. subsect. 6.3.4.7).

Further approximations:

Internal gravity motions occur in the frequency range from the local Coriolis frequency f to the maximum of the buoyancy frequency N. This interval generally spans two to three decades. Depending on the frequencies of interest further approximations can be made:

(a) the high-frequency approximation ($\omega \gg f$): For high-frequency motions the Coriolis force may be omitted (formally, $f=0$). As shown in subsect. 6.3.5. rotation becomes important when the horizontal scale exceeds the vertical scale by a factor N/f which generally is large in the ocean (of order 10^2).

(b) the low-frequency or long-wave approximation ($\omega \ll N$): From the third equation of eqs. (1a) and the buoyancy equation (1b) one finds that $\partial u_3/\partial t$ may be dropped for low-frequency motions. Then the third component of the momentum balance becomes a diagnostic relation between the buoyancy field and the vertical gradient of the pressure field

$$-b + \frac{\partial p}{\partial x_3} = 0 \tag{9}$$

if the remaining nonlinear terms are small. This hydrostatic balance is also obtained for motions with a small aspect ratio, i.e. for waves with a horizontal scale L which is large compared to the vertical scale H of the motion (the term $\partial u_3/\partial t$ is of order $(H/L)^2$ relative to the pressure gradient $\partial p/\partial x_3$).

A further simplification which is frequently used in internal wave theory is the *rigid lid approximation* which replaces the free sea surface by a rigid horizontal boundary so that the kinematic boundary condition, eq. (5a), is replaced by the no-flux condition $u_3=0$, at $x_3=0$. The dynamic boundary condition, eq. (7a), must then be omitted. The implication of this approximation will be discussed in subsect. 6.3.4.5.

It should further be emphasized that the concept of wave motion implies that the nonlinearities in the equations are small. Formally, *linearization* requires that the particle speed u is small compared to the phase speed $c=\omega/k$ of the wave motion, i.e. $u \ll c$ or $ak \ll 1$ where a is the wave amplitude. This *small-amplitude approximation* is generally assumed to hold in the ocean: internal wave theory so far mainly adresses linearized problems and solves nonlinear problems by perturbation expansion. The validity of this assumption will be addressed in subsect. 6.3.7.

6.3.4.2 Motions relative to a large-scale mean flow

In the ocean wave motion generally occurs in the presence of a larger scale mean flow. The equations discussed in subsect. 6.3.4.1 may be generalized to include interaction with a shear flow U_j, buoyancy field B, pressure field P, and sea level variation Z associated with larger scale motions. Formally the appropriate equations of motion may be derived by separating flow components by a suitably defined space-time averaging procedure $\langle\,\rangle$ in the Reynolds sense. The total flow is decomposed as the sum $u_j + U_j$, $b + B$, etc., of the small-scale wave component u_j, b, etc., with $\langle u_j \rangle = 0$, $\langle b \rangle = 0$, etc., and the large-scale component $U_j = \langle U_j \rangle$, $B = \langle B \rangle$, etc.

The averaging procedure introduces for the large-scale components the familiar Reynolds stress terms $\langle u_i u_j \rangle$, $\langle u_j b \rangle$, etc., induced by the small-scale motion. The small-scale motion is governed by

$$\frac{\partial u_\alpha}{\partial t} - f\varepsilon_{\alpha\beta} u_\beta + \frac{\partial p}{\partial x_\alpha} = S_\alpha \quad (\alpha = 1, 2), \tag{10a}$$

$$\frac{\partial u_3}{\partial t} - b + \frac{\partial p}{\partial x_3} = S_3, \tag{10b}$$

$$\frac{\partial b}{\partial t} + u_3 N^2 = S_4, \tag{10c}$$

$$\frac{\partial u_j}{\partial x_j} = 0, \tag{10d}$$

$$\frac{\partial \zeta}{\partial t} - u_3 = S_5, \quad \text{at} \quad x_3 = 0, \tag{10e}$$

$$p - g\zeta = S_6, \quad \text{at} \quad x_3 = 0, \tag{10f}$$

$$u_3 = S_7, \quad \text{at} \quad x_3 = -H, \tag{10g}$$

where the nonlinear terms and the terms describing the interaction with larger scale components are comprized into the source terms S_n, $n = 1, \ldots, 7$, given by

$$S_i = \frac{\partial}{\partial x_j}(\langle u_i u_j \rangle - u_i u_j) - \frac{\partial}{\partial x_j}(U_i u_j + U_j u_i) \quad (i = 1, 2, 3)$$

$$S_4 = \frac{\partial}{\partial x_j}(\langle u_j b \rangle - u_j b) - \frac{\partial}{\partial x_j}(u_j B + U_j b)$$

$$S_5 = \left\langle u_\alpha \frac{\partial \zeta}{\partial x_\alpha} \right\rangle - u_\alpha \frac{\partial \zeta}{\partial x_\alpha} - U_\alpha \frac{\partial \zeta}{\partial x_\alpha} - u_\alpha \frac{\partial Z}{\partial x_\alpha} + Z \frac{\partial u_3}{\partial x_3} + \zeta \frac{\partial U_3}{\partial x_3} + \zeta \frac{\partial u_3}{\partial x_3} - \left\langle \zeta \frac{\partial u_3}{\partial x_3} \right\rangle + \cdots \tag{11}$$

$$S_6 = (p_a - \langle p_a \rangle) - \zeta \frac{\partial p}{\partial x_3} + \left\langle \zeta \frac{\partial p}{\partial x_3} \right\rangle - N^2 Z \zeta - \frac{1}{2} N^2 (\zeta^2 - \langle \zeta^2 \rangle) - Z \frac{\partial p}{\partial x_3} - \zeta \frac{\partial P}{\partial x_3} - \cdots$$

$$S_7 = (U_\alpha + u_\alpha) \frac{\partial \eta}{\partial x_\alpha} - \eta \frac{\partial}{\partial x_3}(U_3 + u_3) + \cdots.$$

To obtain the last relation it was assumed that $\langle \eta \rangle = 0$. The set of eqs. (10) and (11) are the most general form of evolution equations of internal wave theory considered in this section.

Viscous and diffusive terms:

The equations of motion (10) and the source terms (11) do not contain the effect of viscosity and diffusion on the wave field. Generally, these terms are assumed to be negligible. In some problems, however, they must be retained, e.g. when describing the coupling of the atmospheric wind stress or buoyancy flux to the wave field. There are essentially two approaches to this problem. In the first, the atmospheric fluxes are assumed to be distributed over a small surface layer, e.g. in the case of forcing by the wind stress τ_α, the effect of the stress is modelled by the body force $S'_\alpha = \tau_\alpha / d$ in a layer of thickness d. Alternatively, the common parametrization of turbulent fluxes by austausch coefficients is used which enter the source term S_α and S_4 in the form

$$S'_\alpha = \frac{\partial}{\partial x_3}\left(\nu \frac{\partial u_\alpha}{\partial x_3}\right), \qquad S'_4 = \frac{\partial}{\partial x_3}\left(\varkappa \frac{\partial b}{\partial x_3}\right), \tag{12}$$

where ν and $\varkappa$ are the eddy-induced diffusion coefficients for vertical turbulent transports. Corresponding parametrization of horizontal turbulent transport may also be included. This parametrization scheme by austausch coefficients then requires additional boundary conditions which equates the turbulent fluxes of momentum and buoyancy through the boundary with the externally imposed fluxes of these quantities.

6.3.4.3 Energy equations

The energy equations

$$\frac{\partial}{\partial t}\left(\frac{1}{2}u_j^2\right)+\frac{\partial}{\partial x_j}(u_j p)-u_3 b=u_j S_j$$
$$\frac{\partial}{\partial t}\left(\frac{1}{2}N^{-2}b^2\right)+u_3 b=N^{-2}bS_4 \tag{13}$$

may be derived from eqs. (10a–d). Apparently, $(1/2)u_j^2$ is the kinetic energy associated with the motion and $(1/2)N^{-2}b^2$ can be identified the perturbation potential energy: a fluid particle displaced by ξ from its equilibrium level z in a density profile $\bar{\varrho}(z)$ induces a perturbation in the density field, $\bar{\varrho}(z)-\bar{\varrho}(z+\xi)\simeq -\xi \mathrm{d}\bar{\varrho}/\mathrm{d}z$, so that $b=-N^2\xi$ and thus $(1/2)N^{-2}b^2=(1/2)N^2\xi^2$.

Vertical integration of eq. (13) and use of eqs. (10e–g) yield the conservation equation for the total energy per unit surface area in the form

$$\frac{\partial}{\partial t}\left[\int_{-H}^{0}\mathrm{d}z\left(\frac{1}{2}u_j^2+\frac{1}{2}N^{-2}b^2\right)+\frac{1}{2}g\zeta^2\right]+\frac{\partial}{\partial x_\alpha}\left[\int_{-H}^{0}\mathrm{d}z u_\alpha p+p\eta(U_\alpha+u_\alpha)|_{-H}\right]$$
$$=\int_{-H}^{0}\mathrm{d}z(u_j S_j+N^{-2}bS_4)+g\zeta S_5-\frac{\partial\zeta}{\partial t}S_6+S_5 S_6 \tag{14}$$

which identifies the form of the vertically averaged energy flux. The energy $(1/2)g\zeta^2$ is associated with free motion of the surface.

6.3.4.4 The *w*-equation

The complex set of evolution equations (10) may be reduced by standard manipulations to an evolution equation for the vertical velocity $w=u_3$ in the form

$$\left\{\frac{\partial^2}{\partial z^2}\left(\frac{\partial^2}{\partial t^2}+f^2\right)+\frac{\partial^2}{\partial x_\alpha \partial x_\alpha}\left(\frac{\partial^2}{\partial t^2}+N^2\right)\right\}w=Q_1 \tag{15}$$

with boundary conditions

$$\left(\frac{\partial^2}{\partial t^2}+f^2\right)\frac{\partial w}{\partial z}-g\frac{\partial^2 w}{\partial x_\alpha \partial x_\alpha}=Q_2,\ \text{at}\quad z=0,$$
$$w=Q_3,\quad \text{at}\quad z=-H. \tag{16}$$

In terms of the S_{n} the source terms appearing in these equations are

$$Q_1=\frac{\partial^2}{\partial x_\alpha \partial x_\alpha}\left(\frac{\partial S_3}{\partial t}+S_4\right)-\frac{\partial^2}{\partial z \partial x_\alpha}\left(\frac{\partial S_\alpha}{\partial t}+f\varepsilon_{\alpha\beta}S_\beta\right)$$
$$Q_2=\frac{\partial^2}{\partial x_\alpha \partial x_\alpha}\left(\frac{\partial S_6}{\partial t}+gS_5\right)-\frac{\partial}{\partial x_\alpha}\left(\frac{\partial S_\alpha}{\partial t}+f\varepsilon_{\alpha\beta}S_\beta\right) \tag{17}$$
$$Q_3=S_7.$$

For the unforced linearized case (i.e. $Q_{\mathrm{n}}=0$, $\mathrm{n}=1,2,3$) eqs. (15) and (16) represent a closed problem for w which is second order in time. As shown below the w-equation describes the two internal gravity wave branches. The geostrophic branch is filtered out. However, for nonlinear problems or interaction with a mean flow the w-equation represents an ill-posed problem since the remaining field variables u_α, p, b, and ζ cannot entirely be expressed by w, i.e. w does not completely determine the state of the motion.

6.3.4.5 Separation into vertical normal modes

For $Q_n = 0$ (n = 1, 2, 3) the w-equation is solved by a linear superposition of free internal gravity waves of the form

$$w(\boldsymbol{X}, z, t) = Re\{a_{\boldsymbol{K}}^{\nu} \varphi_{\boldsymbol{K}}^{\nu}(z) e^{i(\boldsymbol{K} \cdot \boldsymbol{X} - \omega t)}\} \tag{18}$$

with arbitrary amplitudes $a_{\boldsymbol{K}}^{\nu}$.

The vertical eigenvalue problem for w:

The real vertical normal mode $\varphi_{\boldsymbol{K}}^{\nu}(z)$ is solution of the eigenvalue problem

$$\begin{gathered} \frac{d^2}{dz^2}\varphi + K^2 \frac{N^2 - \omega^2}{\omega^2 - f^2}\varphi = 0, \\ \frac{d}{dz}\varphi - \frac{gK^2}{\omega^2 - f^2}\varphi = 0, \quad \text{at} \quad z = 0, \\ \varphi = 0, \quad \text{at} \quad z = -H. \end{gathered} \tag{19}$$

Solution of this Storm-Liouville type problem yields an infinite, complete and orthogonal set of modes $\varphi_{\boldsymbol{K}}^{\nu}(z)$, $\nu = 0, 1, 2, \dots$ with eigenvalues $\omega^2 = (\omega_{\boldsymbol{K}}^{\nu})^2$ if the wavenumber K is prescribed, or vice versa. The numbering may be arranged such that the ν-th mode has $\nu - 1$ zero-crossings in the interior $0 > z > -H$. Then $(\omega_{\boldsymbol{K}}^{\nu})^2$ is a monotonically decreasing sequence. The lowest mode $\nu = 0$ (the barotropic or surface wave mode) has a frequency $|\omega_{\boldsymbol{K}}^{0}| \gg |\omega_{\boldsymbol{K}}^{1}|$ and vanishes only at the bottom. It is mainly associated with the displacement of the free surface and represents surface gravity waves. The baroclinic or internal modes $\nu \geqq 1$ exist only in the frequency range $f < |\omega_{\boldsymbol{K}}^{\nu}| < N_{max}$ where N_{max} is the maximum buoyancy frequency in the water column. These modes represent internal gravity waves. Waves in the branch of positive frequencies propagate in the direction of the horizontal wavevector $\boldsymbol{K}$, and waves in the branch of negative frequencies have opposite propagation direction. The modes are orthogonal with respect to the scalar product

$$\int_{-H}^{0} dz (N^2 - f^2) \varphi_{\boldsymbol{K}}^{\nu} \varphi_{\boldsymbol{K}}^{\mu} + g \varphi_{\boldsymbol{K}}^{\nu} \varphi_{\boldsymbol{K}}^{\mu}|_{z=0} = \delta_{\nu\mu}\{(\omega_{\boldsymbol{K}}^{\nu})^2 - f^2\}. \tag{20}$$

The normalization has been chosen in order to simplify the expression for the total wave energy eq. (28).

Long-wave approximation:

For low frequencies, ω^2 may be neglected compared to N^2 and eq. (19) may be written

$$\begin{gathered} \frac{d^2\varphi}{dz} + \lambda^2 \frac{N^2}{f^2}\varphi = 0, \\ \frac{d\varphi}{dz} - \lambda^2 \frac{g}{f^2}\varphi = 0, \quad \text{at} \quad z = 0 \\ \varphi = 0, \quad \text{at} \quad z = -H \end{gathered} \tag{21}$$

with the eigenvalue $\lambda^2 = f^2 K^2/(\omega^2 - f^2)$. Then

$$\omega = \pm f(1 + K^2/\lambda^2)^{1/2}. \tag{22}$$

The low-frequency eigenvalue problem, eq. (21), has discrete eigenvalues λ_{ν}^2, $\nu = 0, 1, 2, \dots$, and eigenfunctions $\varphi_{\nu}(z)$ which, in contrast to eq. (19), are independent of the wavevector $\boldsymbol{K}$.

Rigid-lid approximation:

For the rigid-lid boundary condition ($w = 0$ at $z = 0$) the upper boundary condition of eq. (19) becomes $\varphi = 0$ at $z = 0$. By this approximation the barotropic or surface wave mode is lost, but the modifications of the baroclinic modes are negligible (see example below). Formally, the rigid-lid approximation is the lowest order of the expansion of eq. (19) with respect to the small parameter HN_0^2/g, which is of order 10^{-3}.

Solutions of the eigenvalue problems:

Solution of eqs. (19) or (21) for constant or piece-wise constant $N(z)$ is straightforeward. As an example, for constant $N=N_0$ the low-frequency problem is solved by

$$\varphi_\nu(z)=C\left\{\sin\left(\lambda_\nu\frac{N_0}{f}z\right)+\frac{N_0 f}{g\lambda_\nu}\cos\left(\lambda_\nu\frac{N_0}{f}z\right)\right\}, \tag{23}$$

where λ_ν follows from

$$\frac{N_0}{gf\lambda_\nu}=\tan\left(\lambda_\nu\frac{N_0}{f}H\right), \tag{24}$$

with the approximate solutions $\lambda_0\approx f/(gH)^{1/2}$, $\lambda_\nu\approx\nu\pi f/(N_0H)$ for $\nu\geqq 1$. Thus $\lambda_0^2/\lambda_\nu^2=\frac{N_0^2H}{g}\left(\frac{1}{\nu\pi}\right)^2\ll 1$ and $\omega_0^2\gg\omega_\nu^2$, $\nu\geqq 1$, for nonzero K. The normalization constant may be determined from eq. (20). Further, the eigenfunctions may be written in the approximate form

$$\varphi_0=C\left(\frac{N_0^2H}{g}\right)^{1/2}\left\{1+\frac{z}{H}-\frac{1}{2}\frac{N_0^2H}{g}\left(\frac{z}{H}\right)^2+\cdots\right\},$$
$$\varphi_\nu=C\left\{\sin\frac{\nu\pi z}{H}+\frac{N_0^2H}{g\nu\pi}\cos\frac{\nu\pi z}{H}\right\},\quad \nu\geqq 1, \tag{25}$$

so that the barotropic mode varies, effectively, in a linear way from zero at the bottom to its maximum at the surface whereas the baroclinic modes have their maxima in the water column and negligible surface values of order $N_0^2H/g\ll 1$. For these latter modes, describing internal wave motions, the rid-lid approximation, $\varphi=0$ at $z=0$ may be applied which neglects the second term in the curly brackets of eqs. (23) and (25) and changes the eigenvalue condition eq. (24) to $\sin\left(\lambda\frac{N_0H}{f}\right)=0$. Note that by this procedure the barotropic mode is lost.

The general problem eq. (19) yields corresponding results. For more complex stratifications, in particular for observed N profiles, the eigenfunctions must be determined by numerical solution techniques. The Figs. 12b and 12c display the dispersion relations and eigenfunctions, respectively, for the first five baroclinic modes computed for the profile of the stability frequency shown in Fig. 12a. This consists of a high-frequency waveguide in the upper ocean and a low-frequency waveguide at greater depths separated by a broad minimum at intermediate depths with N around 1 cph. The high-frequency modes nicely illustrate the trapping of energy in the region $\omega<N(z)$: here the modes have a sinusoidal appearance whereas they fade away exponentially beyond the turning points. Notice the distortion of the dispersion curves close to the minimum of $N(z)$ separating the upper and lower thermoclines. Here the wavelength suddenly decreases for a slight increase of frequency (this causes reduction of the waveguide to the extent of the seasonal thermocline). The phase velocity drops and the group velocity attains low values in this frequency range. Also the dispersion curves come very close to each other by pairs.

The eigenfrequencies of the baroclinic modes are confined to the range from f to N_{max} but the barotropic mode extends beyond N_{max}. For $\omega_0\gg N_{max}$ the eigenvalue problem eq. (19) takes the form

$$\frac{d^2\varphi_0}{dz^2}-K^2\varphi_0=0,$$
$$\frac{d\varphi_0}{dz}-\frac{gK^2}{\omega_0^2}\varphi_0=0,\quad\text{at}\quad z=0, \tag{26}$$
$$\varphi_0=0,\quad\text{at}\quad z=-H$$

which describes high-frequency surface wave motion (see sect. 6.2).

Representation of field variables:

The eigenfunction $\varphi_{\boldsymbol{K}}^\nu(z)$ reflects the vertical structure of the vertical velocity. The complete vector of field variables is obtained from the linearized equations (10). One finds for an ensemble of waves

$$\begin{pmatrix}u_\alpha\\ w\\ b\\ p\\ \xi\end{pmatrix}=\sum_{\pm\nu,\boldsymbol{K}}\frac{a_{\boldsymbol{K}}^\nu}{\omega_{\boldsymbol{K}}^\nu K^2}\begin{pmatrix}(\mathrm{i}\omega_{\boldsymbol{K}}^\nu K_\alpha-f\varepsilon_{\alpha\beta}K_\beta)\frac{d}{dz}\\ \omega_{\boldsymbol{K}}^\nu K^2\\ -\mathrm{i}K^2N^2\\ \mathrm{i}((\omega_{\boldsymbol{K}}^\nu)^2-f^2)\frac{d}{dz}\\ \mathrm{i}K^2\end{pmatrix}\varphi_{\boldsymbol{K}}^\nu(z)\ \mathrm{e}^{\mathrm{i}(\boldsymbol{K}\cdot\boldsymbol{X}-\omega_{\boldsymbol{K}}^\nu t)}, \tag{27}$$

where, for notational convenience, negative mode numbers have been introduced and $a_{\boldsymbol{K}}^{-\nu}=(a_{-\boldsymbol{K}}^{\nu})^*$, $\omega_{\boldsymbol{K}}^{-\nu}=-\omega_{-\boldsymbol{K}}^{\nu}$, $\varphi_{\boldsymbol{K}}^{-\nu}=\varphi_{-\boldsymbol{K}}^{\nu}$. Notice that $\xi(z=0)=\zeta$ is the surface displacement.

Energy and energy flux:

The total energy (averaged horizontally which is denoted by an overbar) of the linear wave field is given by

$$\frac{1}{2}\int_{-H}^{0} dz(\overline{u_j^2}+N^{-2}\overline{b^2})+\frac{1}{2}g\overline{\zeta^2} = \sum_{\pm\nu,\boldsymbol{K}} |a_{\boldsymbol{K}}^{\nu}|^2 \tag{28}$$

so that each mode ν carries a total energy $2|a_{\boldsymbol{K}}^{\nu}|^2$ at the wave vector $\boldsymbol{K}$. The horizontal energy flux becomes

$$\int_{-H}^{0} dz\overline{u_\alpha p}=\sum_{\pm\nu,\pm\mu,\boldsymbol{K}} \frac{a_{\boldsymbol{K}}^{\nu}a_{-\boldsymbol{K}}^{-\mu}}{\omega_{\boldsymbol{K}}^{\mu}K^2}K_\alpha\left\{g\varphi_{\boldsymbol{K}}^{\nu}\varphi_{\boldsymbol{K}}^{\mu}\Big|_{z=0}+\int_{-H}^{0} dz[N^2-(\omega_{\boldsymbol{K}}^{\mu})^2]\varphi_{\boldsymbol{K}}^{\nu}\varphi_{\boldsymbol{K}}^{\mu}\right\}. \tag{29}$$

Unlike the energy this quantity is not diagonal in the mode number.

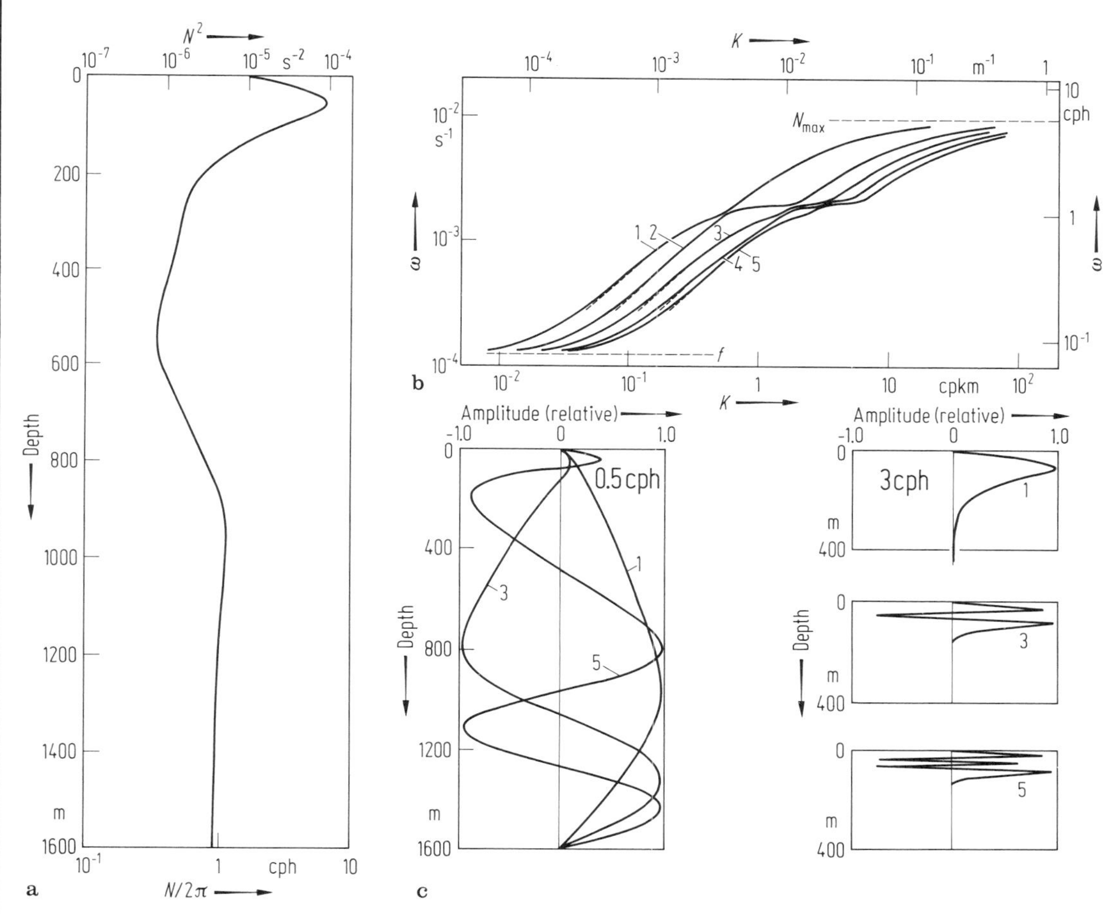

Fig. 12a. Buoyancy frequency profile for the JASIN area (Olbers [83 O]).

Fig. 12b. Dispersion relation of the first five internal wave modes computed for the buoyancy frequency shown in Fig. 12a (Olbers [83 O]).

Fig. 12c. Modes 1, 3, and 5 at a low frequency ($\omega=0.5$ cph) and a high frequency ($\omega=3.0$ cph) (Olbers [83 O]).

6.3.4.6 The *w*-equation in the presence of a horizontal shear flow

The concept of a *w*-equation can be extended to wave motions relative to a horizontal mean current $U_\alpha(z)$ with vertical shear. From eq. (10) one derives, omitting all source terms and taking $f=0$ for simplicity,

$$\mathrm{D}^2 \frac{\partial^2 w}{\partial x_i \partial x_i} + N^2 \frac{\partial^2 w}{\partial x_\alpha \partial x_\alpha} - \mathrm{D} \frac{\partial^2 U_\alpha}{\partial z^2} \frac{\partial w}{\partial x_\alpha} = 0 \tag{30}$$

with boundary conditions

$$\begin{aligned} &\mathrm{D}^2 \frac{\partial w}{\partial z} - \mathrm{D} \frac{\partial U_\alpha}{\partial z} \frac{\partial w}{\partial x_\alpha} - g \frac{\partial^2 w}{\partial x_\alpha \partial x_\alpha} = 0, \quad \text{at} \quad z=0, \\ &w=0, \quad \text{at} \quad z=-H \end{aligned} \tag{31}$$

and $\mathrm{D} = \frac{\partial}{\partial t} + U_\alpha \frac{\partial}{\partial x_\alpha}$.

Taylor-Goldstein equation:
The normal mode problem

$$\begin{aligned} &(\hat{U}-c)^2 \left[\frac{\mathrm{d}^2}{\mathrm{d}z^2} - K^2\right] \chi - \left[(\hat{U}-c)\frac{\partial^2 \hat{U}}{\partial z^2} - N^2\right] \chi = 0, \\ &(\hat{U}-c)^2 \frac{\mathrm{d}\chi}{\mathrm{d}z} - \left[(\hat{U}-c)\frac{\partial \hat{U}}{\partial z} + g\right] \chi = 0, \quad \text{at} \quad z=0, \\ &\chi = 0, \quad \text{at} \quad z=-H \end{aligned} \tag{32}$$

derived from eqs. (30) and (31) for $w \sim \chi(z)\exp\{\mathrm{i}(\boldsymbol{K}\cdot\boldsymbol{X} - \omega t)\}$ is called the Taylor-Goldstein equation after Taylor [31 T] and Goldstein [31 G]. In eq. (32) $c=\omega/K$ is the phase speed and

$$\hat{U}(z,\alpha) = \boldsymbol{K}\cdot\boldsymbol{U}/K = U_1(z)\cos\alpha + U_2(z)\sin\alpha \tag{33}$$

with $\boldsymbol{K} = K(\cos\alpha, \sin\alpha)$.

The Taylor-Goldstein equation is usually formulated for a parallel mean flow (i.e. $U_2=0$, e.g. LeBlond and Mysak [78 L 1]) but this form is retained for a spiraling current by the definition of $\hat{U}(z)$. Note that eq. (19) with $f=0$ is recovered from eq. (32) with $\hat{U}=0$. However, the character of the eigenvalue problem is severely changed by the inclusion of the shear current. This does not only lead to shear modification of the internal wave modes but the singularity of eq. (32) at depths where $\hat{U}(z)=c$ (the critical levels) is responsible for additional modes. These are either unstable or damped (i.e. the eigenvalue $c=c_r+\mathrm{i}c_i$ is complex) or they have real eigenvalues but, in contrast to the internal wave modes, the eigenfunctions are discontinuous at the critical levels. Unstable modes are confined to $\hat{U}_{min}<c_r<\hat{U}_{max}$ with c_i bounded by Howard's semicircle theorem and do not occur if the Richardson number (based on $\hat{U}$) is above 1/4 everywhere in the water column (Miles' theorem) (cf. LeBlond and Mysak [78 L 1]). A complete classification of the modes of the Taylor-Goldstein equation has only recently been found (Banks et al. [76 B]). The class of unstable modes is finite. Stable modes include (a) a finite set of damped modes (those conjugate to the unstable modes), (b) a discrete set of modes which essentially are internal waves modified by the shear (called shear modes; these have $c>\hat{U}_{max}$ or $c<\hat{U}_{min}$), (c) a finite set of modes with c_r inside the range of the velocity distribution (these have branch points at the critical levels), and (d) a continuous set of modes which also have c_r inside the range of $\hat{U}(z)$ (these have discontinuities at the critical levels).

6.3.4.7 Complete representation in terms of the state vector

It is possible to derive a well-posed problem for a three-dimensional state vector corresponding to the three wave branches contained in the original equations. We describe here the specification of the flow state in terms of the horizontal divergence, the vertical component of the vorticity, and the horizontal Laplacian of the pressure. Other representations are also possible (see sect. 4.1; note, however, that there the representation does not include the displacement field ζ).

State vector:

Thus, we define the state vector ψ_a, a = 1, 2, 3, by

$$\psi_1 = \frac{\partial u_\alpha}{\partial x_\alpha}$$
$$\psi_2 = \varepsilon_{\alpha\beta}\frac{\partial u_\beta}{\partial x_\alpha} \tag{34}$$
$$\psi_3 = \frac{1}{f}\mathscr{L}p,$$

where $\mathscr{L} = \frac{\partial^2}{\partial x_\alpha \partial x_\alpha}$ is the horizontal Laplacian operator. The vector completely specifies the state of the motion since the six field variables are uniquely determined by the relations

$$u_1 = \mathscr{L}^{-1}\left(\frac{\partial\psi_1}{\partial x_1} - \frac{\partial\psi_2}{\partial x_2}\right),$$
$$u_2 = \mathscr{L}^{-1}\left(\frac{\partial\psi_2}{\partial x_1} + \frac{\partial\psi_1}{\partial x_2}\right),$$
$$u_3 = -\int_{-H}^{x_3} \mathrm{d}x_3'\psi_1 + S_7,$$
$$p = f\mathscr{L}^{-1}\psi_3, \tag{35}$$
$$\zeta = (f\mathscr{L}^{-1}\psi_3 - S_6)/g,$$
$$b = f\mathscr{L}^{-1}\frac{\partial\psi_3}{\partial x_3} - \int_{-H}^{x_3}\mathrm{d}x_3'\left[f(\psi_2 - \psi_3) + \frac{\partial S_\alpha}{\partial x_\alpha}\right] - S_3 + \frac{\partial S_7}{\partial t}.$$

The expressions for u_α and p are immediately obvious; u_3 follows from eq. (10d) and eq. (10g), ζ from eq. (10f), b follows from (10c).

The relations, eq. (35), involve either vertical differentiation or integration, or solution of a Poisson equation with appropriate horizontal boundary conditions. This is simple for a horizontally homogeneous problem.

Evolution equation:

The evolution of the state vector, ψ_a is governed by

$$\frac{\partial\psi_a}{\partial t} + \mathrm{i}H_{ab}\psi_b = q_a[\psi] \quad (a = 1, 2, 3), \tag{36}$$

where the operator H_{ab} arises from the left-hand side of eq. (10) and the functionals q_a from the right-hand side:

$$H_{ab} = -\mathrm{i}f\begin{pmatrix} 0 & -1 & 1 \\ 1 & 0 & 0 \\ \mathscr{N}^{-1}\mathscr{L}\mathscr{M} & 0 & 0 \end{pmatrix} \tag{37}$$

and

$$q_1 = \frac{\partial S_\alpha}{\partial x_\alpha},$$
$$q_2 = \varepsilon_{\alpha\beta}\frac{\partial S_\beta}{\partial x_\alpha}, \tag{38}$$
$$q_3 = -\frac{1}{f}\mathscr{N}^{-1}\mathscr{L}\left\{\int_{x_3}^{0}\mathrm{d}x_3'\left[S_4 - \left(\frac{\partial^2}{\partial t^2} + N^2\right)S_7 + \frac{\partial S_3}{\partial t} + \int_{-H}^{x_3'}\mathrm{d}x_3''\left(\frac{\partial^2 S_\alpha}{\partial t\partial x_\alpha} + f\varepsilon_{\alpha\beta}\frac{\partial S_\beta}{\partial x_\alpha}\right)\right]\right.$$
$$\left. - \frac{\partial S_6}{\partial t} - g(S_5 + S_7)\right\}$$

with

$$\mathcal{N} = 1 - \mathcal{L} \int_{x_3}^{0} dx'_3 \int_{-H}^{x'_3} dx''_3$$

and

$$\mathcal{M} = \frac{1}{f^2}\left\{\int_{x_3}^{0} dx'_3 (N^2 - f^2) \int_{-H}^{x'_3} dx''_3 + g \int_{-H}^{0} dx''_3\right\}.$$

We note that the operator $\mathcal{N}$ reduces to unity when the long-wave approximation is made. Also some terms in the functionals q_a disappear (see Olbers [81 O 2]).

The three wave branches:

The operator H_{ab} possesses separable eigenfunctions

$$\beta_a(\boldsymbol{X}, z) = \varepsilon_a \phi(z) e^{i\boldsymbol{K}\cdot\boldsymbol{X}} \tag{39}$$

with real eigenvalue ω, i.e.

$$H_{ab}\beta_b = \omega\beta_a \tag{40}$$

The spatial dependence $\phi(z)\exp[i\boldsymbol{K}\cdot\boldsymbol{X}]$ is the eigenfunction of the operator $\mathcal{N}^{-1}\mathcal{L}\mathcal{M}$ with the eigenvalue $-K^2/\lambda^2$, and ω and the polarization vector are found to be

$$\begin{aligned} \omega = \omega^s = sf(1+K^2/\lambda^2)^{1/2} &= s\Omega, \\ \varepsilon_1^s &= is\Omega/f, \\ \varepsilon_2^s &= 1, \\ \varepsilon_3^s &= 1-(s\Omega/f)^2, \end{aligned} \tag{41}$$

where s = +, −, 0 counts the three wave branches. Apparently, for s = 0 one finds $\omega = 0$, $\varepsilon_2 = \varepsilon_3$, and $\varepsilon_1 = 0$ which specifies a stationary geostrophic flow with vanishing horizontal divergence, as can be inferred from eq. (20). Further, for linearized motion this branch is hydrostatically balanced, as can be inferred from eq. (35). The cases $s = +, -$ represent the two internal gravity wave branches, which may be shown by comparing the polarisation vector of eq. (41) with eq. (27) using eq. (34).

The eigenvectors ε_a^s are not orthogonal. Orthogonal vectors follow from the adjoint problem, which yields

$$\begin{aligned} \begin{Bmatrix} \tilde{\varepsilon}_1^s \\ \tilde{\varepsilon}_2^s \\ \tilde{\varepsilon}_3^s \end{Bmatrix} &= \tfrac{1}{2}(\Omega/f)^{-2}\begin{Bmatrix} -is\Omega/f \\ 1 \\ -1 \end{Bmatrix}, \quad s = +, - \\ \begin{Bmatrix} \tilde{\varepsilon}_1^0 \\ \tilde{\varepsilon}_2^0 \\ \tilde{\varepsilon}_3^0 \end{Bmatrix} &= (\Omega/f)^{-2}\begin{Bmatrix} 0 \\ (\Omega/f)^2 - 1 \\ 1 \end{Bmatrix}. \end{aligned} \tag{42}$$

These satisfy the relations

$$\begin{aligned} \tilde{\varepsilon}_a^s \varepsilon_a^r &= \delta^{sr}, \\ \varepsilon_a^s \tilde{\varepsilon}_b^s &= \delta_{ab}. \end{aligned} \tag{43}$$

It can be shown that the vertical eigenvalue problem of this decomposition is completely equivalent to eq. (19) with $\varphi(z) = \int_{-H}^{z} dz' \phi(z')$.

Evolution of the normal mode amplitudes:

The vector eigenfunctions $\beta_a = \beta_{a\nu\boldsymbol{K}}^s(\boldsymbol{X}, z)$ are complete so that any state ψ_a can be expanded in the form

$$\psi_a(\boldsymbol{X}, z, t) = \sum_{s,\nu,\boldsymbol{K}} A_{\nu\boldsymbol{K}}^s(t)\beta_{a\nu\boldsymbol{K}}^s(\boldsymbol{X}, z), \tag{44}$$

where the normal mode amplitudes

$$A^s_{\nu \boldsymbol{K}}(t) = \tilde{B}^s_{a\nu \boldsymbol{K}} \psi_a \tag{45}$$

are obtained by applying the decomposition operator

$$\tilde{B}^s_{a\nu \boldsymbol{K}} = \left(\frac{\lambda_\nu}{Kf}\right)^2 \tilde{\varepsilon}^s_a \int_{-H}^{0} dz \{(N^2 - f^2) + g\delta(z)\} \varphi^\nu_{\boldsymbol{K}}(z) \int_{-H}^{z} dz' \frac{1}{(2\pi)^2} \int d^2X \, e^{-i\boldsymbol{K}\cdot\boldsymbol{X}} \tag{46}$$

to the state ψ_a. The structure of this projection operator follows from eqs. (43), (20), and the orthogonality properties of the functions $\exp(i\boldsymbol{K}\cdot\boldsymbol{X})$. From eq. (36) we find that the normal mode amplitudes evolve according to

$$\frac{d}{dt} A^s_{\nu \boldsymbol{K}} + i\omega^s_{\nu \boldsymbol{K}} A^s_{\nu \boldsymbol{K}} = \tilde{B}^s_{a\nu \boldsymbol{K}} q_a, \qquad s = +, -, 0. \tag{47}$$

The source function on the r.h.s. of eq. (47) may be expressed in terms of the A's by means of eqs. (44), (38), (35), and (11). This reduces the entire problem, eq. (10), of the evolution of the wave motion to a set of ordinary differential equations. We return to these equations when discussing internal wave dynamics in subsect. 6.3.7.

Evolution of the wave branches:

The procedure above describes the complete decomposition of the state of the motion into the normal mode amplitudes. For some purposes it may be convenient to separate the state ψ_a only into the three wave branches

$$\psi^s(\boldsymbol{X}, z, t) = \sum_{\nu, \boldsymbol{K}} A^s_{\nu \boldsymbol{K}}(t) \beta^s_{a\nu \boldsymbol{K}}(\boldsymbol{X}, z). \tag{48}$$

This is most easily done by formal replacement of $-K^2/\lambda^2$ in eqs. (41) and (42) by the operator $\mathcal{N}^{-1}\mathcal{L}\mathcal{M}$ so that

$$\Omega = f(1 - \mathcal{N}^{-1}\mathcal{L}\mathcal{M})^{1/2} \tag{49}$$

becomes an operator with eigenfunctions $\phi(z)e^{i\boldsymbol{K}\cdot\boldsymbol{X}}$ and eigenvalues $\omega^s = sf(1 + K^2/\lambda^2)^{1/2}$. Also, ε^s_a and $\tilde{\varepsilon}^s_a$ become operators which satisfy

$$\begin{aligned} H_{ab}\varepsilon^s_b &= s\Omega\varepsilon^s_a \\ \tilde{\varepsilon}^s_a H_{ab} &= s\Omega\tilde{\varepsilon}^s_b, \qquad s = +, -, 0. \end{aligned} \tag{50}$$

The operator $\tilde{\varepsilon}^s_a$ projects a state ψ_a onto the wave branch s, i.e.

$$\tilde{\varepsilon}^s_a \psi_a = \psi^s, \tag{51a}$$

while ε^s_a transforms back to ψ_a, i.e.

$$\psi_a = \varepsilon^s_a \psi^s. \tag{51b}$$

Further, one finds that ψ^s satisfies

$$\frac{\partial \psi^s}{\partial t} + s\Omega\psi^s = \tilde{\varepsilon}^s_a q_a, \qquad s = +, -, 0, \tag{52}$$

where again the source term $\tilde{\varepsilon}^s_a q_a$ may entirely be expressed in terms of the wave branch variables ψ^s.

6.3.5 Kinematics of linear small-scale waves

Waves are by definition an essentially linear disturbance of the wave-carrying medium. Strongly nonlinear effects such as breaking occur only as very localized events in space and time. This picture is certainly true for surface gravity waves and generally accepted for internal waves.

6.3.5.1 The geometrical optics or WKB approximation

In the subsect. 6.3.4.5 we derived the concept of decomposing the motion into vertical normal modes. From a mathematical point of view a representation of a linear internal wave field in terms of standing modes is always possible but physically it may be inappropriate. A well-developed standing mode for the full water column appears a priori rather questionable at least for high mode numbers (i.e. small vertical wavelengths) since typical vertical propagation times are comparable to typical relaxation times of nonlinear interactions (Olbers [76 O], McComas and Bretherton [77 M 2]). Phases then will be randomized and energy will be exchanged among the waves before a mode can be formed. In this instance the motion is better described by vertically progressive waves. The kinematical structure of these waves can formally be derived from the equations of motion by performing a WKB expansion with respect to a small parameter (usually the ratio of vertical wavelength to the vertical scale of ambient background fields). This procedure is also known as the geometrical optics approximation.

In this framework a linear wave is characterized by an amplitude $a(\boldsymbol{k})$, a 3-dimensional wavevector $\boldsymbol{k}$ and a frequency ω which are related by a dispersion relation $\omega=\Omega(\boldsymbol{k})$. Large scale inhomogeneities (compared to period and wavelength) of the wave-carrying background can be treated by WKB methods. Waves then appear in form of slowly varying wave trains which may locally be represented by wave groups characterized by a local dispersion relation

$$\omega=\Omega(\boldsymbol{k},\boldsymbol{x},t)\,. \tag{53}$$

A mean current $\boldsymbol{U}$ is included in Ω as a Doppler shift $\boldsymbol{k}\cdot\boldsymbol{U}$ so that $\omega-\boldsymbol{k}\cdot\boldsymbol{U}$ represents the intrinsic frequency.

A wave group propagates with the group velocity

$$\dot{\boldsymbol{x}}=\frac{\partial}{\partial \boldsymbol{k}}\Omega\,. \tag{54}$$

On the trajectory (ray) wavevector and frequency change according to

$$\dot{\boldsymbol{k}}=\frac{\partial}{\partial \boldsymbol{x}}\Omega \tag{55}$$

$$\dot{\omega}=\frac{\partial\Omega}{\partial t}\,. \tag{56}$$

Notice that $\partial/\partial\boldsymbol{x}$ acts on the explicit $\boldsymbol{x}$-dependence whereas ∇ below in eq. (57) also acts on the spatial variations of $\boldsymbol{k}$. Changes in amplitude are conveniently expressed in the form of action conservation (Whitham [70 W], Bretherton and Garrett [68 B])

$$\frac{\partial}{\partial t}\left(\frac{\varepsilon}{\omega-\boldsymbol{k}\cdot\boldsymbol{U}}\right)+\nabla\cdot\left(\dot{\boldsymbol{x}}\frac{\varepsilon}{\omega-\boldsymbol{k}\cdot\boldsymbol{U}}\right)=0\,. \tag{57}$$

The wave energy density $\varepsilon\sim|a(\boldsymbol{k})|^2$ is a quadratic functional of the amplitude. Equation (57) states that the wave action $\int\varepsilon/(\omega-\boldsymbol{k}\cdot\boldsymbol{U})\mathrm{d}^3x$ is an adiabatic invariant (see e.g. Landau and Lifshitz [70 L]) for slowly varying linear wave groups. The application of this general framework of wave kinematics to internal waves is the aim of this subsection.

6.3.5.2 Description of the wave guide

The kinematical structure of internal waves is determined by the mean buoyancy and current fields and the bottom topography. Only variations of these mean oceanic fields have an essential influence: the reflection properties of a flat horizontal bottom are fairly simple, a constant current merely implies a Doppler shift, and the waves feel the buoyancy field only through its gradients. Horizontal gradients of buoyancy and currents are generally small compared to the vertical gradients and therefore are neglected in the traditional kinematical models based on a horizontally homogeneous buoyancy frequency $N(z)$ and shear current $\boldsymbol{U}(z)$. A typical profile of the mean current may be extracted from Fig. 4 averaging by eye the wiggles of the internal wave motion. The buoyancy frequency usually has maxima below the upper mixed layer and in the deeper ocean where strong gradients of temperature, and thus density, exist. Peak values of N are about $10^{-2}\,\mathrm{s}^{-1}$ (period 10 min) in the upper (seasonal) thermocline and $10^{-3}\,\mathrm{s}^{-1}$ (period 1.5 h) in the lower (main) thermocline. Typical profiles of $N(z)$ are displayed in Fig. 6.

A further external parameter which is essential for the kinematic description of internal waves, in particular for waves of low frequency and large wavelength, is the local Coriolis frequency f. In most kinematical models f is assigned a constant local value (the f-plane approximation). However, if a wave group travels substantial distances on the globe the variation of f with latitude must be taken into account.

6.3.5.3 Properties of linear small-scale waves

We restrict the discussion to small-scale waves in an ambient mean flow that is geostrophically balanced. In such a state the horizontal density gradient affects the WKB approximation only in higher orders (Olbers [81 O 1]). Thus, the kinematics are determined by the slowly varying buoyancy frequency $N(\boldsymbol{x})$ and current $\boldsymbol{U}(\boldsymbol{x})$.

Dispersion relation:

Applying the WKB approximation to the linearized equations (10) one finds that a wave with a phase factor $\exp[i(\boldsymbol{k}\cdot\boldsymbol{x}-\omega t)]$ obeys the local dispersion relation

$$\omega=\Omega(\boldsymbol{k},\boldsymbol{x})=\omega_0+\boldsymbol{k}\cdot\boldsymbol{U}(\boldsymbol{x}), \tag{58}$$

where the intrinsic frequency is given by

$$\omega_0=\Omega_0\,\boldsymbol{k},\boldsymbol{x})=\left\{N^2(\boldsymbol{x})\frac{K^2}{k^2}+f^2\frac{k_3^2}{k^2}\right\}^{1/2}. \tag{59}$$

Equivalently,

$$(k_3/K)^2=\tan^2\theta=\frac{N^2(\boldsymbol{x})-\omega_0^2}{\omega_0^2-f^2}, \tag{60}$$

which relates the inclination θ of the three-dimensional wavevector $\boldsymbol{k}=(k_1,k_2,k_3)$ to the intrinsic frequency (see Fig. 14). In these expressions k is the modulus of $\boldsymbol{k}$ and K is the modulus of the horizontal wavevector $\boldsymbol{K}=(k_1,k_2)$. The intrinsic frequency ω_0 is restricted to $f<\omega_0<N$ and independent of the magnitude of $\boldsymbol{k}$. The dependence of ω_0 on wavelength can be inferred from Fig. 13.

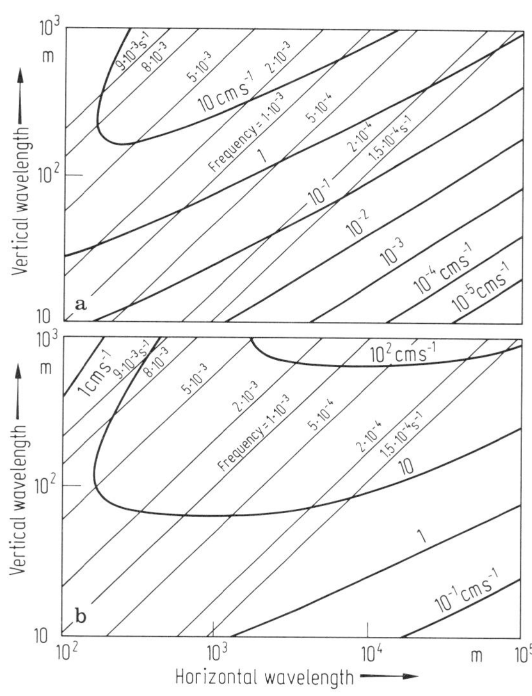

Fig. 13a. Vertical group velocity v_3 as function of vertical and horizontal wavelength and frequency ($N/f=10^2$, $f=10^{-4}\,\mathrm{s}^{-1}$). Thick lines: velocity contours, thin lines: frequency contours.

Fig. 13b. Modulus of the horizontal group velocity as function of vertical and horizontal wavelength and frequency ($N/f=10^2$, $f=10^{-4}\,\mathrm{s}^{-1}$). Thick lines: velocity contours, thin lines: frequency contours.

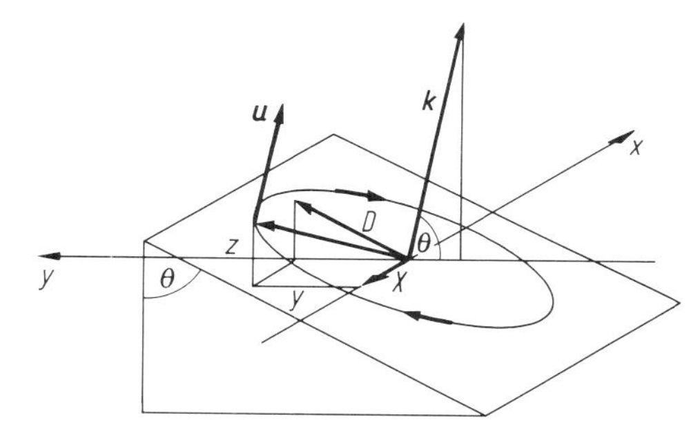

Fig. 14. The particle motion is in the plane perpendicular to the wavevector $\boldsymbol{k}$. The orbit is elliptical with the major axis along the line of maximum slope and in an anticyclonic sense (here shown for the northern hemisphere). The wavevector $\boldsymbol{k}=k(0,\cos\theta,\sin\theta)$ is in the (y,z)-plane, (X,Y,Z) = particle position vector, $\boldsymbol{u}$ = velocity vector.

Group velocity and rate of refraction:

The ray equations (54), (55), and (56) take the form

$$\dot{x}_i=\frac{\partial\Omega}{\partial k_i}=v_i+U_i,$$
$$\dot{k}_i=-\frac{\partial\Omega}{\partial x_i}=r_i-k_j\frac{\partial U_j}{\partial x_i}\quad(i=1,2,3), \tag{61}$$
$$\omega=\text{constant},$$

where $\boldsymbol{v}$ and $\boldsymbol{r}$ are the intrinsic group velocity and rate of refraction, respectively,

$$\boldsymbol{v}=\frac{\partial}{\partial \boldsymbol{k}}\Omega_0=\frac{(N^2-\omega_0^2)(\omega_0^2-f^2)}{\omega_0 K^2(N^2-f^2)}\left(k_1, k_2, -\frac{\omega_0^2-f^2}{N^2-\omega_0^2}k_3\right),$$
$$\boldsymbol{r}=-\frac{\partial}{\partial \boldsymbol{x}}\Omega_0=-\frac{N}{\omega_0}\frac{N^2-\omega_0^2}{N^2-f^2}\nabla N. \tag{62}$$

Well-known properties of the wave-group propagation are that the intrinsic stretching or shrinking of the wave vector $\boldsymbol{k}$ always occurs along the gradient of the buoyancy frequency, and that phase and group velocity of a wave are orthogonal ($\boldsymbol{v}\cdot\boldsymbol{k}=0$). Note further that upward phase propagation ($k_3>0$) implies downward propagation of the wave group and vice versa. The dependence of the group velocity on frequency and wavenumber is given in Fig. 13.

Particle orbits:

The compressibility condition constrains the particle motion of a wave to the plane orthogonal to the wave vector $\boldsymbol{k}$. To describe the particle motion in detail it is instructive to express the equations of motion (10a–d) in terms of the displacement vector (X, Y, Z) and transform to the component X and $D=Y/\sin\theta$ describing the displacement in this plane (see Fig. 14). This yields

$$\ddot{X}-f\sin\theta\dot{D}=0,$$
$$\ddot{D}+f\sin\theta\dot{X}+N^2\cos^2\theta D=0 \tag{63a}$$

which expresses the balance between inertia, Coriolis force (the component of rotation perpendicular to the plane is $\frac{1}{2}f\sin\theta$), and buoyancy force (which is the product of gravity component $g\cos\theta$ parallel to the plane and the density change $\frac{\mathrm{d}\bar{\varrho}}{\mathrm{d}z}\cos\theta$ along the plane). The solution for the particle orbit is

$$D=D_0\cos\omega_0 t,$$
$$X=D_0\frac{f\sin\theta}{\omega_0}\sin\omega_0 t \tag{63b}$$

which describes an ellipse with ratio of along-slope to up-slope axes of $(f/\omega_0)\sin\theta$ where ω_0 is given by eq. (59). A sketch of the particle orbit is shown in Fig. 14. At near-inertial frequencies $\omega_0\approx f$ (where $\sin\theta\approx 1$ and $k_3^2\gg K^2$) the motion is almost horizontal and circular, at higher frequencies the ellipse tends towards the vertical and becomes eccentric with almost up- and downward motion at $\omega_0\approx N$ (where $\sin\theta\approx 0$ and $K^2\gg k_3^2$). The motion around the orbit is or anticyclonic, i.e. clockwise in the northern hemisphere.

Field representation:

The velocity field $\boldsymbol{u}$, vertical displacement field $\xi=-b/N^2$, and pressure field p of the wave are represented by

$$\begin{pmatrix}\boldsymbol{u}(\boldsymbol{x},t)\\ \xi(\boldsymbol{x},t)\\ p(\boldsymbol{x},t)\end{pmatrix}=a(\boldsymbol{k})\begin{pmatrix}\boldsymbol{\mathcal{U}}(\boldsymbol{k})\\ \mathrm{i}\mathcal{U}_3/\omega_0\\ C(\omega_0^2-f^2)\end{pmatrix}\mathrm{e}^{\mathrm{i}(\boldsymbol{k}\cdot\boldsymbol{x}-\omega t)}+\text{complex conjugate}, \tag{64a}$$

where

$$\boldsymbol{\mathcal{U}}(\boldsymbol{k})=C\begin{pmatrix}\omega_0 k_1+\mathrm{i}fk_2\\ \omega_0 k_2-\mathrm{i}fk_1\\ -\omega_0\frac{\omega_0^2-f^2}{N^2-\omega_0^2}k_3\end{pmatrix} \tag{64b}$$

is the polarisation vector of the velocity and $a(\boldsymbol{k})$ the wave amplitude.

Energy and energy flux:

The normalization factor C is conveniently chosen as

$$C=(\omega_0 K)^{-1}\{(N^2-\omega_0^2)/(N^2-f^2)\}^{1/2}.$$

Then the total local energy density of the wave component (averaged over a wave period, denoted by the overbar) is

$$\varepsilon=\tfrac{1}{2}(\overline{u_j u_j}+N^2\overline{\xi^2})=2|a(\boldsymbol{k})|^2\,, \tag{65}$$

and its energy flux vector is

$$\boldsymbol{\phi}=\overline{p\boldsymbol{u}}=2|a(\boldsymbol{k})|^2\boldsymbol{v}=\varepsilon\boldsymbol{v}\,. \tag{66}$$

These equations show that the wave energy flux becomes the product of the energy density and the group velocity, a relation which is true for many wave types.

For future reference we give the expressions for the horizontal and vertical kinetic energy and potential energy

$$\begin{aligned}\varepsilon^{\mathrm{h}}_{\mathrm{kin}}&=\frac{1}{2}\overline{u_\alpha u_\alpha}=\frac{\omega_0^2+f^2}{\omega_0^2}\,\frac{N^2-\omega_0^2}{N^2-f^2}|a(\boldsymbol{k})|^2\,,\\ \varepsilon^{\mathrm{v}}_{\mathrm{kin}}&=\frac{1}{2}\overline{u_3^2}=\frac{\omega_0^2-f^2}{N^2-f^2}|a(\boldsymbol{k})|^2\,,\\ \varepsilon_{\mathrm{pot}}&=\frac{1}{2}N^2\overline{\xi^2}=\frac{\omega_0^2-f^2}{N^2-f^2}\,\frac{N^2}{\omega_0^2}|a(\boldsymbol{k})|^2\,.\end{aligned} \tag{67}$$

Apparently, near-inertial waves carry their energy mainly in kinetic form associated with horizontal motion while near-buoyancy waves carry their energy in potential and vertical kinetic form at about the same amount.

Reflection at solid boundaries:

The WKB theory is completed by reflection conditions for the rays at the sea surface and at the bottom. These are given, e.g., in the textbooks of Phillips [77 P] and LeBlond and Mysak [78 L 1]. For a rigid boundary sloping at an angle α with respect to the horizontal (i.e. $x_3=x_1\tan\alpha$) the reflected wavevector (l_1, l_2, l_3) is given by

$$\begin{aligned}l_1&=\frac{(1+\tan^2\alpha\sin^2\theta)k_1+2k_3\tan\alpha}{1-\tan^2\alpha\sin^2\theta}\\ l_2&=k_2\,,\\ l_3&=-\frac{(1+\tan^2\alpha\sin^2\theta)k_3+2k_1\tan\alpha\sin^2\theta}{1-\tan^2\alpha\sin^2\theta}\,,\end{aligned} \tag{68}$$

where (k_1, k_2, k_3) is the incident wavevector with the inclination angle θ. If $\tan^2\theta\tan^2\alpha<1$, i.e. $\omega^2>\omega_s^2$ with

$$\omega_s^2=N^2\sin^2\alpha+f^2\cos^2\alpha\,, \tag{69}$$

the reflection process is horizontally transmissive in the sense that the group velocities of the incident and the reflected wave point in the same horizontal direction. For $\omega^2<\omega_s^2$ the horizontal group velocities are opposite to each other and the reflection process is said to be horizontally reflective. The frequency ω_s thus is an important parameter of the wave field close to sloping bottoms. Waves with a frequency below ω_s will be reflected backward if they impinge on a sloping bottom. Waves with $\omega>\omega_s$ must proceed up the slope, which may lead to accumulation of high-frequency energy at topographic features. Consequences of this process will be considered in subsect. 6.3.7.

6.3.5.4 Turning levels and critical levels

For a single monochromatic wave the wave pattern at a fixed position is stationary and the action conservation, eq. (57), takes the form

$$\nabla\cdot\left\{(\boldsymbol{v}+\boldsymbol{U})\frac{\varepsilon}{\omega_0}\right\}=0 \tag{70}$$

stating that wave action flux through any cross-section of a ray tube is constant. This equation and the ray equations (61) determine changes of the wave parameters along the ray. Solutions of the ray equations (61) and the action conservation eq. (70) are discussed in this subsection for a horizontal mean flow $\boldsymbol{U}$.

Turning level:

In a horizontally homogeneous ocean with a constant horizontal mean current (the traditional kinematical model) integration of eqs. (61) and (70) yields the constancy of the horizontal wave vector $\boldsymbol{K}$ and the intrinsic frequency ω_0 whereas the vertical wave number and energy change according to

$$k_3(z) \sim \{N^2(z) - \omega_0\}^{1/2},$$
$$\varepsilon(z) \sim \frac{1}{v_3} \sim \frac{N^2(z) - f^2}{(N^2(z) - \omega_0^2)^{1/2}}. \tag{71}$$

Thus, when a wave group propagates towards a region of lower $N(z)$ the vertical wavenumber and the group velocity tend to zero but the wave group still reaches the depth where $N(z) = \omega_0$ (turning depth) in a finite time and internal reflection occurs. Near this depth the WKB solution becomes invalid and should be replaced by solutions in terms of Airy functions (Desaubies [75 D]). These show that the energy $\varepsilon(z)$ possesses a finite maximum at the turning depth rather than the weak singularity given by eq. (71).

Critical levels:

By allowing the current to have a vertical shear, i.e. $\boldsymbol{U} = \boldsymbol{U}(z)$, another important kinematical feature is introduced. Here $\boldsymbol{K}$ is still constant but the intrinsic frequency $\omega_0(z) = \omega - \boldsymbol{K} \cdot \boldsymbol{U}(z)$ now varies with depth and eqs. (60) and (70) yield

$$k_3(z) \sim \left\{\frac{N^2(z) - \omega_0^2(z)}{\omega_0^2(z) - f^2}\right\}^{1/2},$$
$$\varepsilon(z) \sim \frac{\omega_0}{v_3} \sim \frac{\omega_0^2(z)\,(N^2(z) - f^2)}{(\omega_0^2(z) - f^2)^{3/2}(N^2(z) - \omega_0^2)^{1/2}}. \tag{72}$$

A wave group propagating towards increasing $\boldsymbol{K} \cdot \boldsymbol{U}(z)$ and thus decreasing ω_0 may encounter a level where ω_0 approaches f. Here k_3 and ε tend to infinity, the wave group shrinks, its vertical shear increases, and the group velocity tends towards the horizontal. However, in contrast to the turning depth behavior, the group never reaches the level where $\omega_0(z) = f$ (Bretherton [66 B]). Dynamical considerations suggest that the near this critical layer where the wave shear becomes very large there will be substantial dissipation, and the wave will be absorbed by the mean flow (Booker and Bretherton [67 B]).

Effect of horizontal inhomogeneities:

Horizontally homogeneous models are of limited value, in particular when discussing internal waves in the upper ocean where N and $\boldsymbol{U}$ are known to vary in the horizontal direction on a broad range of scales. The effect of horizontal inhomogeneities of large scale on the oceanic internal wave field has been investigated only for very special cases. The propagation of internal waves in a geostrophic current with straight, sloping isopycnals was analyzed by Olbers [81 O 1]. For this configuration, which covers the cases discussed above, the refraction equations still can be integrated analytically. Taking the current into the x_2-direction it is found that the frequency of encounter ω, the wavenumber component k_2 along the current U_2, and the wavenumber component $k_\parallel$ parallel to the isopycnals remain constant whereas the normal component $k_\perp$ changes along the ray. The wave guide is the region defined by

$$\omega_c^2(x_1, x_3) \leqq \omega_0^2 = (\omega - k_2 U_2(x_1, x_3))^2 \leqq N^2(x_1, x_3) \tag{73}$$

with

$$\omega_c^2 = \frac{L^2 k_2^2 + f^2 k_\parallel}{k_2^2 + k_\parallel^2},$$
$$L^2 = N^2 \varrho_1^2 + f^2 \varrho_3^2, \tag{74}$$

where $\boldsymbol{\varrho} = (\varrho_1, \varrho_3)$ is the normal to the isopycnals in the (x_1, x_3)-plane. At $\omega_0 = \pm\omega_c$ and $\omega_0 = \pm N$ waves are reflected, and at $\omega_0 = \pm L$ (which is inside the wave guide) waves encounter a critical layer with a valve type behavior: these surfaces can be penetrated from one side while incidence from the other side results in absorption. A sketch of this behavior is shown in Fig. 15. For a horizontally homogeneous ocean ($\varrho_1 = 0$) one recovers the critical layer at $\omega_0^2 = \omega_c^2 = L^2 = f^2$, and for a vertically homogeneous ocean ($\varrho_3 = 0$) the critical layer shifts to $\omega_0^2 = L^2 = N^2$. The conservation of action takes the form

$$v_\perp \frac{\varepsilon}{\omega_0} = \text{constant}, \tag{75}$$

where $v_\perp$ is the component of the intrinsic group velocity normal to the isopycnals. It is then shown in agreement with the concept of internal reflection and critical layer absorption that the energy possesses an integrable singularity at the reflecting surfaces $\omega_0 = \pm\omega_c$ and $\omega_0 = \pm N$ while at $\omega_0 = \pm L$ a nonintegrable singularity appears for waves arriving from the nonpenetrative side.

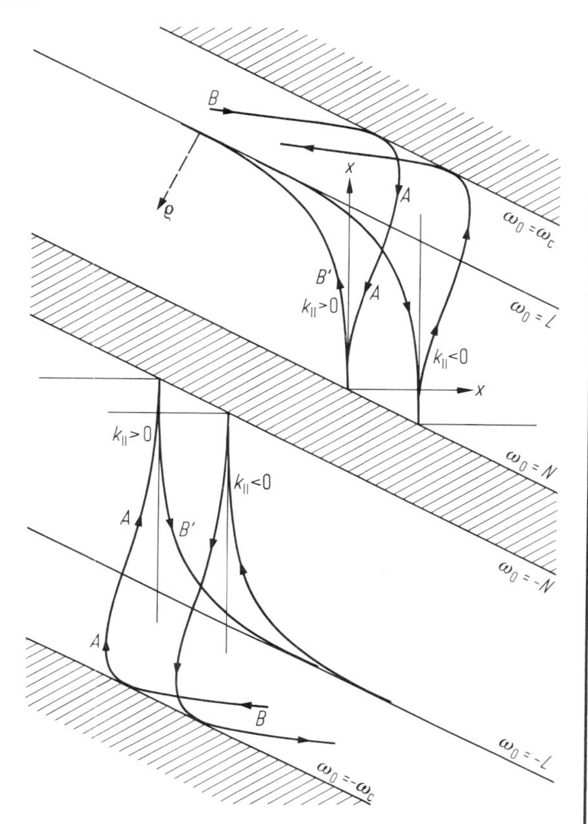

Fig. 15. Sketch of rays for a wave $(\omega, k_2, k_\perp)$ for different $k_\parallel$, projected onto the (x_1, x_3)-plane. The current is normal to this plane and the vector $\boldsymbol{\varrho}$ is normal to the sloping isopycnals. The region forbidden for wave propagation is hatched. Reflection occurs at $\omega = \pm\omega_c$ and $\omega = \pm N$. Wave A is allowed to penetrate whereas B' runs into a critical layer at $\omega_0 = \pm L$ (Olbers [81 O 1]).

Limitations of the WKB theory:

The WKB approximation presented so far is valid for large values of the Richardson number $Ri = N^2/(\partial U/\partial z)^2$. The behavior at critical layers is quite different for small Richardson numbers. For wave propagation in a horizontally homogeneous shear flow Brooker and Bretherton [67 B] found that very little energy and momentum is transmitted or reflected. The wave energy flux is attenuated by a factor $\exp\{2\pi(Ri - 1/4)\}$ across the critical layer which in the limit of large Ri recovers the WKB prediction of complete absorption. Jones [68 J] extends the theory to Richardson numbers in the range $0 < Ri < 1/4$ and found that there substantial reflection may occur. At any given wavenumber and intrinsic frequency there is a critical value of Ri below which the wave is able to extract energy and momentum from the mean flow so that the reflected wave is actually larger than the incident wave. The intimate relation of such an "overreflection" process to the instability of the ambient shear flow at $Ri < 1/4$ was analyzed by Lindzen and Rosenthal [76 L 2].

6.3.5.5 The planetary wave guide

The kinematical models presented so far have neglected the latitudinal dependence of the Coriolis frequency. This may be incorporated and leads to latitudinal bounds of the wave guide. Waves of a frequency ω travelling polewards encounter reflection at the latitude ϕ given by $\omega = 2\Omega_e \sin\phi$. As demonstrated by Munk [80 M] the behavior at these turning latitudes is closely analogeous to the reflection process of vertically progressive waves in an ocean with a variable $N(z)$-profile. Kroll [75 K 3] presented a WKB-theory of internal waves on a β-plane (i.e. taking $\beta = \mathrm{d}f/\mathrm{d}x_2$ constant). Munk [80 M] considers a β-plane with turning point solutions in terms of Airy functions, and Munk and Phillips [68 M] consider Airy function solutions on a sphere. A slightly more complete analysis is provided by Fu [81 F]. Again, as in the vertical direction, the problem of propagating versus standing waves appears.

6.3.6 Spectral models

In subsect. 6.3.3 we have introduced the energy distribution of internal waves by one-dimensional spectra of either frequency or wavenumber, which we directly obtained by applying spectral analysis techniques to the observed series of data points. The complete energy distribution, however, is determined by a three-dimensional spectrum which describes the distribution of energy with respect to the three-dimensional wave-vector (or two-dimensional wave-vector and mode number in case of a modal description). The modeling and description of the complete energy spectrum is subject of this subsection.

6.3.6.1 The cross-spectral matrix

Oceanic internal waves are a phenomenon best treated statistically. The wave motion may be represented by a continuous superposition of linear waves with random amplitudes and phases each of which moves in $(\boldsymbol{x}, \boldsymbol{k})$-space subject to the kinematical constraints discussed in the subsects. 6.3.4 and 6.3.5 and weakly affected by dynamical processes which will be discussed in subsect. 6.3.7.

Spectral matrix for small-scale waves:
A superposition of vertically progressive waves [cf. eqs. (64a) and (64b)]

$$\boldsymbol{u}(\boldsymbol{x}, t) = \int d^3k a(\boldsymbol{k}) \mathcal{U}(\boldsymbol{k}) e^{i(\boldsymbol{k} \cdot \boldsymbol{x} - \omega t)} + \text{complex conjugate} \tag{76}$$

adequately describes the state of the wave field away from turning points in the deep ocean. The remaining field variables may be expressed corresponding to eq. (64). It is reasonable to assume statistical independence of WKB waves with different wavevectors far from reflecting boundaries, i.e.

$$\begin{aligned} \langle a(\boldsymbol{k}) a(\boldsymbol{k}') \rangle &= 0\,, \\ \langle a(\boldsymbol{k}) a^*(\boldsymbol{k}') \rangle &= \tfrac{1}{2} E(\boldsymbol{k}) \delta(\boldsymbol{k} - \boldsymbol{k}')\,. \end{aligned} \tag{77}$$

The statistical properties of the wave field are then condensed in the spectrum $E(\boldsymbol{k})$ which is the density in $\boldsymbol{k}$-space of the total local energy density. The spectrum represents a complete description of the statistical state when the wave field is Gaussian.

Provided that the observed fluctuations are caused merely by internal waves the relation of observed one-dimensional cross-spectra and the energy spectrum $E(\boldsymbol{k})$ is fairly simple and may be computed from the representation in eq. (76) and the statistical properties in eq. (77). Thus, e.g., the one-sided cross-spectrum between moored sensors with separation $\boldsymbol{r}$ becomes

$$A_{ij}(\boldsymbol{r}, \omega) = \frac{1}{\pi} \int_{-\infty}^{+\infty} d\tau \langle u_i(\boldsymbol{x}, t) u_j(\boldsymbol{x} + \boldsymbol{r}, t + \tau) \rangle e^{-i\omega\tau} = \int d^3k \mathcal{U}_i \mathcal{U}_j^* E(\boldsymbol{k}) \delta(\omega - \Omega(\boldsymbol{k})) e^{-i\boldsymbol{k} \cdot \boldsymbol{r}} \tag{78}$$

which represents a weighted projection of the spectral density $E(\boldsymbol{k})$ onto the dispersion surface $\omega = \Omega(\boldsymbol{k})$. Cross-spectra of towed and dropped sensor pairs may be expressed as similar one-dimensional projections. Recovering the complete spectrum $E(\boldsymbol{k})$ by direct methods – e.g. by Fourier transformation of eq. (78) – appears to be impossible in view of the sparse (in space) data base so that indirect methods must be employed.

Spectral matrix for a modal wave field:
For high-frequency waves in the upper ocean a modal representation

$$\boldsymbol{u}(\boldsymbol{x}, t) = \sum_{\nu} \int d^2K a^{\nu}(\boldsymbol{K}) \mathcal{U}^{\nu}[\boldsymbol{K}, \varphi_{\boldsymbol{K}}^{\nu}(z)] e^{i(\boldsymbol{K} \cdot \boldsymbol{X} - \omega t)} + \text{complex conjugate} \tag{79}$$

seems to be more appropriate. Here $\mathcal{U}^{\nu}[\boldsymbol{K},]$ is a differential operator acting on the vertical mode $\varphi_{\boldsymbol{K}}^{\nu}$. It can be inferred from eq. (27) which also yields the representation for the remaining field variables. The basic difference between vertically progressive and vertically standing waves lies in the statistical concept associated with these representations: in eq. (77) up- and downward propagating waves are statistically independent while in a modal field up- and downward propagating waves with the same frequency and horizontal wavevector must have equal amplitude and deterministic phase relation to form the standing mode. Thus we assume

$$\begin{aligned} \langle a^{\nu}(\boldsymbol{K}) a^{\nu'}(\boldsymbol{K}') \rangle &= 0\,, \\ \langle a^{\nu}(\boldsymbol{K}) (a^{\nu'}(\boldsymbol{K}'))^* \rangle &= \tfrac{1}{2} E^{\nu}(\boldsymbol{K}) \delta_{\nu\nu'} \delta(\boldsymbol{K} - \boldsymbol{K}') \end{aligned} \tag{80}$$

to form the spectral matrix for a modal field

$$A_{ij}(\boldsymbol{R}, z, z', \omega) = \sum_{\nu} \int d^2K E^{\nu}(\boldsymbol{K}) \mathcal{U}_i^{\nu}[\boldsymbol{K}, \varphi_{\boldsymbol{K}}^{\nu}(z)] \mathcal{U}_j^{\nu}[\boldsymbol{K}, \varphi_{\boldsymbol{K}}^{\nu}(z')] \delta(\omega - \omega_{\boldsymbol{K}}^{\nu}) e^{-i\boldsymbol{K}\cdot\boldsymbol{R}}, \quad (81)$$

where $E^{\nu}(\boldsymbol{K})$ is the spectrum of total wave energy per unit surface area in the mode ν.

6.3.6.2 Consistency tests

There are relations between the different components of the spectral matrix which do not depend on the specific form of the energy spectrum, e.g., the ratio of the potential and horizontal kinetic energy spectra is simply

$$\frac{N^2 P_{\xi\xi}(\omega)}{P_{11}(\omega) + P_{22}(\omega)} = \frac{\omega^2 - f^2}{N^2 - \omega^2} \frac{N^2}{\omega^2 + f^2}, \quad (82)$$

in agreement with the expressions in eq. (67) for a single wave component. Another relation fixes the ratio of the energies of the clockwise (−) and anticlockwise (+) rotating current components $u_{\pm} = u_1 \pm iu_2$ in the form

$$\frac{P_{++}(\omega)}{P_{--}(\omega)} = \left(\frac{\omega - f}{\omega + f}\right)^2, \quad (83)$$

where $P_{++}(\omega)$ and $P_{--}(\omega)$ is the power spectrum of u_+ and u_-, respectively. Such relation may be utilized to check for the consistency of observed fluctuations with internal wave kinematics. The above energy tests are shown in Fig. 16 for data from the IWEX experiment. Obviously, for frequencies above the tidal frequency M_2, the relation (83) is fairly well satisfied whereas eq. (83) is systematically violated (due to contamination of the displacement estimates).

There are also consistency relations between cross-spectra for observations at different positions. A complete set of such relations for vertical progressive waves as well as modal waves has been worked out by Müller and Siedler [76 M 2] and applied to the IWEX data set (Müller et al. [78 M]).

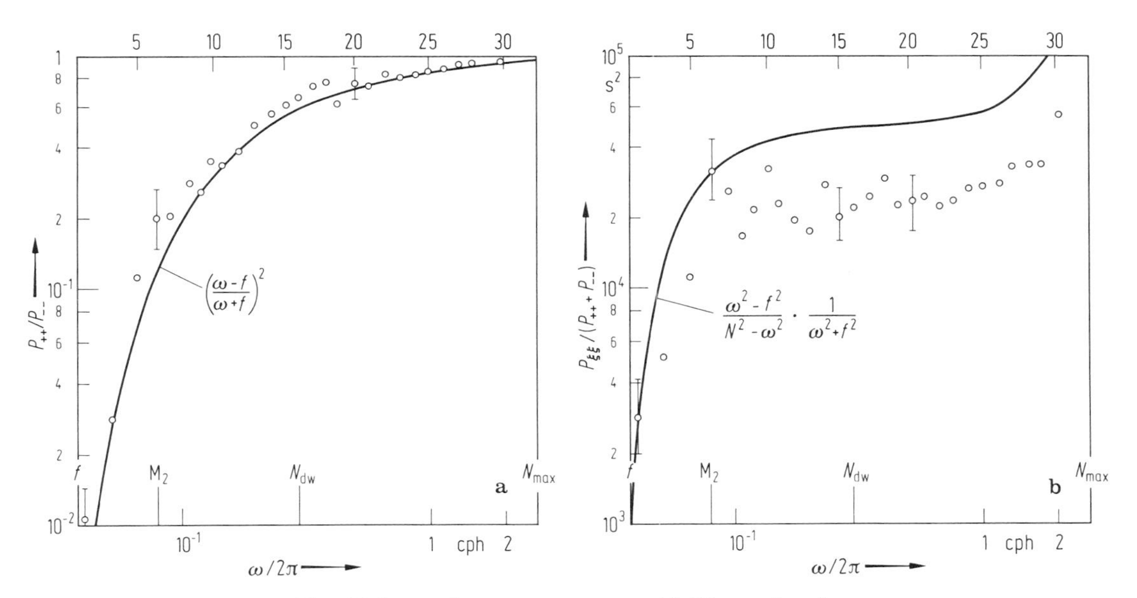

Fig. 16. Internal wave energy tests. (a) Observed ratio P_{++}/P_{--} of average rotary spectra and theoretical curve. (b) Observed ratio $P_{\zeta\zeta}/(P_{++} + P_{--})$ and theoretical curve (Müller et al. [78 M]).

6.3.6.3 The GM model for the deep ocean wave field

The first attempt to provide a unified picture of the internal wave field was made by Garrett and Munk [72 G] who synthesized a model of the complete wavenumber-frequency spectrum of the motion in the deep ocean on the basis of linear theory and the available observations (GM model). Except for inertial internal waves and tides this model is believed to reflect the spectral features of the internal wave climate in the deep ocean and to possess a certain global validity. Most data were in good agreement with the model or could be incorporated by slight modifications (Garrett and Munk [75 G], Müller et al. [78 M]).

The GM spectral form:

The basic features of the GM model are
(a) horizontal isotropy of the energy distribution
(b) vertical symmetry of the energy distribution
(c) -2 slope in frequency domain with a cuspy increase at the inertial frequency
(d) -2 decrease in wavenumber
(e) bandwidth of equivalently about 10 vertical modes at each frequency.

Because of horizontal isotropy and vertical symmetry the spectrum is completely specified by its density $E(\alpha, \beta)$ in (α, β)-space where α and β are the moduli of horizontal wavevector and vertical wavenumber, respectively. Likewise, the densities $E(\alpha, \omega)$ or $E(\beta, \omega)$ may be used. These distributions of energy in wavenumber-frequency space are displayed in Fig. 17. The analytical form of the spectrum will briefly be described.

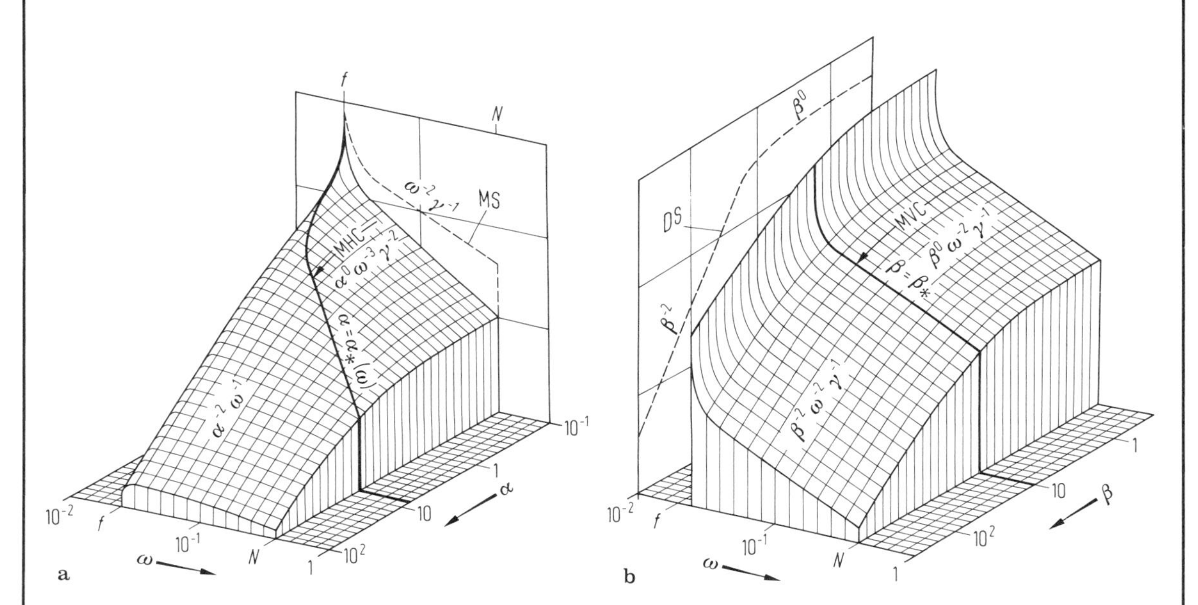

Fig. 17. The Garrett and Munk model for the energy spectrum of internal waves. (a) displays $E(\alpha, \omega)$, (b) displays $E(\beta, \omega)$ with $\gamma=(1-f^2/\omega^2)^{1/2}$. Coordinates are dimensionless and plotted logarithmically, so that plane surfaces represent power laws, as designated. The moored spectrum MS is a projection on a vertical plane, as shown in (a), and the dropped spectrum DS is displayed similarly. Coherences (MHC, MVC) are related to the bandwidths, as indicated (after Garrett and Munk [75 G]).

The spectrum is factorized in the form

$$E(\alpha,\omega)=\varepsilon B(\omega)A(\alpha,\omega)\,, \tag{84}$$

where the frequency distribution $B(\omega)$ and the wavenumber distribution $A(\alpha,\omega)$ of each frequency are normalized so that ε represents the total local energy. The wavenumber distribution was chosen such that $A(\alpha,\omega)$ has the same shape at each frequency and only one scale parameter $\alpha_*(\omega)$ characterizing its width. Then

$$A(\alpha,\omega)=A(\alpha/\alpha_*(\omega))/\alpha_*(\omega)\,. \tag{85}$$

The form of the wavenumber scale $\alpha_*(\omega)$ was assumed to correspond roughly to a constant mode number ν_* which is obtained if the continuum of the (α,ω)-domain is resolved in equivalent modes. Solving the modal eigenvalue problem eq. (19) by WKB approximation yields the modal dispersion relation

$$\alpha_\nu \int_{-H}^{0} \mathrm{d}z \left(\frac{N^2(z)-\omega^2}{\omega^2-f^2}\right)^{1/2}=\nu\pi \tag{86}$$

relating ω to the discrete wavenumbers α_ν, $\nu=1,2,\ldots$. Thus a choice

$$\alpha_*(\omega)=\frac{\nu_*\pi}{bN_0}(\omega^2-f^2)^{1/2} \tag{87}$$

corresponds to a constant number ν_* of equivalent modes at low frequencies and therefore also to a scale of the local vertical wavenumbers

$$\beta_*(\omega,z)=\frac{\nu_*\pi}{bN_0}\{N^2(z)-\omega^2\}^{1/2}\approx\frac{\nu_*\pi N(z)}{bN_0} \tag{88}$$

which is almost constant for $\omega \ll N$. The scales α_* and β_* are related by the dispersion relation eq. (60) of progressive waves. The parameter $bN_0=\int \mathrm{d}z N(z)$ is an integral scale of the stratification. A typical value for the mid-ocean is $bN_0=6.5\,\mathrm{m\,s^{-1}}$ with $b=1300\,\mathrm{m}$ and $N_0=5\cdot 10^{-3}\,\mathrm{s^{-1}}$.

The spectral shape in the frequency domain was inferred from the moored energy spectrum and chosen as

$$B(\omega)=\frac{2}{\pi}\frac{f}{\omega}(\omega^2-f^2)^{-1/2}\,. \tag{89}$$

The shape in wavenumber domain $A(\lambda)$ came from the (one-sided) towed and dropped spectra of vertical displacement. A choice

$$A(\lambda)=\frac{2}{\pi}(1+\lambda^2)^{-1} \tag{90}$$

agrees with most observations of these spectra. The shape of the spectrum is then fixed and the scale parameters ε and ν_* remain to be specified. The form

$$\varepsilon(z)=\frac{\varepsilon_0}{N_0}N(z) \tag{91}$$

with $\varepsilon_0=3\,\mathrm{J/m^3}$ agreed with the WKB scaling, eq. (71), and (within a factor of 3) with most observed internal waves rms velocities (≈ 5 cm/s) and rms vertical displacements (≈ 7 m). The mean square quantities derived from the spectrum are

$$\langle\zeta^2\rangle=\frac{1}{2}\frac{\varepsilon_0}{N_0}N^{-1}(z)\,,$$

$$\langle u_1^2\rangle=\langle u_2^2\rangle=\frac{3}{4}\frac{\varepsilon_0}{N_0}N(z)\,. \tag{92}$$

Integrating of eq. (91) yields $b\varepsilon_0$ for the energy per unit surface area amounting to $4\cdot 10^3\,\mathrm{J/m^2}$.

Information on the wavenumber scale was taken from the moored coherence. Estimates of ν_* may conveniently be obtained independently from moored vertical and horizontal coherences using relations between ν_* and the distance at which the coherences as function of separation drops below a certain value. In terms of the bandwidth

$$\beta_e(\omega) = \left(\int_0^\infty d\beta E(\beta, \omega) \right)^2 \Big/ \int_0^\infty d\beta E^2(\beta, \omega) \tag{93}$$

of the spectrum the moored vertical coherence drops below 1/2 at $\Delta z_{1/2} \approx 2/\beta_e$ rather independently of the shape of $A(\lambda)$. For the form in eq. (90) $\beta_e = \pi\beta_*$. Observations then yield a mode number scale ν_* of about 3 corresponding to mode number bandwidth $\nu_e = \pi\nu_*$ of 10 modes.

Model one-dimensional cross-spectra:

The GM spectrum may be used to recover the shape of all one-dimensional spectra and cross-spectra which apply for the mean oceanic condition. Thus. e.g. the analytical expression for the frequency moored spectrum (MS), dropped spectrum (DS), towed spectrum (TS), and moored coherences for vertical (MVC) and horizontal (MHC) separations are given by

$$\begin{aligned}
\mathrm{MS}^{\mathrm{h}}_{\mathrm{kin}}(\omega) &= \varepsilon(z)\frac{\omega^2+f^2}{\omega^2}B(\omega) \sim N(z)\frac{2\varepsilon_0}{\pi N_0}\frac{f}{\omega^2}, \quad \omega \gg f,\\
\mathrm{MS}_\xi(\omega) &= \frac{\varepsilon(z)}{N^2}\frac{\omega^2-f^2}{\omega^2}B(\omega) \sim N^{-1}(z)\frac{2\varepsilon_0}{\pi N_0}\frac{f}{\omega^2}, \quad \omega \gg f,\\
\mathrm{DS}_\xi(\beta) &= \frac{1}{2}\frac{\varepsilon(z)}{N^2}A\left(\frac{\beta}{\beta_*}\right) \sim N^{-1}(z)\frac{\varepsilon_0}{\pi N_0}\left(\frac{\beta}{\beta_*}\right)^{-2}, \quad \beta \gg \beta_*,\\
\mathrm{TS}_\xi(k_1) &= \frac{\varepsilon(z)}{N^2}\left(\frac{2}{\pi}\right)^3\frac{f\nu_*\pi}{bN_0}\left\{\ln\frac{N}{f}-\frac{1}{2}\left(1-\frac{f^2}{N^2}\right)\right\}k_1^{-2}, \quad k_1 \gg \alpha_*,\\
\mathrm{MVC}(\omega, \Delta z) &= \exp[-\beta_*\Delta z],\\
\mathrm{MHC}(\omega, R) &= I_0(\alpha_* R) - L_0(\alpha_* R)
\end{aligned} \tag{94}$$

where h and ξ refer to horizontal kinetic energy and displacement, respectively. Further, I_0 and L_0 are the modified Bessel and Struve function, respectively (see Abramowitz and Stegun [64 A]). Observe the scaling of the spectra with N mentioned in subsect. 6.3.3.1. The shape of the GM models for the MS, DS, and TS spectra are inserted in the Figs. 7 and 8.

Limitations:

There is an obvious necessity for a spectral cutoff at high wavenumbers to avoid an infinite mean shear. Experimental evidence of such a wavenumber bound for internal waves is masked by the presence of finestructure. As shown in Fig. 8b the -2 wavenumber slope continues to about 10^{-1} cpm. However, it is not clear at present if the following -3 part can still be attributed (or partly attributed) to internal waves. With respect to this problem we refer to Holloway [83 H]. The GM-model does not reproduce the bump in the sub-buoyancy range caused by turning point effects. This feature can be modelled by extensions of the WKB-theory (Desaubies [75 D]). The inertial peak in the spectrum is generally estimated too low by the shape function in eq. (89). This problem will be considered in subsect. 6.3.6.5.

Extensions:

The most comprehensive efforts to get insight into the precise local structure of the wave field have been made with the IWEX (Internal Wave Experiment) performed 1973 in the Sargasso Sea. The experiment consisted of a three-dimensional array of 20 current meter and temperature sensors (to estimate vertical velocity) deployed in the main thermocline with a spatial resolution ranging from 2 m to about 1000 m. A detailed description of the experiment and the data is given by Briscoe [75 B 2], the modelling of the spectrum was performed by Müller et al. [78 M]. The analysis basically confirmed the universal GM spectrum but revealed a great deal of variability of the spectral parameter with frequency which is not of statistical but of dynamical origin. Further, the spectrum was asymmetric at low frequencies with more energy going down than up, and also showed some degree of anisotropy.

6.3.6.4 Upper ocean models

Spectral modelling of upper ocean spectra in the spirit of Garrett and Munk has not yet far advanced. Käse and Clarke [78 K] presented a model of the superenergetic waves at high frequencies which is based on the peculiar response function of the stratification in the upper ocean. The energy is mainly contained in the first mode with wavelength of about 1 km. Levine et al. (1983) applied spectral modelling to temperature and velocity spectra from the Mixed-Layer-Experiment (MILE) which was performed in the North-East Pacific Ocean. The low-frequency part of the spectra (0.1···1.0 cph) could successfully be interpreted by a WKB model whereas for the high-frequency shoulder in the frequency range 1.0···5.0 cph a modal picture was more appropriate. Again the first mode was predominant. Peters [83 P] conceived a spectral model based on the shear modes computed from the actual shear current and stratification (see subsect. 6.3.4.6). A significant fraction of the observed anisotropy of the wave field could be attributed to the ambient shear. The hump at high frequencies, however, could only be modelled by a corresponding distortion in the wavenumber distribution. Peters thus concluded that the superenergetic waves are of dynamical origin and not related to the shear current. Roth et al. [81 R] attempted by a comparison of various upper ocean spectra to substantiate the idea that the stationary universal Garrett and Munk spectrum is still present in the upper ocean but masked by a variable spectral contribution generated at the surface (see Fig. 10).

6.3.6.5 Near-inertial waves

Most of the features of the near-inertial wave field described in subsect. 6.3.3.3 could be modelled (Fu [81 F]) by a superposition of a remotely generated contribution and a locally generated contribution to the inertial peak. Waves which are generated closer to the equator and propagate polewards approach their turning latitudes (see subsect. 6.3.5.5) to form here part of the inertial wave field. This global wave field is identified with the observations in dynamically inactive regions. A model based on latitudinal Airy solution (Munk and Phillips [68 M]) and the frequency-wavenumber model of Garrett and Munk at lower latitudes is able to reproduce the main features of the global part: the magnitude of the peak height, the latitudinal dependence of the bandwidth. and the correlation of the blueshift with the height of the peak. The excess of inertial energy above the spectral level of the global model is interpreted as the result of the local forcing. This contribution then is responsible for the enhanced peaks and the downward propagation of energy in the observations above 2000 m depth.

6.3.7 Spectral dynamics

The state of the oceanic internal wave field may be affected by a great variety of processes due to the coupling of the waves with external fields and due to interactions among the waves themselves. For a large ensemble of waves the dynamics reduce to the problem of the spectral balance, which requires the derivation of the radiative transfer equation describing the evolution of the wave spectrum, and its solution. At present we are only at the initial stages of this complex problem, trying to shed light onto the different generation, dissipation, and internal transfer mechanisms and sort out the most important contributions by experimental and theoretical investigations. This subsection presents the present state of this attempt.

6.3.7.1 The radiative transfer equation

The dynamics of internal waves is governed by the conservation equations for momentum and buoyancy which in various stages of approximation are discussed in subsect. 6.3.4. External forcing fields enter through the boundary conditions (e.g., wind stress, atmospheric pressure, and buoyancy flux) or are introduced by suitable spacetime averaging (e.g., large scale mean flow, stationary thermal finestructure) or modal decomposition (e.g., surface gravity waves). Coupling between wave components arises from the nonlinearities in the equations. These may be reduced to an equation which describes the evolution of the wave amplitude $a(\boldsymbol{k})$ in wavenumber space [cf. eq. (47), here for simplicity for vertically progressive wave

$$\left\{\frac{\partial}{\partial t} + \mathrm{i}\Omega_0(\boldsymbol{k})\right\} a(\boldsymbol{k}) = F[a, e; \boldsymbol{k}], \tag{95}$$

where the source term $F(a, e; \boldsymbol{k}]$ derives from nonlinearities and external forcing fields denoted here by e.

Radiative transfer equation:

An evolution equation for the spectrum defined by eq. (77) is readily obtained

$$\frac{\partial}{\partial t}\langle a(\boldsymbol{k})a^*(\boldsymbol{k}')\rangle = \mathrm{Re}\langle a^*(\boldsymbol{k}')F[a,\mathrm{e};\boldsymbol{k}]\rangle. \tag{96}$$

Slow spatial variations (in the WKB sense) of the spectrum may be included defining $E(\boldsymbol{k},\boldsymbol{x},t)$ such that $E(\boldsymbol{k},\boldsymbol{x},t)\mathrm{d}^3k$ is the total energy density at the position $\boldsymbol{x}$ in the wavenumber band d^3k at $\boldsymbol{k}$.

For purpose of spectral transfer it is more convenient to consider the action spectrum

$$A(\boldsymbol{k},\boldsymbol{x},t) = E(\boldsymbol{k},\boldsymbol{x},t)/\Omega_0(\boldsymbol{k}) \tag{97}$$

which, loosely, may be interpreted as a number density of waves in the $(\boldsymbol{k},\boldsymbol{x})$-space. Its evolution is covered by a radiative transfer equation

$$\left\{\frac{\partial}{\partial t} + \dot{\boldsymbol{x}}\frac{\partial}{\partial \boldsymbol{x}} + \dot{\boldsymbol{k}}\frac{\partial}{\partial \boldsymbol{k}}\right\} A(\boldsymbol{k},\boldsymbol{x},t) = S(\boldsymbol{k},\boldsymbol{x},t) \tag{98}$$

which is the generalization of the action conservation, eq. (57), to a random wave field. Interaction processes described by the r.h.s. of eq. (96) have been accounted for by the source function $S(\boldsymbol{k},\boldsymbol{x},t)$ which determines the local change of action of the wave groups due to coupling between them and with external fields as they propagate along their rays. The resemblance between an ensemble of interacting wave groups and an ensemble of interacting particles is apparent: indeed, the ray equations (54) and (55) are the Hamiltonian equations with a Hamiltonian $\Omega(\boldsymbol{k},\boldsymbol{x},t)$ for a particle with generalized coordinate $\boldsymbol{x}$, momentum $\boldsymbol{k}$, and energy ω. Action conservation is then conservation of particle number, and the radiative transfer equation is the analogue of transport equations governing the particle distribution function. The radiative transfer equation (98) has to be augmented by radiation conditions at the surface and the bottom stating that the difference of up- and downward flux of action equals the flux through the boundary

$$\dot{x}_3(\boldsymbol{K},k_3)A(\boldsymbol{K},k_3) + \dot{x}_3(\boldsymbol{K},-k_3)A(\boldsymbol{K},-k_3) = \phi(\boldsymbol{k}), \tag{99}$$

where the boundary source function $\phi(\boldsymbol{k})$ is the net flux of action through the surface or the bottom of the ocean due to coupling with external fields. In a modal description volume and boundary forcing appear both in the modal source function $S^\nu(\boldsymbol{K},\boldsymbol{X},t)$ in the radiation balance of the spectrum $E^\nu(\boldsymbol{K},\boldsymbol{X},t)$ of the vertical mode ν.

Closure hypothesis:

The framework of spectral evolution presented so far is not complete. As a consequence of non-linearities in the equation of motion the source term $\langle a^*F[a]\rangle$ involves triple correlations of the wave amplitudes. The spectral treatment of wave-wave interactions thus requires a closure hypothesis. This dilemma, which is fundamental in turbulence theory, is much reduced in the theory of random wave fields. Because of the dispersive nature of wave propagation linear random wave fields are in a Gaussian state. This is a state in which wave amplitudes are mutually statistically independent so that any wave correlation can be expressed in terms of the spectrum. Weakly nonlinear wave fields never depart much from a Gaussian state. It has been shown by Prigogine [62 P] that in the limit of infinitely weak nonlinear coupling the correlations $\langle a^*F[a]\rangle$ can be determined under the assumption that the lowest order amplitudes are elements of a Gaussian ensemble. If, in addition, the free wave amplitudes are uncorrelated with external fields the radiation balance, eqs. (98) and (99), is a closed equation for the wave spectrum. These ideas form the weak interaction theory which has found wide applications in geophysical wave problems (e.g., Hasselmann [66 H], Olbers [79 O 1]).

Limitations:

Formally, the concept of a wave requires that the growth time of the amplitude due to wave-wave coupling is small compared to the wave period, i.e. $|\partial a/\partial t| \ll \omega a$. The validity of this condition can, however, no longer be assessed from the radiation balance. As pointed out by Holloway [80 H, 82 H] the condition $|S_{\mathrm{ww}}/A| \ll \omega$ on the spectral growth appears to be necessary but not sufficient for the validity of weak coupling between the waves which constitute the spectrum. The source term S_{ww} for wave coupling as derived in the weak interaction limit and evaluated for the Garrett and Munk model is comparable or even larger than ωA in some part of the spectral domain (see below). For this reason the applicability of weak interaction theory to the oceanic internal wave field has recently been questioned (e.g., Holloway [80 H, 82 H]) and theories for strong interactions have been proposed (a survey can be found in the AIP Conference Proceeding [81 A]). These approaches have not yet advanced as to prove consistency or even utility for tackling the energy balance of the oceanic internal wave field. The reader is referred to the exhaustive summary by Holloway [81 H] in the AIP Conference report.

6.3.7.2 Observational evidence of dynamical relations

The energy level of deep-sea internal waves varies only very little: a factor of 2 to 4 as between IWEX and the Garrett and Munk model (with total energy $4 \cdot 10^3$ J/m^2) is typical. The energy in the surface wave field can change by a factor of 10^3 between periods of calm and stormy weather. A relation to the wind as driving agent is quite obvious and a phenomenology involving wave height and fetch and duration of the wind can be found by rather crude observations even if the generation mechanism remains unknown (see, e.g., sect. 6.2 and Phillips [71 P]). In view of the universal character of the deep-sea wave spectrum and the large number of external fields interacting with the internal wave field it seems difficult to establish such a phenomenology in this case.

Topography:

Indeed, the search of Wunsch [76 W] and Wunsch and Webb [79 W] for deviations from the canonical wave spectrum was only moderately successful. These authors investigated the spectral level and slope and the isotropy at medium frequencies of moored measurements in the North Atlantic, the Mediterranean, and the equatorial Indian Ocean. Significant increases of the energy level (Fig. 18) and anisotropy of the spectra were observed near pronounced topographic features (Muir Seamount, canyons), which, however, are lost very rapidly with the distance from these regions.

But the identification of topographic features as a source is not imperative. As shown by Eriksen [82 E] linear inviscid theory of wave reflection can account for much of the observations close to a sloping bottom. Indeed, from eqs. (60), (68), and (70) we find that an incident wave with total energy E_i and vertical wavenumber k_3 is reflected into a wave with wavenumber l_3 and total energy $E_r = E_i l_3/k_3 \sim E_i/(\omega^2 - \omega_s^2)$ where ω_s is the critical frequency, eq. (69), associated with the bottom slope. The spectrum of the local energy density will thus be enhanced in the vicinity of the bottom. As apparent in Fig. 18 the perturbation of the spectrum disappears within a few hundred meters, presumably due to redistribution by nonlinear wave interactions and frictional dissipation, and topographic features would rather be a sink for wave energy and a source for mixing than an energy source.

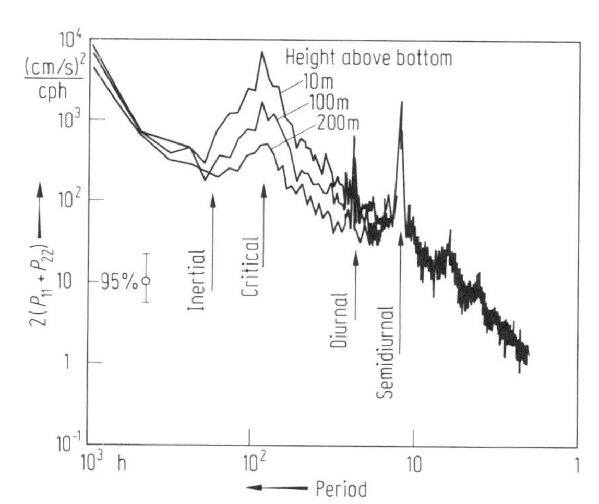

Fig. 18. Spectra of horizontal kinetic energy, $P_{11} + P_{22}$, at 10 m, 100 m, and 200 m height above the bottom. The critical frequency is given by eq. (69) (Eriksen [82 E]).

Surface forcing:

Relation of the upper ocean wave field to surface forcing could be established at low frequencies. Evidence for wind-driven near-inertial waves in the upper thermocline has been reported, e.g., by Käse and Olbers [79 K 2] and Weller [81 W]. Correlations between the local wind and inertial currents down to depths of about 200 m were found. The enhancement of the inertial peak in spectra above 2000 m with respect to the global inertial wave field (Fu [81 F]) was already mentioned above. To interpret this as the consequence of surface forcing seems plausible, in particular in view of the overwhelming evidence of the downward direction of energy propagation in the inertial band. Estimates of the energy flux at near-inertial frequencies by different authors agree roughly within an order of magnitude: values range from 0.1 mW/m^2 to 3 mW/m^2 (e.g., Leaman [76 L 1], Käse and Olbers, [79 K 2]), but show no obvious correlation with environmental conditions. Briscoe [83 B] searched for correlations between high-frequency wave energy and various forcing candidates. Integrating the high-frequency energy over frequency (above the tidal frequency) and depth (over the upper 1500 m) this bulk measure of internal wave energy showed time variations of a few mW/m^2. Correlations were found with the deep near-inertial energy with a lag of about 2 days which makes a transfer of energy from high to low frequencies plausible. Unique correspondences with surface forcing could, however, not be detected.

Other sources:

The identification of other dominant sources or sinks of the wave energy by experimental means has so far been rather limited. In particular, no unique relationships were found between the spectral level in the wave continuum and tidal or inertial energy in the deep ocean. Also, extensive experimental search for empirical relations between the mean shear and the momentum flux of the wave field was only marginally successful (Ruddick and Joyce [79 R], Brown and Owens [81 B]).

6.3.7.3 Theoretical estimates of transfer rates

Only a few processes have been evaluated in a spectral representation. Many investigations consider the behavior of discrete internal waves forced by a deterministic external field or a Fourier component. Spectral growth rates are inherently smaller than growth rates of deterministically forced waves. Studies of deterministic models are useful for clarifying the dynamics but they are useless for estimating time scales of a random forcing process and give no information on the form or magnitude of the source function. Studies of deterministic models will not be reviewed here when spectral treatments exist. The reader is referred to the review of Thorpe [75 T].

The following survey of interaction processes is formally divided into generation processes (i.e., mechanism with excite waves or enhance those which already exist), dissipation processes (i.e., mechanisms which destroy waves), and internal transfer processes (i.e., mechanisms which shift energy within the wave spectrum while conserving the total amount). This separation is not a strict one: depending upon wavenumber and frequency or upon conditions of external fields a process as, e.g., wave-mean flow interaction may either enhance or attenuate waves.

6.3.7.3.1 Generation processes

Sources of internal waves exist in the interior of the ocean as well as at its boundaries. Energy may be extracted from the mean flow, from the tides, from atmospheric and mixed layer turbulence, and from surface waves.

Atmospheric forcing:

At the surface, internal waves can be generated by the atmosphere through resonant coupling to travelling pressure fields and fluctuations of the buoyancy flux and the wind stress. An atmospheric disturbance with wavevector $\boldsymbol{K}$ and frequency ω will generate an internal wave with the same horizontal wavevector and frequency, the mode number (or the vertical wavenumber) adjusts to satisfy the resonance condition (and the surface radiation condition). The source terms are of the form

$$S_{\mathrm{a}}^{\nu}(\boldsymbol{K}) = T_{\mathrm{a}}^{\nu}(\boldsymbol{K}) F^{\mathrm{a}}(\boldsymbol{K}, \omega_{\boldsymbol{K}}^{\nu}), \tag{100}$$

where $F^{\mathrm{a}}(\boldsymbol{K}, \omega)$ is the spectrum of the atmospheric forcing field and $T_{\mathrm{a}}^{\nu}(\boldsymbol{K})$ a transfer function which may be calculated from the equations of motion (see Olbers [83 O] for detailed expressions). Insufficient knowledge of the wavenumber structure of the spectra $F^{\mathrm{a}}(\boldsymbol{K}, \omega_{\boldsymbol{K}}^{\nu})$ in the internal wave range at present prevents a detailed theoretical analysis of these source terms, but it is generally accepted that generation by wind stress is the dominant process. The magnitude of the energy transfer by the wind has been estimated by Käse [79 K 1]. Integration of eq. (100) over low wavenumbers approximately yields a total transfer rate to near-inertial waves

$$\phi_{\tau}^{\nu} = \int \mathrm{d}^2 K S_{\tau}^{\nu}(\boldsymbol{K}) \approx 2\pi F^{\tau}(f)/L_{\nu}, \tag{101}$$

where $F^{\tau}(f)$ is the density of the stress spectrum at low frequencies and L_{ν} a measure of the vertical length scale of the normal mode ν. With typical oceanic parameters transfer rates of about $1\,\mathrm{mW/m^2}$ are obtained. This magnitude of the transfer determined by a dynamical theory agrees favorably with the experimental estimates of the low-frequency energy flux reported above which are merely derived from kinematical properties of the wave field (basically the phase propagation and the blueshift of the inertial peak).

Surface waves:

By resonant interaction a pair of surface waves with wavevectors $\boldsymbol{K}_1$ and $\boldsymbol{K}_2$ and frequencies ω_1 and ω_2 can generate an internal wave with wavevector $\boldsymbol{K}$ and frequency ω satisfying

$$\begin{aligned} \omega &= \omega_1 - \omega_2, \\ \boldsymbol{K} &= \boldsymbol{K}_1 - \boldsymbol{K}_2, \end{aligned} \tag{102}$$

where ω equals the eigenfrequency $\omega_{\boldsymbol{K}}^{\nu}$ of a mode ν. The energy transfer must bridge the large gap in the frequency and wavenumber domain between the two wave modes. This gap causes one of the prominent signatures of the interaction: since ω is much less than ω_i, one gets also $K \ll K_i$ so that the surface wave components propagate almost parallel to each other and perpendicular to the generated internal wave. The transfer to internal waves in the deep ocean is insignificant but the resonance mechanism may very efficiently generate high-frequency internal waves in the upper ocean (Olbers and Herterich [79 O 2]). The spectral transfer from a surface wave spectrum $F^{\mathrm{g}}(\boldsymbol{K})$ to the ν-th mode is given by Hasselmann [66 H].

$$S_{\mathrm{g}}^{\nu}(\boldsymbol{K}) = \int \mathrm{d}^2 K_1 \delta(\omega_1 - \omega_2 - \omega) T_{\mathrm{g}}^{\nu} F^{\mathrm{g}}(\boldsymbol{K}_1) F^{\mathrm{g}}(\boldsymbol{K}_1 - \boldsymbol{K}). \tag{103}$$

The scattering cross-section T_{g}^{ν} depends on the overlapping of the surface and internal wave modes. The analysis shows predominant generation of the first mode at high frequencies and horizontal wavelengths of order 1 km. The total transfer to the first mode is given by

$$\phi_{\mathrm{g}} = \omega_{\mathrm{m}} \varepsilon_{\mathrm{g}} \frac{K_{\mathrm{m}}^2 \langle \zeta_{\mathrm{g}}^2 \rangle}{2\sigma_{\omega}\sigma_{\chi}} \left(\frac{N_0}{\omega_{\mathrm{m}}}\right)^4 \left(\frac{\Delta}{d}\right)^2 \tag{104}$$

for a three layer model of the stability frequency: a mixed layer with depth d, a thermocline of thickness Δ and stability frequency N_0, and a much less stratified deep ocean. Further, $\omega_{\mathrm{m}} = (gK_{\mathrm{m}})^{1/2}$ is the peak frequency of the surface wave spectrum with total energy $\varepsilon_{\mathrm{g}} = g\varrho\langle\zeta_{\mathrm{g}}^2\rangle$ and band widths σ_{ω} and σ_{χ} in the frequency and directional distributions, respectively. In extreme situations – a rough sea and a shallow, strongly stratified thermocline – this rate may attain values comparable to the rate 1 mW/m² of generation by wind stress.

Radiation of internal waves from the surface mixed layer:

There are several mechanisms which may excite internal waves from turbulence or inertial shear currents in the mixed layer. Bell [78 B] considers the advection of corrugations of the mixed layer base by inertial oscillations. The resulting pressure forces excite waves in the stratified layer below. If the inertial current U_0 is moderately large ($U_0 \gtrsim 5$ cm/s) waves are generated at a rate

$$\phi_{\mathrm{m}l} = \frac{2}{3\pi} l N_0^3 \langle \zeta_0^2 \rangle \left\{1 + O\left(\frac{N_0 l}{U_0}\right)\right\}, \tag{105}$$

where the mixed layer turbulence is characterized by an internal scale l and a rms displacement $\langle\zeta_0^2\rangle^{1/2}$. Further N_0 is the stability frequency below the mixed layer. The flux, eq. (105), may attain values as high as 1 mW/m².

Mean flow:

Two mechanisms by which internal waves may draw energy from the mesoscale mean flow have been investigated in a spectral concept. Bell [75 B 1] considered the generation of lee waves by the mean flow over abyssal hills and Müller [76 M 1, 77 M 3] studied the interaction of waves with a mesoscale shear flow.

A steady bottom current $\boldsymbol{U}_{\mathrm{b}}$ interacting with a Fourier component $\boldsymbol{K}$ of the bottom roughness η generates a lee wave of frequency $\boldsymbol{K} \cdot \boldsymbol{U}_{\mathrm{b}}$. Wavenumbers of upward propagating waves are in the range $f/U_{\mathrm{b}} < K < N_{\mathrm{b}}/U_{\mathrm{b}}$ corresponding to wavelengths between 400 m and 4000 m for a bottom current of 4 cm/s. The energy flux may be parameterized as

$$\phi_{\mathrm{b}} = 2 f N_{\mathrm{b}} U_{\mathrm{b}} \langle \eta^2 \rangle \left\{1 + O\left(\frac{f^2}{N_{\mathrm{b}}^2}\right)\right\} \tag{106}$$

and amounts to about 1 mW/m². It is likely that these waves encounter critical layers in the bottom kilometer of the ocean and do not much contribute to the deep ocean wave field.

The interaction of an internal wave with a mean shear flow is described by the ray equations (61) and conservation of action, eq. (70). The wave exchanges energy with the mean flow such that action $\varepsilon/(\omega - \boldsymbol{K} \cdot \boldsymbol{U}(z))$ remains constant. However, the system is reversible: a wave that would gain energy on its way would loose the same amount if it is reflected back to its initial level provided other processes (as e.g., critical layer effects) can be disregarded. A symmetrical field of freely propagating waves would thus remain unaffected by a mean shear. This is not true in the presence of processes which try to relax local distortions of the spectrum by the shear to some shear-independent equilibrium shape. As shown by Müller [76 M 1, 77 M 3], the balance between the tendency of the shear to distort the spectrum and the internal relaxation tendency of the wave field leads to asymmetries in the wave field such that the field exerts a stress which opposes the mean shear. The effect may be parameterized by eddy viscosities which depend on the equilibrium wave spectrum. The magnitude of the theoretically predicted wave-induced viscosities is still in controversy with experimental evidence (see e.g., Ruddick and Joyce [79 R], Brown and Owens [81 B]).

Tides:

The conversion of barotropic tidal energy to baroclinic tides and the coupling to the internal wave field has been proposed as a source of internal wave energy. The total input into the barotropic tide is about $5 \cdot 10^{12}$ W (Kaula and Harris [75 K 2]) which amounts to 12 mW/m^2. This would be more than enough to feed the internal wave field. However, it has been shown (Bell [75 B 1]) that only a small fraction (about 10%) enters the internal tides. Further, spreading of their energy over the internal wave frequency band by nonlinear coupling seems to be entirely negligible (Olbers and Pomphrey [81 O 3]).

6.3.7.3.2 Dissipation mechanisms

A survey of the many mechanism by which internal waves can dissipate has been given by Thorpe [75 T]. As yet it is still a matter of speculation which of these actually work in the ocean to limit the growth of the observed spectrum. Processes suggested in this respect are wave breaking by gravitational (Orlanski and Bryan [69 O]) and shear (Phillips [66 P]) instability, and critical larger absorption (Bretherton [66 B]).

Instabilities:

Gravitational overturning of a wave occurs if the fluid particle velocity exceeds the phase velocity. Shear instability requires the local Richardson number to be less than 1/4. Obviously, both criteria do not only depend on the wave properties (slope and shear) of the wave which actually breaks but also on the ambient flow including other waves which might be present. In tank experiments Thorpe [78 T 1, 78 T 2] has demonstrated how these mechanisms work separately, i.e., overturning in the absence of ambient shear and shear instability in the absence of ambient wave disturbances, and how each may be enhanced if the constituents of the other mechanism are present. In these experiments the breaking can be attributed to a particular wave in the fluid. In the many-wave environment of the ocean there may be no identifiable breaking internal wave, there is only a breakdown in the fluid due to internal waves. Breaking is a more amorphous, unrecognizable process which occurs locally in the fluid and thus affects a broader band in the spectrum. It is still unclear if wave breaking in the ocean prefers one of the two modes of instability.

Critical layers:

The properties of wave propagation near critical layers have been discussed in subsect. 6.3.5. The value of the ambient Richardson number controls whether an incident wave is absorbed, reflected or even overreflected. Critical layers represent a sink of wave energy for large values of the ambient Richardson number. In the ocean critical layers are likely to exist for internal waves with short vertical wavelengths. Roughly, the vertical wavenumber must satisfy

$$\beta(z) > \beta_c(z) = N(z)/\Delta U \,. \tag{107}$$

This vertical wavenumber may attain typical values of $10^{-3}\,\mathrm{cm}^{-1}$ in the upper ocean ($N = 10^{-2}\,\mathrm{s}^{-1}$ and $\Delta U = 10\,\mathrm{cm\,s}^{-1}$) and the main thermocline ($N = 10^{-3}\,\mathrm{s}^{-1}$ and $\Delta U = 1\,\mathrm{cm\,s}^{-1}$). Thus waves with vertical wavelengths shorter than about 60 m (i.e., a substantial part of the wave spectrum) are susceptible to critical layer effects.

Dissipation rates:

Little is known about the rate at which wave breaking extracts energy from the spectrum. A direct estimate of the dissipation rate in observed Kelvin-Helmholtz billows in the Mediterranean seasonal thermocline was derived by Woods (personal communication). From the billow height L, overturning speed U, and intermittency factor I, Woods gets for the dissipation rate $\varepsilon \sim IU^3/L$ some mW/m^2 over 100 m depth. Some information can be obtained from indirect considerations. The kinetic energy released by breaking will partly be used to mix the fluid and thereby increase the mean potential energy. If this turbulent mixing is parameterized by an eddy diffusity $\varkappa$ the local rate of increase is

$$\varepsilon_p = \varkappa N^2 \,. \tag{108}$$

The remaining part of the energy is converted to smaller scales at a rate ε_{sc} where it will eventually be dissipated by molecular action. The ratio $\varepsilon_p/(\varepsilon_p + \varepsilon_{sc})$ gives the efficiency of converting kinetic wave energy to mean potential energy. Following Thompson's [80 T] arguments this ratio should just be the critical Richardson number

$$\varepsilon_p/(\varepsilon_p + \varepsilon_{sc}) = Ri_c = 1/4 \tag{109}$$

which agrees well with estimates by Thorpe [73 T] who studied breaking events in a tank. The relations in eqs. (108) and (109) can be used to evaluate the total dissipation rate due to breaking. If internal waves were responsible for mixing the ocean with the "classical" diffusivity $10^{-4}\,\mathrm{m}^2\,\mathrm{s}^{-1}$ (Munk [66 M]) this would require a total local dissipation of $4 \cdot 10^{-6}\,\mathrm{W/m}^3$ (using a main thermocline $N = 3 \cdot 10^{-3}\,\mathrm{s}^{-1}$) which amounts to some mW/m^2 by vertical integration. Recent measurements of temperature finestructure do not support a value as high as $10^{-4}\,m^2\,s^{-1}$ for the vertical diffusivity (see, e.g., Garrett [79 G 1]). Values of the order $10^{-6}\,m^2\,s^{-1}$ were found which reduces the dissipation rate in the main thermocline to some $10^{-8}\,W/m^3$ locally, or $10^{-5}\,W/m^2$ as vertical integral. Estimates of ε_{sc} can be obtained from velocity microstructure measurement (e.g., Osborn [78 O] and Garrett et al. [81 G]) if one assumes that all of the variance in these scales (less than 1 m) derives from wave dissipation. Observed values for ε_{sc} range from $10^{-6}\,W/m^3$ to $10^{-5}\,W/m^3$ in the upper few hundred meters of ocean which supports Woods' direct estimate mentioned above. Deeper values of ε_{sc} have not been reported. The relationship between wave dissipation and oceanic finestructure has been reviewed by Gregg and Briscoe [79 G 3] and Munk [81 M 3], the aspects of mixing due to internal waves by Garrett [79 G 1].

6.3.7.3.3 Internal transfer processes

There are processes which redistribute energy within the spectrum but conserve its total amount. These processes effect a cascade of energy through the spectrum and may play an important role in the interplay of generation and dissipation in the shaping of the spectrum. The most prominent transfer process is the weak resonant coupling among the spectral components. Other possible mechanisms for transfer are resonant coupling of internal waves to external steady fields (Bragg scattering), at, e.g., bottom inhomogeneities or stationary finestructure in the density field.

Resonant wave-wave interactions:

Due to the nonlinear terms in the equations of motion a triad of internal waves with frequencies ω, ω_1, and ω_2 and wavevectors $\boldsymbol{k}$, $\boldsymbol{k}_1$, and $\boldsymbol{k}_2$ may interact resonantly and redistribute energy among themselves if

$$\begin{aligned} \omega \pm \omega_1 \pm \omega_2 &= 0\,, \\ \boldsymbol{k} \pm \boldsymbol{k}_1 \pm \boldsymbol{k}_2 &= 0\,. \end{aligned} \tag{110}$$

The form of the spectral transfer rate of this process

$$S_{ww}(\boldsymbol{k}) = \int d^3k_1 \int d^3k_2 \{T^+(A_1A_2 - AA_1 - AA_2)\delta(\omega-\omega_1-\omega_2) + 2T^-(A_1A_2 + AA_1 - AA_2)\delta(\omega+\omega_1-\omega_2)\} \tag{111}$$

has been worked out by Hasselmann [66 H] who also pointed out its formal similarity with Boltzmann's collision integral for interacting particles. Intensive study and evaluation of the transfer integral began when Garrett and Munk [72 G] published their first version of the wave spectrum. Olbers [76 O] found that wave energy is systematically transferred from the intermediate frequency range to $f<\omega<2f$ and towards smaller vertical wavelengths with a delivery time scale of the order of days. McComas and Bretherton [77 M 2] found a strong sensitivity of the spectral transfer to the wavenumber slope. This was confirmed by Pomphrey et al. [80 P] by systematic variation of the spectral slope parameter. Following their analysis a spectrum with wavenumber slope -2 is closest to equilibrium in the high-wavenumber region but the major flow of energy is still from low to high vertical wavelengths and low frequencies at rate $0.6\,mW/m^2$.

McComas and Bretherton [77 M 2] were able to identify three classes of resonant wave coupling which are responsible for much of the complex transfer in a spectrum with the form of the Garrett and Munk model where the energy is mainly in low frequencies and large vertical scales and the shear mainly in small vertical scales. A class may dominate the transfer in some region either because the cross-section of the triad is large or because the triad has one small-wavenumber and low-frequency component with a very large action. The resonant triads of the three classes which satisfy eqs. (110) are sketched in Fig. 19. The Garrett and Munk models are in approximate equilibrium with two of these mechanisms (elastic scattering and induced diffusion) whereas the third process (parametric subharmonic instability) controlls the energy flow to the low-frequency, high-wavenumber region with the transfer rates given above. The processes are also responsible for a rapid relaxation of spectral perturbations to the equilibrium form (McComas [77 M 1], McComas and Müller [81 M 2]). In the induced diffusion mechanism (Fig. 19a) a high-frequency, high-wavenumber component interacts with a wave component of much lower frequency and wavenumber to generate another high-frequency, high-wavenumber component. The process has some similarity with diffusion of particles in physical space. Spectra with wavenumber slope -2 or -3 are unchanged by the induced diffusion so that the GM spectrum is unaffected by this mechanism. In the elastic scattering mechanism (Fig. 19b) a high-frequency wave is scattered into another one with approximately reversed vertical wavenumber by interaction with a low-frequency component of approximately twice the vertical wavenumber. This Bragg scattering thus tends to eliminate vertical asymmetries in the wave field. The process is very efficient with exclusion of the very low frequencies. The asymmetries of near-inertial waves do not relax in agreement with the observations of the wave field asymmetry reported above. The last mechanism, the parametric subharmonic instability (Fig. 19c), describes the decay of a low-vertical-wavenumber component into two high-vertical-wavenumber components of about half the frequency. This process accounts for much of the transfer to the near-inertial band at high vertical wavenumbers in the Garrett and Munk models. The growth time is about 100 inertial periods at 100 m vertical wavelength and decreases to about 1/3 inertial period at 1 m.

Other transfer processes:

The scattering of internal waves at horizontally layered irreversible finestructure of the stratification was considered by Mysak and Howe [76 M 3]. The process affects a decay of the asymmetric part of the spectrum. The effect of random bottom inhomogeneities on internal wave modes was studied by Cox and Sandstrom [62 C] in attempt to explain the conversion of the surface tide into internal tides. However, for the internal waves bottom scattering is of minor importance.

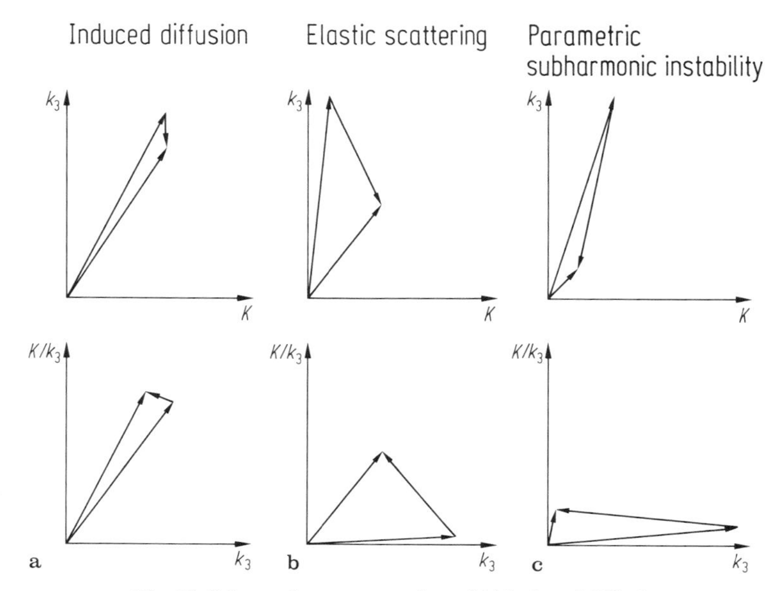

Fig. 19. Schematic representation of (a) induced diffusion, (b) elastic scattering, and (c) parametric subharmonic instability triads, displayed in (top) wavenumber space of vertical wavenumber k_3 vs. horizontal wavenumber K and (bottom) a stretched frequency-vertical wavenumber space. The aspect ration K/k_3 is equivalent to a fixed frequency [see eq. (60)] (McComas [77 M 2]).

6.3.7.4 A perspective of the spectral balance

The discussion of individual processes affecting the oceanic internal wave field is summarized in Fig. 20. There are many interactions which have been left out because their effect on the wave spectrum could not be estimated or turned out to be small. Scattering at finestructure has been mentioned to cause symmetry of the wave field. Generation by atmospheric pressure and buoyancy flux fluctuation is presumably negligible. Shear instability of the mean flow might be important (see Thorpe [75 T]) as well as scattering of internal waves at meso-scale fronts in the upper ocean (Olbers [81 O 2]).

Surprisingly many of the energy fluxes associated with generation and transfer processes of internal waves turn out to be of the same order of magnitude, 1 mW/m^2. Such a magnitude has been found as typical of the energy input in other oceanic motions. Gill et al. [74 G] estimate the energy input into the Sverdrup flow as 1 mW/m^2. Frankignoul and Müller [79 F] give a lower bound of 0.1 mW/m^2 for the generation of baroclinic Rossby waves by stochastic atmospheric forcing. Bryden [82 B] estimates the net local conversion of mean to eddy energy as $2 \cdot 10^{-3}$ mW/m^3. These examples demonstrate that internal waves play an active dynamical role in the interplay of oceanic motion. A tentative picture of this interplay has recently been put forward by Woods [80 W].

The oceanic wave field may draw energy from many sources. Forcing as well as dissipation may be weak. It may well be that characteristics of the forcing have only marginal control on the energy content and even less control on the shape of the spectrum. The wave field may be in such a state where any additional weak input of energy is taken by nonlinear transfer to the spectral domain where dissipation can handle it. There are analytical as well as numerical attempts to verify these ideas (Orlanski and Cerasoli [81 O 4], McComas and Müller [81 M 1]). However, important problems remained unsolved. It is unclear if the upper ocean wave field is dynamically independent from the wave field in the deep ocean. Also the fate of the downward near-inertial wave flux is unknown. Complete transfer of the asymmetric low-frequency part to the wave continuum would represent a large and presumably steady energy input which is hardly compatible with the low dissipation rates. Dissipation of near-inertial energy in the benthic boundary layer is weak (Fu [81 F], D'Asaro [82 D]), so that the question arises if the near-inertial flux may be overestimated (of course this question may as well apply to other energy fluxes).

Aside from such particular items of the energy balance a concept must be put forward which explains the universality of the spectrum. The known sources of energy are not sufficiently steady and uniformly distributed to account for the universality. A simple but important idea was presented by Cox and Johnson [79 C] and discussed by Garrett and Munk [79 G 2]. If internal waves do not dissipate quickly as suggested by the low dissipation rates in the deep ocean they can travel large distances. The low-frequency, energy-containing waves with nonlinear relaxation times of order 100 days can propagate 1000 km within their relaxation time of some days. Beyond this mean free path of the wave groups, wave energy may diffuse and spread over even larger regions (in the same way as phonon propagate in a lattice and interact resonantly leading to a diffusion of heat). This would make the energy level almost uniform.

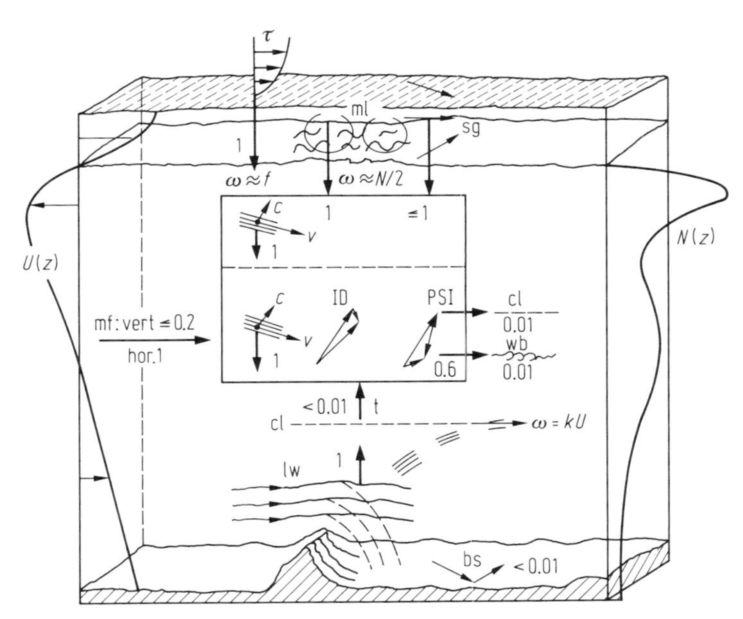

Fig. 20. Sketch of interaction processes affecting the internal wave field in the upper and the deep ocean. Energy fluxes are in units of [mW/m²]. Abbreviations as follow. τ: wind stress, ml: mixed layer turbulence, sg: surface gravity waves, cv: near inertial waves, mf: large scale mean flow, t: baroclinic tides, lw: lee waves, bs: bottom scattering, cl: critical layers, wb: wave breaking, ID: induced diffusion, and PSI: parametric subharmonic instability (Olbers [83 O]).

6.3.8 References for 6.3

1847 S Stokes, G.G.: Trans. Cambridge Philos. Soc. **8** (1847) 441.
1883 R Rayleigh, Lord: Proc. London Math. Soc. **14** (1883) 170.
31 G Goldstein, S.: Proc. R. Soc. London Ser. A **132** (1931) 524.
31 T Taylor, G.I.: Proc. R. Soc. London Ser. A **132** (1931) 524.
61 D Defant, A.: Physical Oceanography, Vol. 2, New York: Pergamon Press **1961**, 598 pp.
62 C Cox, C., Sandstrom, H.: J. Oceanogr. Soc. Japan 20th anniv. vol. **1962**, 499.
62 L Lafond, E.C., in: The Sea, Vol. **1**, part 1 (Hill, M.N., ed.), New York: Wiley-Interscience **1962**, 731.
62 P Prigogine, I.: Non equilibrium statistical mechanics, New York: Wiley-Interscience **1962**.
64 A Abramowitz, M., Stegun, E.A., eds.: Handbook of mathematical functions, Dover **1964**.
66 B Bretherton, F.P.: Q. J. R. Meteorol. Soc. **92** (1966) 466.
66 H Hasselmann, K.: Rev. Geophys. Space Phys. **4** (1966) 1.
66 M Munk, W.H.: Deep Sea Res. **13** (1966) 707.
66 P Phillips, O.M.: The Dynamics of the upper ocean, Cambridge: Cambridge University Press **1966**, 261 pp.
67 B Booker, J.R., Bretherton, F.P.: J. Fluid Mech. **27** (1967) 513.
68 B Bretherton, F.P., Garrett, C.J.R.: Proc. R. Soc. London Ser. A **302** (1968) 529.
68 J Jones, W.L.: J. Fluid Mech. **34** (1968) 609.
68 M Munk, W.H., Phillips, O.M.: Rev. Geophys. **6** (1968) 44.
69 O Orlanski, I., Bryan, K.: J. Geophys. Res. **74** (1969) 6975.
70 L Landau, L.D., Lifschitz, E.M.: Mechanik, Berlin: Akademie-Verlag **1970**.
70 W Whitham, G.B.: J. Fluid Mech. **44** (1970) 373.
71 F Fofonoff, N.P., Webster, F.: Philos. Trans. R. Soc. London Ser. A **279** (1971) 423.
71 P Phillips, O.M.: J. Phys. Oceanogr. **1** (1971) 1.
72 G Garrett, C.J.R., Munk, W.H.: Geophys. Fluid Dyn. **2** (1972) 225.
73 T Thorpe, S.A.: Boundary-Layer-Meteorology **5** (1973) 95.
74 G Gill, A.E., Green, J.S.A., Simmons, A.J.: Deep Sea Res. **21** (1974) 499.
75 B 1 Bell, T.H., jr.: J. Geophys. Res. **80** (1975) 320.
75 B 2 Briscoe, M.G.: J. Geophys. Res. **80** (1975) 3872.
75 B 3 Briscoe, M.G.: Rev. Geophys. Space Phys. **13** (1975) 591, 636.
75 C Cairns, J.L.: J. Geophys. Res. **80** (1975) 299.
75 D Desaubies, Y.J.F.: J. Geophys. Res. **80** (1975) 895.
75 G Garrett, C.J.R., Munk, W.H.: J. Geophys. Res. **80** (1975) 291.
75 K 1 Katz, E.J.: J. Geophys. Res. **80** (1975) 1163.
75 K 2 Kaula, W.M., Harris, A.W.: Rev. Geophys. Space Phys. **13** (1975) 363.
75 K 3 Kroll, J.: J. Mar. Res. **33** (1975) 15.
75 M Müller, P., Olbers, D.J.: J. Geophys. Res. **80** (1975) 3848.
75 P Pinkel, R.: J. Geophys. Res. **80** (1975) 3892.
75 S Sanford, T.B.: J. Geophys. Res. **80** (1975) 3861.
75 T Thorpe, S.A.: J. Geophys. Res. **80** (1975) 328.
75 W Wunsch, C.: Rev. Geophys. Space Phys. **13** (1975) 167.
77 G Gregg, M.C.: J. Phys. Oceanogr. **7** (1977) 436.
76 B Banks, W.H.H., Drazin, P.G., Zaturska, M.B.: J. Fluid Mech. **75** (1976) 149.
76 L 1 Leaman, K.D.: J. Phys. Oceanogr. **6** (1976) 894.
76 L 2 Lindzen, R.S., Rosenthal, A.J.: J. Geophys. Res. **81** (1976) 1561.
76 M 1 Müller, P.: J. Fluid Mech. **77** (1976) 789.
76 M 2 Müller, P., Siedler, G.: Deep Sea Res. **23** (1976) 613.
76 M 3 Mysak, L.A., Howe, M.S.: Dyn. Atmos. Oceans **1** (1976) 3.
76 O Olbers, D.J.: J. Fluid Mech. **74** (1976) 375.
76 U U.S. Polymode Organizing Committee: U.S. Polymode program and plan, Cambridge, Mass. **1976**.
76 W Wunsch, C.: J. Phys. Oceanogr. **6** (1976) 471.
77 G Gregg, M.C.: J. Phys. Oceanogr. **7** (1977) 436.
77 M 1 McComas, C.H.: J. Phys. Oceanogr. **7** (1977) 836.
77 M 2 McComas, C.H., Bretherton, F.P.: J. Phys. Oceanogr. **11** (1977) 139.
77 M 3 Müller, P.: Dyn. Atmos. Oceans **2** (1977) 49.
77 P Phillips, O.M.: The Dynamics of the upper ocean, 2nd edition, London: Cambridge University Press **1977**, 336 pp.

77 S Stern, M.E.: J. Mar. Res. **35** (1977) 479.
78 B Bell, T.H., jr.: J. Fluid Mech. **88** (1978) 289.
78 K Käse, R.H., Clarke, R.A.: Deep Sea Res. **25** (1978) 815.
78 L 1 LeBlond, P.H., Mysak, L.A.: Waves in the ocean, New York: Elsevier Scientific Publishing Co. **1978**, 560 pp.
78 L 2 Lighthill, J.: Waves in Fluids, London: Cambridge University Press **1978**, 504 pp.
78 M Müller, P., Olbers, D.J., Willebrand, J.: J. Geophys. Res. **83** (1978) 479.
78 O Osborn, T.R.: J. Geophys. Res. **83** (1978) 2939.
78 T 1 Thorpe, S.A.: J. Fluid Mech. **85** (1978) 7.
78 T 2 Thorpe, S.A.: J. Fluid Mech. **88** (1978) 623.
79 C Cox, C.S., Johnson, C.L.: Inter-relations of microprocesses internal waves, and large scale ocean features. Unpublished manuscript **1979**.
79 F Frankignoul, C., Müller, P.: J. Phys. Oceanogr. **9** (1979) 194.
79 G 1 Garrett, C.: Dyn. Atmos. Oceans **3** (1979) 239.
79 G 2 Garrett, C., Munk, W.H.: Annu. Rev. Fluid Mech. **11** (1979) 339.
79 G 3 Gregg, M.C., Briscoe, M.G.: Rev. Geophys. Space Phys. **17** (1979) 1524.
79 K 1 Käse, R.H.: Deep Sea Res. **26** (1979) 227.
79 K 2 Käse, R.H., Olbers, D.J.: Deep Sea Res., Suppl. to Vol. **26** (1979) 191.
79 O 1 Olbers, D.J.: Z. Angew. Math. Mech. **59** (1979) 10.
79 O 2 Olbers, D.J., Herterich, K.: J. Fluid Mech. **92** (1979) 349.
79 P Pinkel, R.: J. Phys. Oceanogr. **9** (1979) 675.
79 R Ruddick, B.R., Joyce, T.M.: J. Phys. Oceanogr. **9** (1979) 498.
79 W Wunsch, C., Webb, S.: J. Phys. Oceanogr. **9** (1979) 235.
80 H Holloway, G.: J. Phys. Oceanogr. **10** (1980) 906.
80 K Käse, R.H., Siedler, G.: Deep Sea Res., Suppl. I to Vol. **26** (1979) **1980**, 161.
80 M Munk, W.H.: J. Phys. Oceanogr. **10** (1980) 1718.
80 P Pomphrey, N., Meiss, J.D., Watson, K.M.: J. Geophys. Res. **85** (1980) 1085.
80 T Thompson, R.O.R.Y.: J. Geophys. Res. **85** (1980) 6631.
80 W Woods, J.D.: Nature **288** (1980) 219.
81 A AIP Conference Proceedings (West, B., ed.): Nonlinear properties of internal waves, New York: American Institute of Physics **1981**.
81 B Brown, E.D., Owens, W.B.: J. Phys. Oceanogr. **11** (1981) 1474.
81 F Fu, L.-L.: Rev. Geophys. Space Phys. **19** (1981) 141.
81 G Gargett, A.E., Hendricks, P.J., Sanford, T.B., Osborn, T.R., Williams III, A.J.: J. Phys. Oceanogr. **11** (1981) 1258.
81 H Holloway, G.: Theoretical approach to interactions among internal waves. Nonlinear properties of internal waves, Conference Proceedings (West, B., ed.), New York: American Institute of Physics, **1981**, p. 47–77.
81 M 1 McComas, C.H., Müller, P.: J. Phys. Oceanogr. **11** (1981) 970.
81 M 2 McComas, C.H., Müller, P.: J. Phys. Oceanogr. **11** (1981) 139.
81 M 3 Munk, W.H.: Internal waves and small-scale processes. Evolution of physical oceanography scientific surveys in honor of Henry Stommel (Warren, B.A., Wunsch, C., eds.), Cambridge, Mass.: MIT Press **1981**, 264.
81 O 1 Olbers, D.J.: J. Phys. Oceanogr. **11** (1981) 1224.
81 O 2 Olbers, D.J.: J. Phys. Oceanogr. **11** (1981) 1078.
81 O 3 Olbers, D.J., Pomphrey, N.: J. Phys. Oceanogr. **11** (1981) 1423.
81 O 4 Orlanski, I., Cerasoli, C.P.: J. Geophys. Res. **86** (1981) 4103.
81 P Pinkel, R.: Deep Sea Res. **28** (1981) 269.
81 R Roth, M.W., Briscoe, M.G., McComas III, C.H.: J. Phys. Oceanogr. **11** (1981) 1234.
81 W Weller, R.A.: J. Geophys. Res. **86** (1981) 1969.
82 B Bryden, H.L.: J. Mar. Res. **40** (1982) 1047.
82 D D'Asaro, E.: J. Phys. Oceanogr. **12** (1982) 323.
82 E Eriksen, C.C.: J. Geophys. Res. **87** (1982) 525.
82 G Gill, A.E.: Atmosphere-ocean dynamics, New York and London: Academic Press **1982**.
82 H Holloway, G.: J. Phys. Oceanogr. **12** (1982) 293.
83 A Alpers, W., Salusti, E.: J. Geophys. Res. **88** (1983) 1800.
83 B Briscoe, M.G.: Philos. Trans. R. Soc. London Ser. A **308** (1983) 427.

83 H Holloway, G.: Atmosphere-Ocean **21** (1983) 107.
83 L 1 Levine, M.D.: Rev. Geophys. Space Phys. **21** (1983) 1206.
83 L 2 Levine, M.D., de Soeke, R.A., Niiler, P.P.: J. Phys. Oceanogr. **13** (1983) 240.
83 L 3 Levine, M.D., Paulson, C.A., Briscoe, M.G., Weller, R.A., Peters, H.: Philos. Trans. R. Soc. London Ser. A **308** (1983) 389.
83 O Olbers, D.J.: Rev. Geophys. Space Phys. **21** (1983) 1567.
83 P Peters, H.: Deep Sea Res. **30** (1983) 119.

6.4 Astronomical tides

6.4.0 List of symbols

A_K, A_P	amplitudes of sea-surface elevation of Kelvin wave and of Poincaré wave, respectively, in [m]
A_h, A_z	kinematic eddy viscosity for the horizontal and vertical directions, respectively, in [$m^2 s^{-1}$]
$A_z^{(1)}$	A_z at unit level above sea bottom
A_n, B_n	expansion coefficients of Poincaré wave, in [$m s^{-1}$]
A_n^m, B_n^m	time dependent coefficients of spherical harmonic of degree n and order m in gravitational potential, in [m]
a	half width of rectangular basin, in [m]
a, b, c	speeds of tidal constituent and corresponding indices, in [$deg\,h^{-1}$]
$\boldsymbol{a}$	column vector consisting of η and $\bar{\boldsymbol{v}}_h$
a_n^m	normalization coefficients, in [m^{-1}]
a_1, a_2	minor and major axis of tidal current ellipse, respectively, in [$m s^{-1}$]
b	$((\omega+iR')\omega/(gh))^{1/2}$, in [$m^{-1}$]
b_1, b_2	dimensionless sea-surface amplitudes
C	phase velocity of tidal wave, in [$m s^{-1}$]
$c, \bar{c}$	dimensionless quadratic bottom friction coefficients
D	sense of rotation of tidal current vector, in [$m^2 s^{-2}$]
d	distance of tide-generating body from earth's mass center, in [m]
dF	element of area, in [m^2]
$\boldsymbol{e}_x, \boldsymbol{e}_y, \boldsymbol{e}_z$	unit vectors in the x, y, z directions
$\boldsymbol{F}$	vector of frictional forces, in [$m s^{-2}$]
$\bar{\boldsymbol{F}}'$	depth-averaged vector of frictional forces
$\boldsymbol{F}^{(e)}$	lateral eddy dissipation vector, in [$m s^{-2}$]
F_4, F_6	ratios for determining quarter(sixth)-diurnal harmonic constants, in [m^{-1}], [m^{-2}], respectively
f	Coriolis parameter, in [s^{-1}]
$\boldsymbol{f}$	forcing term
f'	lunar nodal factor
f_4, f_6	angles for determining quarter(sixth)-diurnal harmonic constants, in [deg]
$\boldsymbol{f}_j, \boldsymbol{g}_j$	biorthogonal sets of eigenfunctions
G	surface loading Green's function
g	surface gravity of a spherical earth, in [$m s^{-2}$]
g^0	harmonic constant (phase) referring to the zonal time meridian, in [deg]
g_u, g_v	harmonic constants (phase) for u and v, respectively, in [deg]
H	instantaneous water depth, in [m]
H	harmonic constant (amplitude), in [m], or in [$m s^{-1}$]
H_i, H_{im}	amplitudes of harmonic component, in [m]
H_a, H_{ab}, H_{abc}	harmonic constants H of semidiurnal, quarter-diurnal, and sixth-diurnal tidal constituents with speeds $a, a+b, a+b+c$, respectively, in [m]
h	mean longitude of the sun, in [deg] or [rad]
h	height of the benthic boundary layer, in [m]
h	undisturbed water depth, in [m]
h_1	dimensionless water depth on the shelf
h_1, h_2	water depth on and off the shelf, respectively, in [m]
h_2, k_2	Love numbers
h_n', k_n'	load Love numbers
K	harmonic constant (phase) referring to Greenwich meridian, in [deg] or [rad]
k, k_n	wave numbers, in [m^{-1}]
L	shelf width; characteristic length; length of channel; in [m]
L	spatial hydrodynamic differential operator
$\bar{\mathsf{L}}$	operator adjoint to L
l	half length of rectangular basin, in [m]
N	mean longitude of ascending node of lunar orbit, in [deg] or [rad]
N	Brunt-Väisälä frequency, in [s^{-1}]
M	mass of tide-generating celestial body, in [kg]

P_n	(normalized) Legendre polynomial of degree n
P_n^m	(normalized) associated Legendre functions of degree n and order m
p	sea pressure, in [N m^{-2}]
p	mean longitude of perigee, in [deg] or [rad]
p'	mean longitude of perihelion, in [rad]
p_r, p_{-r}	coefficients of spherical harmonics in Φ and in Ψ, respectively, in [m^3]
R	geocentric radial distance, in [m]
R'	coefficient of linear friction, in [kg m^{-2} s^{-1}], or in [s^{-1}]
R^*	$R'-i\omega$, in [s^{-1}]
R_e	mean radius of the earth, in [m]
R_i	amplitude of harmonic component
Ro	Rossby number
Q_j	Brown's variables, in [rad]
S	longitude of zonal time meridian, in [deg]
S_j, S_{jm}	secular arguments, in [rad]
s	mean longitude of the moon, in [deg] or [rad]
T	tidal interference period, in [d]
t	time, in [s^{-1}]; dimensionless time
t', t^*	solar mean time at observation point and at zonal time meridian, respectively, in [h]
t_a	tide age, in [d]
t_0	Greenwich Mean Time, in [d]
t_0	epoch of maximum current, in [s]
U	characteristic velocity, in [m s^{-1}]
u	x-component of current velocity vector, in [m s^{-1}]
$\bar{u}$	u, depth-averaged, in [m s^{-1}]
u^*	friction velocity, in [m s^{-1}]
u'	lunar nodal phase, in [deg]
u', v'	complex time-independent current velocity components, in [m s^{-1}]
u_1	current velocity 1 m above bottom, in [m s^{-1}]
u_K, u_P	dimensionless component u of Kelvin wave and of Poincaré wave, respectively
u_1, u_2	u at Greenwich passage and at a quarter of tidal period later, respectively, in [m s^{-1}]
u_1, u_2	u on the shelf and off the shelf, respectively, in [m s^{-1}]
V	total tide-generating potential, in [m^2 s^{-2}]
$V^{(2)}$	principal second degree part of V, in [m^2 s^{-2}]
V'	potential due to surface load, in [m^2 s^{-2}]
V_0, V_0'	astronomical argument of tidal potential at Greenwich meridian and at observation point, respectively, in [deg]
v	y-component of current velocity vector, in [m s^{-1}]
$\bar{v}$	v, depth-averaged, in [m s^{-1}]
$\boldsymbol{v}$	current velocity vector, in [m s^{-1}]
$\bar{\boldsymbol{v}}_h$	depth-averaged horizontal current velocity vector, in [m s^{-1}]
v_K, v_P	dimensionless component v of Kelvin wave and of Poincaré wave, respectively
v_1, v_2	v at Greenwich passage and at a quarter of tidal period later, respectively, in [m s^{-1}]
v_1, v_2	v on and off the shelf, respectively, in [m s^{-1}]
W_1	rate of divergence of energy flux, in [W m^{-2}]
W_2	rate of work per unit area by tide-generating forces, in [W m^{-2}]
w	z-component of current velocity vector, in [m s^{-1}]
x, y, z	rectangular coordinates, in [m]
x, y	dimensionless horizontal rectangular coordinates
Z_0	mean sea level above chart datum, in [m]
z_0	level of roughness, in [m]
α	right ascension of tide-generating body, in [rad]
α	$2(\omega h/A_z^{(1)})^{1/2}$
α_1, α_2	dimensionless wavenumbers
α_{1K}, α_{2K}	α_1, α_2 of Kelvin wave, respectively
α_{1P}, α_{2P}	α_1, α_2 of Poincaré wave, respectively
α_n	normalized density ratio
α_n'	$(1+k_n'-h_n')\alpha_n$

β	dimensionless wavenumber
β_K, β_P	β of Kelvin wave and of Poincaré wave, respectively
$\beta_{\pm r, \pm s}$	gyroscopic coefficients, in $[m^{-2}]$
γ	gravitational constant, in $[m^3 s^{-2} kg^{-1}]$
δ, δ'	codeclination and declination of tide-generating body, respectively, in [rad]
δ	$(2\nu/\omega)^{1/2}$
δ	elevation of sea bottom, in [m]
δ_b, δ_s	elevation of sea bottom due to body tide of solid earth and due to surface load, respectively, in [m]
δ_{ik}	Kronecker symbol
ε	ellipticity of tidal current ellipse
ε	obliquity of the ecliptic, in [rad]
ε'	$4\Omega^2 R_e^2/(gh)$
η	sea-surface elevation, in [m]
η_n	n-th degree spherical harmonic constituent of η, in [m]
η_1, η_2	η at Greenwich passage and at a quarter of tidal period later, respectively, in [m]
η_1, η_2	dimensionless η on the shelf and off the shelf, respectively
η_L	η at open end of a channel, in [m]
η_K, η_P	dimensionless sea-surface elevations of Kelvin wave and of Poincaré wave, respectively
$\bar{\eta}$	total lift of equipotential surface of gravity by tide-generating potential, in [m]
$\bar{\eta}_2$	2nd degree spherical harmonic constituent of $\bar{\eta}$, in [m]
$\bar{\eta}_r$	coefficient of Φ_r in $\bar{\eta}$, in $[m^2]$
ϑ	zenith angle of tide-generating body relative to observation point, in [rad]
Θ	geocentric colatitude of observation point, in [deg]
$\varkappa$	harmonic constant (phase) referring to observation point, in [deg]
$\varkappa_a, \varkappa_{ab}, \varkappa_{abc}$	κ of semi-diurnal, quarter-diurnal, and sixth-diurnal tidal constituents with speeds a, $a+b$, and $a+b+c$, respectively, in [deg]
λ, λ'	geographic east longitude, in [deg] or [rad]
μ	$\cos\vartheta$
μ	frequency of external force, in $[s^{-1}]$
μ_i	eigenfrequencies of adjoint operator $\bar{\mathbf{L}}$, in $[s^{-1}]$
μ_r, ν_r	eigenvalues of horizontal Laplacean operator on a hemisphere, in $[m^{-2}]$
ν	kinematic viscosity, in $[m^2 s^{-1}]$
ϱ	density of sea water, in $[kg\,m^{-3}]$
ϱ	ecliptic longitude, in [rad]
$\bar{\varrho}$	mean density of sea water, in $[kg\,m^{-3}]$
ϱ_e	density of solid earth, in $[kg\,m^{-3}]$
τ_B	bottom stress, in $[N\,m^{-2}]$
ϕ, ϕ'	geographic latitude, in [deg] or [rad]
Φ	function analogous to a potential, in $[m^2]$
Φ_r	basis functions for expanding potential Φ, in $[m^{-1}]$
$\Phi, \Phi', \Phi_1, \Phi_2$	symbols for sea-surface elevation and current velocity components, in [m] or $[m\,s^{-1}]$, respectively
χ	von Kármán's constant
χ	direction of major axis of tidal current ellipse, in [deg]
ψ_K, ψ_P	phases of Kelvin wave and Poincaré wave, respectively, in [rad]
Ψ	function analogous to stream function, in $[m^2]$
Ψ_r	basis functions for expanding Ψ, in $[m^{-1}]$
ω	circular frequency, in $[s^{-1}]$
ω	circular frequency of tidal harmonic, in $[deg\,h^{-1}]$
ω	dimensionless circular frequency
$\bar{\omega}$	$\omega/(2\Omega)$
Ω	rotation angular velocity of the earth, in $[s^{-1}]$
$\boldsymbol{\Omega}$	vector of angular velocity of earth's rotation, in $[s^{-1}]$

6.4.1 The tide-generating potential

The astronomical tide-generating forces, present in the whole universe, are acting as inertia forces in all parts of the earth, viz. in the atmosphere, in the oceans, and in the solid earth, producing periodic phenomena which obey the dynamical equations. Whereas the centers of mass of the celestial bodies, including planets and moons, are moving, determined by Newton's law of gravitation, to a very good approximation as if the bodies' mass were concentrated in these centers, properties of the real bodies, e.g. the shape of their surfaces, do depend on the variation of the gravitational forces over the extension of the bodies. These variations give rise to residual forces to which the tidal phenomena are due to. See also LB, NS vol. V/2, sect. 2.5, for details on the tidal forcing field and the tides of the solid earth.

The tide-generating forces *per unit mass* can be given by the spatial gradients of a tide-generating potential V, and it is convenient generally to refer to this scalar potential. V may be defined by

$$V=\gamma M/r-\gamma MR\cos\vartheta/d^2-\gamma M/d\,, \tag{1}$$

where γ is the gravitational constant, M the mass of the celestial body B, r the distance of its center of gravity from a given point P on the earth's surface, d the corresponding distance BC from the earth's center, R the distance PC, and ϑ the zenith angle of B relative to P (see Fig. 1). Considering the earth as a sphere of radius R_e, which is permitted in view of the unimportant corrections for a spheroid, V can be expanded in terms of ϑ:

$$V=\gamma MR_e^2d^{-3}(P_2(\cos\vartheta)+R_ed^{-1}P_3(\cos\vartheta)+O(R_e^2d^{-2}))\,, \tag{2}$$

with the Legendre polynomials $P_2(\mu)=(3\mu^2-1)/2$ and $P_3(\mu)=(5\mu^3-3)/2$. The first term of eq. (2), to be denoted $V^{(2)}$ in the following, is the principal part of the tide-generating potential, for the sun it is in magnitude on an average 0.46 times that of the moon. Geometrically, the tidal acceleration vectors belonging to this term yield a symmetrical oval with major axis along CB (see Fig. 2) where B′ denotes the sublunar (subsolar) point on the earth's surface. R_ed^{-1} taking about 1/60 for the moon and about 10^{-4} for the sun, the contribution from the sun to the P_3-term is realized to be quite negligible.

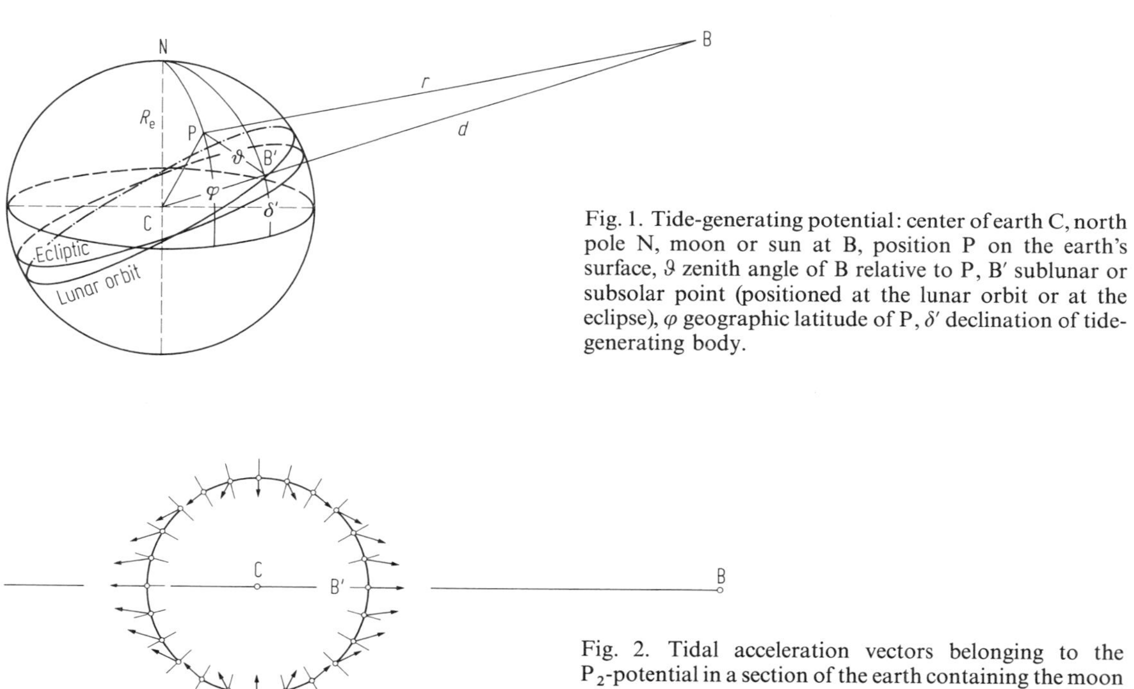

Fig. 1. Tide-generating potential: center of earth C, north pole N, moon or sun at B, position P on the earth's surface, ϑ zenith angle of B relative to P, B′ sublunar or subsolar point (positioned at the lunar orbit or at the eclipse), φ geographic latitude of P, δ' declination of tide-generating body.

Fig. 2. Tidal acceleration vectors belonging to the P_2-potential in a section of the earth containing the moon (sun) positioned at B and the center of earth, C. B′: sublunar (subsolar) point.

Making use of the relations between ϑ, the colatitude Θ of P, the codeclination δ of the body, and the angle PNB$'=\lambda-\alpha+h+2\pi t-\pi$ in the spherical triangle NPB$'$ (Fig. 1), where λ is the east longitude of P, α the right ascension of the body, h the mean longitude of the sun, and t dimensionless Greenwich mean time, i.e. $t=t_0/T$ with t_0 Greenwich mean time in days, and period $T=1$ d, the principal term of V/g writes

$$\frac{V^{(2)}}{g}=\frac{3\gamma M R_e^2}{4gd^3}\Bigg[\frac{16}{9}P_2(\cos\Theta)P_2(\cos\delta)$$

$$-\frac{16}{27}P_2^1(\cos\Theta)P_2^1(\cos\delta)(\cos\lambda\cos(\alpha-h-2\pi t)+\sin\lambda\sin(\alpha-h-2\pi t))$$

$$+\frac{4}{27}P_2^2(\cos\Theta)P_2^2(\cos\delta)(\cos 2\lambda\cos 2(\alpha-h-2\pi t)+\sin 2\lambda\sin 2(\alpha-h-2\pi t))\Bigg]$$

$$=A_2^0(t)P_2(\cos\Theta)+\sum_{m=1}^{2}(A_2^m(t)\cos m\lambda+B_2^m(t)\sin m\lambda)P_2^m(\cos\Theta) \qquad (3)$$

where $P_2^1(\cos\Theta)=3/2\sin 2\Theta$, $P_2^2(\cos\Theta)=3\sin^2\Theta$ and the equilibrium tide $\bar{\eta}=V^{(2)}/g$ is used instead of the potential $V^{(2)}$. g denotes gravity.

The term with m$=0$ varies with the monthly or yearly frequencies of $R_e d^{-1}$ and δ. The terms with $m=1,2$ primarily vary with approximately one day and half day period, respectively, with a low frequency modulation from $R_e d^{-1}$ and δ. Now δ and α depend by simple sine- and cosine-relations on the ecliptic longitude ϱ and the obliquity ε of the ecliptic, and in case of the moon, on the moon's latitude additionally. ε being nearly constant with secular trend only, for the sun the variations with time of $R_e d^{-1}$ and ϱ are given by h and $h-p'$ in the main, where p' is the mean longitude of the perihelion. The influence of the sun on the motion of the moon leads to a more complicated situation in case of the moon making the moon's parallax, longitude, and latitude depend on Q_j, j$=1\cdots4$, i.e. on $Q_1=s-p$, $Q_2=h-p'$, $Q_3=s-N$, and $Q_4=s-h$, respectively, where s, p, and N are the mean longitudes of, respectively, the moon, the lunar perigee, and the ascending node of the lunar orbit on the ecliptic.

This dependence can approximately be given by expressions of the form

$$\sum_i R_i\begin{Bmatrix}\cos\\ \sin\end{Bmatrix}\left(\sum_{j=1}^{4}\mathrm{M}_j^{(i)}Q_j\right)$$

with integers $\mathrm{M}_j^{(i)}$, j$=1\cdots4$, and constants R_i for the tidal constituents i. Introducing these expressions for $R_e d^{-1}$, δ, and α into $A_2^m(t)$ and performing a harmonic expansion yields expressions of the form

$$A_2^m(t)=\sum_i H_i\begin{Bmatrix}\cos\\ \sin\end{Bmatrix}\left(\sum_{j=1}^{6}\mathrm{N}_j^{(i)}S_j\right)\quad\text{if}\quad\begin{Bmatrix}\text{m even}\\ \text{m odd}\end{Bmatrix} \qquad (4)$$

containing some hundred or more constituents with constants H_i, integers $\mathrm{N}_j^{(i)}$ and the arguments $S_1=2\pi t-\pi+h-s$, $S_2=s$, $S_3=h$, $S_4=p$, $S_5=-N$, and $S_6=p'$ representing periods of a lunar day (1.035050 mean solar days), a tropical month (27.321582 mean solar days), a tropical year (365.242199 mean solar days), 8.847 tropical years, 18.613 tropical years, and 20940 tropical years, respectively. For $\mathrm{B}_2^m(t)$ one gets the same expressions as in eq. (4) except for a $\pi/2$ phase change, i.e. from $A_2^m=H_s\cos\omega t$ follows $B_2^m=H_s\sin\omega t$. Although some hundred terms are necessary in eq. (4) to approximate the potential to an accuracy of 1%, a few terms contribute to the potential significantly more than the others. The most important terms are diurnal ($\mathrm{N}_1=1$) and semi-diurnal ($\mathrm{N}_1=2$) ones, whereas long-period terms ($\mathrm{N}_1=0$) are of relatively minor importance. In Table 1 a list of the more important terms is given. See also LB, NS vol. V/2, subsect. 2.5.1.6.

All terms with the same values of N_1 constitute a species, the species being separated from each other by about one cycle per lunar day. All terms with the same (N_1, N_2) constitute a group, the groups being separated by one cycle per month. The terms with the same (N_1, N_2, N_3) form a constituent, constituents are separated by one cycle per year.

A detailed presentation of the tide generating potential including reference to secular trends in the variation of the amplitudes of the harmonic expansion can be found in [71 C].

Table 1. Astronomical tidal constituents according to [71 C].

Origin	Symbol	N_1	N_2	N_3	N_4	N_5	N_6	ω °/h	Amplitude m	Geodetic factor
Moon/Sun	M_0+S_0	0	0	0	0	0	0	0.0000000	0.19839	
Sun	Sa	0	0	1	0	0	−1	0.0410667	0.00311	
Sun	Ssa	0	0	2	0	0	0	0.0821373	0.01953	
Moon	MSm	0	1	−2	1	0	0	0.4715211	0.00425	$1/2(1-3\cos^2\Theta)\cos\omega t$
Moon	Mm	0	1	0	−1	0	0	0.5443747	0.02219	
Moon	MSf	0	2	−2	0	0	0	1.0158958	0.00367	
Moon	Mf	0	2	0	0	0	0	1.0980331	0.04648	
Moon	σ_1	1	−3	2	0	0	0	12.9271398	0.00309	
Moon	Q_1	1	−2	0	1	0	0	13.3986609	0.01939	
Moon	ϱ_1	1	−2	2	−1	0	0	13.4715145	0.00368	
Moon	O_1	1	−1	0	0	0	0	13.9430356	0.10129	
Moon	NO_1	1	0	0	1	0	0	14.4966939	−0.00796	$-\sin 2\Theta \sin(\omega t+\lambda)$
Sun	P_1	1	1	−2	0	0	0	14.9589314	0.04716	
Moon/Sun	K_1	1	1	0	0	0	0	15.0410686	−0.14246	
Moon	J_1	1	2	0	−1	0	0	15.5854433	−0.00796	
Moon	OO_1	1	3	0	0	0	0	16.1391017	−0.00436	
Moon	$2N_2$	2	−2	0	2	0	0	27.8953548	0.00618	
Moon	μ_2	2	−2	2	0	0	0	27.9682084	0.00746	
Moon	N_2	2	−1	0	1	0	0	28.4397295	0.04672	
Moon	ν_2	2	−1	2	−1	0	0	28.5125831	0.00887	
Moon	M_2	2	0	0	0	0	0	28.9841042	0.24407	$\sin^2\Theta \cos(\omega t+2\lambda)$
Moon	L_2	2	1	0	−1	0	0	29.5284789	−0.00690	
Sun	T_2	2	2	−3	0	0	1	29.9589333	0.00665	
Sun	S_2	2	2	−2	0	0	0	30.0000000	0.11356	
Moon/Sun	K_2	2	2	0	0	0	0	30.0821373	0.03089	

6.4.2 The tidal dynamical equations

For studying the field of motion which is due to the astronomical tidal potential, the dynamical equations are basic. These are defined by the conservation laws of mass and momentum, where the momentum equations represent the dynamical equilibrium between the astronomical forces and the internal forces in the sea, like inertial, pressure gradient, frictional and Coriolis forces, and moreover buoyancy forces when the density stratification is considered:

$$\frac{\partial \varrho}{\partial t} + \nabla \cdot (\varrho \boldsymbol{v}) = 0 \tag{5}$$

$$\frac{\partial \boldsymbol{v}}{\partial t} + \boldsymbol{v}\nabla\boldsymbol{v} + 2\boldsymbol{\Omega} \times \boldsymbol{v} + \boldsymbol{F} = -\varrho^{-1}\nabla p + g\nabla\bar{\eta} - g\boldsymbol{e}_z \tag{6}$$

where $\boldsymbol{v} = (u, v, w)$ denotes the current velocity vector, $\bar{\eta}$ the total lift of equipotential surface of gravity by tide-generating potential (equilibrium tide), ϱ the fluid density, p the pressure, $\boldsymbol{\Omega} = (0, \Omega\cos\phi, \Omega\sin\phi)$ the earth's angular velocity, ϕ the geographic latitude, and $\boldsymbol{F}$ the vector of frictional forces. The unit vectors $\boldsymbol{e}_x$ and $\boldsymbol{e}_y$ are directed eastwards and northwards, respectively, and $\boldsymbol{e}_z$ vertically upwards. For long-wave tidal motions simplifications are introduced. Controversies about the justification for the simplified equations have been settled fairly well by [74 M].

Assuming incompressibility, eq. (5) yields after vertical integration

$$\frac{\partial \eta}{\partial t} + \nabla_h \cdot (H\bar{\boldsymbol{v}}_h) = 0 \tag{7}$$

with $\bar{\boldsymbol{v}}_h = (\bar{u}, \bar{v})$ the depth-averaged horizontal current velocity vector, $H = h + \eta$ the actual depth, h the undisturbed depth, and η the actual water elevation. In view of the predominance of the terms left, the terms involving the vertical components of $\boldsymbol{v}$ and $2\boldsymbol{\Omega} \times \boldsymbol{v}$ are generally omitted when investigating barotropic tidal motions. Allowing for density stratification and baroclinic tidal motions, the equation

$$\frac{\partial \varrho}{\partial t} - wN^2\varrho/g = 0 \tag{8}$$

arising from eq. (5) is used additionally. Here N is the Brunt-Väisälä frequency. After vertically integrating also eq. (6) and neglecting the inertia terms $\boldsymbol{v}\nabla\boldsymbol{v}$, the equations governing barotropic tidal motions result:

$$\frac{\partial \eta}{\partial t} + \nabla_h \cdot (H\bar{\boldsymbol{v}}_h) = 0 \tag{9}$$

$$\frac{\partial \bar{\boldsymbol{v}}_h}{\partial t} + 2\boldsymbol{\Omega} \times \bar{\boldsymbol{v}}_h + \bar{\boldsymbol{F}}' = -g\nabla \cdot (\eta - \bar{\eta}) \tag{10}$$

where the hydrostatic approximation $p = \bar{\varrho}g(\eta - z)$ with the mean density $\bar{\varrho}$ is included and $\bar{\boldsymbol{F}}' = \bar{c}|\bar{\boldsymbol{v}}_h|\bar{\boldsymbol{v}}_h/H + \boldsymbol{F}^{(e)}$ denotes the sum of quadratic bottom friction vector and lateral eddy dissipation vector $\boldsymbol{F}^{(e)}$. $\bar{c}$ is a nondimensional constant known from measurements in shallow water to take about 0.0025 (see [56 B]. $\boldsymbol{F}^{(e)}$ has the spherical components

$$F_\lambda^{(e)} = -A_h\left(\Delta\bar{u} + R^{-2}\left(-\bar{u}(1 + \tan^2\phi) - 2\tan\phi\cos^{-1}\phi\frac{\partial \bar{v}}{\partial \lambda}\right)\right)$$

$$F_\phi^{(e)} = -A_h\left(\Delta\bar{v} + R^{-2}\left(-\bar{v}(1 + \tan^2\phi) + 2\tan\phi\cos^{-1}\phi\frac{\partial \bar{u}}{\partial \lambda}\right)\right)$$

with constant eddy viscosity coefficient. Concerning variable A_h, see [78 S]. Linearizations are generally well justified for open ocean tides contrary to tides in shallow water. Eqs. (9) and (10) with $\bar{\boldsymbol{F}}' = 0$ are known as *Laplace's tidal equations* (LTE). Eqs. (9) and (10) with modifications have been taken as a basis for investigating ocean tide dynamics. Although since long known in theory, the elastic yielding of the earth's crust to the astronomical forces and to the loading of the ocean tide itself as well as the ocean self-attraction effect have not been included in the LTE for tidal computations before the last decennium.

As the corresponding normal modes of the earth's oscillation, different from those of the ocean, have frequencies at least an order of magnitude higher than tidal frequencies, the response is virtually static.

$$\delta_b(\phi, \lambda, t) = h_2 \bar{\eta}(\phi, \lambda, t) \tag{11}$$

holds for the bottom deformation δ_b with the elastic constant (Love number) $h_2 = 0.61$. As a result of this redistribution of mass the potential V is increased by $k_2 g \bar{\eta}$ with the Love number $k_2 = 0.3$. In eq. (9) η has to be understood relative to the moving sea bottom. The pressure gradient is referred to the geocentric elevation $\eta + \delta_b$, whence together with the change of potential the right hand side of eq. (10) now writes $-g\nabla \cdot (\eta - (1 + k_2 - h_2)\bar{\eta})$, thus the tidal potential is provided with the additional factor 0.69. η firstly gives rise to the ocean self-attraction potential $g\alpha_n \eta_n$ with $\alpha_n = \frac{3}{2n+1} \frac{\varrho}{\varrho_e}$, η_n the n-th degree spherical harmonic constituent of η, and ϱ_e the density of the solid earth. The surface load $\varrho g \eta_n$ yields the solid earth deformation

$$\delta_s = h'_n \alpha_n \eta_n \tag{12}$$

due to the weight of the additional column of water, and due to the opposing but smaller effect of the gravitational attraction of the solid earth by the shell. The potential due to the surface load finally amounts to

$$V' = g(1 + k'_n)\alpha_n \eta_n \,. \tag{13}$$

Values of different origin of Love numbers k'_n and h'_n defining the solid earth response are put together in [78 m].

Thus, including loading and self-attraction eq. (10) turns to

$$\frac{\partial \bar{\boldsymbol{v}}_h}{\partial t} + 2\boldsymbol{\Omega} \times \bar{\boldsymbol{v}}_h + \bar{\boldsymbol{F}}' = -g\nabla \cdot \left(\eta - (1 + k_2 - h_2)\bar{\eta} - \sum_n (1 + k'_n - h'_n)\alpha_n \eta_n \right). \tag{14}$$

By means of the expansion of η into constituents η_n, eq. (14) includes the integral

$$\sum_n (1 + k'_n - h'_n)\alpha_n \eta_n = \iint_B \eta G \,\mathrm{d}\lambda' \mathrm{d}\phi' \cos\phi'$$

where B is the surface of the globe with unit radius and

$$G(\phi, \lambda, \phi', \lambda') = \frac{1}{4\pi} \sum_n \alpha'_n \sum_m P_n^m(\sin\phi) P_n^m(\sin\phi') \cos(m(\lambda' - \lambda)),$$

$$\alpha'_n = (1 + k'_n - h'_n)\alpha_n$$

with normalized associated Legendre functions P_n^m. Eq. (14) thus represents integro-differential equations (see e.g. [78 Z].

6.4.3 Time dependence of the tidal field of motion

6.4.3.1 The linear case and general considerations

In accordance with the expansion of the tidal potential (equilibrium tide)

$$\bar{\eta} = \sum_i H_{i0} P_2(\cos\Theta) \left\{ \cos\left(\sum_{j=1}^{6} N_{j0}^{(i)} S_j \right) \right\} + \sum_{m=1}^{2} \sum_i H_{im} P_2^m(\cos\Theta) \begin{Bmatrix} \cos \\ \sin \end{Bmatrix} \left(\sum_{j=1}^{6} N_{jm}^{(i)} S_j + m\lambda \right) \Bigg\}, \quad \text{if} \begin{Bmatrix} \text{m even} \\ \text{m odd} \end{Bmatrix} \tag{15}$$

in tidal constituents (see eqs. (3), (4)), the elevation η and the current velocity components u and v can be given, as long as nonlinear effects are negligible, also by six-dimensional Fourier series for every fixed position P as solutions to a forced linear differential equation problem. The constituents of this expansion of η, u, and v can be written in the following way

$$H\cos(V_0' + \omega t' - \varkappa) = H\cos(V_0 + \omega t^* - g^0), \tag{16}$$

where the *harmonic constants* H and $\varkappa$ (amplitude and phase, respectively) or H and g^0 depend on the position $P = P(\Theta, \lambda)$. t' and t^* denote solar mean time at λ and at the zonal time meridian S, respectively, V_0' and V_0 the astronomical argument of the tidal potential at λ and at the Greenwich meridian at 00.00 hours local mean time, respectively, whence $V_0' = V_0 + p\lambda - (\omega\lambda/15°)$ and $g^0 = \varkappa - p\lambda + (\omega S/15°)$ are valid with $p = 1$ for diurnal tides and $p = 2$ for semi-diurnal tides. ω denotes the frequency of the tidal constituent. When making reference to the passage of the corresponding tide-generating body, i.e. of the maximum of the equilibrium tide, at the Greenwich meridian, the expressions in eq. (16) are rewritten as $H\cos(\omega t - K)$, $K = \varkappa + p\lambda$, where t denotes the time after passage at Greenwich and K the phase delay at P referring to the maximum of the equilibrium tide in Greenwich. A selection of harmonic constants H and $\varkappa$ for η is given in Table 2. All tidal harmonic constants which become known are included in a computerized data bank, the IHO Tidal Constituent Bank operated by the Canadian Hydrographic Service, this data base replacing the IHO Special Publication No. 26 produced by the International Hydrographic Bureau, Monaco. Analyzed deep-sea tidal measurements have been published by [71 F, 72 I, 75 P, 75 Z], a collection of such data (108 stations) is found in [79 C]. Relevant methods for obtaining harmonic constants from measured time series are presented in [72 g]. When deriving harmonic constants from observations extending over a one year interval, the harmonic constants belonging to different groups can be separated.

In view of longer durations being rare, the influence of the nonresolved constituents is considered by introducing correction factors f' and arguments u' into the harmonic constants thus derived:

$$f'H\cos(V_0' + \omega t' - (\varkappa - u')) = f'H\cos(V_0 + \omega t^* - (g^0 - u')). \tag{17}$$

f' and u' are obtained by assuming that amplitudes and phases of the minor constituents are related to the major constituents according to the respective astronomical potentials. Neglecting in case of the moon tides those contributions which would make f' and u' depend on $S_4 = p$, one gets the relations in use for calculating f' and u' (see [77 G 1]), those for O_1, K_1, M_2, and K_2 are given in Table 3.

To obtain the complete tidal elevations and current velocities in the open ocean and wherever linearity can be assumed, the constituents (16) or, if necessary, (17) have to be superimposed linearly, considering the actual values of the astronomical arguments. The latter can be taken from the Tide Tables, the most comprehensive of which are referred to in subsect. 6.4.3.3. Dependent on the elevations of which constituents, semi-diurnal or diurnal ones, prevail at a certain place, one of the specific tidal curves given in Fig. 3 in principle results, viz. such of semi-diurnal, mixed, or diurnal tidal form. The superposition of M_2 and S_2 defines spring and neap tides. The combined M_2–S_2 tide is the highest (spring tide) due to the combined potential having a maximum at full and at new moon, and it is the smallest (neap tide) due to the combined potential having a minimum at the quadratures. The phase age at a place is the interval between the time of full or new moon and the time of spring tide at that place, viz.

$$t_a = \frac{\varkappa(M_2) - \varkappa(S_2)}{\omega(M_2) - \omega(S_2)}. \tag{18}$$

The interference period T of the semi-diurnal tide in Fig. 3 (compare with the values in Table 2) is determined by M_2 and S_2 tides yielding $T=14.765$ mean solar days, the interference period of the diurnal tide is governed by O_1 and K_1 tides yielding $T=13.660$ mean solar days. In the diurnal case the maximum of the potential occurs at the maximum declination of the moon. The phase ages amount to $t_a=8.23$ hours (M_2–S_2) and to $t_a=51.00$ hours (O_1–K_1), respectively. Generally, however, not only positive values (delay) occur, but also negative ones, the values of the phase age ranging between $-T/2$ and $T/2$. The average positive phase age (age of the tide) is attributed to the effect of tidal friction. [71 G 2] (see also [73 W]) presented a simple model of this phenomenon based on the expansion of the tidal elevations into an eigenfunction series, eq. (25) assuming that only two eigenoscillations dominate the series. Numerical model computations yield an average age of 42.1 hours (M_2–S_2) and 30.1 hours (O_1–K_1), see [80 B 2], with a distribution across the total time interval as given in Fig. 13.

For single tidal constituents the current velocity vector defines an ellipse within a tidal period, the elements of which are given by (see also [52 H 2]):

minor axis a_1 and major axis a_2:

$$a_{1,2}=1/\sqrt{2}(w^2 \pm (w^4-4(u_1v_2-u_2v_1)^2)^{1/2})^{1/2} \quad \text{where} \quad w=(u_1^2+u_2^2+v_1^2+v_2^2)^{1/2},$$

epochs t_0 of extreme currents:

$$\tan(2\omega t_0)=\frac{2(u_1u_2+v_1v_2)}{u_1^2+v_1^2-u_2^2-v_2^2},$$

directions χ of axes:

$$\tan 2\chi=\frac{2(u_1v_1+u_2v_2)}{u_1^2+u_2^2-v_1^2-v_2^2},$$

sense of rotation D:

$$D=u_1v_2-u_2v_1$$

$D>0$ positive (counterclockwise) rotation

$D<0$ negative (clockwise) rotation,

ellipticity ε:

$$\varepsilon=\operatorname{sign}(D)a_1/a_2,$$

with Φ_1 and Φ_2 in $\Phi=(\Phi_1+i\Phi_2)e^{-i\omega t}$, $\Phi=\eta, u, v$, denoting elevation or current velocity component at $t=0$, e.g. at meridian passage at Greenwich (index 1), and at a quarter of a tidal period later (index 2), respectively.

The real shape of the curve defined by the current velocity vector is more complicated, as it is indicated for Proudfoot Shoal in Fig. 4. The corresponding harmonic constants for this place are given in Table 4. They show that linear superposition of the first seven constituents are sufficient to represent the actual current velocity vector.

Table 2. Harmonic tidal constants (selection from IHO Special Publication No. 26), amplitudes and Z_0 (mean level above chart datum) in [cm], phases $\varkappa$ in degrees.

Place		Z_0	Sa	Ssa	O_1	P_1	K_1	μ_2	N_2	M_2	L_2	K_2	S_2	M_4	MS_4	M_6
Eastern boundary of the Atlantic Ocean																
Ekaterinskaya	H	215	5.6	1.2	2.6	3.3	13.2	3.8	24.8	116.1	1.6	9.9	33.9	2.5	3.3	0.4
69°12′ N, 33°28′ E	$\varkappa$		201	303	98	289	295	98	165	193	210	234	237	238	212	187
Bergen	H	80.1	8.4	5.9	3.2	1.1	3.2	1.7	8.5	43.9	1.6	4.4	15.9	–	1.3	–
60°24′ N, 5°18′ E	$\varkappa$		253	217	18	147	171	292	270	295	326	334	334	–	319	–
Helgoland	H	142.0	12.2	6.8	9.2	2.6	6.4	9.9	17.5	108.6	–	8.4	28.9	7.0	4.3	2.0
54°11′ N, 7°54′ E	$\varkappa$		288	123	243	35	28	56	301	328	–	30	33	179	238	323
Hoek van Holland	H	99	–	–	11.2	3.9	7.9	8.2	11.6	75.3	8.2	5.4	18.6	17.4	10.4	4.4
51°59′ N, 4°07′ E	$\varkappa$		–	–	181	341	351	190	45	71	76	133	131	130	186	62
Immingham	H	355.4	9.7	1.6	15.5	4.6	14.1	8.2	41.3	226.0	11.0	21.1	74.4	2.0	3.5	1.3
53°38′ N, 0°11′ W	$\varkappa$		216	127	116	271	278	221	139	161	172	209	211	165	236	118
Aberdeen	H	198.9	10.3	5.5	12.7	3.3	11.0	2.1	25.7	130.9	4.0	12.5	45.0	2.6	2.7	0.6
57°09′ N, 2°05′ W	$\varkappa$		211	110	50	197	203	309	357	21	46	56	58	163	241	113
Londonderry	H	129.7	8.0	4.5	8.2	2.8	8.9	0.8	14.8	78.6	3.3	8.9	29.5	0.5	2.4	2.3
55°00′ N, 7°19′ W	$\varkappa$		191	55	38	157	186	173	198	217	249	247	241	236	53	313
Cobh	H	180.0	9.9	7.1	3.4	0.7	2.1	4.1	25.5	138.1	5.8	11.6	43.2	5.9	2.1	1.4
51°50′ N, 8°18′ W	$\varkappa$		181	111	37	161	147	210	112	133	139	177	178	208	282	130
Birkenhead	H	462.5	9.9	7.7	12.1	4.5	12.1	6.2	60.4	312.0	16.7	28.9	101.2	23.4	14.1	4.7
53°24′ N, 3°01′ W	$\varkappa$		203	79	41	184	187	28	297	318	317	2	2	202	248	304
Avonmouth	H	658.9	9.2	2.2	7.9	3.0	5.9	43.0	73.3	422.3	33.6	42.7	148.2	34.0	29.6	10.0
51°30′ N, 2°43′ W	$\varkappa$		203	55	4	115	131	260	182	196	189	253	255	337	16	264
Poole Bridge	H	115.8	–	–	4.0	3.7	11.0	4.0	8.2	39.3	3.4	4.6	13.4	14.0	9.8	2.7
50°40′ N, 1°56′ W	$\varkappa$		–	–	14	131	131	163	254	285	293	312	312	55	124	129
Dover	H	369.8	14.9	2.6	5.3	2.3	4.5	8.6	41.0	228.0	11.3	21.0	71.4	25.9	16.8	6.8
51°07′ N, 1°19′ E	$\varkappa$		211	153	171	1	31	54	316	355	348	25	28	227	279	112
Le Havre	H	479	3.7	2.3	5.7	3.3	9.0	7.2	47.8	261.1	11.1	26.3	88.0	24.5	16.1	15.2
49°29′ N, 0°06′ W	$\varkappa$		195	27	4	100	119	352	265	284	285	329	331	75	130	287
Brest	H	465.4	8.5	6.6	6.5	2.6	6.1	8.0	40.6	201.6	5.7	21.3	74.4	5.9	3.3	3.0
48°23′ N, 4°30′ W	$\varkappa$		247	32	327	62	71	95	81	100	96	134	139	87	186	331
Pointe St.-Gildas	H	308	–	4.4	6.5	2.2	6.2	5.8	33.5	167.6	4.1	18.0	62.8	18.8	6.2	–
47°08′ N, 2°15′ W	$\varkappa$		–	39	327	58	73	61	76	96	113	125	129	18	111	–

Table 2 (continued).

Place		Z_0	Sa	Ssa	O_1	P_1	K_1	μ_2	N_2	M_2	L_2	K_2	S_2	M_4	MS_4	M_6
Cascais	H	220	4.5	2.5	5.8	2.1	7.1	3.4	19.3	94.1	2.7	8.9	32.9	1.4	0.9	0.2
38°41′N, 9°25′W	$\varkappa$		201	74	305	42	48	10	33	51	68	80	78	142	175	317
Gibraltar	H	52.8	6.6	1.7	0.8	0.9	2.0	1.2	6.4	29.8	0.9	4.2	10.7	1.7	1.3	0.2
36°08′N, 5°21′W	$\varkappa$		213	156	160	120	125	13	24	35	32	58	61	141	200	95
Venice	H	52	2.2	2.3	5.3	5.8	18.3	0.5	4.1	23.5	0.4	4.0	14.0	0.4	1.0	0.1
45°20′N, 12°15′E	$\varkappa$		188	286	63	67	79	325	292	292	306	294	300	331	294	72
Rhodes	H	10	–	–	1.6	1.0	2.9	–	1.1	5.9	–	1.1	3.9	1.4	1.3	–
36°06′N, 28°06′E	$\varkappa$		–	–	301	340	340	–	318	306	–	326	326	224	164	–
Western boundary of the Atlantic Ocean																
Ponta Delgada	H	100	3.5	4.6	2.5	1.5	4.4	2.1	11.3	49.1	1.1	4.3	17.9	0.4	0.2	–
37°44′N, 25°40′E	$\varkappa$		211	246	292	31	41	329	357	12	162	25	32	14	133	–
Madeira	H	140	2.1	0.4	4.5	2.0	6.1	3.0	13.6	71.2	2.7	7.3	26.7	0.7	0.4	0.1
32°38′N, 16°55′W	$\varkappa$		242	18	285	16	29	313	354	10	17	29	31	50	141	–
Freetown	H	169.8	5.2	2.3	2.5	3.0	9.8	3.0	20.0	97.7	3.9	9.7	32.5	1.4	1.6	1.1
8°30′N, 13°14′W	$\varkappa$		164	138	249	334	334	234	183	201	206	233	234	250	345	354
Simons Bay	H	92.4	3.7	0.8	1.6	1.5	5.8	1.9	10.5	47.1	1.4	5.5	19.6	1.3	1.8	0.1
34°12′S, 18°25′E	$\varkappa$		355	69	247	126	134	67	61	74	133	90	88	79	53	267
Julianehaab	H	–	12.5	6.1	8.5	5.2	15.5	–	17.4	86.6	–	8.8	32.9	0.9	0.1	–
60°43′N, 46°02′W	$\varkappa$		238	217	73	112	112	–	149	175	–	210	210	104	292	–
Father Point	H	232	4.8	5.8	21.7	8.0	23.5	5.4	27.0	127.2	2.7	12.2	41.7	1.3	2.0	0.6
48°31′N, 68°28′W	$\varkappa$		151	156	184	206	208	9	35	62	90	102	103	88	77	115
Anticosti, S.W.-Point	H	104	3.3	2.0	17.2	6.3	18.0	1.6	11.0	50.5	1.4	3.8	14.7	1.1	2.1	1.2
49°24′N, 63°36′W	$\varkappa$		128	194	181	211	207	15	21	41	36	85	81	297	22	258
St. John, N.B.	H	437.0	1.5	4.3	11.5	5.1	15.4	0.9	59.0	302.7	16.1	14.6	48.8	4.3	1.1	3.3
45°16′N, 66°04′W	$\varkappa$		28	108	111	128	128	76	294	325	13	7	3	136	167	193
New York (Sandy Hook)	H	70	10.2	1.2	5.2	2.8	9.7	2.4	14.0	65.4	1.8	4.1	13.8	1.0	–	1.4
40°28′N, 74°01′W	$\varkappa$		126	349	99	104	101	240	205	218	186	249	245	339	–	354
Morehead City	H	43	9.2	6.0	5.7	2.4	7.4	0.8	9.1	41.9	1.9	1.8	7.2	0.4	0.6	0.8
34°43′N, 76°42′W	$\varkappa$		176	76	132	126	123	217	200	215	203	232	239	326	201	8
Galveston	H	15.2	6.4	6.9	11.1	3.2	11.7	0.5	2.3	9.4	0.2	0.3	3.0	0.5	–	0.2
29°19′N, 94°47′W	$\varkappa$		157	54	307	306	317	58	96	111	191	164	112	246	–	116

Table 2 (continued).

Place		Z_0	Sa	Ssa	O_1	P_1	K_1	μ_2	N_2	M_2	L_2	K_2	S_2	M_4	MS_4	M_6
Bermuda	*H*	66.4	10.1	3.4	5.4	2.3	6.9	–	8.6	37.9	0.8	2.0	8.3	0.2	0.2	–
32°19′N, 64°50′W	*ϰ*		238	11	130	129	129	–	216	236	251	266	263	349	59	–
Paramaribo	*H*	150	3.6	3.3	8.4	2.2	10.3	2.8	16.6	85.8	4.4	6.2	24.7	6.8	4.9	3.1
5°50′N, 55°09′W	*ϰ*		83	83	202	232	214	279	135	152	163	176	179	221	263	188
Recife	*H*	116	5.0	3.1	5.1	1.2	4.2	2.9	14.8	75.4	1.9	8.3	27.3	1.0	–	–
8°03′S, 34°52′W	*ϰ*		53	156	146	232	237	113	119	129	129	145	146	267	–	–
Rio de Janeiro	*H*	67	2.3	5.2	10.8	2.3	6.2	1.7	3.4	30.2	1.9	5.0	16.1	3.6	1.8	–
22°56′S, 43°08′W	*ϰ*		52	118	88	135	148	108	121	86	108	85	93	83	184	–
Buenos Aires	*H*	79	10.4	5.1	15.4	4.1	9.4	0.8	11.2	30.3	4.4	1.8	5.7	2.7	1.2	1.1
34°36′S, 58°22′W	*ϰ*		301	51	204	28	25	72	132	176	236	278	254	234	330	262
Puerto Belgrano	*H*	244	9.0	2.3	15.8	4.1	22.5	13.5	19.2	144.3	16.9	2.8	24.3	12.0	5.2	0.2
38°53′S, 62°06′W	*ϰ*		345	202	334	6	35	227	44	135	209	293	274	102	241	252
Orange Bay	*H*	102	4.8	0.4	17.9	5.3	21.5	1.4	15.0	58.9	1.6	2.0	9.2	0.5	–	0.5
55°31′S, 68°05′W	*ϰ*		92	37	347	30	36	74	66	104	109	128	134	197	–	313
Indian Ocean and adjacent places																
Cape Town	*H*	105.2	3.1	1.2	1.6	1.3	6.9	2.1	11.8	53.3	2.0	6.4	23.9	1.4	1.1	0.0
33°54′S, 18°25′E	*ϰ*		14	229	253	128	132	38	62	70	66	89	90	120	144	–
Daressalam	*H*	152.6	3.5	1.9	10.6	4.5	17.1	1.4	19.1	107.4	3.6	14.4	53.6	1.1	0.7	0.4
6°49′S, 39°19′E	*ϰ*		311	69	41	307	36	110	79	103	133	137	144	300	11	258
Aden	*H*	129.5	11.2	3.9	20.1	12.3	39.7	2.2	13.2	47.5	1.3	5.7	20.6	0.2	0.4	0.2
12°47′N, 44°59′E	*ϰ*		346	128	37	32	35	191	222	227	225	239	245	306	162	293
Buschir	*H*	87.8	10.5	3.7	20.4	9.1	30.7	0.5	7.3	33.7	1.6	4.1	12.3	0.7	0.7	0.2
28°54′N, 50°45′E	*ϰ*		142	215	240	269	278	213	185	211	232	257	261	327	23	256
Bombay	*H*	250.8	3.6	4.1	20.1	12.4	42.5	6.3	30.0	122.2	2.5	12.2	48.3	3.5	3.6	0.4
18°55′N, 72°50′E	*ϰ*		285	193	49	43	46	306	315	331	306	338	4	323	30	60
Colombo	*H*	37.8	9.5	4.1	2.9	2.2	7.3	0.5	2.2	17.6	0.8	3.3	11.9	0.5	0.3	0.1
6°56′N, 79°51′E	*ϰ*		308	111	62	26	33	105	34	50	51	90	95	170	253	27
Dublat Saugor	*H*	321.7	27.6	6.3	5.2	4.5	15.0	3.6	27.6	140.2	6.5	19.0	66.1	4.7	4.8	1.1
21°39′N, 88°03′E	*ϰ*		147	124	338	349	350	32	282	292	303	324	326	166	198	213
Singapore	*H*	159.6	9.1	5.5	28.3	8.7	29.3	0.3	14.3	79.6	3.5	8.8	32.5	1.5	1.9	1.1
1°16′N, 103°51′E	*ϰ*		252	226	56	93	100	268	286	303	295	346	349	249	292	46
Soerabaja Harbour	*H*	150	3	7	27	14	47	–	9	44	–	8	26	–	–	–
7°13′S, 112°44′E	*ϰ*		292	165	284	321	318	–	337	351	–	357	355	–	–	–

Table 2 (continued).

Place		Z_0	Sa	Ssa	O_1	P_1	K_1	μ_2	N_2	M_2	L_2	K_2	S_2	M_4	MS_4	M_6
Australian coast and islands in the Pacific Ocean																
Manila	H	48.8	14.0	1.7	28.0	9.2	30.2	0.7	3.9	19.2	0.5	2.2	6.6	0.3	–	0.1
14°35′N, 120°58′E	$\varkappa$		143	352	276	313	321	263	293	304	323	327	334	344	–	274
Darwin	H	410	12.2	6.6	32.9	16.9	58.6	5.4	33.1	182.8	5.1	26.3	94.5	5.5	5.3	2.0
12°28′S, 130°51′E	$\varkappa$		324	15	322	338	331	136	132	153	167	203	201	282	358	203
Port Adelaide	H	148.4	5.0	1.8	15.3	7.6	24.6	7.8	2.3	49.8	2.2	14.0	49.2	0.5	0.5	0.1
34°48′S, 138°28′E	$\varkappa$		138	186	30	46	46	229	207	107	143	166	168	164	30	158
Brisbane River Bar	H	106.7	7.3	2.3	11.8	5.8	21.1	3.2	13.7	68.5	4.9	5.4	18.6	1.6	0.9	1.3
27°19′S, 153°10′E	$\varkappa$		16	67	147	176	174	349	288	292	267	302	310	151	180	197
Auckland	H	175	5.7	2.2	1.6	2.3	7.2	3.1	24.2	116.4	3.3	4.2	17.7	3.2	5.3	0.8
36°51′S, 174°46′E	$\varkappa$		31	165	145	164	169	178	179	206	220	257	268	125	195	314
Proudfoot Shoal	H	177	–	–	30.2	17.1	52.1	–	11.9	68.3	–	4.3	15.5	–	–	–
10°31′S, 141°29′E	$\varkappa$		–	–	128	166	168	–	72	113	–	199	199	–	–	–
Ponape	H	70	8.2	5.1	9.7	5.3	16.1	0.9	4.4	25.1	1.2	5.5	19.5	0.5	0.6	0.4
6°59′N, 158°13′E	$\varkappa$		221	68	200	219	224	71	93	81	86	101	104	150	99	111
Upolu (Samoa)	H	62	3.7	3.1	3.1	1.3	4.7	1.2	9.7	37.7	0.7	2.4	8.8	1.5	0.9	0.5
13°48′S, 171°46′W	$\varkappa$		276	50	246	246	257	156	199	188	199	157	174	297	271	339
Honolulu	H	24.4	3.7	0.9	8.2	4.5	14.9	0.6	2.8	16.0	0.4	1.5	5.1	0.2	0.1	0.1
21°18′N, 157°52′W	$\varkappa$		188	0	58	65	70	64	96	106	114	94	102	276	165	39
Pacific Ocean																
Bangkok	H	149	15.7	1.2	44.9	20.4	68.1	1.3	7.5	55.8	1.5	11.1	28.4	1.1	–	0.3
13°28′N, 100°35′E	$\varkappa$		303	311	120	167	162	75	107	139	171	181	203	190	–	213
Do-Son	H	189.2	–	–	70.0	24.0	72.0	–	0.8	4.4	–	1.0	3.0	–	–	–
20°43′N, 106°48′E	$\varkappa$		–	–	35	91	91	–	99	113	–	140	140	–	–	–
Hongkong	H	137.8	5.9	5.2	28.9	11.1	35.7	1.6	8.4	39.5	0.7	4.4	16.1	2.6	1.7	0.4
22°18′N, 114°11′E	$\varkappa$		213	17	252	287	293	214	247	263	307	284	286	323	10	137
Shanghai	H	231.6	24.7	2.9	14.3	6.4	22.9	5.1	16.1	93.9	5.2	12.1	41.5	16.9	13.7	3.4
31°24′N, 121°29′E	$\varkappa$		131	61	172	222	216	98	10	20	43	60	63	329	11	237
Uno	H	142.7	17.9	2.7	22.9	8.3	31.1	5.0	10.0	67.3	4.6	7.0	21.4	0.9	0.5	0.9
34°30′N, 133°57′E	$\varkappa$		158	270	207	237	231	141	302	323	337	348	348	233	291	19
Kawasaki	H	114	7.6	2.3	19.4	7.8	24.5	1.2	7.0	46.9	1.3	6.2	23.4	1.4	1.4	0.1
35°31′N, 139°45′E	$\varkappa$		178	187	162	177	180	168	151	158	204	205	188	92	109	177

Table 2 (continued).

Place		Z_0	Sa	Ssa	O_1	P_1	K_1	μ_2	N_2	M_2	L_2	K_2	S_2	M_4	MS_4	M_6
Unalaska	H	70	6.37	–	22.5	10.6	33.9	1.6	9.8	26.5	–	–	3.0	–	–	–
55°53′ N, 166°32′ W	$\varkappa$		263	272	139	150	151	358	59	108	–	–	341	–	–	–
Cordova	H	201	14.5	6.6	30.3	15.4	49.1	3.0	28.5	142.7	3.6	13.1	48.0	6.0	3.3	0.9
60°33′ N, 145°46′ W	$\varkappa$		237	121	111	125	128	318	333	359	13	27	31	147	236	342
Vancouver	H	245.3	6.7	5.1	44.8	26.5	84.2	0.7	18.3	90.3	3.7	6.6	21.9	1.3	0.7	1.2
49°18′ N, 123°07′ W	$\varkappa$		260	226	146	169	171	318	129	156	157	179	189	166	177	95
San Francisco	H	91.4	3.0	2.4	23.0	11.4	36.9	0.9	11.6	54.2	1.9	3.7	12.3	2.3	0.9	0.2
37°48′ N, 122°27′ W	$\varkappa$		199	268	88	104	106	226	303	330	351	328	334	24	29	14
San Diego	H	91.4	6.3	1.7	21.2	10.4	33.5	0.9	12.9	54.9	1.2	6.2	22.2	0.4	–	0.6
32°43′ N, 117°10′ W	$\varkappa$		182	273	78	92	93	235	254	275	247	265	272	208	–	117
Balboa	H	259	13.9	9.5	3.6	4.0	13.5	7.5	39.0	184.6	4.5	12.2	48.9	7.0	2.4	1.1
8°57′ N, 79°34′ W	$\varkappa$		167	138	352	341	342	37	57	89	108	139	145	5	64	290
Valparaiso	H	91.4	–	–	9.9	4.8	15.4	–	9.3	42.8	–	4.2	14.1	–	–	–
33°02′ S, 71°38′ W	$\varkappa$		–	–	285	324	329	–	246	278	–	291	299	–	–	–
Antarctica																
Barry Island	H	–	–	–	25.3	11.0	33.2	–	4.3	16.2	–	5.2	19.5	0.9	0.3	–
68°08′ S, 67°05′ W	$\varkappa$		–	–	10	28	28	–	25	132	–	291	291	276	98	–
Gauß Station	H	75	–	–	20.8	6.4	19.4	0.7	5.9	23.0	0.5	3.5	12.0	0.4	–	0.2
66°02′ S, 89°38′ E	$\varkappa$		–	–	0	348	7	303	337	19	96	88	101	138	–	179
Cape Armitage	H	58	–	–	23.5	7.8	23.5	–	2.9	5.0	–	0.8	2.9	–	–	–
77°49′ S, 166°45′ E	$\varkappa$		–	–	0	14	14	–	272	10	–	272	272	–	–	–

Table 3. Correction factors and arguments for O_1, K_1, M_2, K_2 (from [77 G 1]). $N = 259.157° - 19.3282°(Y - 1900) - 0.05295°(D + L)$ where Y = year, D = no. of days elapsed since 00.00 on 1st January in year Y, L = integral part of $0.25(Y - 1901)$.

O_1	$f' = 1.0089 + 0.1871 \cos N - 0.0147 \cos 2N + 0.0014 \cos 3N$
	$u' = 10.80° \sin N - 1.34° \sin 2N + 0.19° \sin 3N$
K_1	$f' = 1.0060 + 0.1150 \cos N - 0.0088 \cos 2N + 0.0006 \cos 3N$
	$u' = -8.86° \sin N + 0.68° \sin 2N - 0.07° \sin 3N$
M_2	$f' = 1.0004 - 0.0373 \cos N + 0.0002 \cos 2N$
	$u' = -2.14° \sin N$
K_2	$f' = 1.0241 + 0.2863 \cos N + 0.0083 \cos 2N - 0.0015 \cos 3N$
	$u' = -17.74° \sin N + 0.68° \sin 2N - 0.04° \sin 3N$

Table 4. Harmonic current velocity constants of Proudfoot Shoal ($\phi = 10°31'$ S, $\lambda = 141°29'$ E), amplitudes H in [m/s], phases $\varkappa$ in degrees (from [52 H 3]). See Fig. 4 for current diagram.

		M_2	S_2	N_2	K_2	K_1	O_1	P_1	M_4	MS_4
North	H	0.123	0.021	0.026	0.005	0.123	0.087	0.041	0.005	0.010
component	$\varkappa$	38	73	345	72	58	337	59	144	86
East	H	0.247	0.129	0.051	0.031	0.205	0.129	0.067	0.005	0.005
component	$\varkappa$	54	172	352	171	118	63	119	290	223

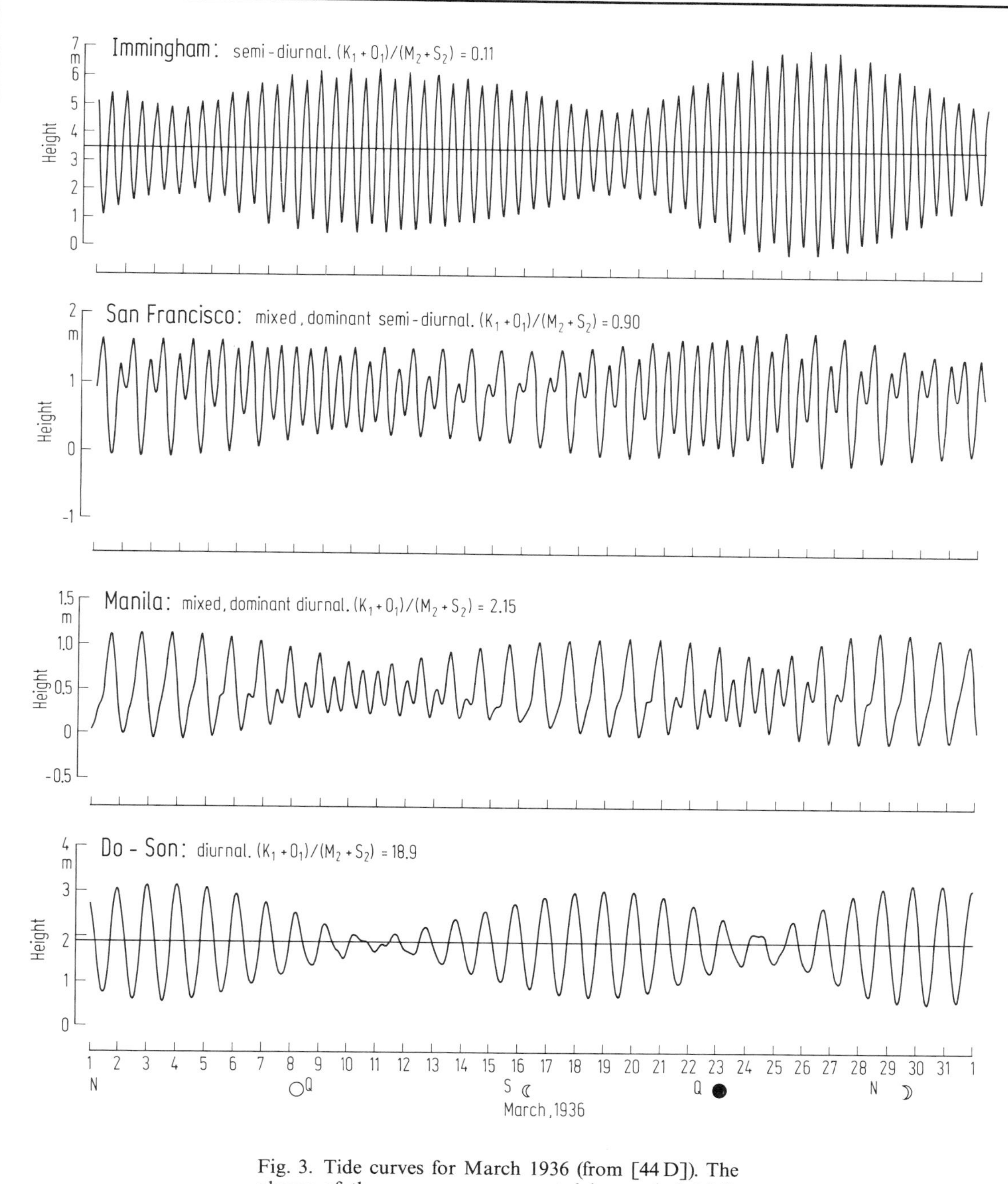

Fig. 3. Tide curves for March 1936 (from [44 D]). The phases of the moon are represented by symbol. N, S: largest northern and southern declination of the moon, respectively, Q: passage of the moon through the equator. The harmonic constants for the above places are given in Table 2. Immingham: semi-diurnal type; San Francisco: mixed, dominant semi-diurnal type; Manila: mixed, dominant diurnal type; Do-Son: diurnal type.

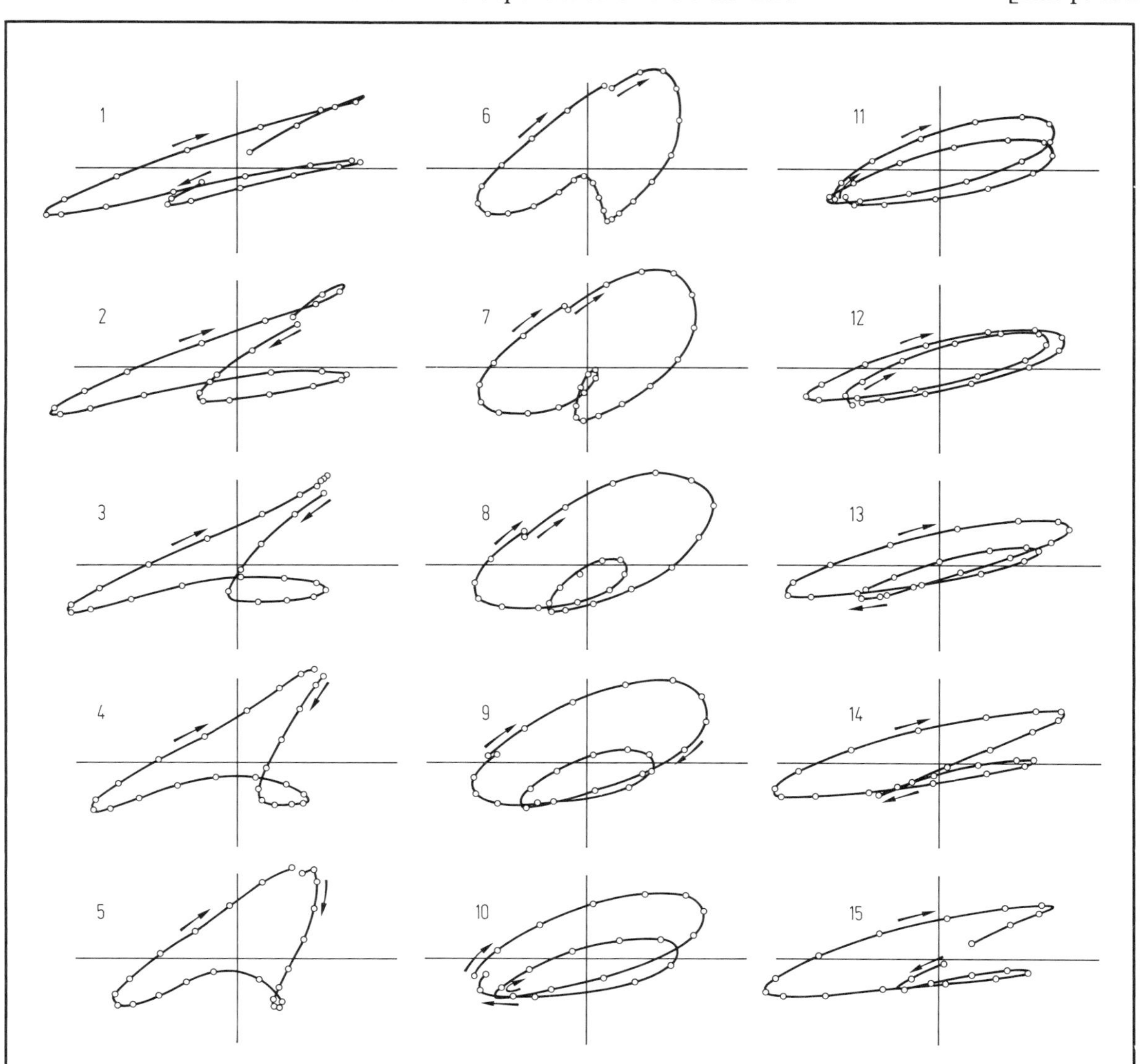

Fig. 4. Tidal current of mixed type: Proudfoot Shoal (10°31′ S, 141°29′ E).
The current diagrams are given for each day from 1–15 March 1936 (from [52 H 3]). For harmonic constants, see Table 4.

6.4.3.2 Shallow water tides

In shallow water linear superposition of the tides due to the astronomical tidal constituents does not represent the real tide due to the appearance of overtides and compound tides, which are caused by nonlinear effects. These shallow water tides have angular speeds which are an exact multiple of the angular speeds of the astronomical tidal constituents (overtides) or which are equal to the linear superposition with integral coefficients of the angular speeds of two or more astronomical tidal constituents (compound tides).

Some important shallow water tides are given in Table 5.

Table 5. List of important short period shallow water tides from [60 H].

Tide	ω [°/h]	V_0	Tide	ω [°/h]	V_0
$2SM_2$	31.0158958	+2s−2h	$2MN_6$	86.4079380	−7s+6h+p
MO_3	42.9271398	−4s+3h−90°	M_6	86.9523127	−6s+6h
SO_3	43.9430356	−2s+ h−90°	MSN_6	87.4238338	−5s+4h+p
MK_3	44.0251729	−2s+3h+90°	$2MS_6$	87.9682085	−4s+4h
SK_3	45.0410686	h+90°	$2MK_6$	88.0503458	−4s+6h
MN_4	57.4238338	−5s+4h+p	$2SM_6$	88.9841042	−2s+2h
M_4	57.9682085	−4s+4h	MSK_6	89.0662415	−2s+4h
SN_4	58.4397295	−3s+2h+p	$3MN_8$	115.3920423	−9s+8h+p
MS_4	58.9841042	−2s+2h	M_8	115.9364169	−8s+8h
MK_4	59.0662415	−2s+4h	$2MSN_8$	116.4079380	−7s+6h+p
S_4	60.0000000	0^0	$3MS_8$	116.9523127	−6s+6h
SK_4	60.0813728	2h	$2(MS)_8$	117.9682085	−4s+4h
			$2MSK_8$	118.0503458	−4s+6h

Assuming that the finite elevation amplitude is small as compared with the depth, it can be derived (see [64 d]) by a perturbation expansion from the one-dimensional tidal equations, when considering in particular the convective term $u\frac{\partial u}{\partial x}$ and the divergence term $\frac{\partial}{\partial x}((h+\eta)u)$, that if $H_a, H_b, H_c, \varkappa_a, \varkappa_b, \varkappa_c$ denote the values of H and $\varkappa$ of any three semi-diurnal constituents of speeds a, b, c, while $H_{\rm aa}, H_{\rm ab}, \varkappa_{\rm aa}, \varkappa_{\rm ab}$ denote the values of H and $\varkappa$ of quarter-diurnal constituents of speeds $2a$ and $a+b$, respectively, and $H_{aaa}, H_{aab}, H_{abc}, \varkappa_{aaa}, \varkappa_{aab}, \varkappa_{abc}$ the values of H and $\varkappa$ of sixth-diurnal constituents of speeds $3a$, $2a+b$, $a+b+c$, respectively, the relations

$$\begin{aligned} H_{aa} &= F_4 H_a^2 & \varkappa_{aa} &= 2\varkappa_a + f_4 \\ H_{ab} &= 2F_4 H_a H_b & \varkappa_{ab} &= \varkappa_a + \varkappa_b + f_4 \\ H_{aaa} &= F_6 H_a^3 & \varkappa_{aaa} &= 3\varkappa_a + f_6 \\ H_{aab} &= 3F_6 H_a^2 H_b & \varkappa_{aab} &= 2\varkappa_a + \varkappa_b + f_6 \\ H_{abc} &= 6F_6 H_a H_b H_c & \varkappa_{abc} &= \varkappa_a + \varkappa_b + \varkappa_c + f_6 \end{aligned} \tag{19}$$

hold. F_4, f_4 and F_6, f_6 are constants at any fixed position for all quarter-diurnal and sixth-diurnal constituents, respectively. Relations (19) are in rather good agreement with the harmonic constants derived from numerous observations. In Table 6 harmonic constants for a place with distinct shallow water tides are given.

The wave amplitude being finite for shallow water tides, the phase speed increases with the surface elevation. Starting from the phase velocity

$$C = (gh)^{1/2}(3(1+\eta/h)^{1/2} - 2) \tag{20}$$

(see [45 l]), [80 F] derives space-dependent laws of interaction between the purely astronomical constituents, [71 G 1] derives shallow water solutions for a single constituent in bounded channels considering also frictional and bottom topography effects. Generally, due to the increase of phase velocity with elevation a wave crest will travel faster than a trough, whence a steepening of the forward face of the wave will occur. Thus the time between low water and high water is expected to be shorter than that one between high water and low water. Fig. 5 shows an example for this situation.

Predicting the distorted tides of shallow water ports requires to consider the shallow water constituents when performing the superposition. Considerable difficulties arise where in view of a great number of relevant shallow water constituents an extremely high resolution is necessary in tidal analysis. For prediction of high and low waters (see tide tables) methods have been developed (see [57 D, 60 H, 77 A 1]) which in case of a dominant M_2-tide can often more effectively be applied and with far less computational expense than the harmonic method yielding the complete tidal profile.

Table 6. Harmonic constants of Cuxhaven (53°52′ N, 8°43′ E) as an example for a detailed analysis of mainly semi-diurnal tides showing shallow water tide phenomena. Duration of data: 5 years. Central time of analysis: 1972. According to Deutsches Hydrographisches Institut.

Tide	H [cm]	$\varkappa$ [°]	Tide	H [cm]	$\varkappa$ [°]
Z_0	167.2		$2MN_2$	12.2	197.8
Sa	12.6	282.3	T_2	1.8	48.1
Ssa	8.0	120.9	S_2	34.4	70.7
Mm	3.3	282.3	K_2	10.2	67.9
MSf	3.4	90.5	$\zeta_2 \sim MSN_2$	2.5	267.8
Mf	2.1	115.9	$2SM_2$	3.1	297.1
$2Q_1$	0.2	147.4	MO_3	1.6	171.9
σ_1	0.3	76.8	M_3	0.3	252.0
Q_1	2.9	199.2	SO_3	0.9	257.8
ϱ_1	0.7	180.2	MK_3	1.0	315.0
O_1	9.3	262.7	SK_3	0.3	34.6
NO_1	0.4	183.8	MN_4	3.8	224.1
π_1	0.6	32.9	M_4	11.4	248.1
P_1	2.9	52.6	SN_4	0.7	358.7
S_1	1.8	266.9	MS_4	7.3	312.7
K_1	6.5	52.0	MK_4	2.1	308.8
ϕ_1	0.4	43.4	S_4	0.8	65.8
J_1	0.6	179.9	SK_4	0.5	67.8
SO_1	0.3	254.4	$2MN_6$	3.4	75.2
OO_1	0.3	164.2	M_6	6.6	106.0
$\varepsilon_2 = MNS_2$	3.2	62.7	MSN_6	1.2	166.0
$2N_2$	2.9	330.4	$2MS_6$	6.6	170.0
μ_2	14.1	87.4	$2MK_6$	1.9	171.1
N_2	21.1	334.5	$2SM_6$	1.1	255.6
ν_2	7.9	320.3	MSK_6	0.8	250.5
$\gamma_2 \sim OP_2$	2.7	342.8	M_8	1.4	13.4
M_2	134.4	1.4			
λ_2	5.1	12.5			

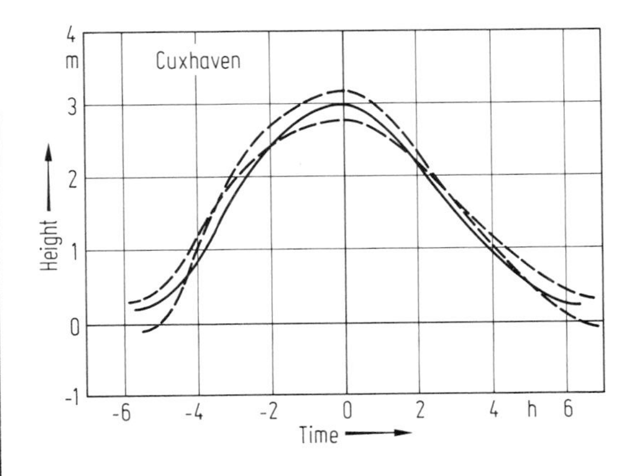

Fig. 5. Mean spring and neap curves (broken lines) for Cuxhaven (53°52′ N, 8°43′ E). Heights refer to Chart Datum (compare Table 6). Tidal curve (solid line) as obtained by a hydrodynamical numerical M_2-tide model (compare Fig. 18).

6.4.3.3 Tide tables

1 Deutsches Hydrographisches Institut, Hamburg: Gezeitentafeln (annually). Band 1: Europäische Gewässer, Band 2: Atlantischer und Indischer Ozean, Westküste Südamerikas. The tables contain: Daily predictions of the times and heights of high and low waters at reference stations (in total 94); tidal curves (mean spring and mean neap tide) for the reference stations (Band 1); data for predictions at many additional stations in the form of time and height differences referred to one of the reference stations, in Band 2 together with mean level Z_0, harmonic constants (M_2, S_2, N_2, K_2, μ_2, K_1, O_1, P_1, M_4, MS_4) and astronomical arguments for prediction by the harmonic method; cotidal charts Southern North Sea, and North Sea with Channel and British and Irish waters.

2 The Hydrographer of the Navy, London: Admiralty Tide Tables (annually). Vol. 1: European Waters (including Mediterranean Sea), Vol. 2: The Atlantic and Indian Oceans (including tidal stream predictions), Vol. 3: The Pacific Ocean and Adjacent Seas (including tidal stream predictions). The tables contain: Daily predictions of the times and heights of high and low waters at together 188 Standard Ports; daily predictions of tidal streams (Vols. 2, 3) at a limited number of places; data for predicting the tide at a large number of Secondary Ports, in Vols. 2, 3 together with mean level Z_0 and harmonic constants (M_2, S_2, K_1, O_1) and astronomical arguments for prediction by the Admiralty Method of Prediction (see N.P. 159, [77 G 1]). Further tidal publications are subdivided in: Admiralty Handbook Series (e.g. Harmonic Tidal Analysis (short periods N.P. 122 (3)), Tidal Stream Atlases, Miscellaneous Publications (e.g. Admiralty Manual of Tides N.P. 120, N.P. 159), Miscellaneous Atlases and Charts (e.g. Persian Gulf, cotidal atlas).

3 U.S. Department of Commerce, National Ocean Survey, Rockville, Md.: Tide Tables (annually) issued in four volumes: Europe and West Coast of Africa (including the Mediterranean Sea), East Coast of North and South America (including Greenland), West Coast of North and South America (including the Hawaiian Islands), Central and Western Pacific Ocean and Indian Ocean. The tables contain daily predictions of the times and heights of high and low waters for together 198 reference stations and differences and other constants for together 6000 subordinate stations.

Further publications relating to tides and tidal currents: Tidal Bench Marks, Tidal Current Tables for Atlantic Coast of North America and for Pacific Coast of North America and Asia (annually, including daily predictions of the times of slack water and the times and velocities of strength of flood and ebb currents for a number of waterways together with differences for numerous other places), Tidal Current Charts (for 12 coastal areas in the U.S.), Tidal Current Diagrams.

Appropriate authorities of many other countries issue tide tables, some of them also for waters beyond the vicinity of their country.

6.4.4 Open ocean tides

6.4.4.1 Tidal oscillations in schematic ocean basins

The family of eigensolutions defines the response of the ocean to the tidal forces by means of the eigenfrequencies and the spatial properties of the eigenoscillations (see eq. (25)). The eigenoscillations divide themselves into gravity waves (class I), which merge into ordinary gravity waves as the earth's rotation rate Ω tends to zero, and into Rossby waves (class II), which tend to steady currents with little associated surface elevation (see sect. 6.1). The complete range of possible eigenfunctions for a spherical ocean was computed by [68 L]. Considering the ocean self-attraction and tidal loading effects, makes the constituent including the effect of gravity $n^2(n+1)^2/(\varepsilon'\bar{\omega})$ (see [68 L 15]), where $\varepsilon' = 4\Omega^2 R_\varepsilon^2/(gh)$ and $\bar{\omega} = \omega/(2\Omega)$, change by the factor $(1-\alpha_n')$, by this way affecting the eigenfunctions of class I for small ε' and those of class II for large ε', realistic ε' belonging to the interval where waves of both classes are affected. For eigenfunctions of spherical harmonic order 2, representing progressive waves with argument $\omega t - 2\lambda$, Fig. 6 shows the dependence of $\bar{\omega}$ on the parameter $(\varepsilon')^{-1/2}$. A numerical value of the abscissa in Fig. 6 for typical oceanic depths is 0.2 and the semi-diurnal tidal frequency gives about ± 1 in the ordinate, these values obviously falling well within the range of eigenoscillations. Thus oceanic resonances to the tide-generating forces are not surprising.

Solutions for a bounded portion of a sphere do reflect a few more characteristic features of oceanic eigenoscillations and oscillations generated by the astronomical tidal forces. The complete set of eigenoscillations for a hemispherical ocean has been computed by [70 L]. Additionally considering linear friction $R'/(\varrho h)\bar{v}_h$ as in [80 W] and ocean self-attraction and tidal loading moreover, the hemispherical K_1-tide and M_2-tide oscillation is depicted in Figs. 7 and 8 as obtained by approximately solving the infinite system of linear algebraic equations (considering constituents up to degree and order 18)

$$-\omega^2 p_r - \frac{iR'\omega}{\varrho h} p_r - 2i\omega\Omega\mu_r^{-1} \sum_{s=-\infty}^{\infty} \beta_{r,s} p_s + gh\mu_r(1-\alpha_r'/2)p_r = g\bar{\eta}_r \tag{21}$$

$$-\omega^2 p_{-r} - \frac{iR'\omega}{\varrho h} p_{-r} - 2i\omega\Omega\nu_r^{-1} \sum_{s=-\infty}^{\infty} \beta_{-r,s} p_s = 0, \qquad r=1,2,\ldots,\infty \tag{22}$$

which has been obtained from eq. (14) with $\bar{F}' = R'/(\varrho h)\bar{v}_h$ by introducing

$$\bar{v}_h = \frac{\partial}{\partial t}(\nabla\Phi + \nabla\Psi \times e_z)$$

(e_z denoting the unit radial vector) and the expansions

$$\Phi = \sum_{r=1}^{\infty} p_r \Phi_r e^{-i\omega t}, \qquad \Psi = \sum_{r=1}^{\infty} p_{-r} \Psi_r e^{-i\omega t}$$

with

$$\Phi_r = a_n^m P_n^m(\sin\phi)\cos(m\lambda), \qquad n=1,2,\ldots,\infty, \qquad m=0,1,2,\ldots,n$$

$$\Psi_r = a_n^m P_n^m(\sin\phi)\sin(m\lambda), \qquad n=1,2,\ldots,\infty, \qquad m=1,2,3,\ldots,n$$

$$\nabla^2\Phi_r + \mu_r\Phi_r = 0 = \nabla^2\Psi_r + \nu_r\Psi_r, \qquad \mu_r = \nu_r = n(n+1)R_e^{-2},$$

$$a_n^m = R_e^{-1}\left(\frac{2n+1}{\pi}\frac{(n-m)!}{(n+m)!}\frac{1}{1+\delta_{m0}}\right)^{1/2}$$

guaranteeing that the current velocities normal to the meridional boundaries are zero by means of Φ_r, Ψ_r fulfilling $\nabla\Phi_r \cdot \boldsymbol{n} = 0$, $\Psi_r = 0$ for $\lambda = 0, \pi$ where $\boldsymbol{n}$ denotes the unit vector normal to the coastline. The coefficients β are given by

$$\beta_{r,s} = -\iint e_z \sin\phi \nabla\Phi_r \times \nabla\Phi_s dF, \qquad \beta_{r,-s} = \iint \sin\phi \nabla\Phi_r \cdot \nabla\Psi_s dF,$$

$$\beta_{-r,s} = -\iint \sin\phi \nabla\Psi_r \cdot \nabla\Phi_s dF, \qquad \beta_{-r,-s} = -\iint e_z \sin\phi \nabla\Psi_r \times \nabla\Psi_s dF$$

while $\bar{\eta}_r = \iint \Phi_r \bar{\eta} dF$, and the equation of continuity yields $\eta = h \sum_{r=1}^{\infty} p_r \mu_r \Phi_r$. The above integrals are over the area of the hemispherical ocean.

An overview of diurnal and semi-diurnal tides in frictionless oceans bounded by meridians and parallels on a rigid earth is given by [58 D] as obtained by series expansions.

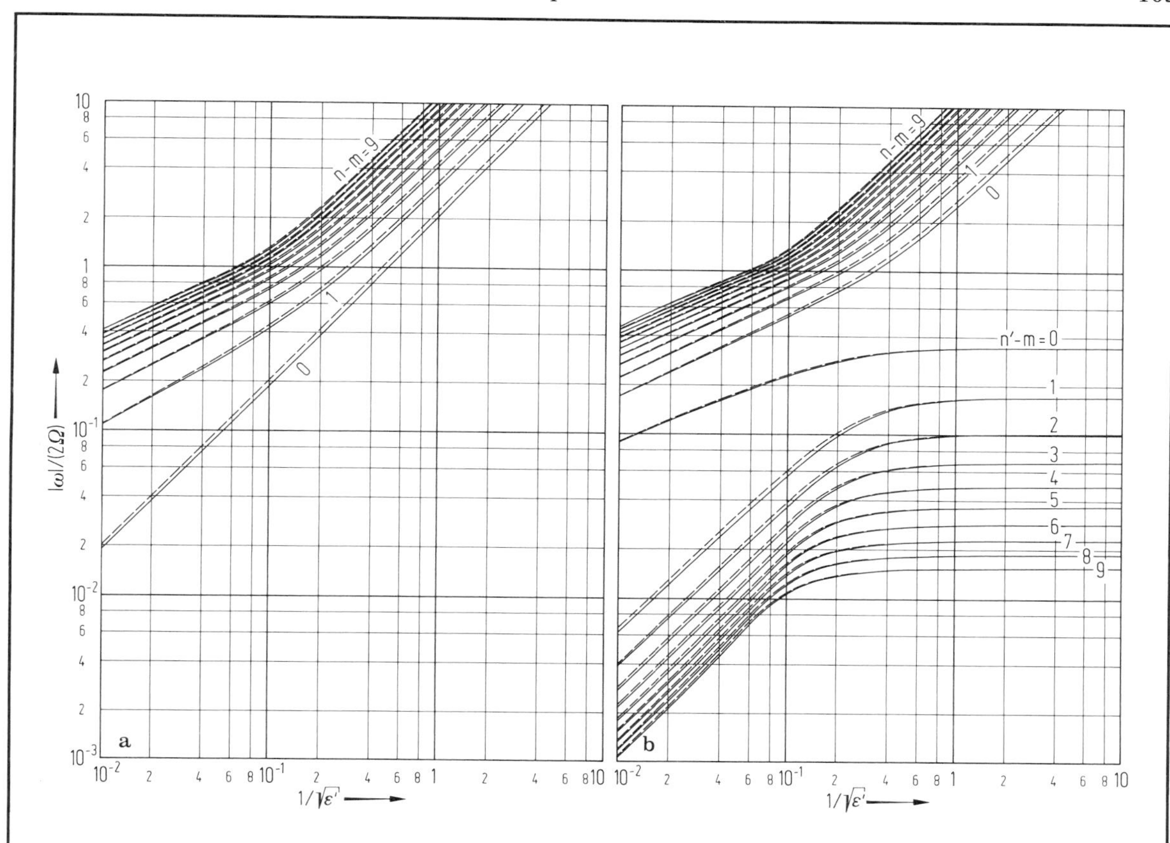

Fig. 6. Eigenfrequencies of free modes of oscillation on a sphere, proportional to $\exp(2i\lambda)$, i.e. m = 2. (a) Modes travelling eastwards, (b) modes travelling westwards. The ordinate gives frequency in units of the inertial frequency at the pole, the abscissa parameter $1/\sqrt{\varepsilon'}$, with $\varepsilon' = 4\Omega^2 R_e^2/(gh)$ and h the ocean basin depth. Degree numbers n refer to gravity modes, degree numbers n' to Rossby modes. Solid lines and dashed lines belong to oscillations considering and neglecting (see also [68 L]) the ocean self-attraction and tidal loading effects, respectively.

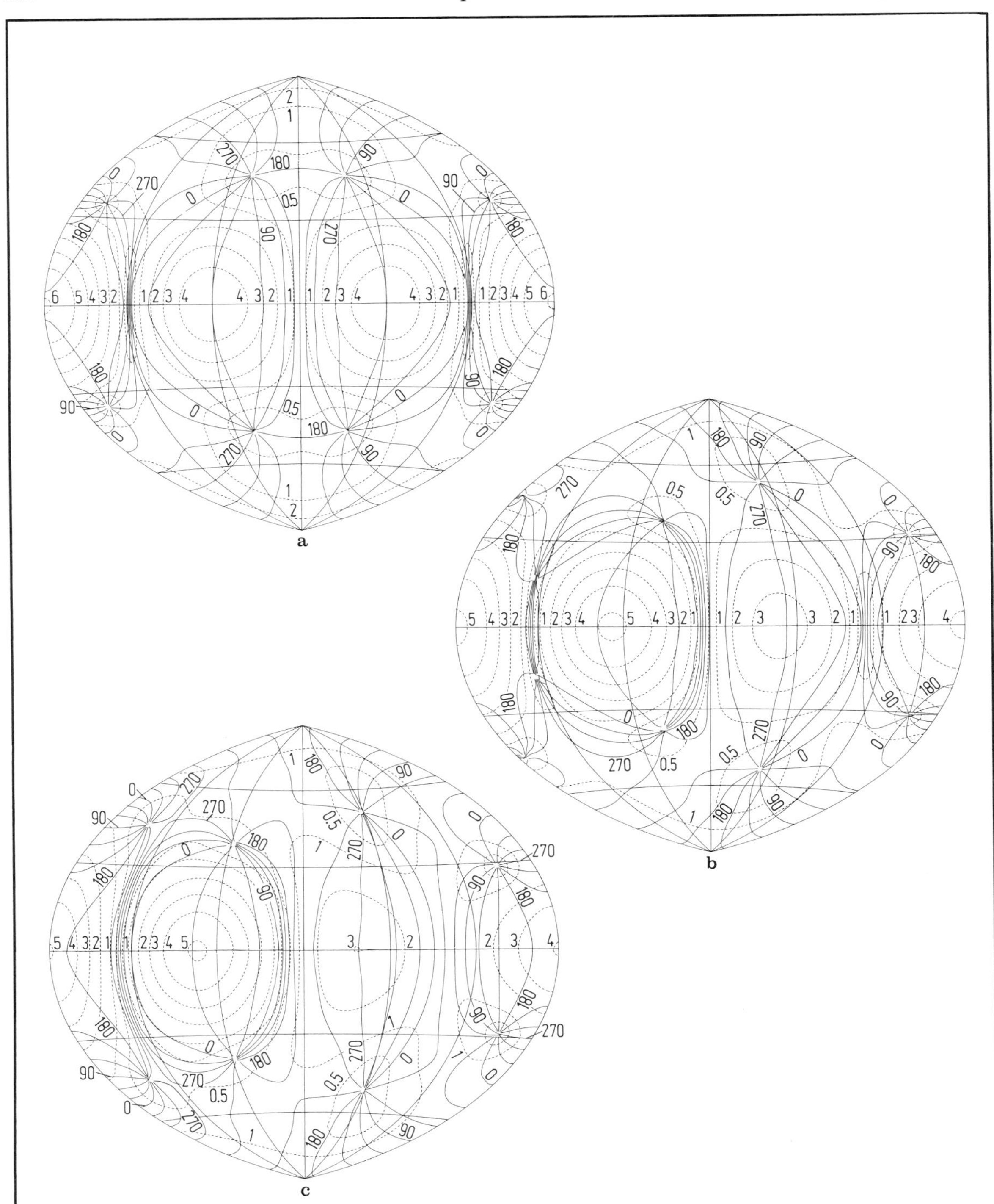

Fig. 7. Diurnal K_1-tide for a hemispherical ocean model: cotidal lines (solid lines) with phases in degrees referred to meridian passage at the western boundary, co-amplitude lines (dashed lines) with amplitudes in units of the maximum equilibrium tide amplitude; near resonance depth $h = 3030$ m, i.e. $1/\sqrt{\varepsilon'} = 0.186$. $\varepsilon' = 4\Omega^2 R_e^2/(gh)$.
a) Frictionless, neglecting ocean self-attraction and tidal loading;
b) frictional decay time 30 h, i.e. $\varrho h/R' = 30$ h, neglecting ocean self-attraction and tidal loading;
c) frictional decay time 30 h, i.e. $\varrho h/R' = 30$ h, considering ocean self-attraction and tidal loading.

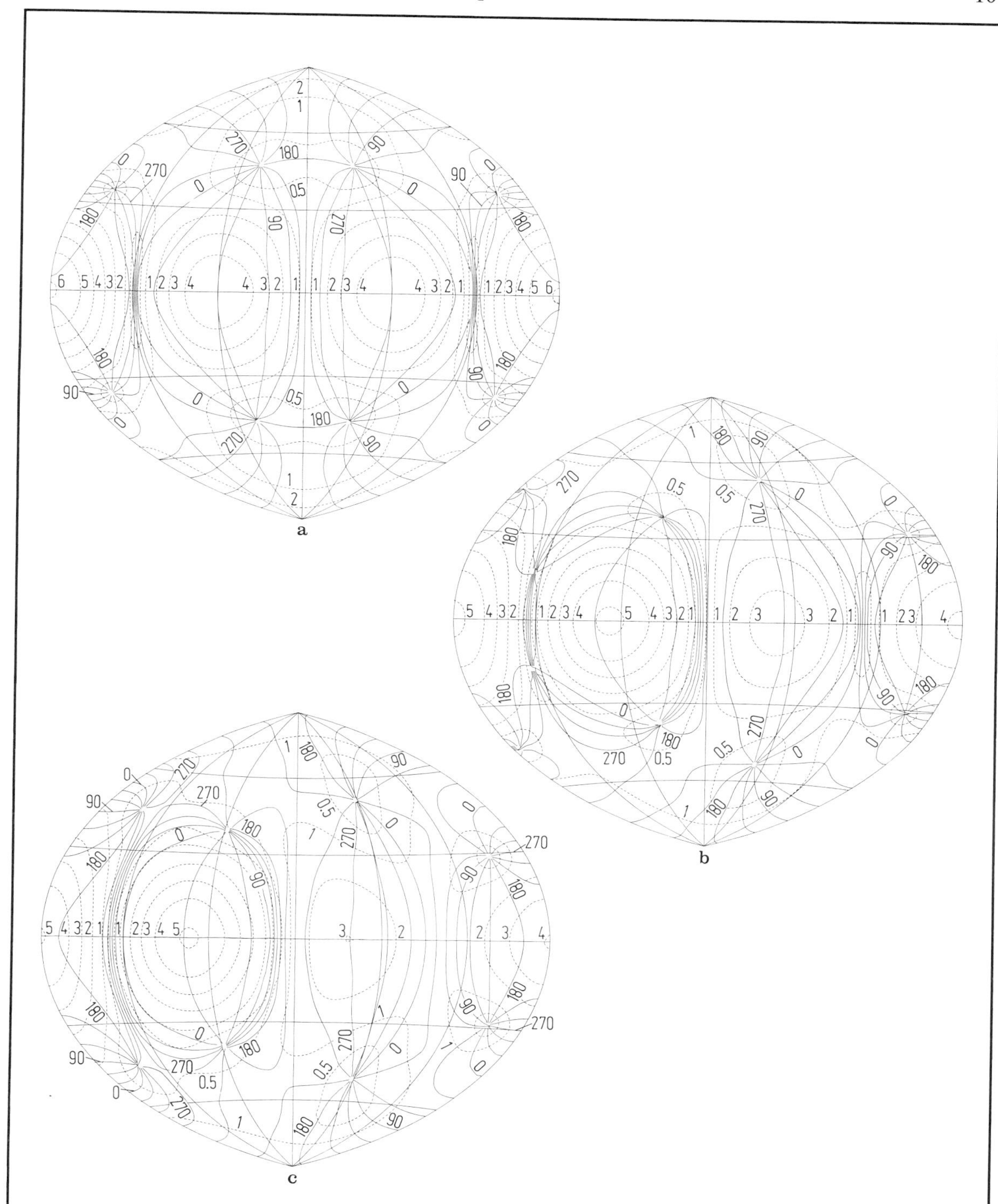

Fig. 8. Semi-diurnal M_2-tide for a hemispherical ocean model: cotidal lines (solid lines) with phases in degrees referred to meridian passage at the western boundary, coamplitude lines (dashed lines) with amplitudes in units of the maximum equilibrium tide amplitude; near resonance depth $h=4420$ m, i.e. $1/\sqrt{\varepsilon'}=0.224$. $\varepsilon'=4\Omega^2 R_e^2/(gh)$.

a) Frictionless, neglecting ocean self-attraction and tidal loading;

b) frictional decay time 30 h, i.e. $\varrho h/R'=30$ h, neglecting ocean self-attraction and tidal loading;

c) frictional decay time 30 h, i.e. $\varrho h/R'=30$ h, considering ocean self-attraction and tidal loading.

6.4.4.2 Tidal oscillations in the real ocean

Due to the failure of obtaining analytical solutions to the tidal equations (9) and (10), or even eqs. (9) and (14) for real oceans, the first attempts to get an overview of the tides in the real oceans were entirely based upon tidal measurements at coasts. By inspection of an increasing number of tidal data, cotidal and, partially, corange maps have been constructed for adjacent seas and partial or world wide oceans (see e.g. [44 D, 52 V, 75 L, 77 M 2, 80 C]).

On the basis of corresponding finite-difference equations numerical solutions have been obtained also for the world ocean as a whole since the availability of suitable computers. Recent global numerical solutions prescribing tidal measurements at the coastal boundaries were presented by [80 P 1] and by [79 S, 81 S, 82 S], where, in the latter series of papers, global charts for the main semi-diurnal, diurnal, and long-period constituents are given. Recent numerical solutions characterized by considering an impermeable coastal boundary and independence of tidal measurements in the sea were presented by [77 G 2, 77 E, 78 A, 77 Z, 78 Z]. An overview of former solutions and the principal numerical methods applied was given by [77 H]. Overviews of charting ocean tides and of the oceanic tides in general were given by [80 S] and [78 C], respectively.

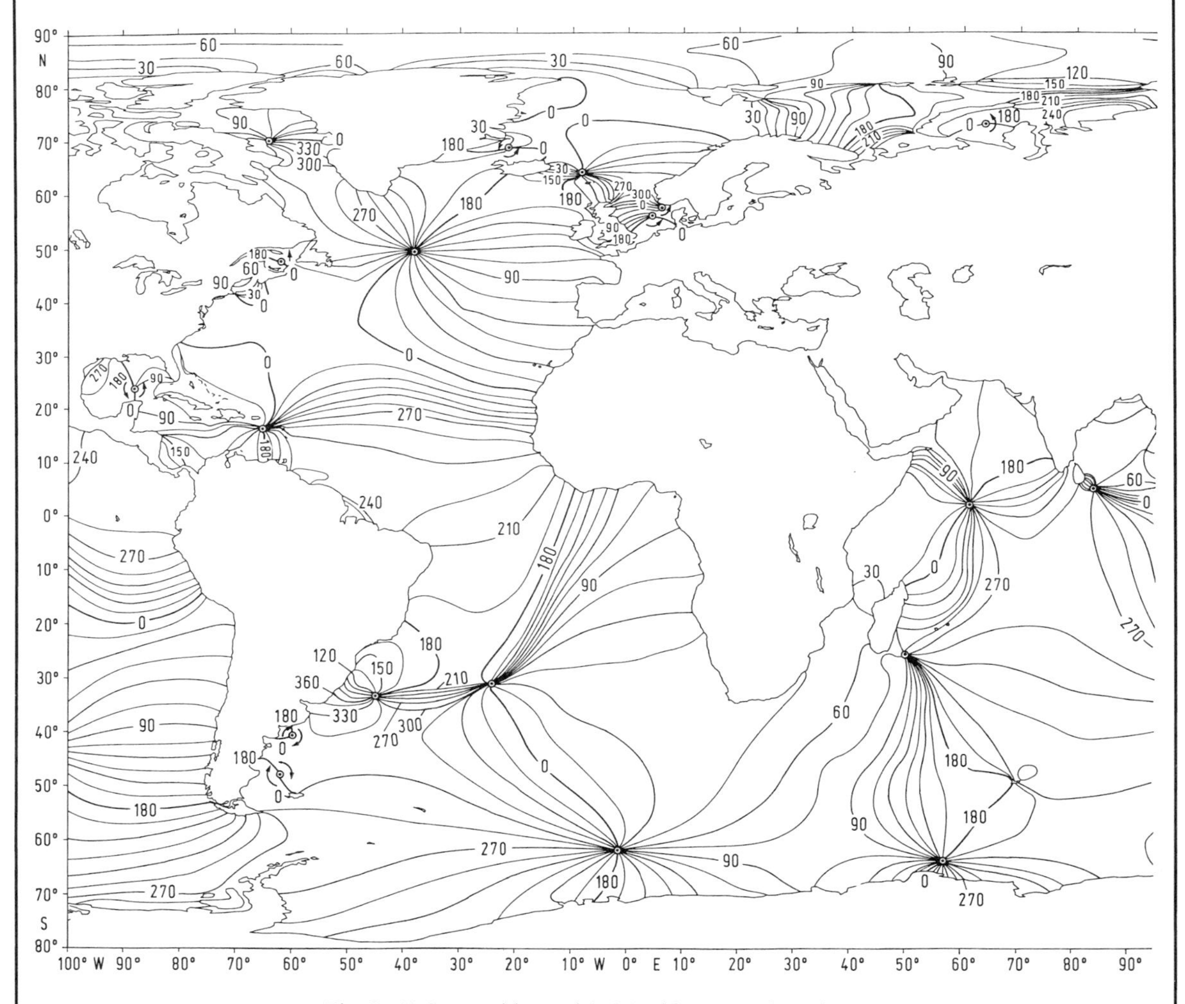

Fig. 9. 1°-Ocean-tide-model: M_2-tide sea-surface elevations. Cotidal map with phases in degrees referred to meridian passage at Greenwich. From [79 S].

In Figs. 9, 10, and 11 cotidal and corange maps for the M_2 tide are given which were obtained by applying a time-stepping procedure independent of tidal measurements ([80 Z]) and by applying a hydrodynamical interpolation technique ([79 S]), thus Figs. 9 and 10 showing at the coastal boundaries values obtained from measurements. The important role of shelf areas (see [67 T, 81 C]) which are poorly resolved in the global tidal models, is going to be considered either by introducing nested grids (see [83 K]) or by a parametrization of the shelf effects (see [81 G 2]).

The nodal points, traditionally denoted as amphidromes, are recognized as centers of zero amplitude round which the cophase lines rotate, the sense of rotation mostly being clockwise in the southern hemisphere and anticlockwise in the northern hemisphere. The other characteristic features are the anti-nodes, zones of locally maximum amplitude with little phase variation. Anti-nodes which in their principle appearance are common to most of the computational results and which are in agreement with the available measurements are those ones in the central Indian Ocean, in the western equatorial Atlantic, and in the eastern and western equatorial Pacific. Amphidromes for which the same is true are the three main amphidromes in the Atlantic, the three ones west of the American continent, the one in the central Pacific, and the amphidrome at the entrance to the Arabian Sea.

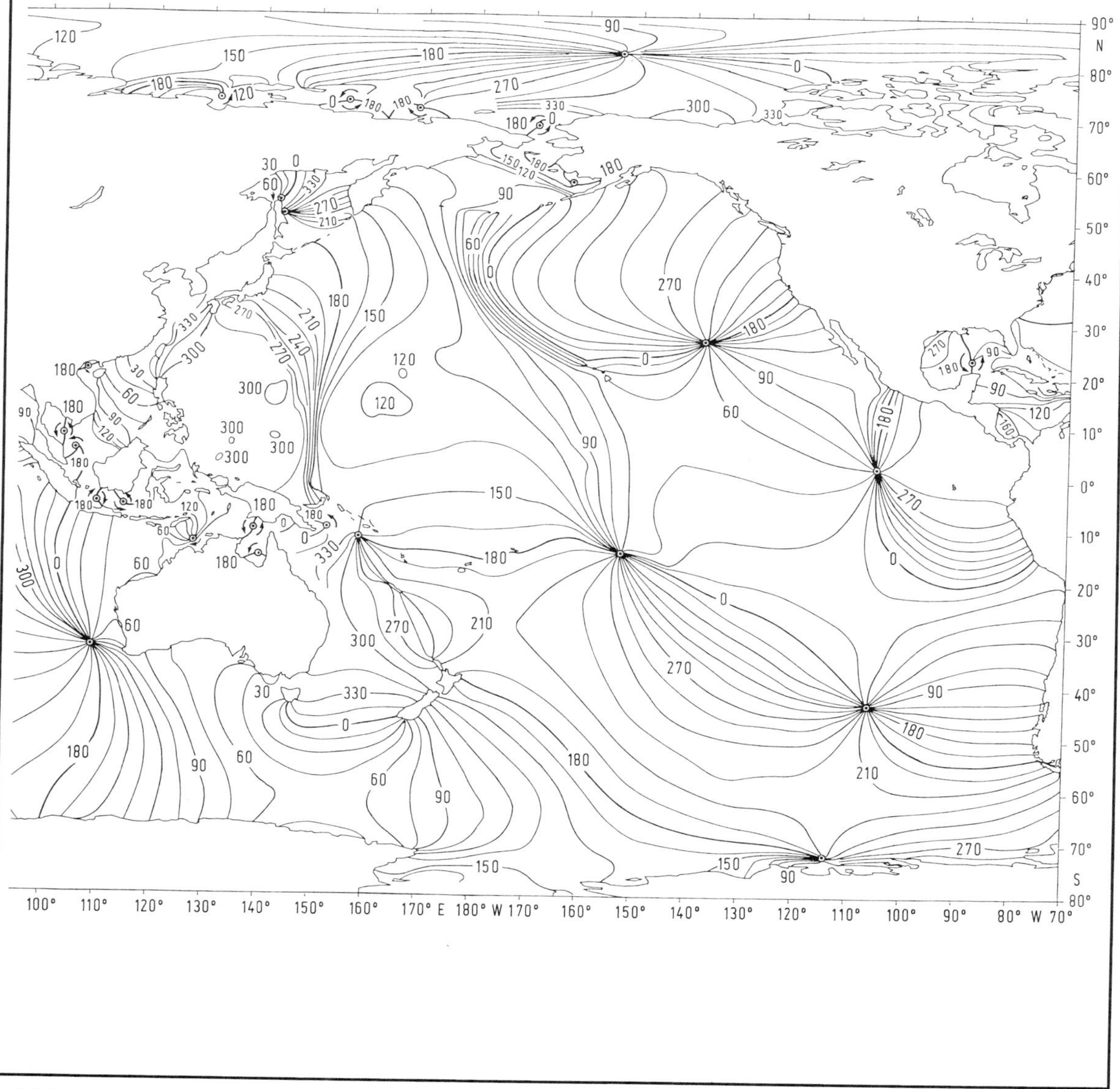

Zahel

The results obtained by a 4°-model and depicted in Figs. 11 and 12 include the tidal loading and ocean self-attraction effect. When compared with the corresponding results obtained neglecting these effects (see comparison for the K_1-tide in Fig. 12 and [78 Z, 80 Z]), it can be realized that the M_2-tide and K_1-tide oscillation systems are not altered significantly, but that a rather uniform phase delay arises when considering these effects, which leads to an improved approximation of the measured tidal wave, in particular at the east sides of the Atlantic and of the Pacific Ocean. Figs. 7 and 8 for a hemispherical ocean show a similar phase delay at the east side.

Establishing and evaluating a tidal energy equation is essential for an understanding of the world ocean's response to the generating forces. This response is determined by the oscillation properties of the oceanic basins, whence the investigation of the eigenoscillations is crucial. Details on the tidal energy equation constituents and their numerical evaluation can be found in [80 Z, 83 S], details on computed global eigenoscillations can be taken from [81 G 1, 81 P]. In Fig. 14 the rate of work per unit area by tide generating forces

$$W_2=(1+k_2)\left(\nabla\cdot(\varrho g H\bar{\eta}\bar{v}_{\mathrm{h}})+\varrho g\bar{\eta}\frac{\partial\eta}{\partial t}\right) \tag{23}$$

is depicted. In the open ocean it is true for each of the oceans that, integrated over a tidal period, tidal energy is fed into the ocean tide regime as a whole. The rate of work pattern is in nearly exact equilibrium with the rate of divergence of energy flux $W_1=\nabla\cdot(\varrho g H(\eta+\delta)\bar{v}_{\mathrm{h}})$, thus the energy is mainly transported into the shelf areas and dissipated there, finally. Concerning deviations from this simple picture, see [77 Z]. The different numerical

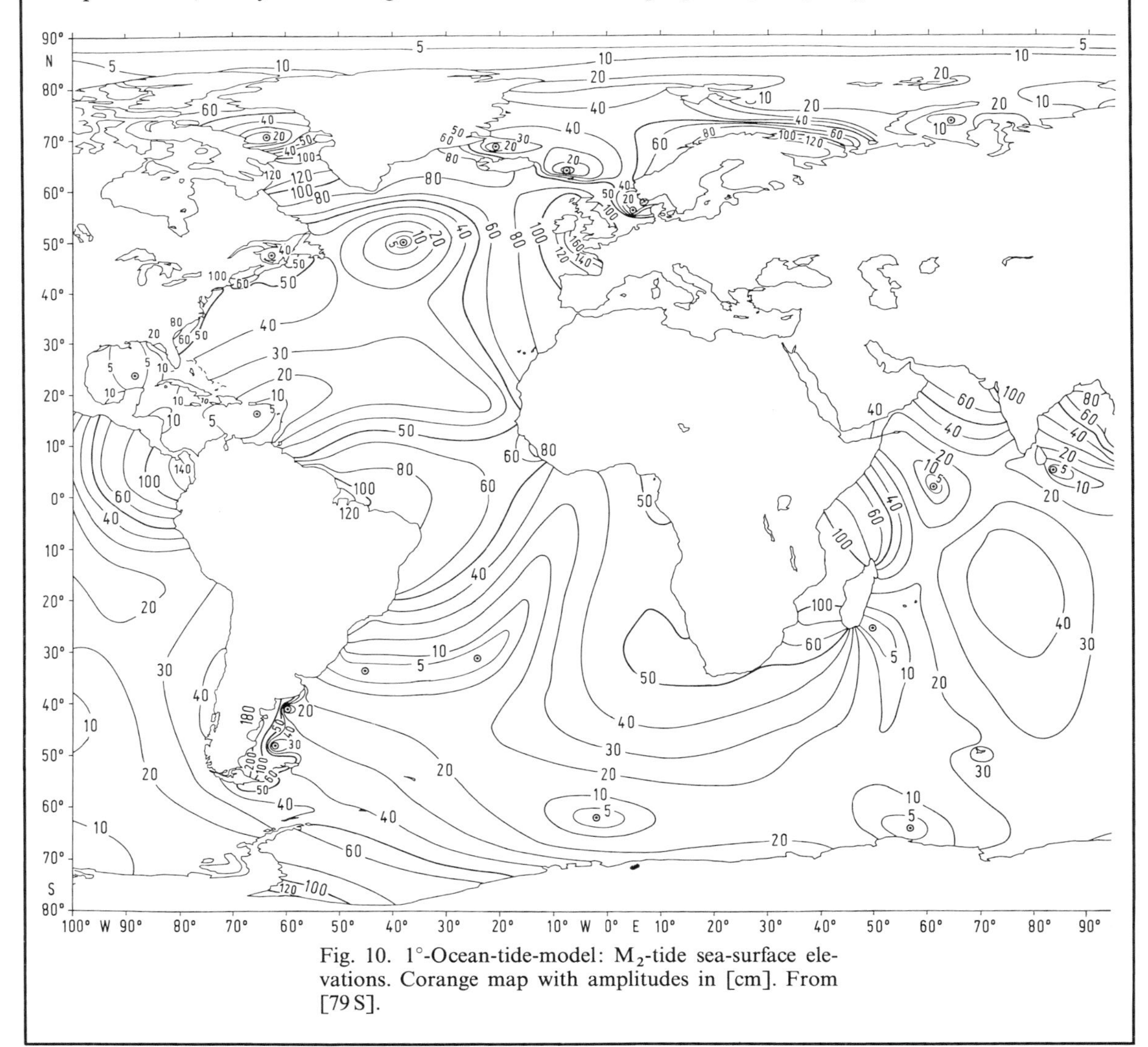

Fig. 10. 1°-Ocean-tide-model: M_2-tide sea-surface elevations. Corange map with amplitudes in [cm]. From [79 S].

evaluations of the tidal energy equation (see [83 K]) yield for the M_2-tide total time-averaged and space-integrated rates of work by tide-generating forces and the moving sea bottom amounting to $(2.2 \cdots 3.8)\,10^{12}$ W. Concerning the relevance of these values for the deceleration of the earth's rotation, see [77 L, 78 b, 82 b].

Linear forced oscillations are uniquely determined by the eigenoscillations and the acting force:

$$\boldsymbol{L}\boldsymbol{a} - \mu \boldsymbol{a} = \boldsymbol{f}, \qquad \boldsymbol{a} = \left[\frac{\eta}{\boldsymbol{v}_{\mathrm{h}}}\right] \tag{24}$$

$$\boldsymbol{a} = -\sum_{\mathrm{j}} \frac{(\boldsymbol{g}_{\mathrm{j}}, \boldsymbol{f})}{(\mu - \mu_{\mathrm{j}})(\boldsymbol{g}_{\mathrm{j}}, \boldsymbol{f}_{\mathrm{j}})} \boldsymbol{f}_{\mathrm{j}} \tag{25}$$

where eq. (24) is defined by eqs. (9) and (10), $\boldsymbol{f}_{\mathrm{j}}$ and $\boldsymbol{g}_{\mathrm{j}}$ are biorthogonal systems of eigenfunctions of $\mathbf{L}$ and of the adjoint operator $\bar{\boldsymbol{L}}$, respectively. $\bar{\boldsymbol{L}}$ arises from $\boldsymbol{L}$ by changing the signs of the dissipative terms when the scalar product (,) is properly defined as a surface overlap integral. μ_{j} denote the eigenvalues of $\mathbf{L}$. Eq. (25) takes the simplified form

$$\boldsymbol{a} = -\sum_{\mathrm{j}} \frac{(\boldsymbol{f}_{\mathrm{j}}, \boldsymbol{f})}{\mu - \mu_{\mathrm{j}}} \boldsymbol{f}_{\mathrm{j}}$$

in the frictionless case with orthonormal eigenfunctions. [81 G 1, 82 G, 81 P] have computed eigenoscillations of the world ocean. Eigenperiods close to semi-diurnal and diurnal tidal periods are given in Table 7.

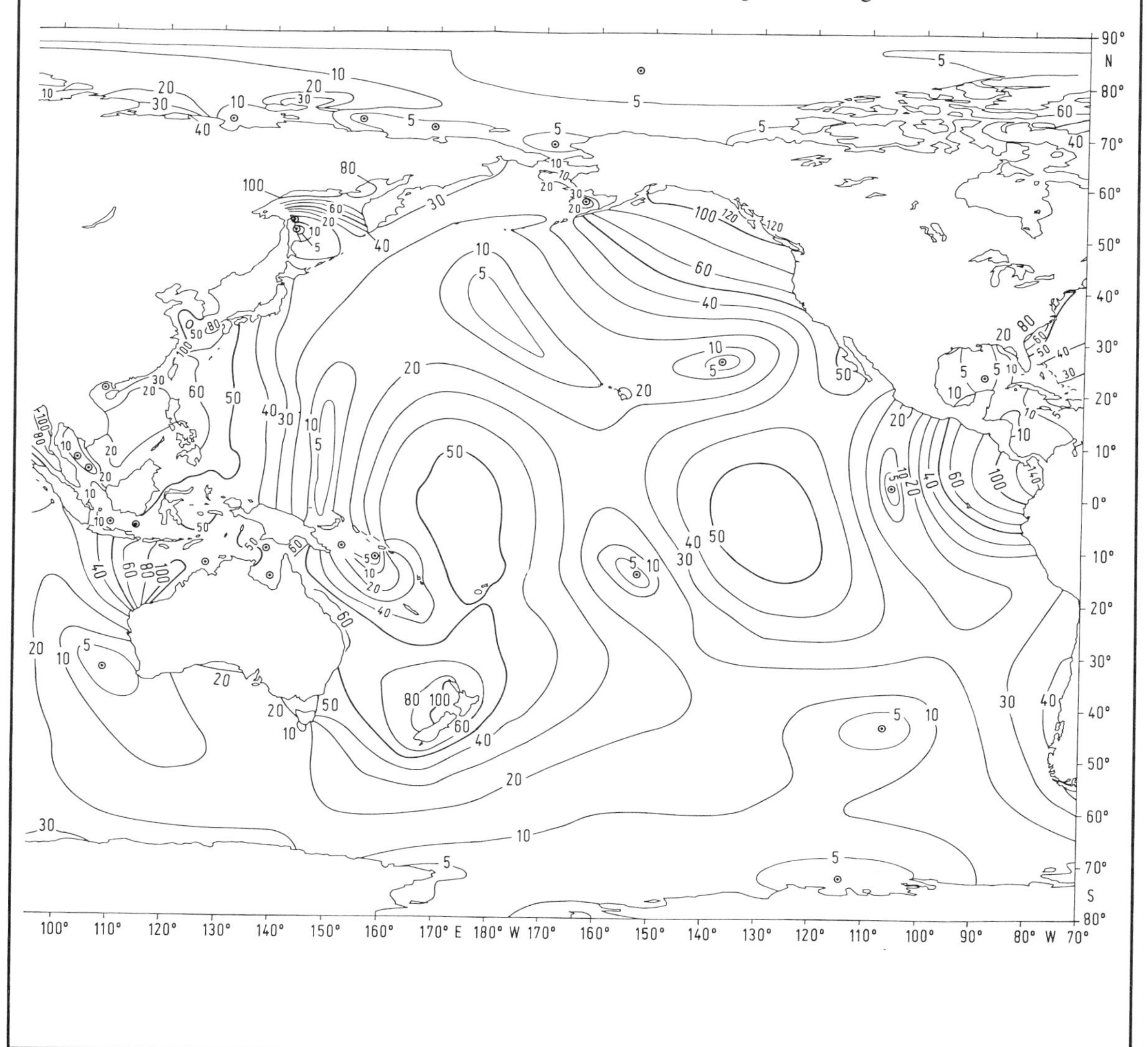

 Zahel

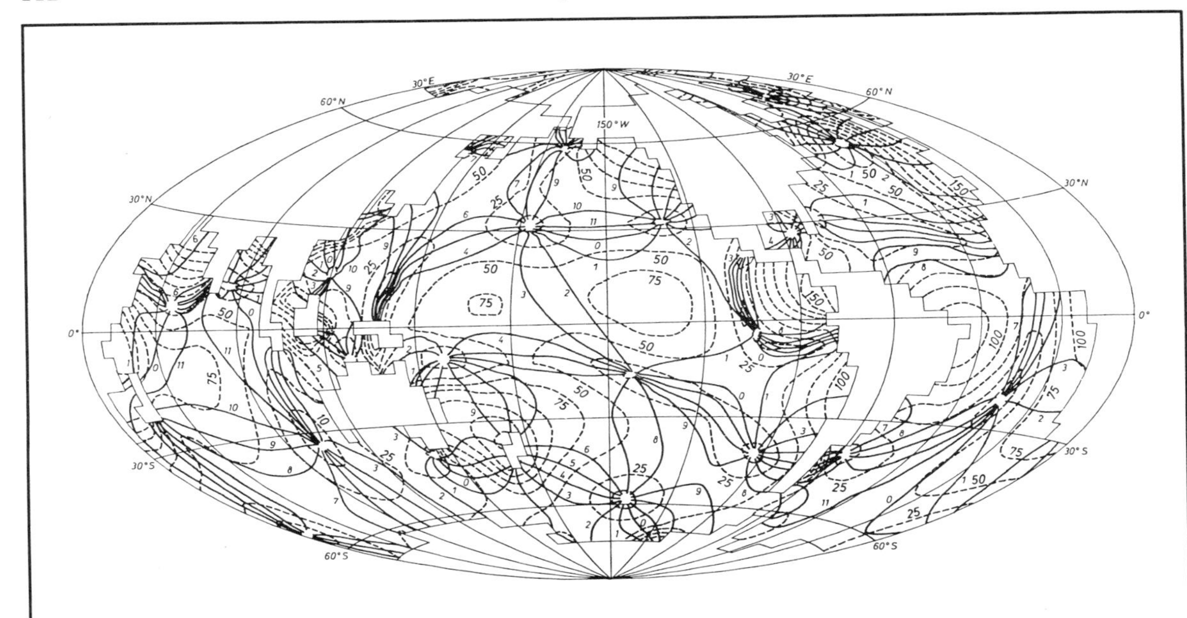

Fig. 11. 4°-Ocean-tide-model: M_2-tide sea-surface elevations considering interaction with earth tides. Cotidal lines solid and corange lines broken. Phases in lunar hours referred to meridian passage at Greenwich, amplitudes in [cm]. From [80 Z].

For Fig. 12, see next page.

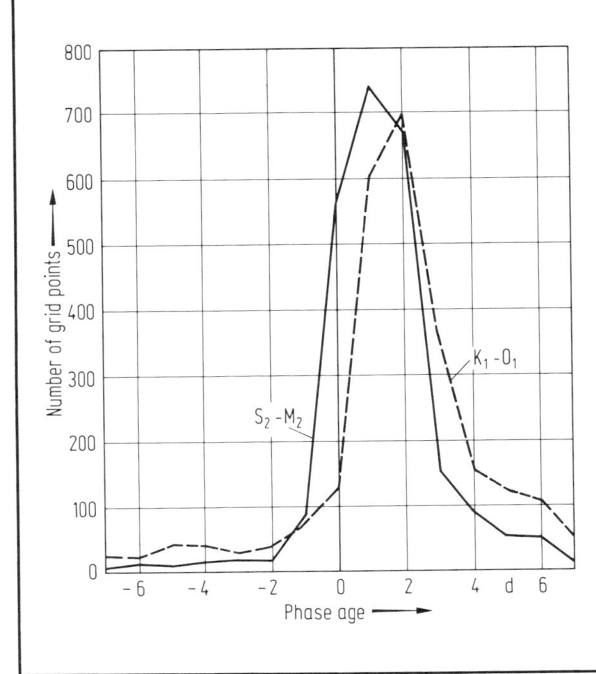

Fig. 13. Number of grid points of a numerical ocean-tide model considering interaction with earth tides at which the age of the tide takes the same values. Solid line refers to the S_2–M_2 age, broken line to K_1–O_1 age. From [80 B 2].

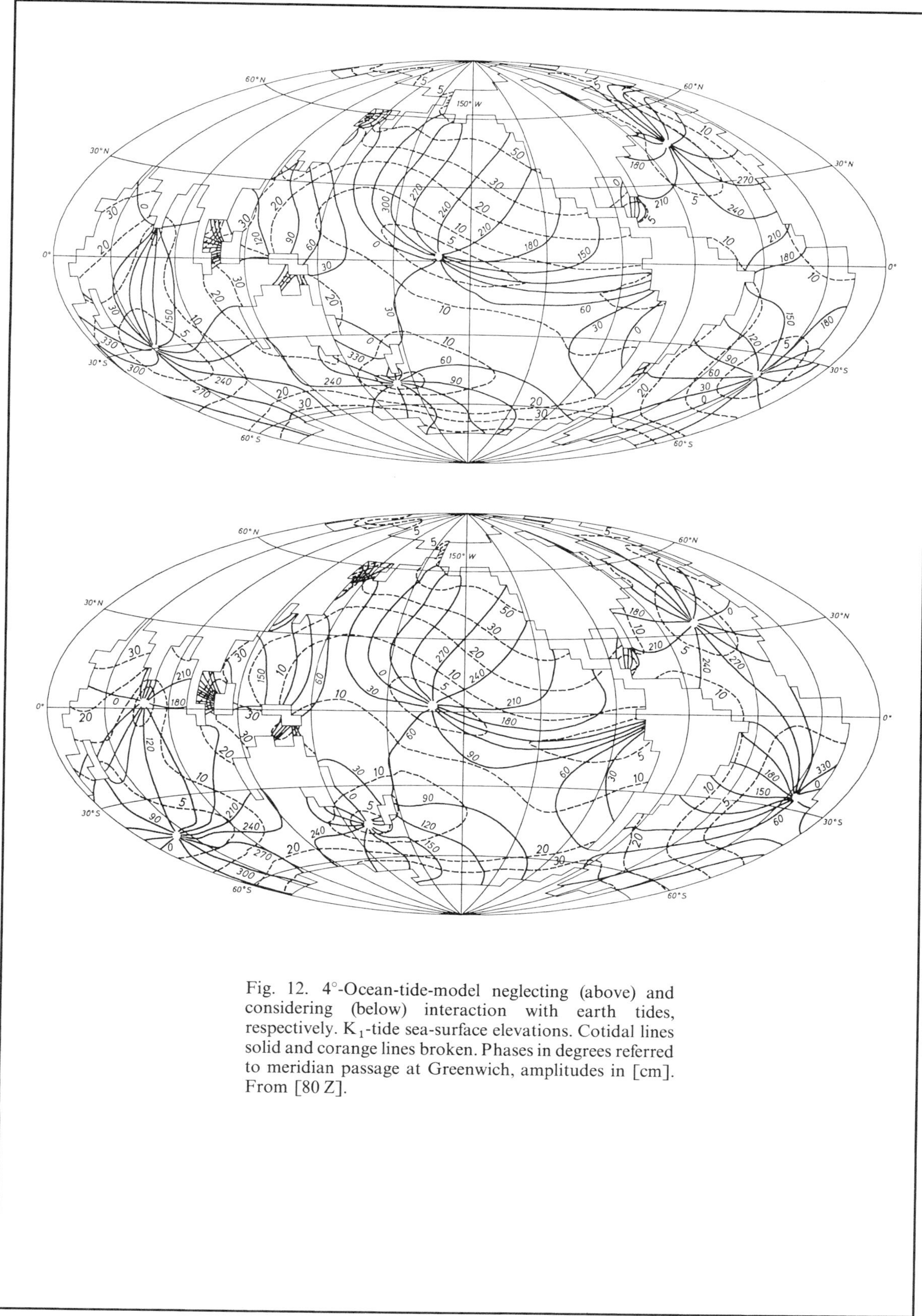

Fig. 12. 4°-Ocean-tide-model neglecting (above) and considering (below) interaction with earth tides, respectively. K_1-tide sea-surface elevations. Cotidal lines solid and corange lines broken. Phases in degrees referred to meridian passage at Greenwich, amplitudes in [cm]. From [80 Z].

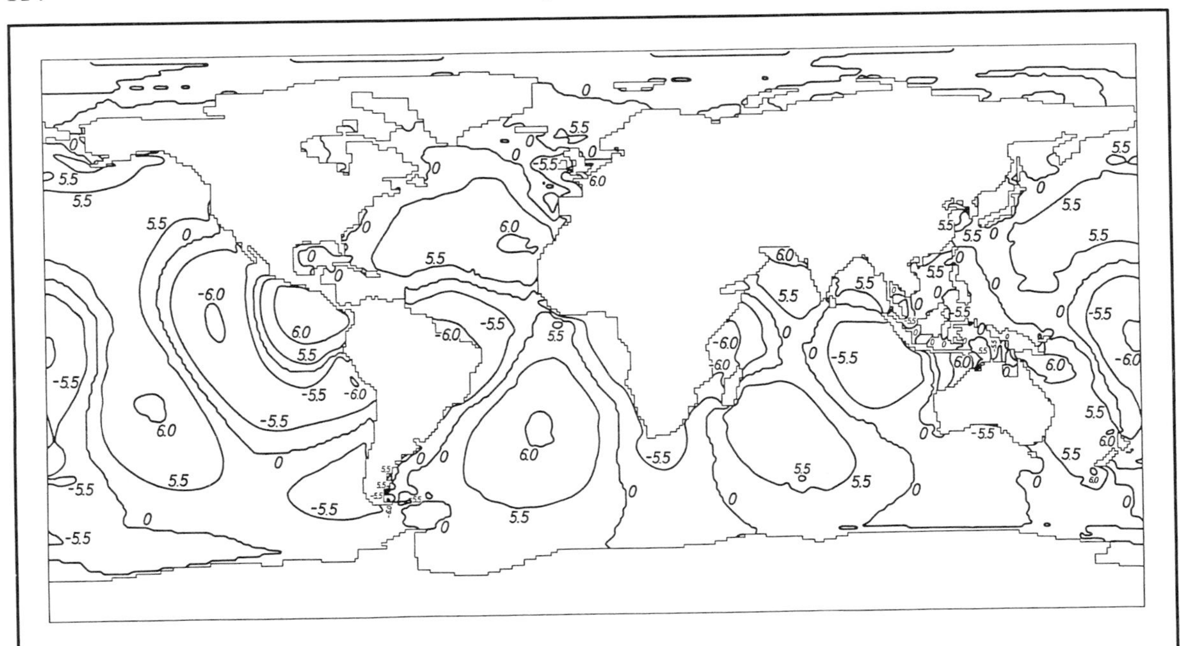

Fig. 14. 1°-Ocean-tide-model: M_2-tide work done by tidal forces and moving sea bottom (mean value over one M_2-period) in 10^n erg/(m^2 sec), lines of equal n (+ gain, − loss). From [77 Z].

Table 7. Important eigenoscillation periods, in hours, of the world ocean in the semi-diurnal and diurnal period range, according to three different references.

81 G 1	82 G 2	81 P
26.0	25.9	25.8
23.5	23.9	23.7
21.9	22.5	21.2
12.8	12.8	12.8
12.7	12.7	–
12.5	12.5	12.5
12.0	12.0	11.9

Fig. 15 demonstrates that the eigenoscillations with periods closest to the K_1 and M_2 tidal periods, respectively, obviously resemble the elevation patterns of the K_1-tide and of the M_2-tide as obtained by the global tidal models. The eigenoscillations of the Atlantic Ocean have also been treated independently of the remainder of the world ocean by [81 G 1]. They do not significantly differ for the above tidal period ranges from the eigenoscillations obtained as part of the global system. This fact suggests in conjunction with the spatial distribution of tidal work (Fig. 14) that the M_2-tide regime in the Atlantic ocean is determined by the local action of the generating forces. This has been verified by tide-model experiments for the M_2-tide as well as for the K_1-tide (see [73 Z]).

Contrary to the open oceans, the adjacent seas extending on the shelves as a rule do receive only a negligible rate of tidal work and prove to be co-oscillating with the open-ocean tides. Eq. (25) points at the insignificance of the independent tide of adjacent seas by means of the smallness of the overlap integrals in the nominators for these areas.

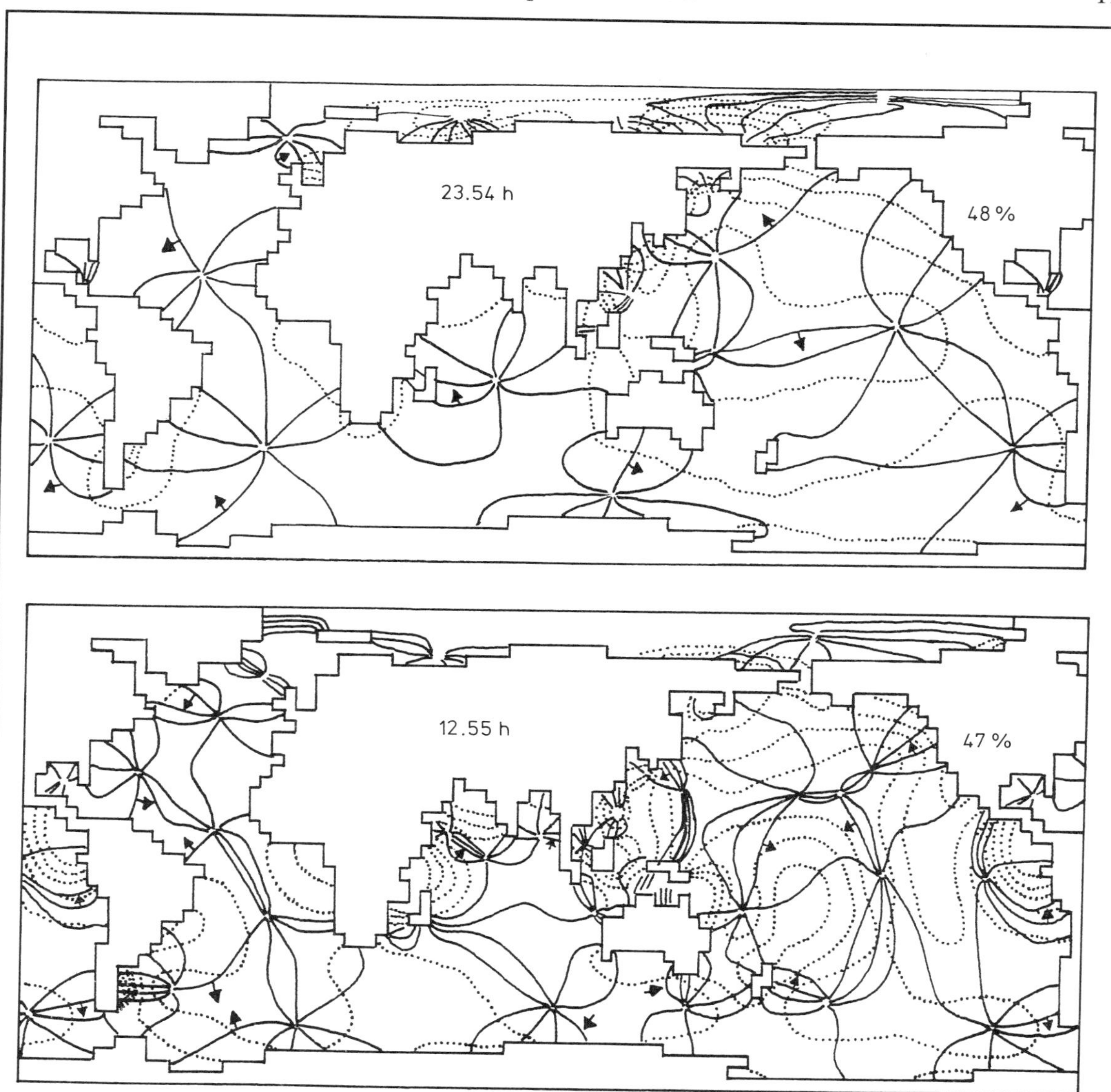

Fig. 15. Cotidal and corange lines for a semi-diurnal and a diurnal free mode in the world ocean. Cotidal lines are solid with 45°-phase difference. Arrows indicate the rotational senses, they are attached to 0°-phase lines. Corange lines are dotted. Percentages indicate the ratio of potential energy to total energy. Eigenperiods are given in hours. From [81 G 1].

6.4.5 Co-oscillating tides

For co-oscillating tides in adjacent seas, Eqs. (9) and (10), with nonlinear terms included, can be simplified by neglecting the tidal potential and by assuming the Coriolis parameter $f=2\omega\cos\phi$ to be constant. For investigations of the three-dimensional tidal field of motion, eqs. (5) and (6) are taken as a point of departure. The problem of co-oscillating tides in real adjacent seas can be treated by applying hydrodynamical numerical methods, also taking into account shallow water areas, the three-dimensional field of motion, and the interaction with the non-tidal field of motion, e.g. surges. Results and methods are described in detail by [76 F, 80 B 1, 80 D, 80 r], coastal phenomena are in the focus of [69 D, 75 D, 78 L, 78 n].

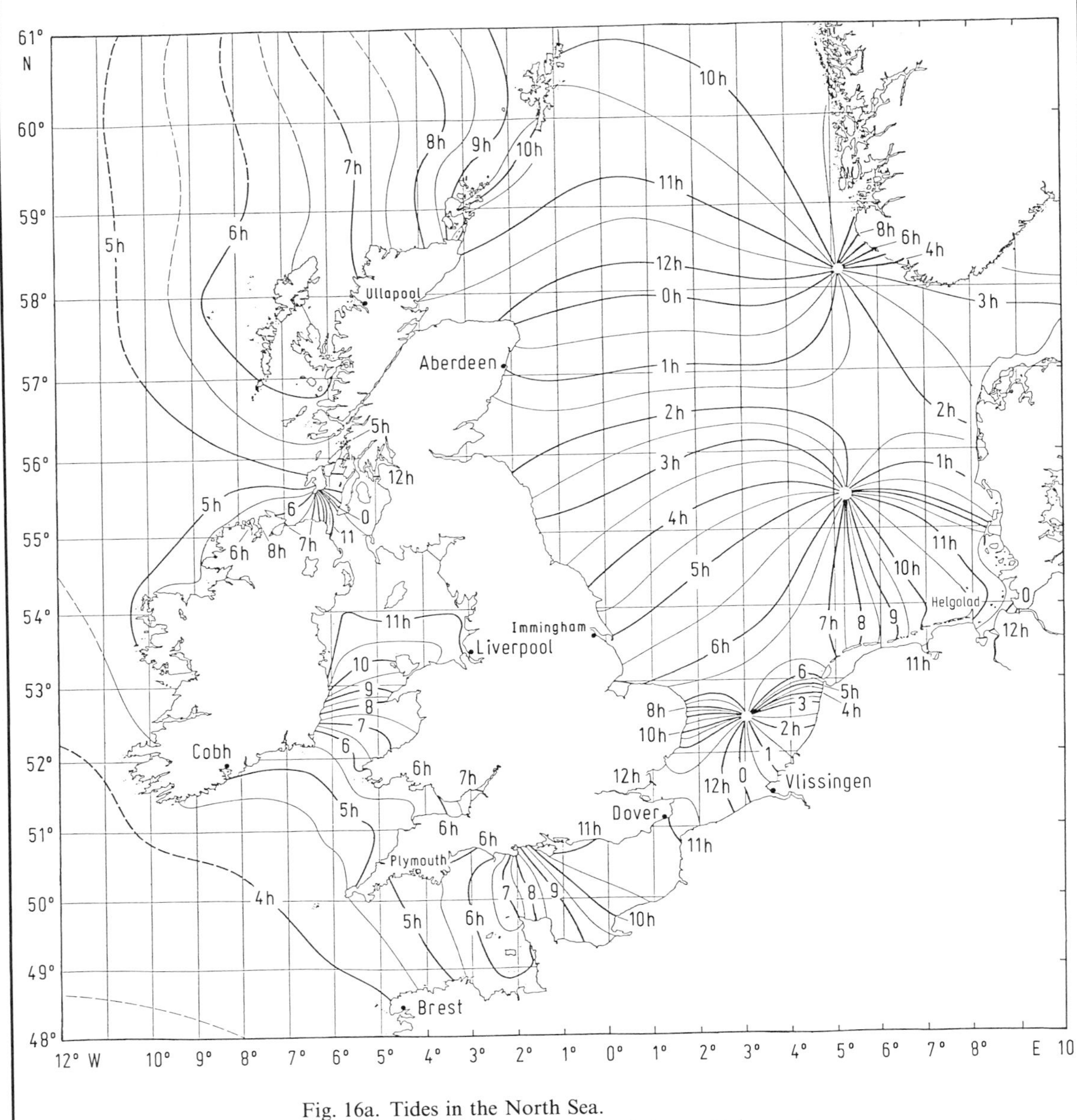

Fig. 16a. Tides in the North Sea.
Lines of equal mean differences between high water and moon's transit of the Greenwich meridian, differences in hours.

A comprehensive collection of tidal charts for adjacent seas was given by [61 d]. Meanwhile for many individual seas charts have been added, e.g. for the Canadian waters by [80 G], the South China Sea by [75 S], the Caribbean Sea by [81 K]. Co-oscillating tides have been computed by means of tidal models for numerous seas, e.g. the Yellow Sea by [77 A 2], the Bering Sea by [77 S 2], the Bay of Fundy and Gulf of Maine by [79 G], the Gulf of Carpentaria and the Arafura Sea by [81 W], and the Gulf of St. Lawrence by [80 P 3]. Some information on the tides in the North Sea (see also [77 M 1, 80 H]) as obtained by inspecting measurements are given in Figs. 16 and 17. Results of the application of a numerical model on the basis of prescribed elevations at the open boundary are shown in Fig. 18.

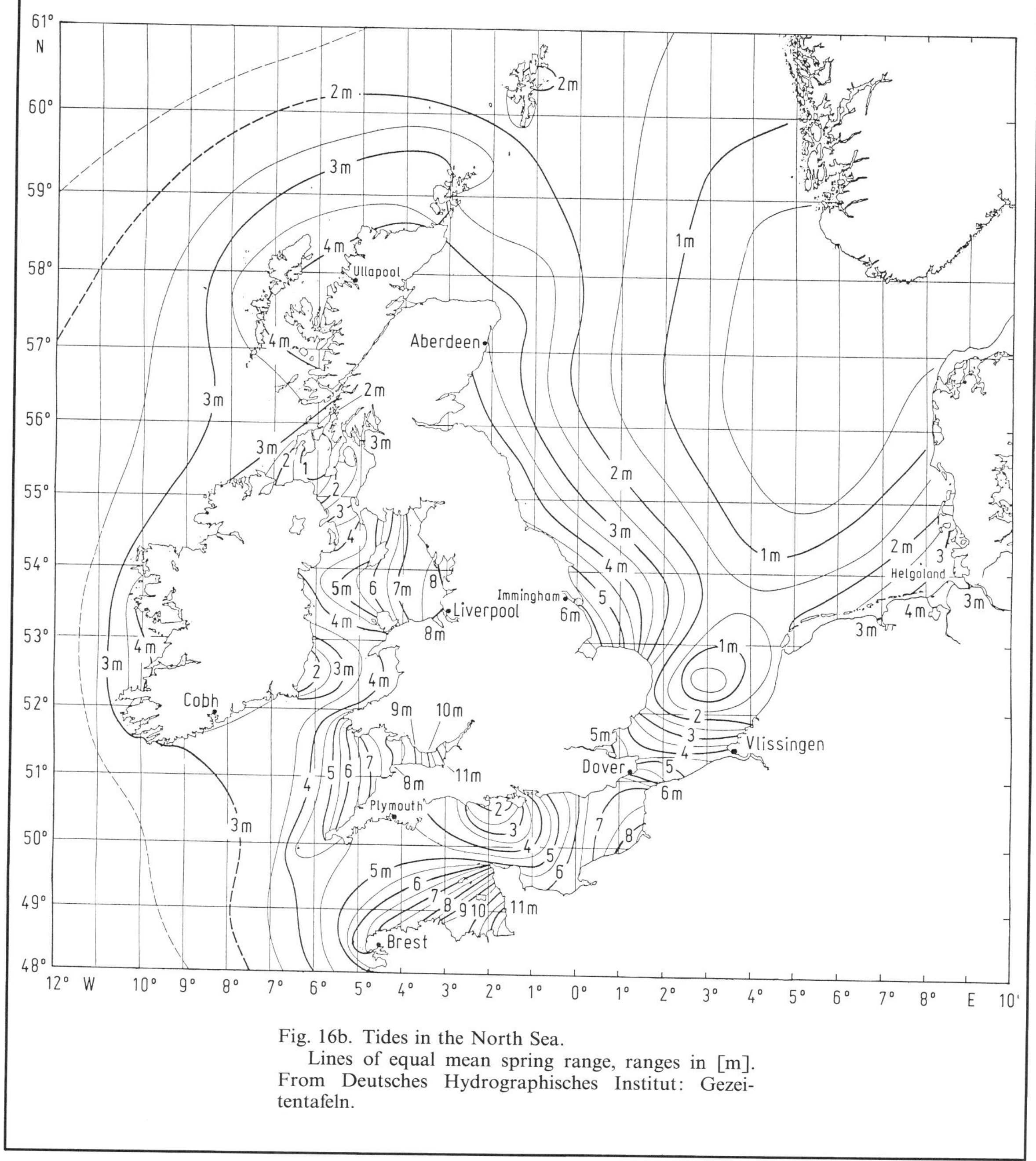

Fig. 16b. Tides in the North Sea.
Lines of equal mean spring range, ranges in [m]. From Deutsches Hydrographisches Institut: Gezeitentafeln.

 Zahel

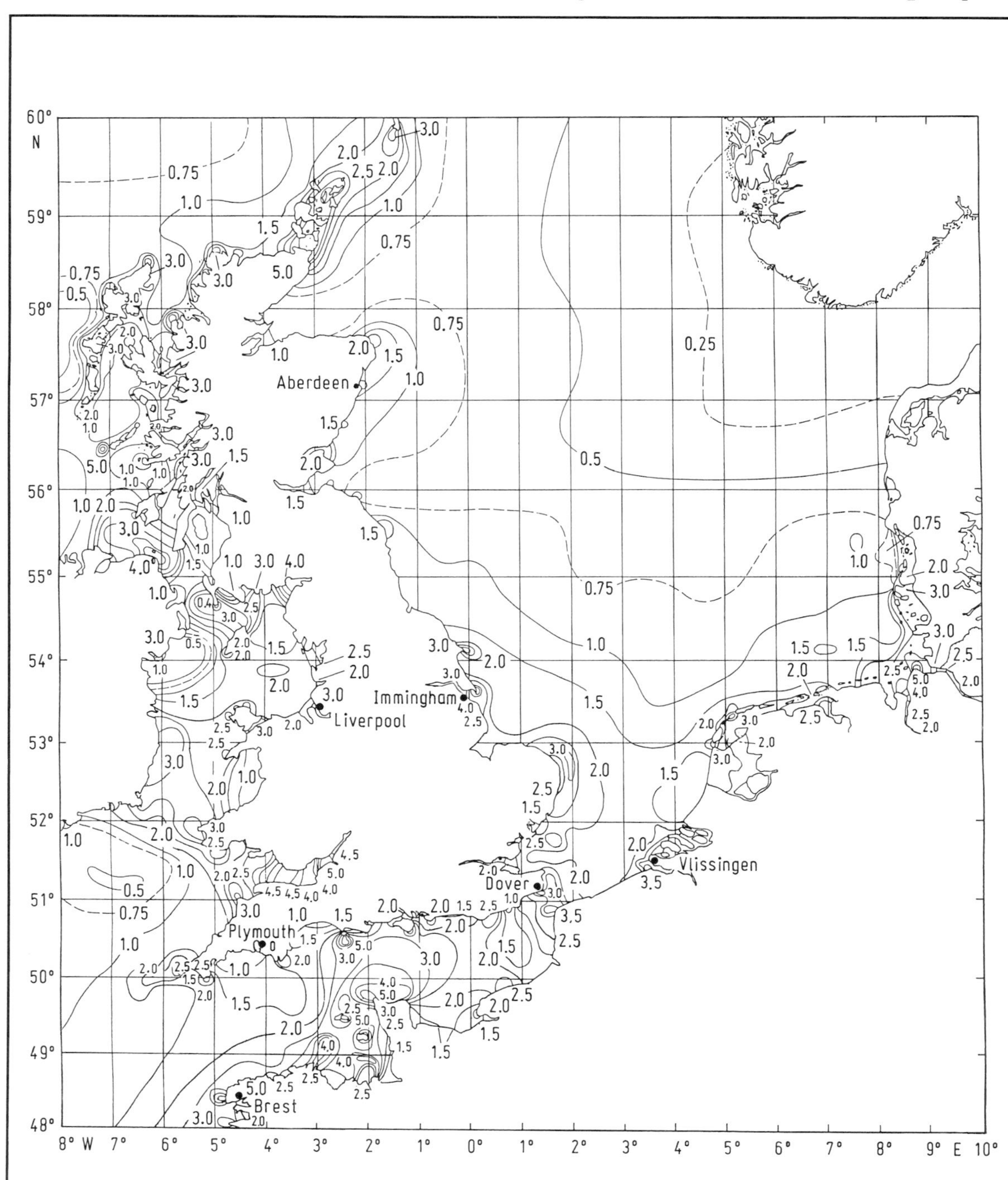

Fig. 17a. Tidal currents in the North Sea.
Lines of equal maximum tidal current velocity at mean spring time, velocities in knots.

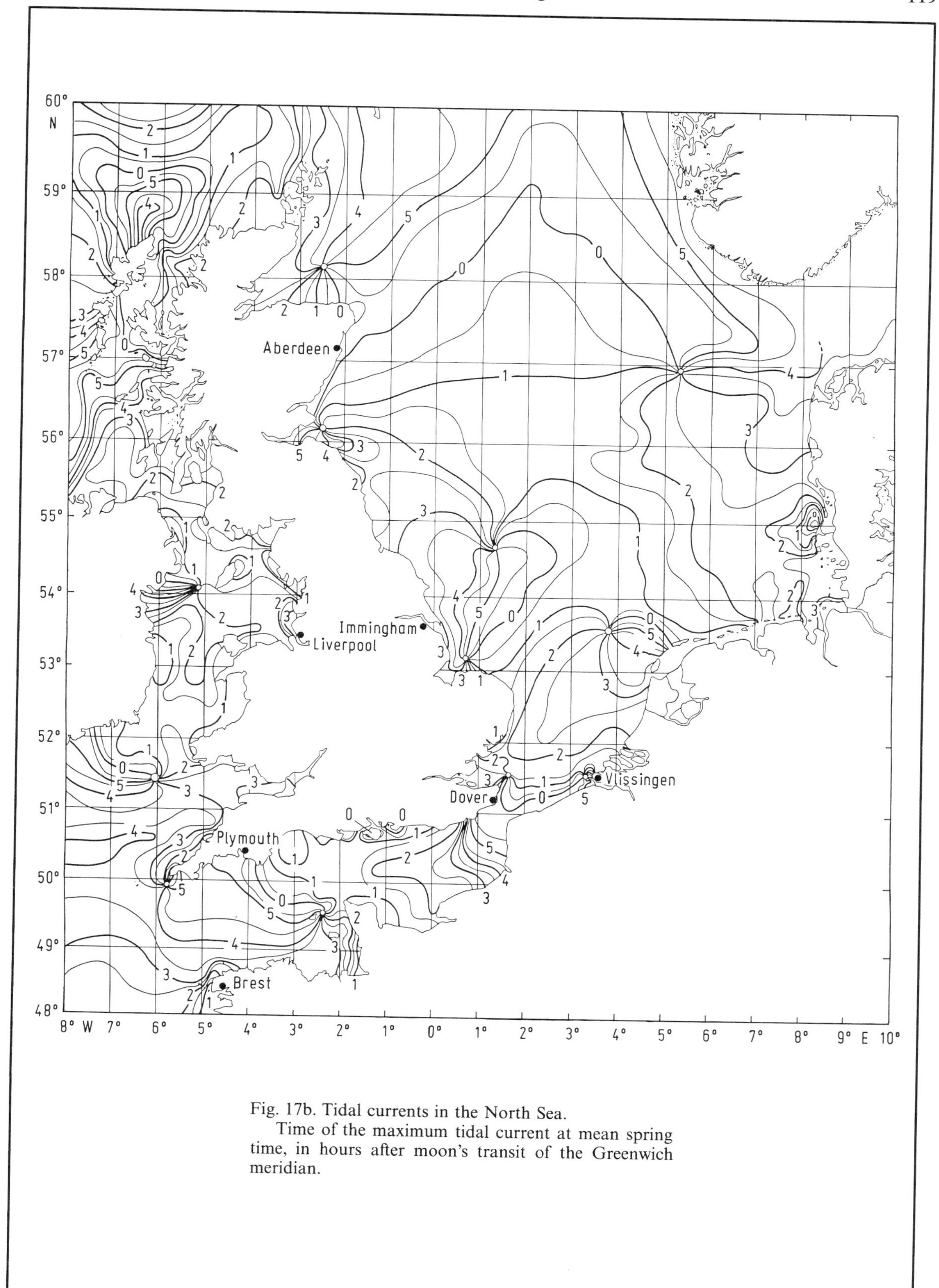

Fig. 17b. Tidal currents in the North Sea.
Time of the maximum tidal current at mean spring time, in hours after moon's transit of the Greenwich meridian.

Zahel

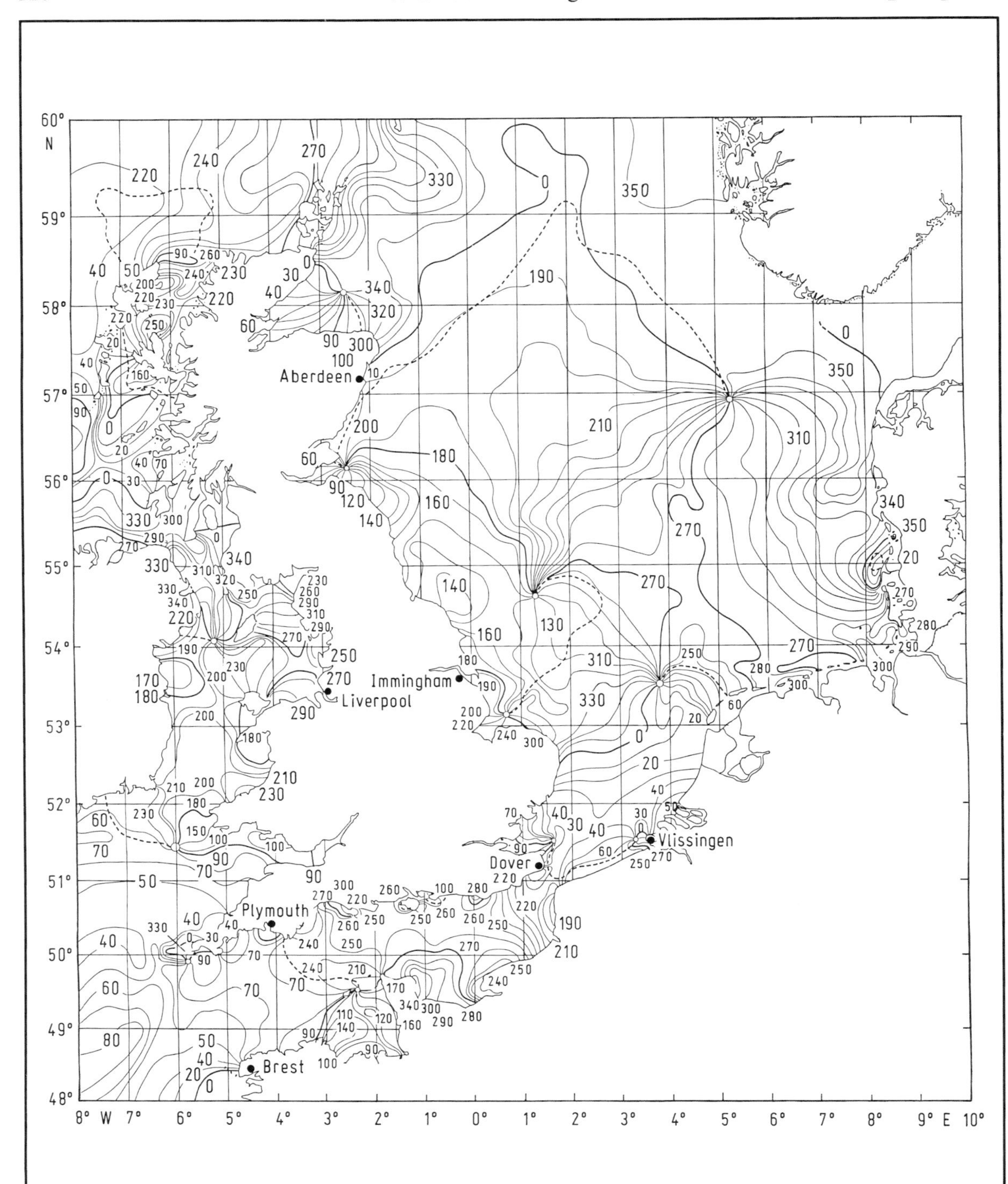

Fig. 17c. Tidal currents in the North Sea.
Lines of equal direction of maximum tidal current at mean spring time. Direction (0° north, 90° east) in degrees.

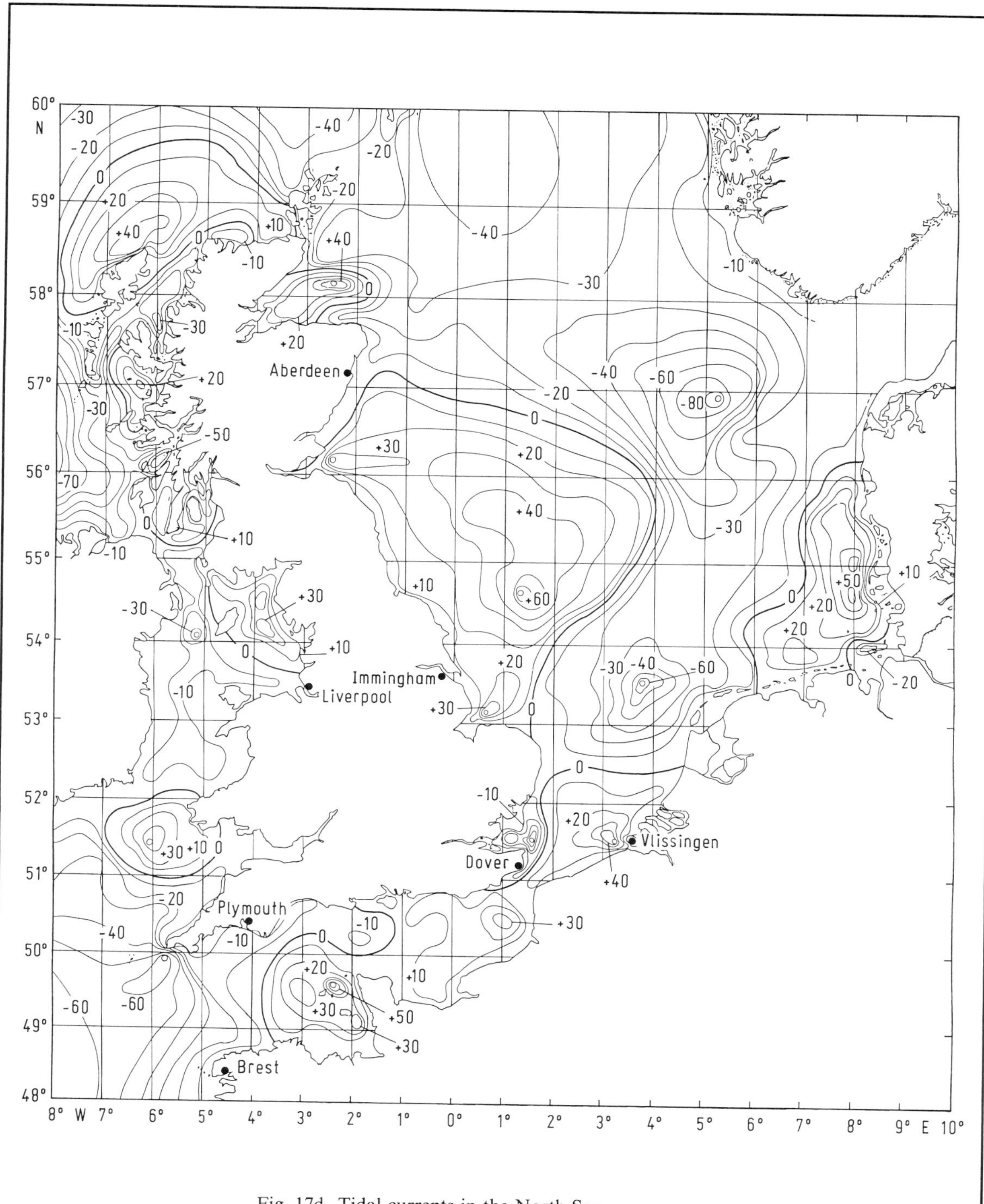

Fig. 17d. Tidal currents in the North Sea.

Lines of equal ratios of minimum to maximum tidal current at mean spring time (−: right-hand rotating current, +: left-hand rotating current). From [68 s].

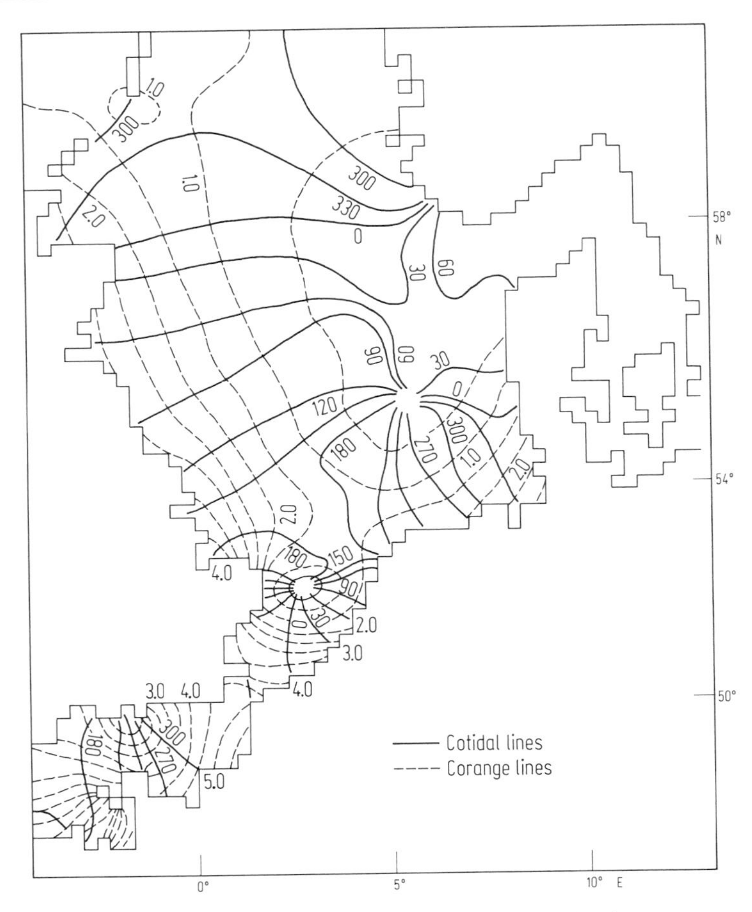

Fig. 18. M_2-tide computed by a North Sea model (Dolata unpublished manuscript): Corange in [m] (dashed line) and cotidal lines (full line) in degrees referred to moon passage at Greenwich.

The classical analytical approach to simplified versions of the problem allows to detect different types of waves playing an important role in the general problem of tidal dynamics (see e.g. [71 P, 78 I]). For analytical investigations of tides in schematic seas the linearized system of equations

$$\begin{aligned} &\frac{\partial u}{\partial t} + \mathrm{R}'u - \mathrm{f}v + g\frac{\partial \eta}{\partial x} = 0 \\ &\frac{\partial v}{\partial t} + \mathrm{R}'v + \mathrm{f}u + g\frac{\partial \eta}{\partial y} = 0 \\ &\frac{\partial \eta}{\partial t} + \frac{\partial (hu)}{\partial x} + \frac{\partial (hv)}{\partial y} = 0 \end{aligned} \tag{26}$$

in Cartesian coordinates is taken as a basis. Here u, v are vertically averaged velocities. Eq. (26) is applicable to the open part of adjacent seas where linearization is allowed due to small Rossby numbers $Ro = U/fL$ and $\eta \ll h$. With the current velocity component normal to the closed boundary being zero, the unknowns u, v, η are uniquely determined in the interior when either the elevation or the normal component of the current velocity is prescribed at the open boundary. By elimination of two of the three unknowns a single equation results:

$$\Delta\Phi' + \frac{1}{h}\frac{\partial h}{\partial x}\frac{\partial \Phi'}{\partial x} + \frac{1}{h}\frac{\partial h}{\partial y}\frac{\partial \Phi'}{\partial y} + \frac{f}{R^*}\left(\frac{1}{h}\frac{\partial h}{\partial x}\frac{\partial \Phi'}{\partial y} - \frac{1}{h}\frac{\partial h}{\partial y}\frac{\partial \Phi'}{\partial x}\right) + \frac{i\omega(R^{*2}+f^2)}{ghR^*}\Phi' = 0 \tag{27}$$

where $\Phi=u, v, \eta$ are supposed to behave simple harmonic in time: $\Phi=\Phi' e^{-i\omega t}$, $\Phi'=\Phi_1+i\Phi_2$. R^* is defined by $R^*=R'-i\omega$. Once the solution for η is obtained, u and v are determined by

$$\begin{bmatrix} u' \\ v' \end{bmatrix} = -\frac{g}{R^{*2}+f^2} \begin{bmatrix} R^*\frac{\partial}{\partial x}+f\frac{\partial}{\partial y} \\ -f\frac{\partial}{\partial x}+R^*\frac{\partial}{\partial y} \end{bmatrix} \eta'. \tag{28}$$

A simple solution for the further simplified ($f=0$) problem of the tides in a channel of length L (or narrow rectangular basin) with constant h is given by

$$\begin{aligned} \eta &= \frac{\eta_L}{\cos bL}\cos(bx)e^{-i\omega t}, \\ u &= \frac{i\omega\eta_L}{bh\cos(bL)}\sin(bx)e^{-i\omega t} \end{aligned} \tag{29}$$

with $b^2=\dfrac{(\omega+iR')\omega}{gh}$, when the elevation $\eta=\eta_L e^{-i\omega t}$ is prescribed at $x=L$, and $u=0$ at the closed end $x=0$. Tide amplification conditions can e.g. be studied by eq. (29) to a first approximation. More general solutions for tides in channels with variable depth and varying sections are given by [45 l, 80 P 3].

A solution for the independent tide corresponding to the solution in eq. (29) for the co-oscillating tide is given by

$$\begin{aligned} \eta &= \frac{1}{b}\frac{\partial\bar{\eta}}{\partial x}(\sin(bx)-\tan(bL)\cos(bx))e^{-i\omega t}, \\ u &= \frac{i\omega}{hb^2}\frac{\partial\bar{\eta}}{\partial x}(1-\cos(bx)-\tan(bL)\sin(bx))e^{-i\omega t} \end{aligned} \tag{30}$$

with $u(0,t)=0=\eta(L,t)$ and $\partial\bar{\eta}/\partial x$ regarded as constant along the channel. The magnitude of $(\partial\bar{\eta}/\partial x)/b$ being less than $0.002\,h^{1/2}$ m (h given in [m]), indicates the independent tide to be negligible as compared with the co-oscillating tide.

Prescribing the normal component of the current velocity at the open side of an arbitrarily shaped rectangular basin of constant depth, the co-oscillating tide in the interior can be obtained as the solution to eq. (29) by superimposing an infinite number of Kelvin and Poincaré type waves (see sect. 6.1) in such a way that the impermeability condition at the closed boundaries is fulfilled and that the solution takes the prescribed values at the open boundary. The Kelvin type wave is given by

$$\begin{aligned} \eta' &= ihk\omega^{-1}\begin{Bmatrix}\cosh\\ \sinh\end{Bmatrix}(-if\omega(ghk)^{-1}x-k(y-l)) \\ v' &= \begin{Bmatrix}\sinh\\ \cosh\end{Bmatrix}(-if\omega(ghk)^{-1}x-k(y-l)) \\ u' &= 0 \end{aligned} \tag{31}$$

with $k^2=-\omega(\omega+iR')(gh)^{-1}$.

The notations of Fig. 19 are used for the geometry of the basin. The Poincaré type waves are given by

$$\begin{aligned} \eta' &= (R'-i\omega)g^{-1}(A_n e^{k_n(y-l)}+B_n e^{-k_n(y+l)})k_n^{-1}\cos\left(\frac{n\pi}{2a}(x+a)\right) \\ &\quad -fg^{-1}(A_n e^{k_n(y-l)}-B_n e^{-k_n(y+l)})\frac{2a}{n\pi}\sin\left(\frac{n\pi}{2a}(x+a)\right), \qquad n=1,2,\dots \end{aligned} \tag{32}$$

with

$$k_n^2=\left(\frac{n\pi}{2a}\right)^2+k^2+\frac{\omega f^2}{gh(\omega+iR')}$$

and k^2 as in eq. (31).

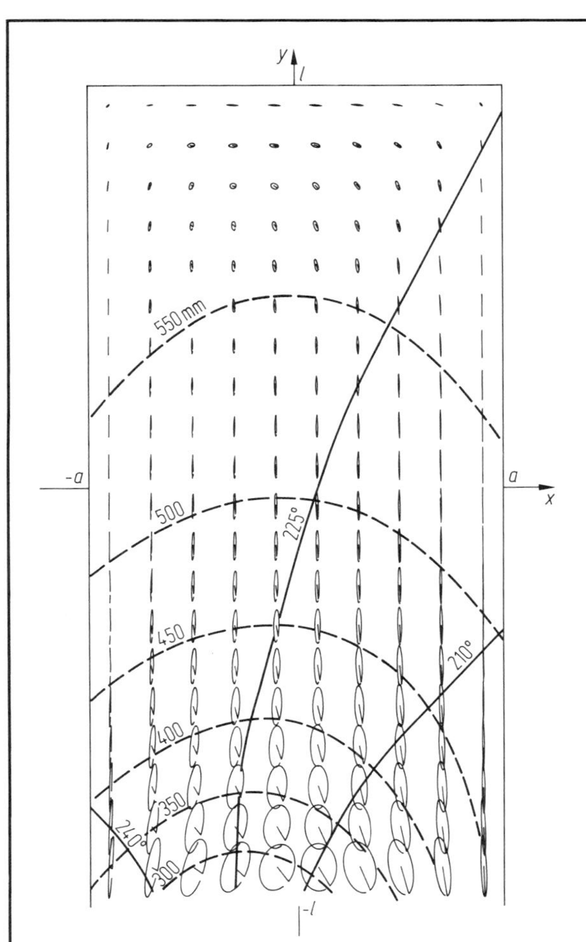

Fig. 19. Analytical solution for co-oscillating diurnal tide in a semi-enclosed rectangular basin with $v(x, -l)$ prescribed and $R' = 10^{-5}\,\mathrm{s}^{-1}$, $\omega = -0.714 \cdot 10^{-4}\,\mathrm{s}^{-1}$, $f = 1.031 \cdot 10^{-4}\,\mathrm{s}^{-1}$, $a = 100\,\mathrm{km}$, $l = 200\,\mathrm{km}$, $h = 100\,\mathrm{m}$. Cotidal lines (solid) with phases in degrees, corange lines (broken) with amplitudes in mm. Current velocity ellipses with current vector at reference time $t = 0$; magnitudes of these vectors given at the two central positions at the open end, are $10.6\,\mathrm{cm\,s^{-1}}$ and $12.9\,\mathrm{cm\,s^{-1}}$ (from left to right). From [70 R].

u' and v' are obtained by applying eq. (28), both components are different from zero in the interior, generally, while $u'(\pm a, y) = 0$.

When prescribing $v'(x, -l)$ (open boundary) the determination of the tidal motion by superimposing waves given by eqs. (31) and (32) reduces to solving an infinite system of linear algebraic equations for the coefficients A_n, B_n. [70 R] presents various solutions for semi-diurnal and diurnal tides, among them the one shown in Fig. 19 for a diurnal tide. The corresponding classical solution of [20 T] for the M_2-tide bases on the frictionless equations and using basin dimensions comparable with those of the North Sea yields a left-hand rotating amphidromic system when prescribing a reflected Kelvin wave at the open boundary. The central amphidrome in the North Sea (see Fig. 18) points at such a reflected Kelvin wave playing a role.

Simple solutions of Kelvin and Poincaré type are also appropriate to give an idea of tidal wave phenomena being affected by the topography, e.g. the transition from the open ocean to the continental shelf. So for a step shelf (see Fig. 20) – other than flat bottom shelves have also been treated (see [80 M]) – parallel to a straight coast, parallel-to-shore progressive wave solutions to the eqs. (26) without the friction terms considered can be obtained:

$$\eta_{1/2} = (\cos(\alpha_{1/2}x) + b_{1/2}\sin(\alpha_{1/2}x))\cos(\beta y - \omega t)$$

$$\begin{pmatrix} u_{1/2} \\ v_{1/2} \end{pmatrix} = \frac{1}{\omega^2 - 1} \begin{pmatrix} -\beta + \omega \dfrac{\partial}{\partial x} \\ \beta\omega - \dfrac{\partial}{\partial x} \end{pmatrix} \eta'_{1/2} \begin{pmatrix} \sin(\beta y - \omega t) \\ \cos(\beta y - \omega t) \end{pmatrix} \tag{33}$$

where quantities with index 1 are valid for $-L \leqq x \leqq 0$ (on the shelf) and those with index 2 for $0 \leqq x$ (off the shelf). The quantities in eqs. (26) have been made dimensionless, depths and elevation by h_2, velocities by $C = (gh_2)^{1/2}$, time and frequency by f^{-1} and f, respectively, distances by Cf^{-1} and wavenumbers by $C^{-1}f$. Eq. (33) and the following relations contain these dimensionless quantities.

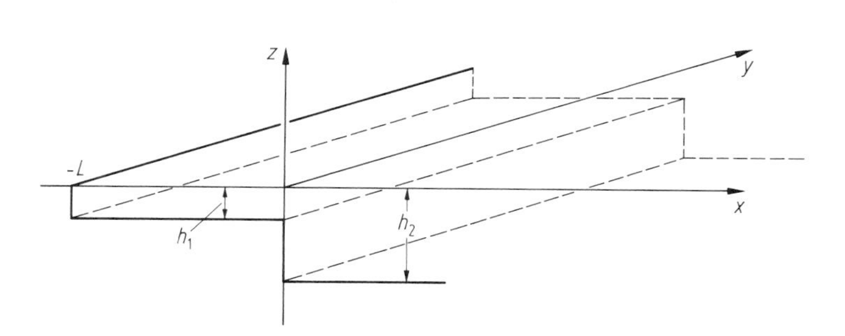

Fig. 20. Normal-to-shore section of a flat-bottom step shelf with infinitely extended coast at $x=-L$. L shelf width, h_1 water depth on the shelf, h_2 water depth off the shelf.

Imposing the conditions of continuity of elevation and normal-to-shore transport at the step, a continuum of Poincaré type waves results being trigonometric on and off the shelf and fulfilling $\omega^2-1-\beta^2>0$. Discrete edge waves result for $\omega^2-1<\beta^2<(\omega^2-1)h_1^{-1}$, they are exponential off the shelf and trigonometric on the shelf. Finally, $h_1\beta^2>\omega^2-1$ defines discrete shelf and edge waves being completely exponential.

The dispersion relations for the discrete waves constitute from

$$\frac{\tan(\alpha_1 L)}{\alpha_1}=\frac{\omega(\beta+i\omega\alpha_2)}{(\omega^2-1)(\omega^2-h_1\beta^2)-\beta(\beta+i\alpha_2)} \tag{34}$$

where $i\alpha_{1/2}$ are positive real, and from

$$h_1(\alpha_1^2+\beta^2)=\omega^2-1=\alpha_2^2+\beta^2, \tag{35}$$

relation (35) holding for all of the above mentioned waves of form (33). The dispersion relations thus determined are displayed in [70 M], corresponding ones for a sloping shelf are displayed in [80 M] (see also sect. 6.1).

In [70 M] shallow water and deep-sea tidal measurements at the Californian coast are fitted by introducing a step shelf approximation and by superimposing free waves of the above types and a forced wave. A free Kelvin like edge wave travelling northwards along the coast and a free Poincaré like wave travelling southwards along the coast dominate this fit of the M_2-tide regime there, which is characterized by a left-hand rotating amphidrome (see Figs. 9, 10, and 11).

These two free waves suffice to approximate the available measurements rather well:

$$\eta(x,y,t)=A_K\eta_K(x)\cos(\beta_K y-\omega t+\psi_K)+A_P\eta_P(x)\cos(\beta_P y-\omega t+\psi_P), \tag{36}$$

their amplitudes at the coast amounting to 62.2 cm (Kelvin wave) and 18.6 cm (Poincaré wave). Normal-to-shore profiles characterizing the free waves and their superposition are given in Fig. 21.

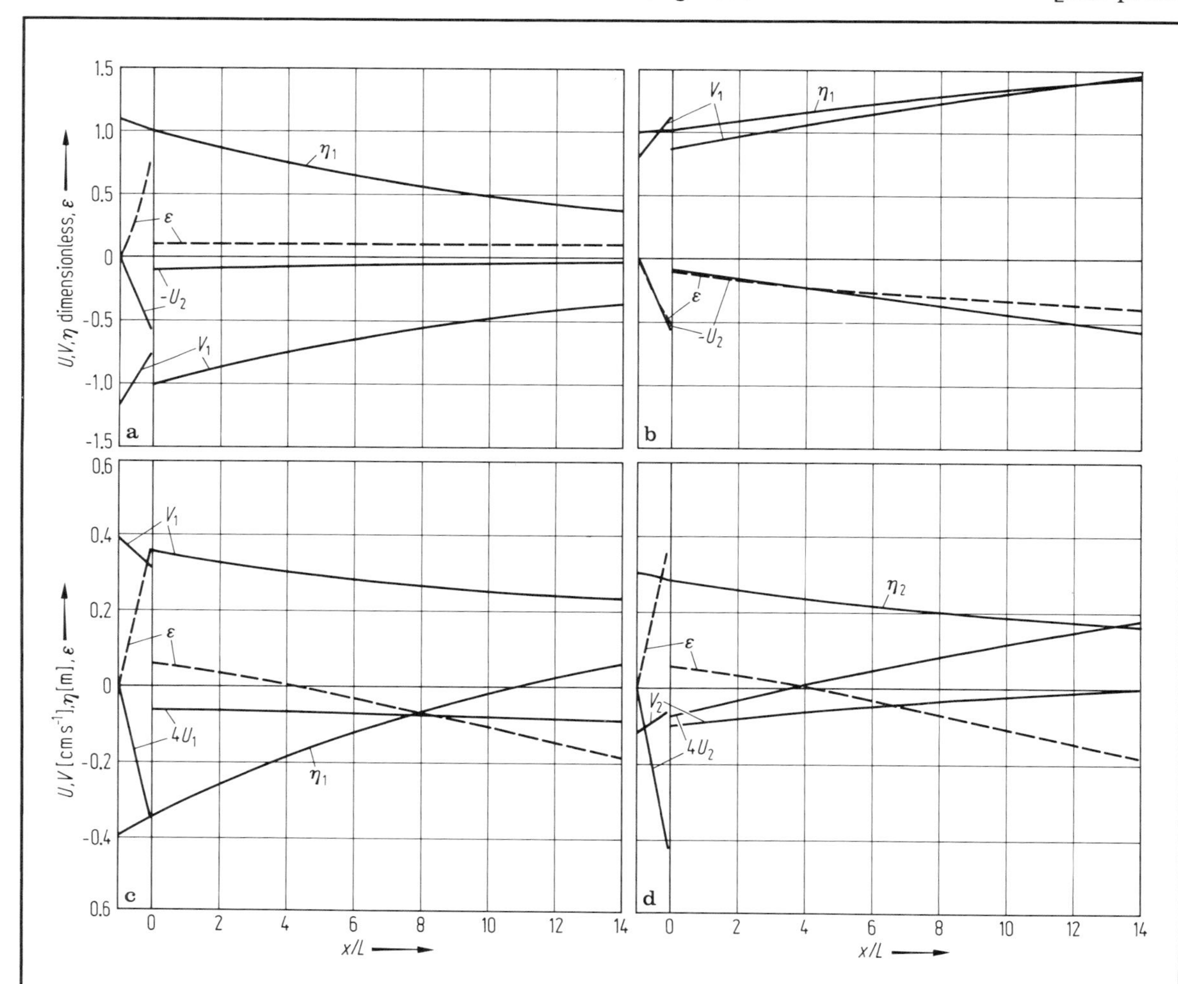

Fig. 21. Normal-to-shore dependence of surface elevation and of current velocity components and their ellipticity. For shelf profile, see Fig. 20. Dimensional quantities: $h_1=600\,\text{m}$, $h_2=3600\,\text{m}$, $f=0.7291\times10^{-4}\,\text{s}^{-1}$, $\phi=29.91°$, $L=155\,\text{km}$. Dimensionless quantities (see text following eq. (33)): $\omega=1.932$, $h_1=1/6$, $\beta_\text{K}=-2.044$, $\alpha_{1\text{K}}=3.496$, $\text{i}\alpha_{2\text{K}}=1.201$, $\beta_\text{P}=1.55$, $\alpha_{1\text{P}}=3.741$, $\alpha_{2\text{P}}=0.575$, $L=0.06$.

a) Kelvin type wave: $\eta_\text{K}(x)$, $u_\text{K}(x)$, $v_\text{K}(x)$, dimensionless, at $y=0$ for $t=0$, $t=3/4\text{T}$, $t=0$, respectively, and ellipticity ε.

b) Poincaré type wave: $\eta_\text{P}(x)$, $u_\text{P}(x)$, $v_\text{P}(x)$, dimensionless, at $y=0$ for $t=0$, $t=3/4\text{T}$, $t=0$, respectively, and ellipticity ε.

c) Superposition of Kelvin and Poincaré type waves with $A_\text{K}\exp(\text{i}\psi_\text{K})=-0.5197\,\text{m}+\text{i}0.2332\,\text{m}$, $A_\text{P}\exp(\text{i}\psi_\text{P})=0.1773\,\text{m}+\text{i}0.0549\,\text{m}$:

$$\eta_1=A_\text{K}\eta_\text{K}(x)\cos\psi_\text{K}+A_\text{P}\eta_\text{P}(x)\cos\psi_\text{P}, u_1, v_1,$$

i.e. η, u, v at $y=0$ for $t=0$, and ellipticity ε.

d) Superposition as in c):

$$\eta_2=A_\text{K}\eta_\text{K}(x)\sin\psi_\text{K}+A_\text{P}\eta_\text{P}(x)\sin\psi_\text{P}, u_2, v_2,$$

i.e. η, u, v at $y=0$ for $t=\text{T}/4$, and ellipticity ε.

6.4.6 Vertical distribution of tidal currents

Atlases of tidal currents, e.g. [68 s, 83 m], show horizontal current vector distributions where these vectors are representative for the upper $(5 \cdots 10)$ m. Tidal models in general refer to the vertically integrated dynamical equations, thus yielding vertically averaged tidal currents, however, meanwhile also results obtained by three-dimensional tidal models, e.g. [80 B 1, 80 D], are available. In fact, due to slowing down of the layer of liquid adjacent to the bottom by adhesion, the velocity of the tidal current in this boundary layer increases from zero at the bottom to its full value in the upper sea layer where a well mixed sea is assumed. The thickness of the boundary layer is of order of 20 m and can be estimated by the parameter $\delta = (2\nu/\omega)^{1/2}$ with viscosity ν. This layer is turbulent and the tidal field of motion is governed by eqs. (26) with the additional turbulent eddy viscosity terms

$$\frac{\partial}{\partial z}\left(A_z \frac{\partial u}{\partial z}\right), \quad \frac{\partial}{\partial z}\left(A_z \frac{\partial v}{\partial z}\right)$$

considered on the right hand sides of the first and the second equation, respectively, and the bottom friction terms omitted. The coefficient A_z decreases linearly to zero towards the bottom within the several meters thick benthic boundary layer (see [62 B]), where the current velocity profile is logarithmic:

$$|u| = A_z^{(1)} \chi^{-2} \ln \frac{z}{z_0}$$

$$\chi = 0.4$$

$$2\sqrt{\frac{\omega h}{A_z^{(1)}}} = \alpha = \text{const.}$$

$$A_z(z) = \begin{cases} A_z^{(1)} z, & z \leqq h \\ A_z^{(1)} h, & z \geqq h \end{cases}$$

z increasing vertically upwards (see sect. 5.5). A_z amounts to some hundreds of $\text{cm}^2\,\text{s}^{-1}$ and decreases again in the upper strata of the boundary layer (see [71 J, 69 B]). On the basis of the modified eqs. (26), using $\alpha = 0.6$, $z_0 = 0.2$ cm, [68 k 1] obtains the vertical current distribution when the level gradients are prescribed (see also the solutions of [26 S] and [70 K] for A_z independent of z, its value having been prescribed and determined, respectively). Fig. 22 shows characteristic profiles of the harmonic velocity constants for the case of the depth exceeding the thickness of the turbulent boundary layer. Fig. 23 shows current ellipses, theoretical and observed, close to the surface and close to the bottom. As predicted by theory, the elements of the current velocity ellipses are observed to change differently with depth in the various tidal regimes and locations. However, as a rather general feature it is found that tidal streams near the bottom reach their maximum values before the surface tidal streams. The M_2-tide current elements in Table 8, as obtained from measurements by [77 P], reflect the indicated properties as well as the tidal current ellipses in Fig. 24 where the depth is comparable with the thickness of the boundary layer (see also [52 H 1]).

Where, e.g. in estuaries, horizontal density gradients appear, the vertical current velocity distribution is modified. Highly saline sea water penetrates below less saline water upstream, thus the flood stream will increase close to the bottom as compared with the surface flood stream, whereas the ebb stream will increase at the surface as compared with the bottom ebb stream. This situation is demonstrated by the observations shown in Fig. 25. At high current velocities due to turbulent mixing ([54 D]) the stratification breaks down, in the example of Fig. 25 this occurs after the flood stream maximum.

Where the stratification of the sea is well developed, currents of tidal frequency strongly depending on the vertical coordinate and having a much larger velocity than the few $\text{cm}\,\text{s}^{-1}$ associated with the barotropic tide may appear due to internal tides. These are also associated with large vertical amplitudes of tens of meters in the interior of the ocean, but the internal tides are not a phenomenon of basin-wide scales. A comprehensive review of the problem of internal tides has been given by [75 W], [77 S 1] reviews experiments in the North Atlantic in view of the tidal energy budget.

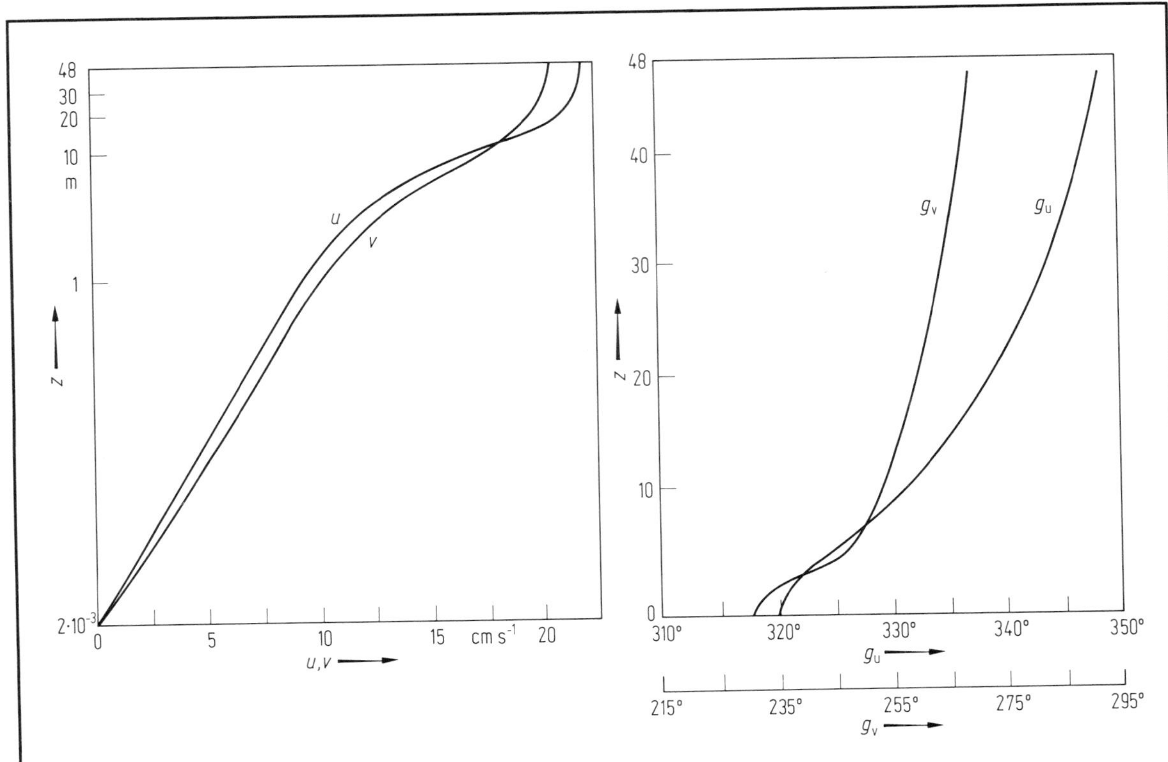

Fig. 22. Computed vertical distribution of harmonic velocity constants of the tidal current at sea. The profiles are computed with tidal sea-surface gradients (M_2-tide, North Sea) prescribed and depth exceeding boundary layer thickness. g_u and g_v denote the phases of u- and v-current velocity component, respectively. From [68 k 1].

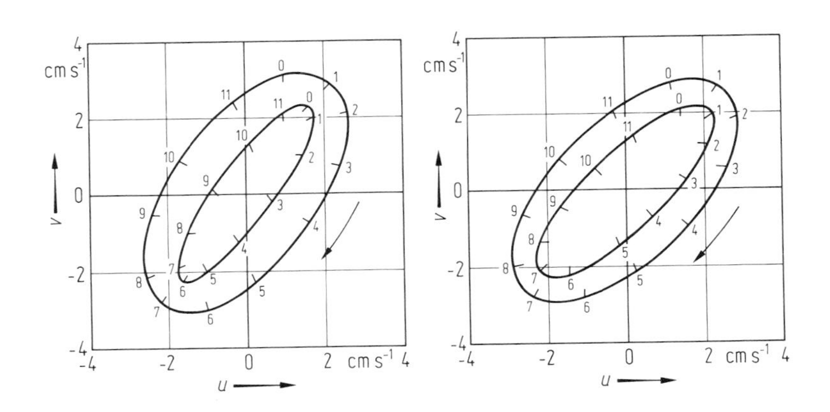

Fig. 23. Comparison of computed (right) and observed (left) semi-diurnal tidal current ellipses. Observation data are given for the layer 0–5 m and 25–30 m below surface, respectively, computational results for 0 m and 25 m, respectively. Current velocities are given in [cm s^{-1}]. From [68 k 1]. Integers represent full hours.

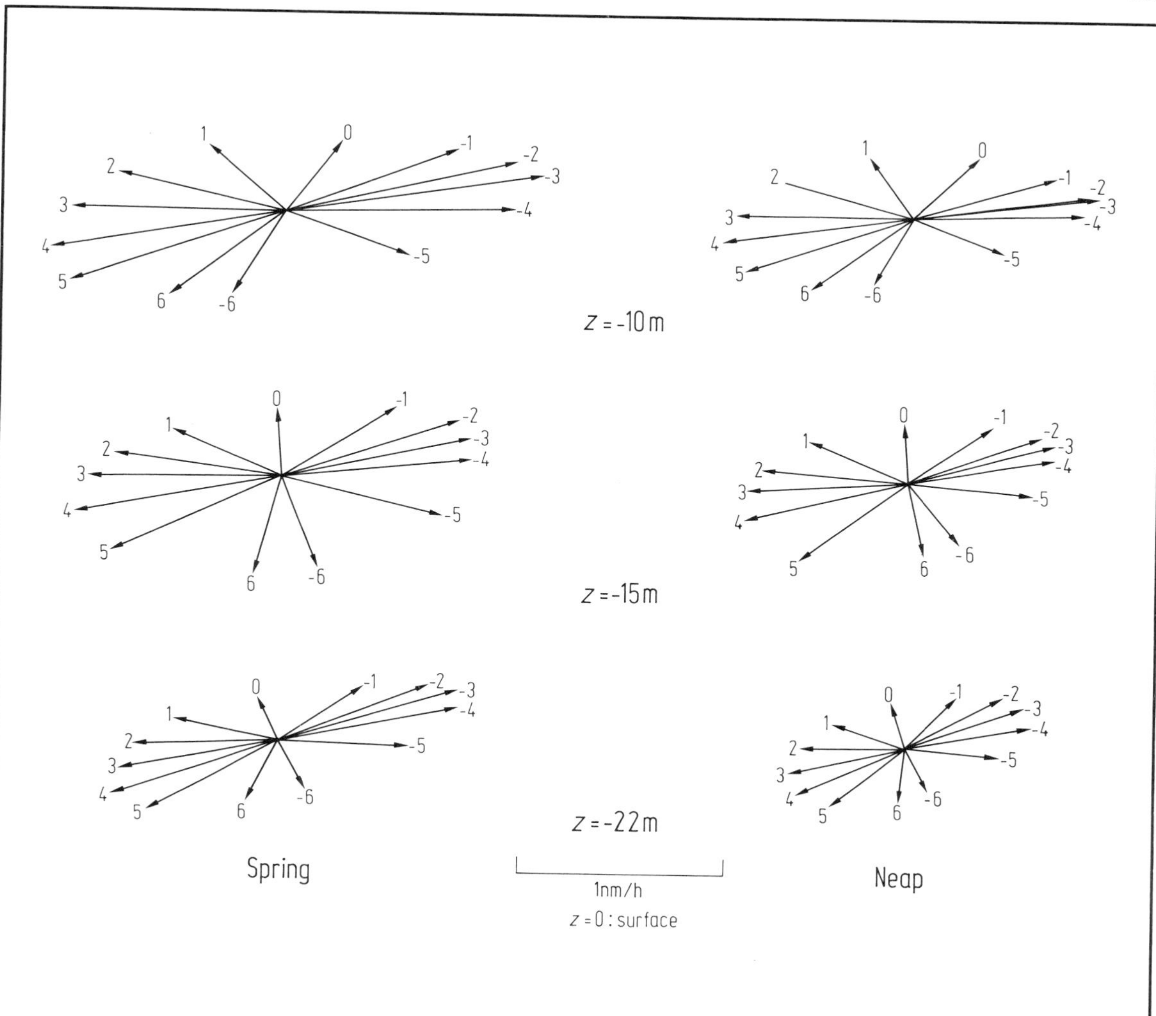

Fig. 24. Tidal current velocity ellipses in three depth levels (10 m, 15 m, and 22 m close to bottom) at 53°48′ N, 6°20′ E in the German Bight at spring tide (left) and at neap tide (right). Integer numbers at the current vectors denote the time difference in hours as against high-water in Helgoland. From [83 m].

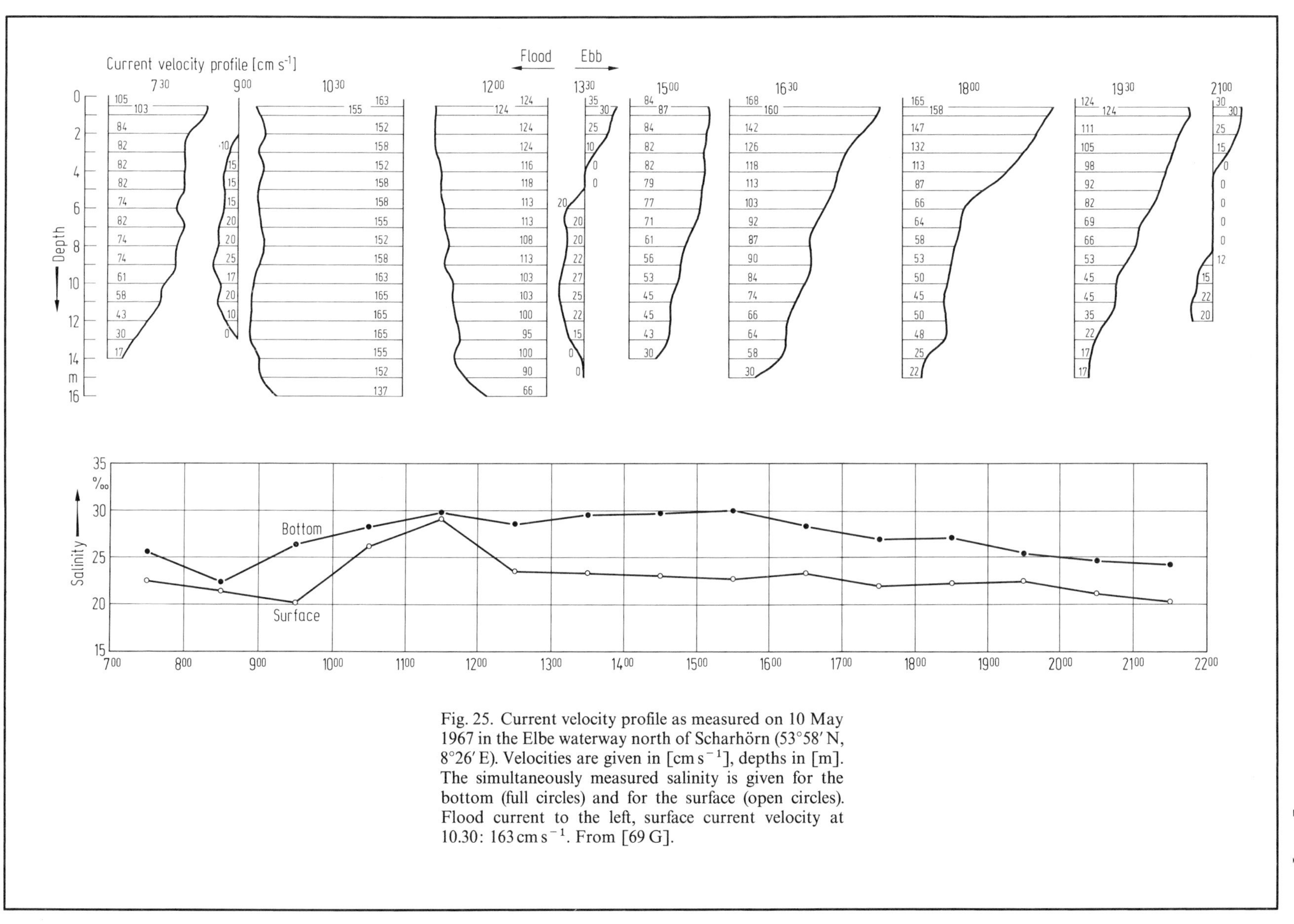

Fig. 25. Current velocity profile as measured on 10 May 1967 in the Elbe waterway north of Scharhörn (53°58′N, 8°26′E). Velocities are given in [cm s^{-1}], depths in [m]. The simultaneously measured salinity is given for the bottom (full circles) and for the surface (open circles). Flood current to the left, surface current velocity at 10.30: 163 cm s^{-1}. From [69 G].

Table 8. Main features of vertical structure of M_2-tide currents in the Celtic Sea and English Channel (from [77 P]). The current phase is with respect to the major axis of the tidal ellipse at the time when the bottom current attains its maximum rate. a_2: M_2 semi-major axis, ε: M_2 ellipticity, u_*: friction velocity of the semi-major axis of the M_2-tide, $\varrho u_*^2 = \tau_B = \varrho c u_1 |u_1|$, where u_1 is the current at 1 m above bottom and τ_B the bottom stress.

Position	Water depth	Height of meter above bottom	% of total kinetic energy in the M_2 constituent	a_2	ε	Orientation of M_2 major axis	Phase of current relative to current at 1 m	u_*	Maximum surface streams at mean springs
	m	m		$cm\,s^{-1}$		°True	degree	$cm\,s^{-1}$	knots
50°02′N, 4°22′W	75	36	71	34.3	−0.19	082	−004		
		3		23.2	−0.06	076	−001		
		2		22.3	−0.05	076	−000	1.2	1.1
		1		20.1	−0.05	076	000		
49°27′N, 4°42′W	90	36	83	55.3	−0.17	076	−005		
		3		39.8	−0.13	073	000		
		2		38.0	−0.13	073	000	1.6	1.6
		1		35.6	−0.11	072	000		
51°10′N, 8°00′W	104	34	73	27.9	−0.21	066	−007		
		3		17.7	0.00	067	−001		
		2		17.2	0.00	068	−001	1.0	0.7
		1		15.1	0.00	068	000		

6.4.7 References

1. Atlases and books

45 l Lamb, H.: Hydrodynamics, New York: Dover **1945**.
61 d Defant, A.: Physical Oceanography, New York: Pergamon Press **1961**.
63 d Deutsches Hydrographisches Institut: Atlas der Gezeitenströme für die Nordsee, den Kanal und die Britischen Gewässer, 2345, Hamburg **1963**.
64 d Dronkers, J.J.: Tidal Computations in Rivers and Coastal Waters, Amsterdam: North-Holland **1964**.
65 u U.S. Naval Oceanographic Office: Oceanographic atlas of the North Atlantic Ocean, Section 1, Tides and Currents, No. 700, **1965**.
68 k 1 Kagan, B.A.: Hydrodynamic Models of Tidal Motions at Sea, Leningrad, Gidrometeorologicheskoye Izdatel'stvo **1968** (Technical Translation DMAAC-TC-2028, St. Louis, 1974).
68 k 2 Keehn, P.A.: Bibliography on Marine Atlases, Washington D.C.: American Met. Soc. **1968**.
68 s Sager, G.: Atlas der Elemente des Tidenhubes und der Gezeitenströme für die Nordsee, den Kanal und die Irische See, Rostock: Seehydrographischer Dienst DDR **1968**.
72 g Godin, G.: The Analysis of Tides, Liverpool: Liverpool Univ. Press **1972**.
78 b Brosche, P., Sündermann, J. (eds.): Tidal Friction and the Earth's Rotation, Berlin, Heidelberg, New York: Springer **1978**.
78 l LeBlond, P.H., Mysak, L.A.: Waves in the Ocean, Amsterdam, Oxford, New York: Elsevier Scientific Publ. Comp. **1978**.
78 m Melchior, P.: The Tides of the Planet Earth, Oxford: Pergamon Press **1978**.
78 n Nihoul, J.C.J. (ed.): Hydrodynamics of Estuaries and Fjords, Amsterdam, Oxford, New York: Elsevier Scientific Publ. Comp. **1978**.
78 s Stommel, H., Fieux, M.: Oceanographic Atlases: a guide to their geographic coverage and contents, Woods Hole: Woods Hole Press **1978**.
80 r Ramming, H.-G., Kowalik, Z.: Numerical Modelling of Marine Hydrodynamics, Amsterdam, Oxford, New York: Elsevier Scientific Publ. Comp. **1980**.
82 b Brosche, P., Sündermann, J.: Tidal Friction and the Earth's Rotation II, Berlin, Heidelberg, New York: Springer **1982**.
83 m Mittelstaedt, E., Lange, W., Brockmann, C., Soetje, K.C.: Die Strömungen in der Deutschen Bucht, Hamburg: Deutsches Hydrographisches Institut (2347) **1983**.

2. Special references

20 T Taylor, G.I.: Proc. London Math. Soc. **20** (2) (1920) 144.
26 S Sverdrup, H.U.: Geophys. Publ. **4** (1926) 1.
44 D Dietrich, G.: Z. Ges. Erdkunde Berlin **3/4** (1944) 69.
52 H 1 Hansen, W.: Landoldt-Börnstein **3** (1952) 516.
52 H 2 Hansen, W.: Dtsch. Hydrogr. Z. Erg.-H. (A) **1** (1952) 3.
52 H 3 Horn, W.: Landoldt-Börnstein **3** (1952) 504.
52 V Villain, C.: Ann. Hydrogr. Paris **3** (1952) 269.
54 D Dietrich, G.: Arch. Meteorol. Geophys. Bioklimatol. Ser. A **7** (1954) 391.
56 B Bowden, K.F., Fairbairn, L.A.: Proc. R. Soc. London Ser. A **237** (1956) 422.
57 D Doodson, A.T.: Int. Hydrogr. Rev. **34** (1957) 85.
58 D Doodson, A.T.: Adv. Geophys. **5** (1958) 117.
60 H Horn, W.: Int. Hydrogr. Rev. **37** (2) (1960) 65.
62 B Bowden, K.F.: J. Geophys. Res. **67** (1962) 3181.
67 T Trepka, L. v.: Mitt. Inst. Meereskd. Univ. Hamburg **9** (1967).
68 L Longuet-Higgins, M.S.: Philos. Trans. R. Soc. London Ser. A **262** (1968) 511.
69 B Bowden, K.F.: Proc. Challenger Soc. **IV** (1969) 25.
69 D Dronkers, J.J.: J. Hydraul. Div., Am. Soc. Civ. Eng. **95** (1969) 29.
69 G Göhren, H.: Hamburger Küstenforschung **6** (1969).
70 K Kraav, V.K.: Oceanology **10** (1969) 195.
70 L Longuet-Higgins, M.S., Pond, G.S.: Philos. Trans. R. Soc. London Ser. A **266** (1970) 193

70 M Munk, W.H., Snodgrass, F., Wimbush, M.: Geophys. Fluid Dyn. **1** (1970) 161.
70 R Röber, K.: Mitt. Inst. Meereskd. Univ. Hamburg **16** (1970).
71 C Cartwright, D.E., Tayler, R.J.: Geophys. J. R. Astron. Soc. **23** (1971) 45.
71 F Filloux, J.H.: Deep-Sea Res. **18** (1971) 275.
71 G 1 Gallagher, B.S., Munk, W.H.: Tellus **23** (1971) 346.
71 G 2 Garrett, C.J.R., Munk, W.H.: Deep-Sea Res. **18** (1971) 493.
71 J Johns, B., Dyke, P.: Geophys. J. R. Astron. Soc. **23** (1971) 287.
71 P Platzman, G.W., in: Mathematical Problems in the Geophysical Sciences (Reid, ed.); Am. Math. Soc. **14** (1971) 239.
72 I Irish, J.D., Snodgrass, F.E.: Am. Geophys. Union Antarctic Res. Ser. **19** (1972) 101.
73 W Webb, D.J.: Deep-Sea Res. **20** (1973) 847.
73 Z Zahel, W.: Pageoph **109** (1973) 1819.
74 M Miles, J.: J. Fluid Mech. **66** (1974) 241.
75 D Dronkers, J.J.: Adv. Hydrosci. **10** (1975) 145.
75 L Luther, D.S., Wunsch, C.: J. Phys. Oceanogr. **5** (1975) 220.
75 P Pearson, C.A.: NOAA (Nat. Oceanic Atmos. Adm.) Tech. Rep. Memo. **17** (1975).
75 S Sager, G.: Beitr. Meereskd., Berlin **36** (1975) 95.
75 W Wunsch, C.: Rev. Geophys. Space Phys. **13** (1975) 167.
75 Z Zetler, B.D., Munk, W.H., Mofjeld, H.O., Brown, W., Dormer, F.: J. Phys. Oceanogr. **5** (1975) 430.
76 F Flather, R.A.: Mem. Soc. R. Sci. Liège Collect. **10** (1976) 141.
77 A 1 Amin, M.: Int. Hydrogr. Rev. Monaco **54** (1) (1977) 87.
77 A 2 An, H.S.: J. Oceanogr. Soc. Japan **33** (2) (1977) 103.
77 E Estes, R.H.: Rep. TR-77-41 Bus. and Technol. Systems Greenbelt, Md. **1977**.
77 G 1 Glen, N.C.: Int. Hydrogr. Rev. Monaco **54** (1) (1977) 73.
77 G 2 Gordeev, R.G., Kagan, B.A., Polyakov, E.V.: J. Phys. Oceanogr. **7** (2) 161.
77 H Hendershott, M.C., in: The Sea, Vol. **6** (Goldberg et al., eds.), New York: Wiley-Interscience **1977**, p. 47.
77 L Lambeck, K.: Philos. Trans. R. Soc. London Ser. A **287** (1977) 545.
77 M 1 Maier-Reimer, E.: Dtsch. Hydrogr. Z. **30** (1977) 69.
77 M 2 McCammon, C., Wunsch, C.: J. Geophys. Res. **82** (37) (1977) 5993.
77 P Pingree, R.D., Griffiths, D.K.: Estuarine Coastal Mar. Sci. **5** (1977) 399.
77 S 1 Schott, F.: Ann. Geophys. **33** (1/2) (1977) 41.
77 S 2 Sündermann, J.: Dtsch. Hydrogr. Z. **30** (1977) 91.
77 Z Zahel, W.: Ann. Geophys. **33** (1/2) (1977) 31.
78 A Accad, Y., Pekeris, C.L.: Philos. Trans. R. Soc. London Ser. A **290** (1978) 235.
78 C Cartwright, D.E.: Int. Hydrogr. Rev. Monaco **55** (1978) 35
78 L Liu, S.K., Leendertse, J.J.: Adv. Hydrosci. **11** (1978) 95.
78 S Schwiderski, E.W.: Naval Surface Weapons Center Dahlgren NSWC/DL TR-3866.
78 Z Zahel, W., in: Tidal Friction and the Earth's Rotation (Brosche, P., Sündermann, J., eds.), Berlin, Heidelberg, New York: Springer **1978**, p. 98.
79 C Cartwright, D.E., Zetler, B.D., Hamon, B.V.: IAPSO Publication Scientifique **30** (1979).
79 G Greenberg, D.A.: Marine Geodesy **2** (2) (1979) 161.
79 S Schwiderski, E.W.: Naval Surface Weapons Center Dahlgren NSWC TR 79-414 (1979).
80 B 1 Backhaus, J.O.: Dtsch. Hydrogr. Z. Erg.-H. (B) **15** (1980).
80 B 2 Berger, R.: Diplomarbeit Inst. Meereskd. Univ. Hamburg, **1980**.
80 C Cartwright, D.E., Edden, A.C., Spencer, R., Vassie, J.M.: Philos. Trans. R. Soc. London Ser. A **298** (1980) 87.
80 D Davies, A.M., Furnes, G.K.: J. Phys. Oceanogr. **10** (1980) 237.
80 F Franco, A.S.: Int. Hydrogr. Rev. Monaco **57** (2) (1980) 139.
80 G Godin, G.: Fish. Mar. Serv., Ms. Rept. Ser. Can. **55** (1980).
80 H Huntley, D.A., in: Elsevier Oceanography Series (Banner, F.T. et al., eds.), **24** B (1980) 301.
80 M Mysak, L.A.: Annu. Rev. Fluid Mech. **12** (1980) 45.
80 P 1 Parke, M.E., Hendershott, M.C.: Marine Geodesy **3** (1/4) (1980) 379.
80 P 2 Pingree, R.D., Griffiths, D.K.: Oceanol. Acta **3** (2) (1980) 221.
80 P 3 Prandle, D., Rahman, M.: J. Phys. Oceanogr. **10** (1980) 1552.
80 S Schwiderski, E.W.: Rev. Geophys. Space Phys. **18** (1980) 243.
80 W Webb, D.J.: Geophys. J. R. Astron. Soc. **61** (1980) 573.
80 Z Zahel, W.: Phys. Earth Planet. Inter. **21** (1980) 202.

81 C Clarke, A., Battisti, D.: Deep-Sea Res. **28A** (7) (1981) 665.
81 G 1 Gaviño, J.H.: Ph.D. thesis Univ. Hamburg **1981**.
81 G 2 Gotlib, V.Y., Kagan, B.A.: Dtsch. Hydrogr. Z. **34** (1981) 273
81 K Kjerve, B.: J. Geophys. Res. **86** (1981) 4243.
81 P Platzman, G.W., Curtis, G.A., Hansen, K.S., Slater, R.D.: J. Phys. Oceanogr. **11** (1981) 579.
81 S Schwiderski, E.W.: Naval Surface Weapons Center Dahlgren NSWC TR 81-122, 142, 144, 218, 220, 222, 224, **1981**.
81 W Webb, D.J.: Aust. J. Mar. Freshwater Res. **32** (1) (1981) 31.
82 G Gotlib, V.Y., Kagan, B.A.: Dtsch. Hydrogr. Z. **35** (1982) 45.
82 S Schwiderski, E.W.: Naval Surface Weapons Center Dahlgren NSWC TR 82-147, 149, 151, **1982**.
83 K Krohn, J.: Ph.D. thesis Univ. Hamburg **1983**.
83 S Schwiderski, E.W.: Marine Geodesy **6** (3–4) (1983) 219.

7 Upwelling regions

7.1 Introduction

The term *upwelling* is used here to describe ascending motions by which water from subsurface layers is brought into the surface layer (Smith [68 S 2]). Upwelling is the result of horizontal divergence in the surface layer and may occur anywhere. But there are areas in the oceans where upwelling is relatively intense over a substantial period of time, e.g. a season or even throughout the year. This causes a typical signal at the surface in terms of horizontal anomalies of the distributions of physical, chemical, and biological properties.

The major upwelling regions being described below are shown in Fig. 1. Basically the regions can be characterized by two types: Open ocean upwelling and coastal upwelling.

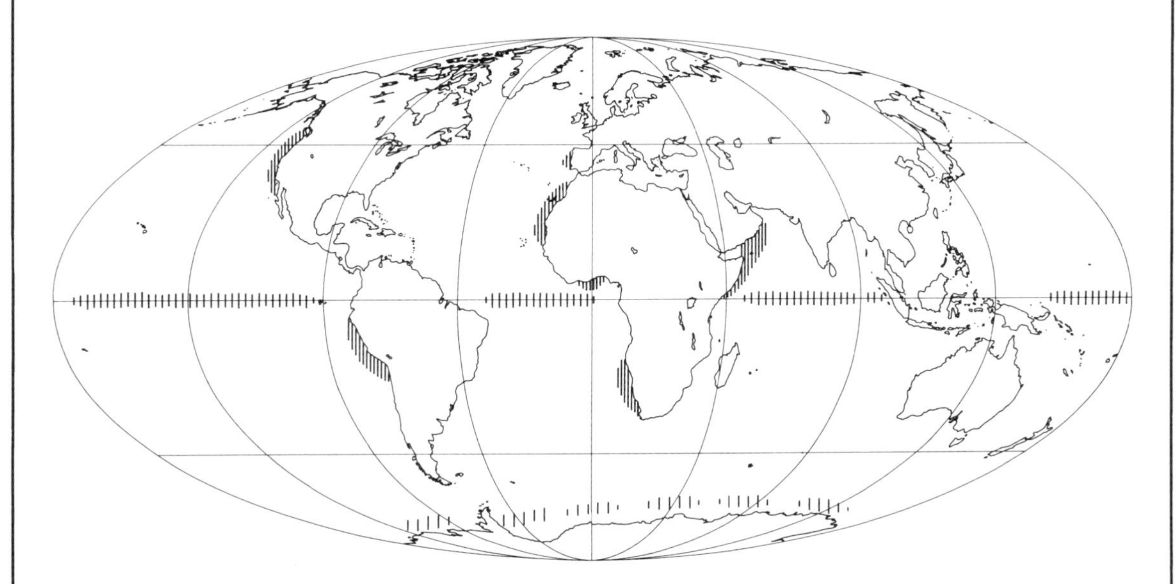

Fig. 1. Upwelling regions (shaded and hatched areas).

7.2 Open ocean upwelling zones

7.2.1 Equatorial upwelling

The divergence of the Ekman flow in the surface layer around the equator is caused by the winds and the varying Coriolis effect across the equator. On average south east trades are prevailing right at the equator over extended distances. Due to the Coriolis force the resulting drift currents are deflected towards the left south of the equator and towards the right north of it producing a divergence of surface currents along the equator. The associated cross-circulation is shown in the schematic representation of Fig. 2. The figure also shows another distinct upwelling area at about 10° N denoting the divergent boundary between the Equatorial Counter Current (ECC) and the North Equatorial Current. In between both upwelling zones the convergence of the trades causes downwelling in the surface layer of the ocean. Corresponding to the seasonal variations of the trades the equatorial divergence and consequently the equatorial upwelling is most intense during summer and fall in the eastern half of the oceans, which also applies for the upwelling along the northern boundary of the ECC.

Mittelstaedt

The Ekman divergence in the surface layer is approximately balanced by a geostrophic convergence in the subsurface layer (Fig. 3). An essential part of the upwelling water comes from the horizontal convergence above the core of the strong eastward flowing equatorial undercurrent. The core raises from about 150 m depths in the western oceans to about 50 m depths in the eastern oceans. The watermasses of the undercurrent represent an important reservoir for the upwelling waters. Because of substantial vertical current shear in the near-surface layer considerable shear-induced turbulence and mixing is to be expected in the equatorial upwelling zone. Occasionally, the undercurrent surfaces which takes place in connection with long equatorial trapped waves, excited by large-scale seasonal wind variations.

Mean ascending motions in the equatorial upwelling zone are estimated to be $1 \cdots 3 \cdot 10^{-5}\,\mathrm{cm\,s^{-1}}$ ($\approx 1 \cdots 3$ cm/day). In actual cases (short-term upwelling events of several days duration) this value will be probably by 2 to 3 magnitudes larger.

Below the core of the undercurrent *downwelling* is assumed which is indicated by the vertical depression of the isolines of various hydrographic parameters such as temperature, salinity, oxygen, etc.

Ekman divergence, however, is only one of the major dynamical principle of equatorial upwelling. By means of models O'Brien et al. [80 O] give a review of the dynamics of the equatorial region and their significance for the variability along the eastern boundaries away from the equator. In particular, they describe theoretically how changes of the equatorial winds in the eastern part of the Pacific excite Kelvin and Rossby waves (see sect. 6.1) causing upwelling and downwelling along the equator.

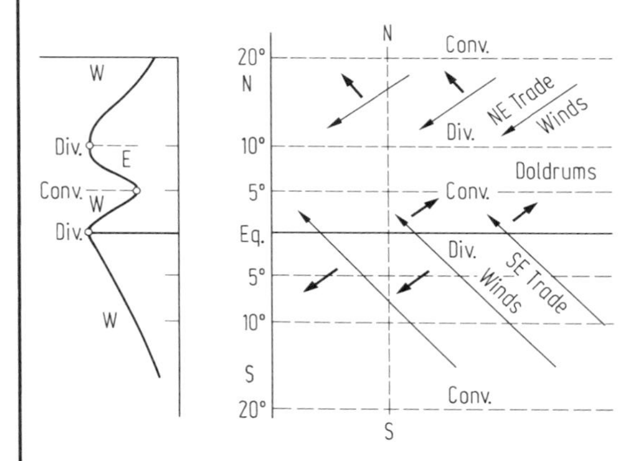

Fig. 2a. Wind distribution in the equatorial region (right) and N–S section (left) showing the idealized topography of the sea surface with indication of currents flowing to the west (W) and to the east (E). Surface water divergences and convergences are indicated by Div. and Conv. (from Neumann and Pierson [66 N]). Wind and current directions are indicated by thin and thick arrows, respectively.

For Fig. 2b, see next page.

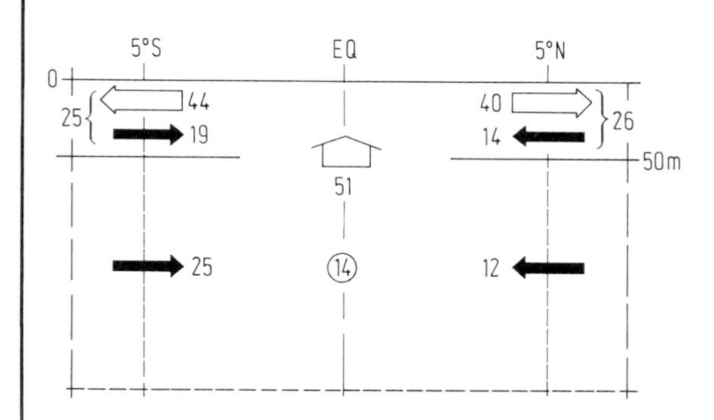

Fig. 3. Water balance in the equatorial upwelling area (from Wyrtki [81 W]). Flows are units of $[10^6\,\mathrm{m^3\,s^{-1}}]$. The open arrows represent surface Ekman flow, the full arrows geostrophic flow. The upward arrow is upwelling and the circled number represents an east-west convergence.

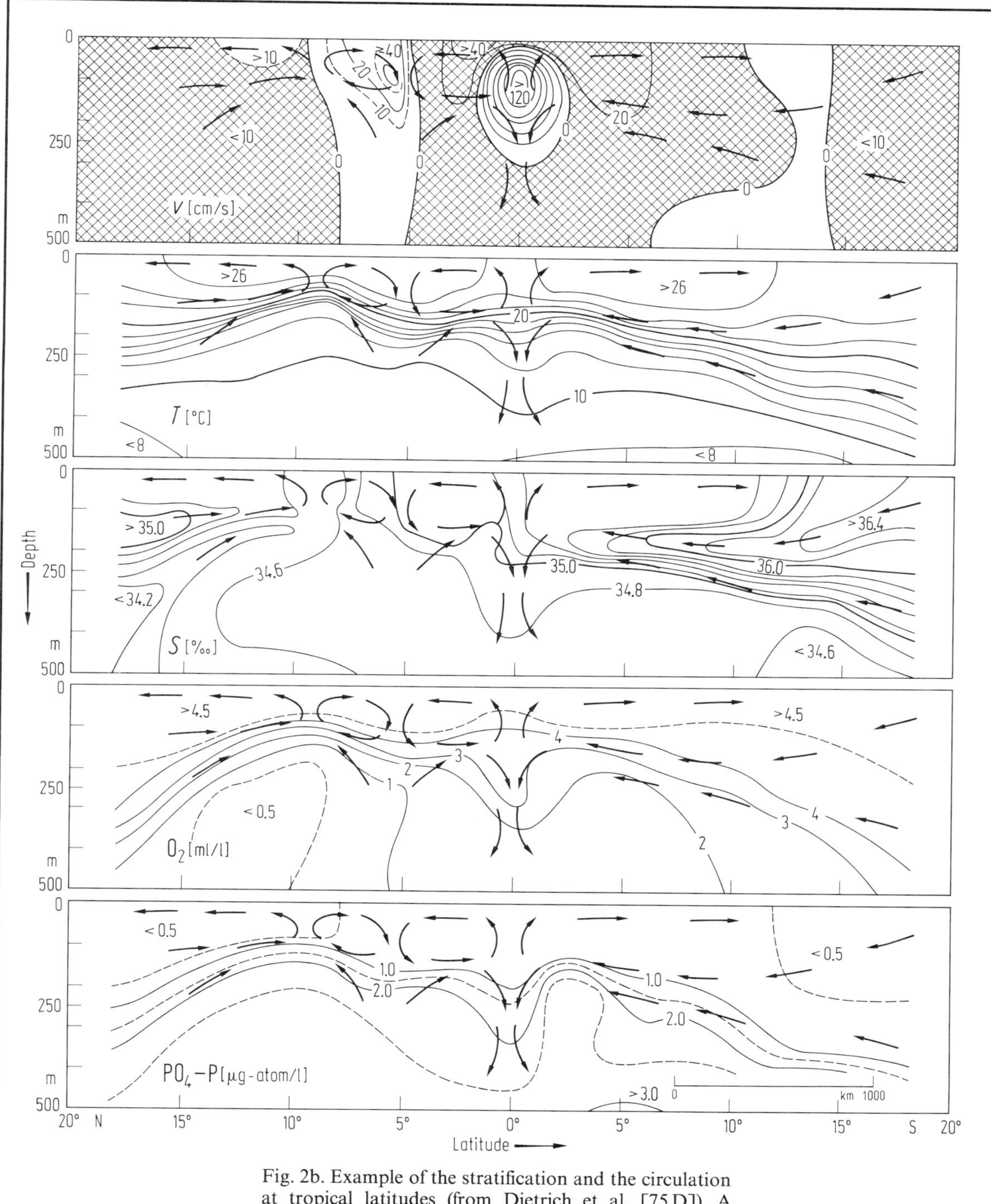

Fig. 2b. Example of the stratification and the circulation at tropical latitudes (from Dietrich et al. [75 D]). A meridional section across the equator in the Pacific at 140° W. Arrows indicate the cross-circulation. From top downwards: Current velocity (to the east, not shaded; to the west, shaded), temperature, salinity, oxygen, and phosphate.

7.2.2 Antarctic upwelling

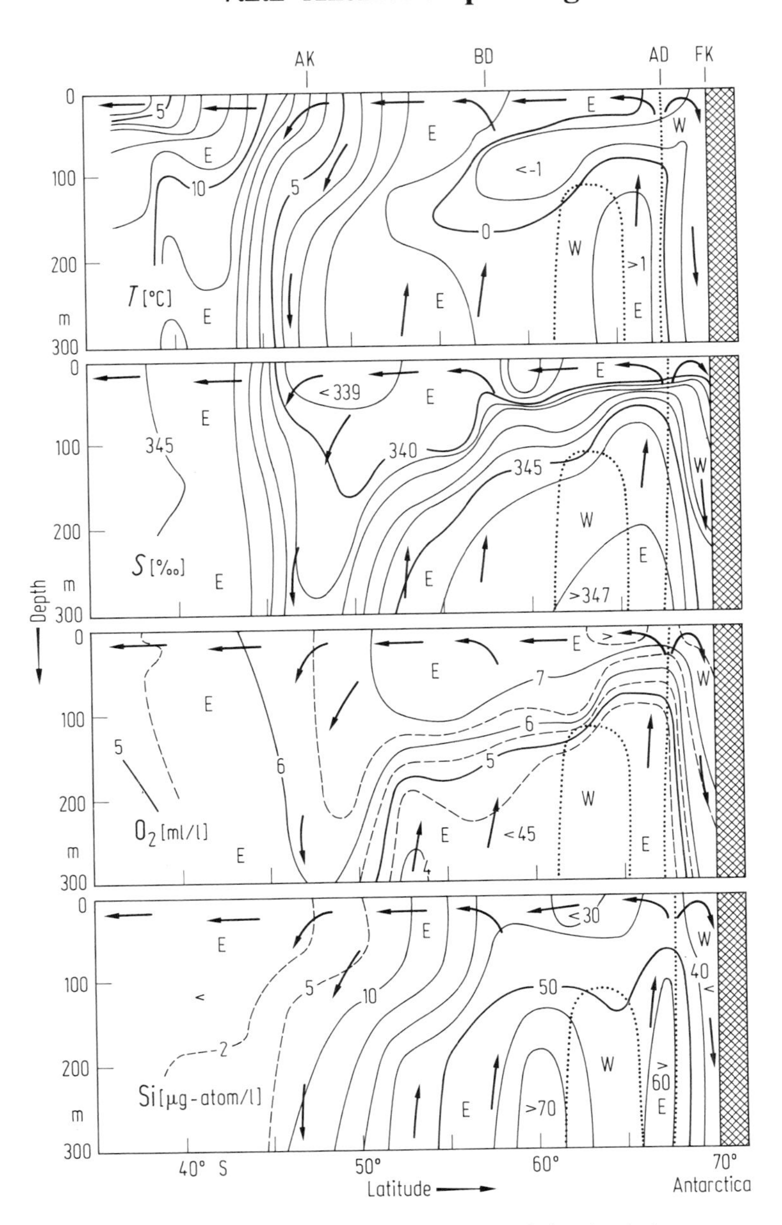

Fig. 4. Example of the stratification and the circulation along a meridional section in the southern Atlantic at 1° E. (from Dietrich et al. [75 D]). Arrows indicate cross-circulation. E: Currents towards east. W: Currents towards west. Dotted lines: Current boundaries. Ak, Fk: Antarctic convergence, coastal convergence, respectively. BD, AD: Bouvet divergence, Antarctic divergence, respectively. Temperature, salinity, oxygen, and silicate.

South of the prevailing west wind belt around Antarctica where at the same time the southern boundary of the mighty eastward flowing Circumpolar Current is found, extends the Antarctic divergence. This divergence occurs between the two oppositely flowing currents: the eastward flowing Circumpolar Current in the north and the westward flowing Antarctic boundary current ("East Wind Drift") in the South along the Antarctic coasts. The effect of the Coriolis force provides the tendency of divergent surface flow between both currents. This causes ascending movements in the divergence zone by which less cool, more saline and nutrient-rich water is brought into the surface layer (Fig. 4). Immediately off the coast, currents tend to be convergent (Antarctic continental front) producing downwelling. The East-Wind-Drift is not completely circumpolar but falls into a number of clockwise gyres in the large embayments of the Wedell, the Ross, and the Bellingshausen Sea, consequently the Antarctic divergence, the region between the Circumpolar Current and East-Wind-Drift, may be broken into segments centered within each gyre (Gordon [71 G]).

The lack of winter data in the ice-covered regions, so far, permits only a one-season study.

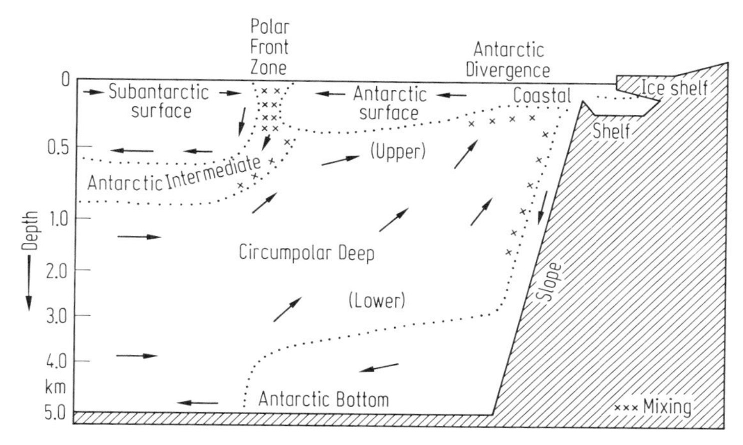

Fig. 5. Schematic representation of antarctic water masses and meridional flow (from Gordon [71 G]).

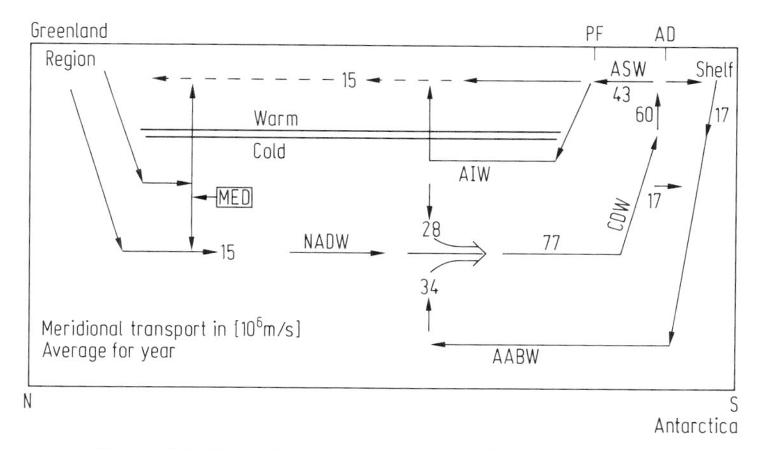

Fig. 6. Meridional transport of the world ocean (from Gordon [71 G]). AABW: Antarctic Bottom Water. AD: Antarctic Divergence. AIW: Antarctic Intermediate Water. ASW: Antarctic Surface Water. CDW: Circumpolar Deep Water. MED: Mediterranean Water. NADW: North Atlantic Deep Water. PF: Polar Front Zone.

The particularity of the upwelling in the Antarctic divergence is the great depth from where parts of the ascending waters come. Apparently the Circumpolar Deep Water between 500 m and 3000 m depth feeds the upwelling (Fig. 5). This water eventually ascends into the surface layer where it converts into fresher, colder, oxygen-rich Antarctic Surface Water. Fig. 6 shows the average meridional transport estimates by Gordon [71 G]. According to the figure $60 \cdot 10^6\,m^3\,s^{-1}$ of Circumpolar Deep Water ascends into the surface layer. Part of the deep water is altered near the coast to a dense cold water mass which sinks and entrains more deep water to produce the Antarctic Bottom Water, which flows northward compensating for the southward flowing Circumpolar Deep Water. Along the polar front part of the Antarctic Surface Water mixes across the front and merges into the slightly less dense Subantarctic Surface Water. Another part of the Antarctic Surface Water sinks along the polar front and entrains some Subantarctic Surface Water and Circumpolar Deep Water to produce the Antarctic Intermediate Water which influences large regions of the oceans in the southern hemisphere. Therefore the conversion of deep water into surface and bottom water and the subsequent transformation of the surface water to deep water farther north represents an oceanic overturning or "breathing" of the deep ocean for which the Antarctic Ocean plays a key role (Gordon [71 G]).

7.3 Coastal upwelling

Along the coasts the signal of upwelling is stronger than in the open ocean upwelling regions. Sea surface temperature in coastal upwelling areas may be 6···10 °C lower than in the open ocean at the same latitudes. The density of the upwelling water over the shelf is remarkably high compared to the density of the surface layer offshore. There are distinct horizontal gradients in the surface layer of physical (e.g. temperature, salinity, density, water color), chemical (e.g. oxygen, nutrients), and biological (e.g. chlorophyll, plankton) parameters. The ascending cool, nutrient-rich and oxygen-poor subsurface waters fertilize the light-flooded (euphotic) surface layer over the upwelling shelf.

The pronounced anomalies and the strong horizontal gradients in those areas indicate intensive upwelling but at the same time downwelling. In fact, upwelling velocities ($10^{-1} \cdots 10^{-3}\,cm\,s^{-1}$) are larger along the coasts than in the open oceans. This is caused by the "guiding effect" of the shelf bottom and the coast acting as a sloped wall.

Theoretically the dynamics of coastal upwelling were described already by Ekamn [05 E]:

The wind-induced equatorwards flowing coastal currents along the western coasts of the continents tend to deviate towards offshore in the surface layer due to the Coriolis force. Along the eastern coasts of the continents coastal currents have to be directed poleward in order to produce upwelling. In those cases the winds blow the light surface waters away from the coast generating a (one-sided) divergence along the coast. For continuity reasons the mass deficit in the surface layer at the coast must be compensated partly by ascending subsurface waters. (Fig. 7). The figures explain Ekman's theory for a homogeneous, nonstratified ocean. According to the theory the onshore-offshore transports are in balance and confined to the upper and lower Ekman layer. In between the subsurface (deep) current flows parallel to the coast.

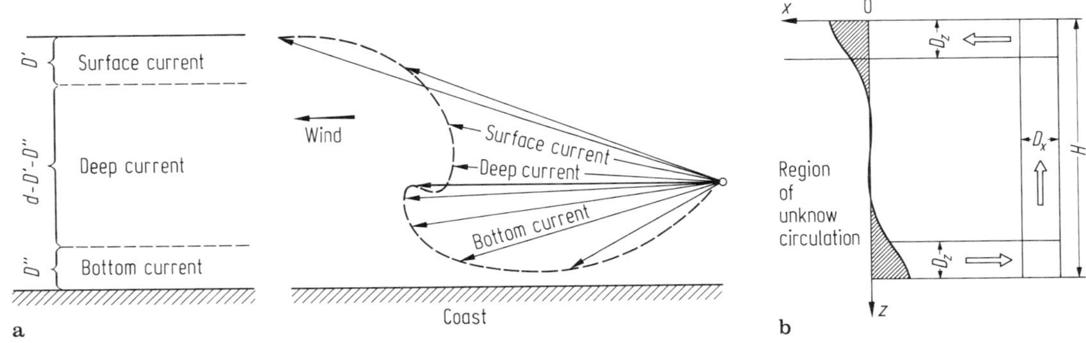

Fig. 7a. Ekman elementary current system (from Defant [61 D]).

Fig. 7b. Model of coastal upwelling (from Tomczak, jr. and Käse [74 T]). O: Origin of coordinate axes. D_x, D_z: Horizontal and vertical Ekman layer width, respectively. The shaded area denotes the vertical profile of the onshore component of the Ekman current at the boundary of the oceanic region, x = 0, and the arrows indicate the integrated Ekman layer mass transport.

The spatial and temporal variability of coastal winds, the irregularities of bottom topography and coastlines, and the interactions of various processes of different scales cause upwelling to be essentially a transient, three-dimensional meso-scale process with distinct variations along the coasts:

– There are seasonal variations of upwelling in the major upwelling regions due to seasonal variations of the largescale wind field.

– Most intense upwelling occurs over the shelf within a narrow strip adjacent to the coast the width of which is of the order of the baroclinic Rossby radius of deformation (10···20 km). Yet, the influence of the upwelling can be observed also at greater distance of the coast (200···300 km and more).

– The bulk of the upwelling water comes from depths not exceeding 200 m.

– An important source of the upwelling water along the western coasts of both Americas, off Northwest- and Southwest Africa is the poleward flowing undercurrent.

– The undercurrent is an eastern boundary phenomenon. Its width appears to be not larger than 100 km. The vertical thickness may extend from near-surface depths down to a level of several hundred meters. Its origin is the subsurface equatorial current system. The flow is maintained by the meridional large-scale pressure gradient in the oceans.

– Local upwelling events with most intense vertical water transports occur at time scales of several days as direct response to corresponding short term variations of the local winds.

– Occasionally, internal Kelvin waves are the dominant or even the only cause of local upwelling.

Yoshida [67 Y] published a comprehensive three-dimensional model with two homogeneous layers, which simulates many important features in coastal upwelling areas along the eastern boundaries of the oceans, including the undercurrent and the potential significance of internal Kelvin waves.

McCreary [81 M 1] developed a linear stratified model of the undercurrent. It can be considered as a generalization of the two-layer model of Yoshida to a continuously stratified ocean. The model predicts a narrow poleward flowing undercurrent along eastern boundaries driven by equatorward winds that has its current maximum 100···300 m below sea surface. Necessary conditions for the existence of the undercurrent are a baroclinic alongshore pressure gradient field and vertical mixing of heat and momentum. In this model the baroclinic alongshore pressure gradient is established by the radiation of Kelvin and Rossby waves.

Recent contributions to the special theory of shelf waves are published by Allan [76 A 1, 76 A 2] and Martell and Allan [79 M].

7.3.1 Peru Current System

On average, the winds are favorable for (Ekman-type) upwelling throughout the year along almost the whole Peruvian and Chilean coasts. The months of relatively strong upwelling in the equatorial region off the Peruvian coast (at about 5° S) are May and September, whereas further south, at about 15° S, intense upwelling occurs during the months June and August (Fig. 8). Upwelling takes place above a depth of 70 m and within 20···40 miles of the coast (Zuta et al. [78 Z]).

In general, coastal upwelling manifests itself by a number of local upwelling centres, rather than by uniform upwelling along the whole coast.

There are three important sources of upwelling waters along the Peruvian-Chilean coast:

– waters from the southward flowing extension of the equatorial undercurrent which predominantly upwells at lower latitudes;

– waters from the poleward flowing undercurrent, which supplies upwelling water along the coast at subtropical latitudes as far as to the northern coast of Chile (Robles et al. [80 R]). This flow is characterized by relative high salinity and temperature and low dissolved oxygen content (Guillen [80 G]). Its southern most extension along the continental slope is traced southwards to a latitude of 48° S (Silva and Neshyba [79 S]).

– Subantarctic waters coming at subsurface depths from the south along the continental margin.

The mean flow as observed during the upwelling experiment JOINT II (Brink et al. [80 B 2]) near 15° S reveals that below a thin Ekman surface layer ($\approx$20 m thick) the poleward flowing undercurrent governs the subsurface layer on the shelf and over the upper continental slope. Between 5° S and 15° S its core (maximum velocity) is at depth of about 100 m. The upper part of the undercurrent reaches up close to the surface layer and may occasionally surface. The poleward subsurface flow seems to be stronger and more stable at low latitudes than at 15° S (Brockmann et al. [80 B 3]). From water mass analysis Robles et al. [80 R] concluded the depth of the undercurrent to be 100···400 m off northern Chile between 18° S and 30° S.

The mean onshore velocities off Peru at 15° S (Fig. 9) show a typical subsurface maximum over the shelf near 15° S reveals indicating the response of the subsurface flow to the divergence at the surface during upwelling.

Fig. 8c gives an example of a vertical hydrographic section across the Peruvian continental slope between 15° S and 16° S. Above about 70 m depth the upwards inclined isolines of temperature and density indicate upwelling over the shelf. The down-tilted isolines there below are the geostrophic signature of the poleward undercurrent.

From field measurements Smith [68 S 2] and Huyer [80 H 2] concluded internal Kelvin waves with phase speeds of 200 km per day to propagate polewards along the Peruvian coast. Heburn [80 H 1] stated by means of a numerical model that the primary causes of the upwelling and its variability are the wind stress and the propagation of Kelvin waves through the region.

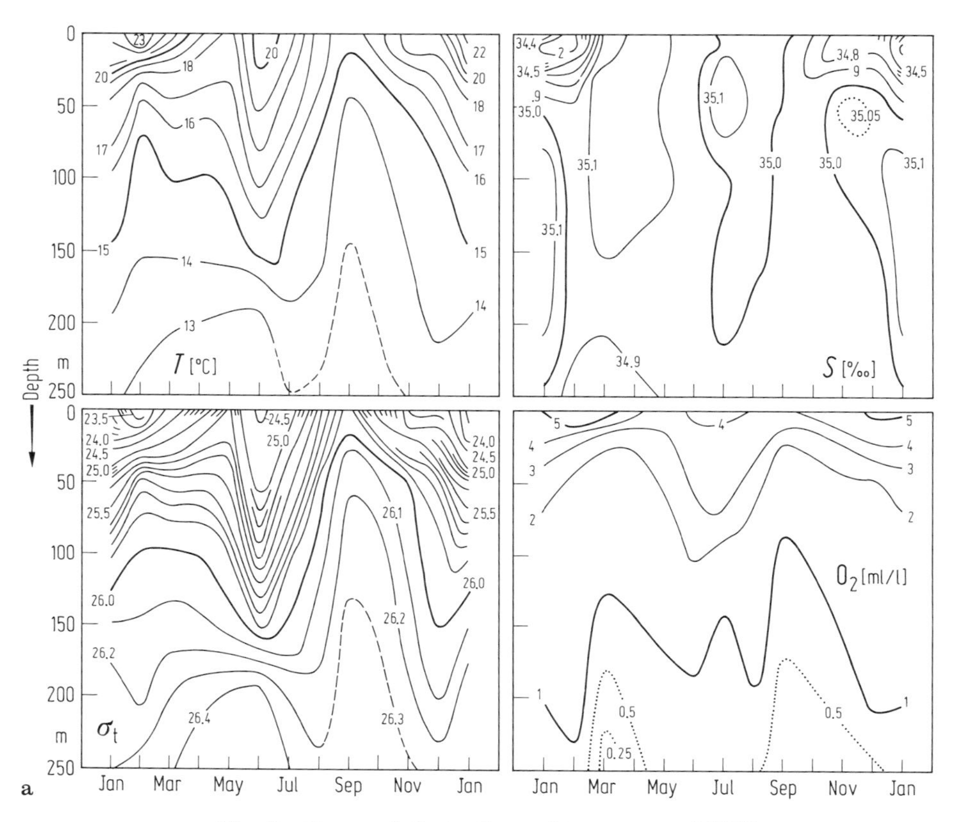

Fig. 8a. Temporal fluctuations of temperature T [°C], salinity S [‰], sigma-t, and oxygen O_2 [ml/l] off the Peruvian upwelling coast (from Zuta et al. [78 Z]) off Paita (5°05′ S).

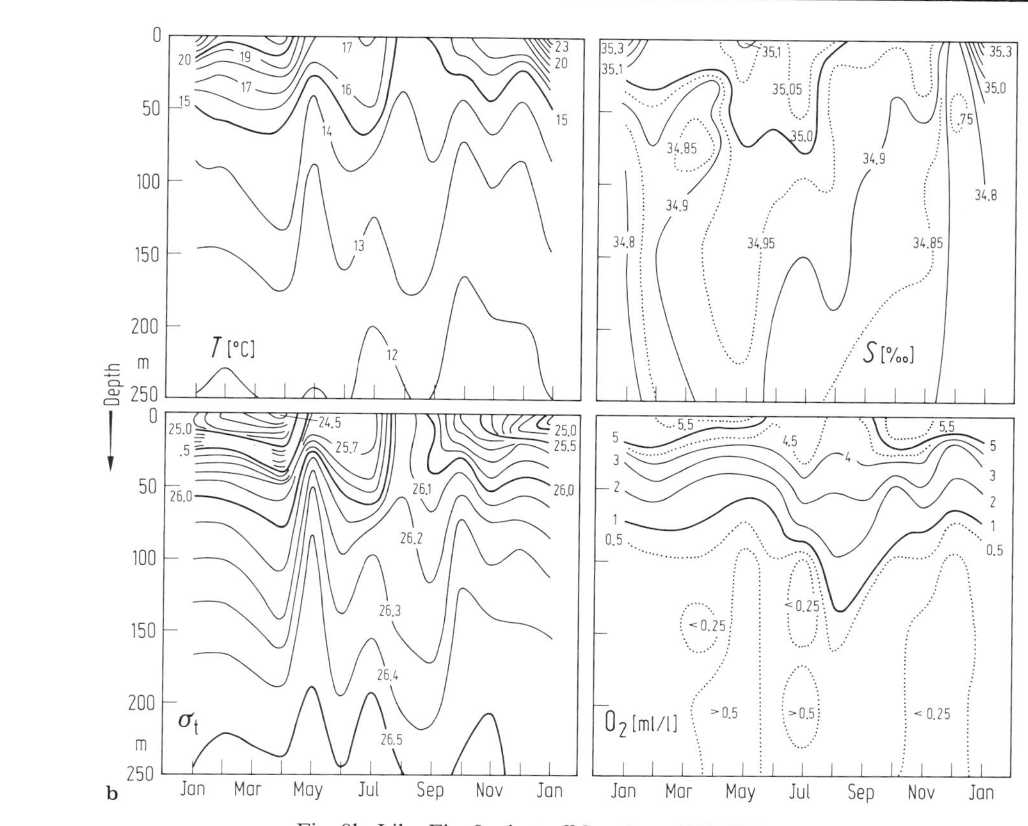

Fig. 8b. Like Fig. 8a, but off San Juan (15°20′ S).

For Fig. 8c, see next page.

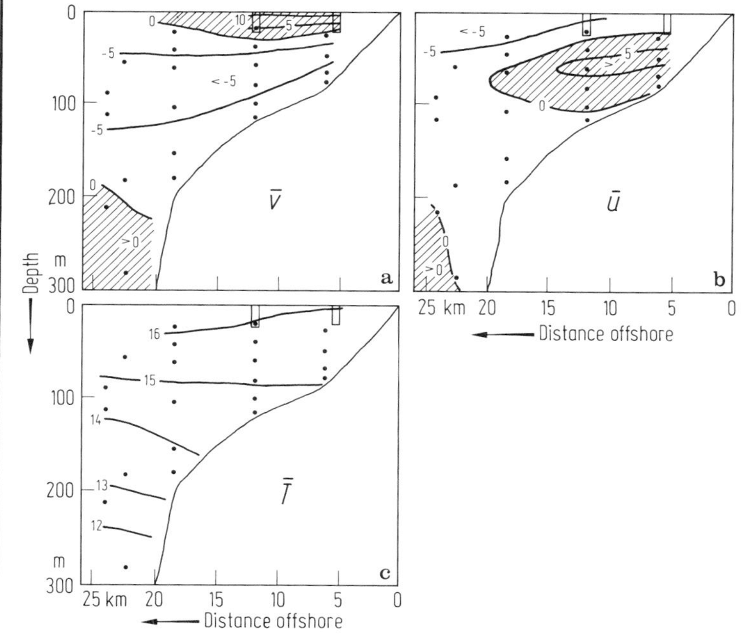

Fig. 9. Vertical sections of mean velocity [cm s^{-1}] and temperature $\bar{T}$ [°C] near 15° S off Peru (from Brink et al. [80 B 2]). Equatorward and onshore flow shaded. $\bar{u}$: cross shelf flow, $\bar{v}$: alongshore flow.

Fig. 8c. A cross section of temperature T [°C], salinity S [‰], and sigma-t off the coast of Peru near 15° S from 18 to 19 March 1977 (from Huyer and Gilbert [79 H 2]).

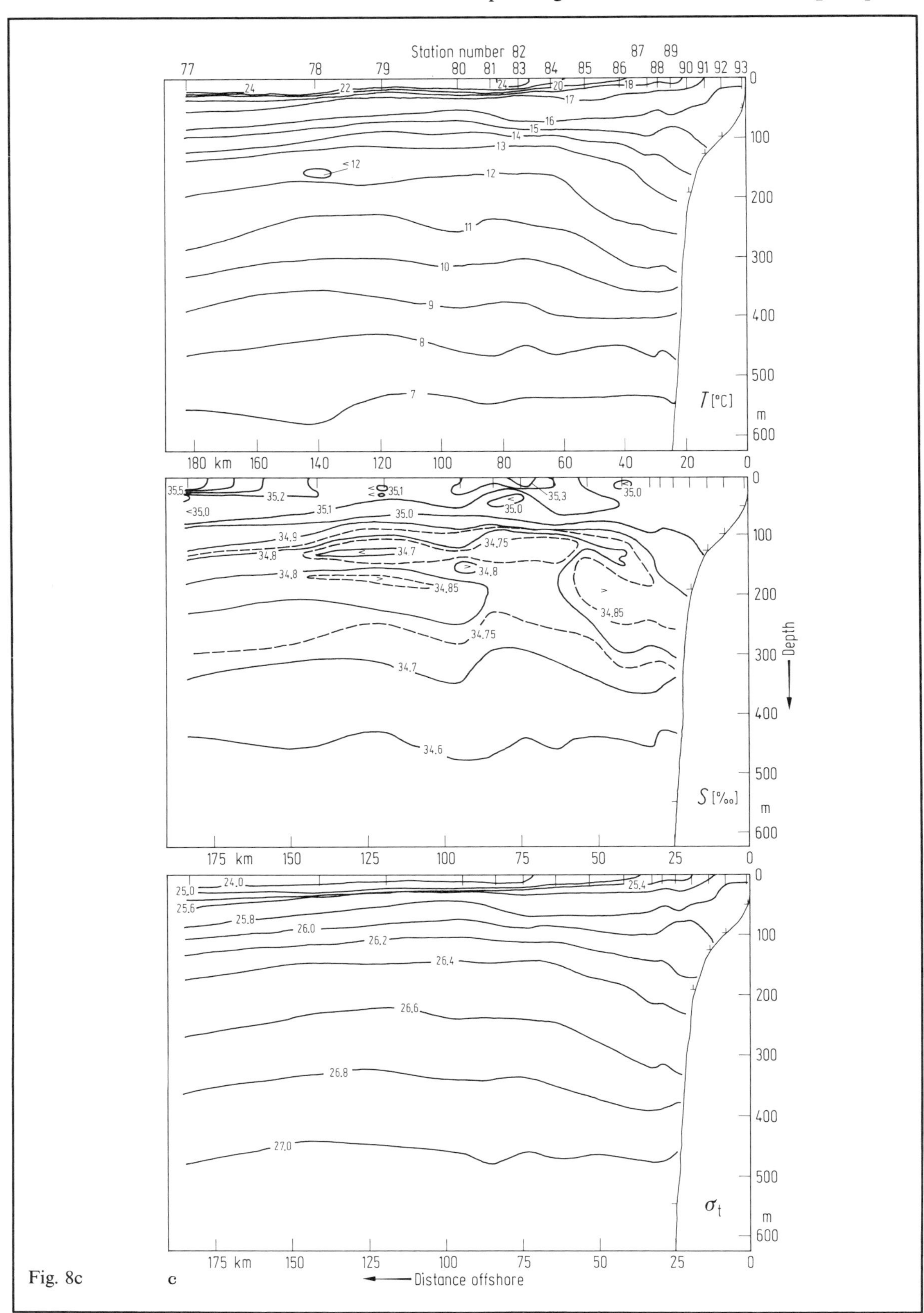
Station number
77 78 79 80 81 82 83 84 85 86 87 88 89 90 91 92 93
24
22
20
18
17
16
15
14
13
<12
12
11
10
9
8
7
T [°C]
180 km 160 140 120 100 80 60 40 20 0
0 100 200 300 400 500 m 600
35.5
35.2
35.1
35.3
35.0
<35.0
34.9
34.8
34.75
34.7
34.85
34.6
S [‰]
175 km 150 125 100 75 50 25 0
Depth
24.0
25.0
25.4
25.6
25.8
26.0
26.2
26.4
26.6
26.8
27.0
σ_t
Distance offshore

Fig. 8c c

El Niño

Around Christmas there is the occasional appearance of abnormally warm coastal surface water off Peru. This water of equatorial origin advances poleward along the Peruvian coast and is called "El Niño" (cf. Zuta et al. [80 Z]). It causes a great dying of local fish stocks and inflicts heavy damages to the local fishing industry.

Although the onset of El Niño is coincident with the normal seasonal warming, the magnitude and extent of the warm equatorial water go well beyond the usual seasonal limits.

El Niño occurs despite of upwelling favorable winds along the coast of Peru. But the coastal upwelling during these events brings to the surface only warm and nutrient-poor water from subsurface depths.

Quinn [74 Q] found that El Niño is preceded by pronounced peaks in the 12-month running means of the Southern Oscillation Index, which represents the atmospheric pressure difference between Darwin (Australia) and Easter Island. When the index is high, the trades blow strongly towards the equator. When it is low the trades are weak. The most intense El Niño events occur when strong southeasterly trades over the central Pacific relax (Wyrtki [75 W]). During the two years preceeding El Niño, excessively strong southeast trades are present in the central Pacific. As soon as the wind stress in the central Pacific relaxes, the accumulated water flows eastward in form of an internal Kelvin wave. The Kelvin wave leads to an accumulation of warm water along the eastern boundary and to a depression of the usually shallow thermocline. The combination of the eastward propagating Kelvin wave and the reflected westward propagating Rossby waves causes downwelling at the eastern boundary (O'Brien et al. [80 O]). The downwelling response at the equator is propagated poleward by coastally trapped Kelvin waves. Hence, along the eastern boundary a poleward flow is generated which advects warm equatorial water along the coasts of Ecuador and Peru. Because of the deepening of the thermocline during the El Niño events the Ekman divergence does not reach down far enough to bring cool and nutrient-rich water from below the thermocline to the surface.

7.3.2 California Current System

The prevailing winds are from the north or northwest throughout the year except in the northern region, north of about 35° N (Nelson [77 N]), where equatorwards blowing winds (favorable for upwelling), in general, only occur during spring and summer. The northerly winds are stronger off Baja California in May and June, off northern California in June and July, and off Oregon and Washington in July and August. Hence, the upwelling season propagates northward as spring and summer progress (Smith [68 S 2]). In late fall and winter upwelling ceases north of about Point Conception (35° N) due to westerly or southwesterly winds in this region. South of 35° the winds are favorable for upwelling along the coast year-round.

Measurements of sea level and currents off Oregon suggest the existence of propagating forced and free continental shelf waves contributing to the observed variability of the coastal water (Mooers and Smith [68 M]; Cutchin and Smith [73 C]; Halpern et al. [78 H]; Kundu and Allen [76 K 2]).

In contrast to other major coastal upwelling regions the salinity of the coastal waters increases with depth, especially in the north (Fig. 10). Upwelling in the California Current System, therefore, means an increase of salinity at the surface towards the coast.

The surface waters of the California Current are of subarctic origin and have relatively low salinity compared to the subsurface waters which are of equatorial origin. The vertical salinity gradient in the coastal waters is additionally enhanced by considerable amounts of fresh water discharged by rivers, especially in the north. The greatest fresh water runoff provides the Columbia river.

During upwelling season in the north (summer), the coastal flow there is southward driven by the equatorward blowing winds. But the subsurface flow is northward over the shelf. The southward surface flow has typically a maximum speed (coastal jet) over the mid shelf (Huyer [75 H]). The coastal jet has its maximum along the maximum offshore density gradient at the surface. A dynamical computation by Mooers et al. [76 M 2] of the alongshore flow during upwelling off Oregon is shown in Fig. 11.

When the characteristic upwelling circulation across the shelf has developed during an upwelling event the surface Ekman layer with offshore drift currents extends downward less than 20 m (Halpern [76 H 1]). The bottom Ekman layer is about 10···15 m thick (Kundu [76 K 1]). Most intense onshore flow (Fig. 12), compensating for the waters upwelling within a narrow inshore strip of 10···20 km along the coast, occurs at intermediate depths, in between both Ekman layers (Huyer [76 H 3]). The northward subsurface flow generates a weak offshore flow component near the bottom (Ekman veering). When the flow is southward throughout the whole water column over the shelf the near-bottom flow is onshore. Estimates of the vertical velocities yield a magnitude of 10^{-2} cm s^{-1} (Halpern [73 H]; Johnson [77 J]; Bryden [78 B]).

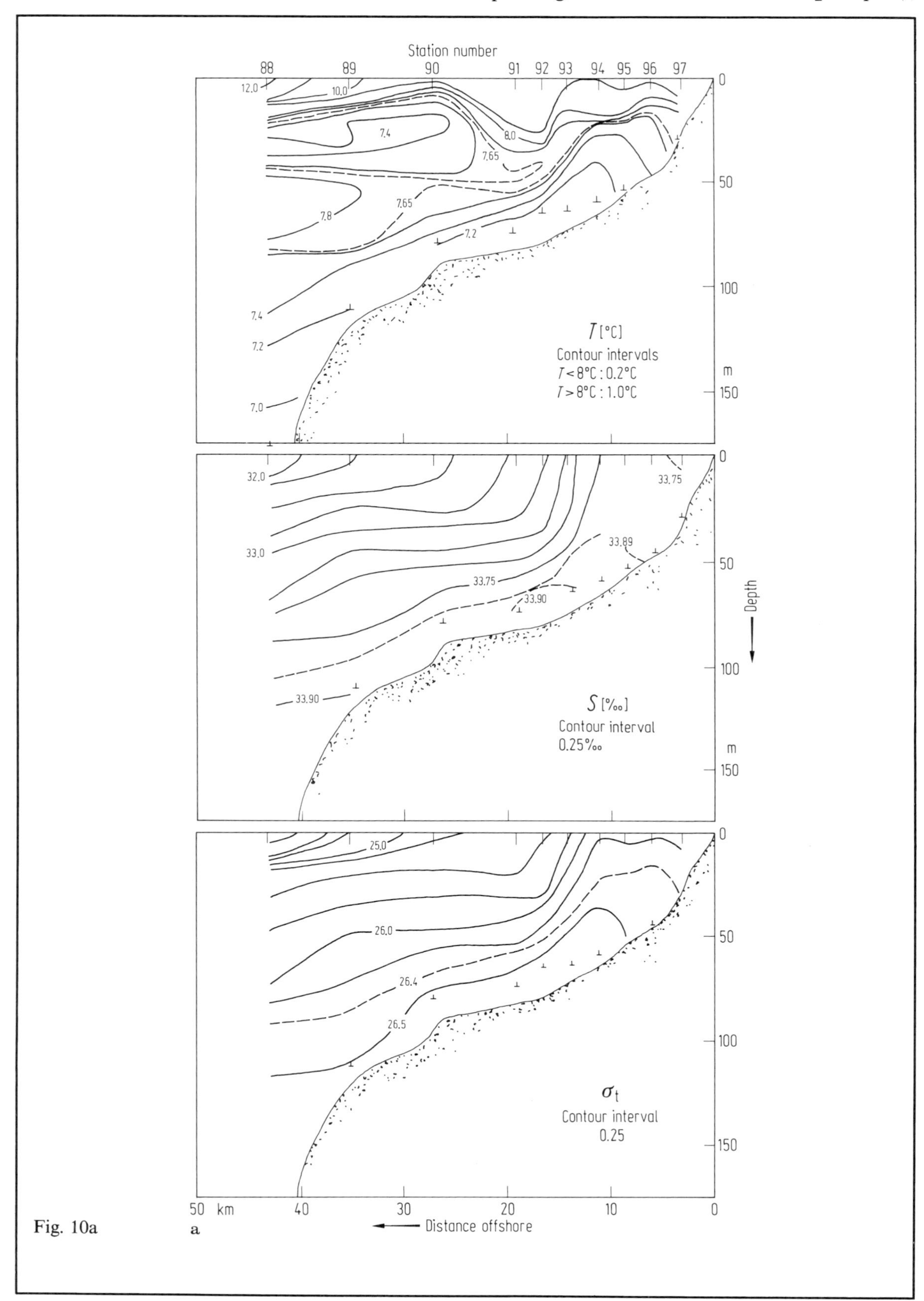
Station number
88
89
90
91
92
93
94
95
96
97
12.0
10.0
7.4
8.0
7.65
7.65
7.8
7.2
7.4
7.2
7.0
T [°C]
Contour intervals
T < 8°C : 0.2°C
T > 8°C : 1.0°C
32.0
33.0
33.75
33.89
33.75
33.90
33.90
S [‰]
Contour interval
0.25‰
25.0
26.0
26.4
26.5
σt
Contour interval
0.25
0
50
100
m
150
Depth
50 km
40
30
20
10
0
Distance offshore

Fig. 10a a

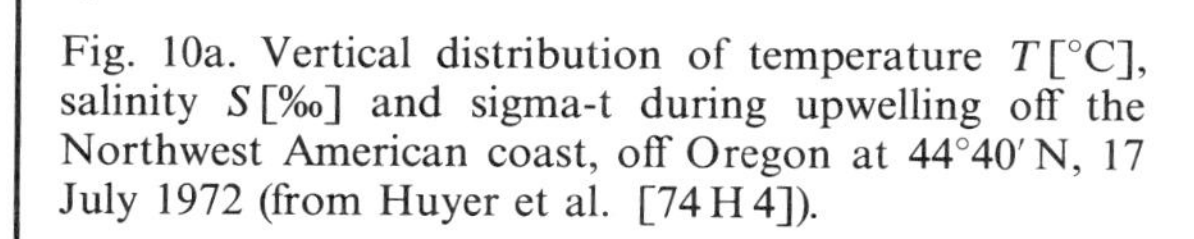

Fig. 10a. Vertical distribution of temperature T [°C], salinity S [‰] and sigma-t during upwelling off the Northwest American coast, off Oregon at 44°40′ N, 17 July 1972 (from Huyer et al. [74 H 4]).

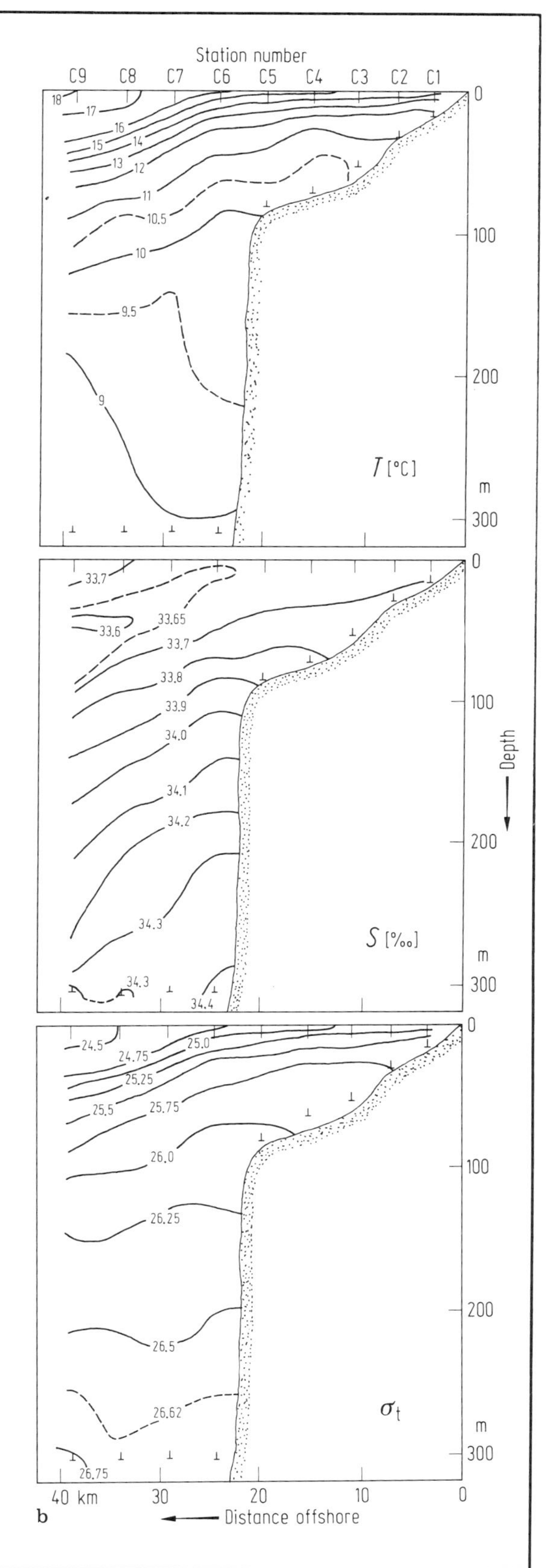

Fig. 10b. Like Fig. 10a, but off Baja California, near 31° N. Average of 5 to 11 repeated surveys (from Barton and Argote [80 B 1]).

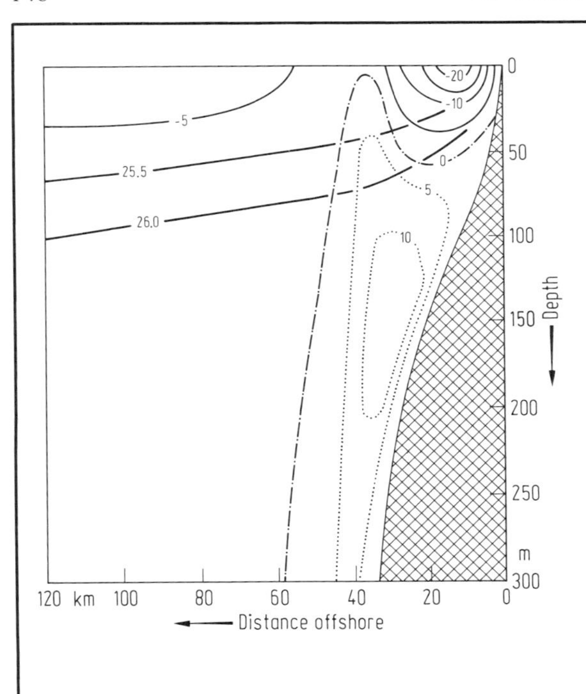

Fig. 11. Geostrophic alongshore flow, in [$cm\,s^{-1}$], off Oregon during upwelling (from Mooers et al. [76 M 2]). Positive isotachs (dotted lines) indicate poleward flow; negative isotachs (solid lines) indicate equatorward flow. The thick lines 25,5 and 26,0 represent isopycnals (σ_t).

Fig. 12. Measured cross-shelf (left) and alongshore current components (right) off Oregon [$cm\,s^{-1}$] during upwelling season (from Huyer [76 H 3]). Above: at weak winds. Below: at strong winds. Positive velocities: onshore and poleward flow, respectively. Negative velocities (shaded): offshore and equatorward flow, respectively.

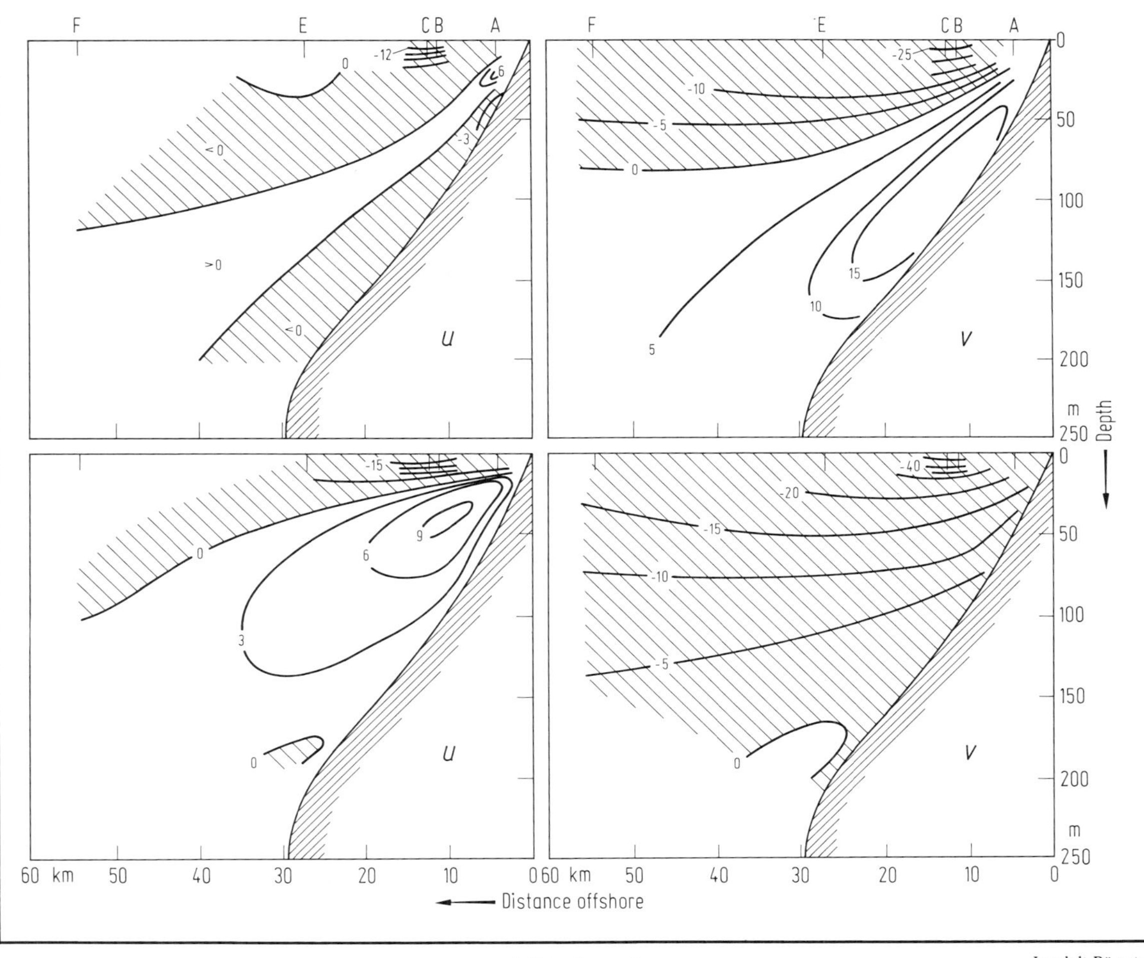

Offshore, along the continental slope is the subsurface poleward flowing California undercurrent, which is present year-round with maximum observed velocities of 30 cm s^{-1} off Baja California (Wooster and Jones [70 W]), of 20···25 cm s^{-1} off central California (Reid [62 R]), and of about 10···20 cm s^{-1} off Washington (Hickey [79 H 1]) in the north (Fig. 13).

The undercurrent is characterized by high temperature, salinity, and phosphate, and low dissolved oxygen concentration. It carries water from equatorial latitudes all along the west coast as far as Vancouver Island (Hickey [79 H 1]). Salinity and temperature of the core of the undercurrent generally decrease from about 34.6‰ and 9.5 °C off southern Baja California to about 33.9‰ and 7.0 °C, respectively, off Vancouver Island. Off Washington the percentages of equatorial waters increase significantly between July and September, during the local upwelling season.

Just south of Point Conception, in the California Bight, the core of the undercurrent lies at a depth of 200···300 m during early summer but rises to a depth of about 100 m as the season progresses and retreats to 200···300 m in the fall (Pavlova [66 P]).

North of Point Conception the core of the undercurrent appears to surface during late fall and winter, when it is known as the Davidson Current. The core of the undercurrent occurs inshore the 2000 m isobath (Hickey [79 H 1]). It shoals as the season progresses. Maximum northward flow is apparently below 300 m depth in July (Halpern et al. [78 H]), near 300 m in September and October and generally at the surface in November (Ingraham [67 I]). The northward flow off Oregon extending from greater depths upwards rises from >400 m in May to the depth of the shelf edge by June, just before the upwelling season (Huyer and Smith [76 H 4]).

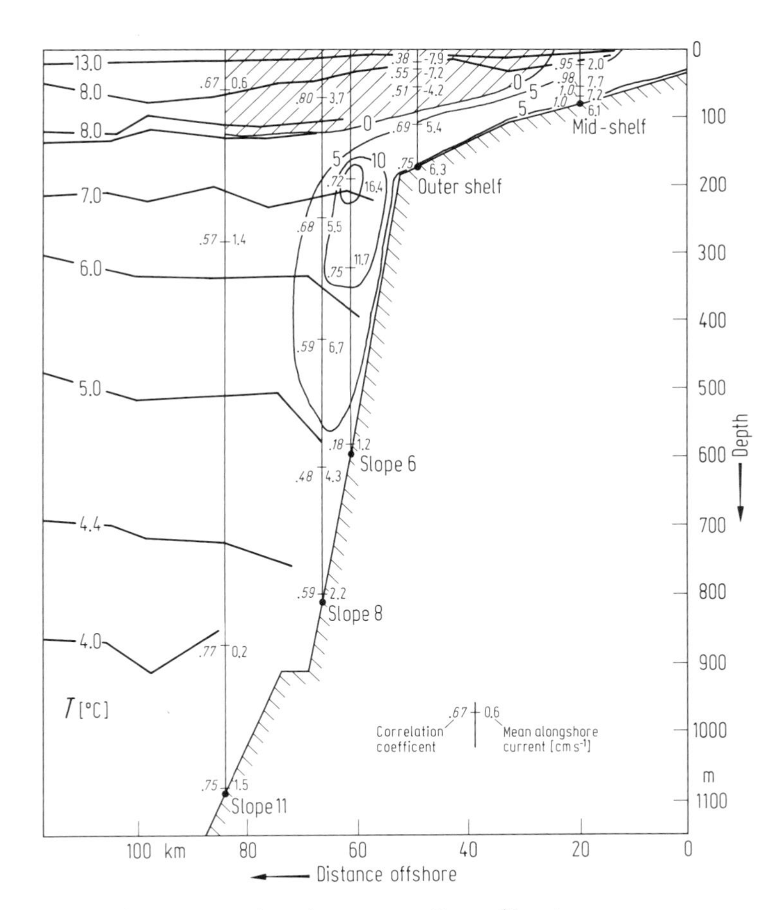

Fig. 13. Mean alongshore current [cm s^{-1}] and temperature [°C] off Washington (from Hickey [79 H 1]). Position of current meter (bar), mean alongshore flow (righthand number) and correlation with an inshore current record (lefthand number) during the summer of 1972. Average of 8 days. Southward flow is shaded.

Mittelstaedt

The example shown in Fig. 13 suggests that during upwelling season off Washington the maximum speed of the undercurrent occurs at a depth of about 190 m. Northward flow was observed below 60 m to at least 80 km offshore and over the entire shelf. Evidently the poleward flowing subsurface undercurrent over the shelf off Oregon and Washington is connected with the California Undercurrent along the continental slope (Halpern et al. [78 H]).

During the winter months (November-February), when no appreciable upwelling occurs, the surface flow is poleward along most of the coast within about 150 km from the coast of California and about 300 km off Oregon and Washington (Hickey [79 H 1]). In the region north of Point Conception this northward surface flow is known as Davidson Current, which might be the surface expression of the California Undercurrent during this season. Within the Channel Islands off Southern California the inshore surface flow is to the northwest (as part of the southern California Eddy) throughout the year oppositely to the prevailing wind stress.

7.3.3 Benguela Current System

Upwelling occurs from about Cape Frio (18° S) southward to Cape Point (34° S). Minimum sea surface temperature due to upwelling occurs immediately along the coast. Maximum upwelling induced sea surface temperature anomalies reach values of 8···10 °C (Fig. 14). The anomalies are greatest off the Cape Peninsula in the south during southern spring and summer (Andrews and Hutching [80 A]). This southernmost upwelling site is bounded by particularly sharp fronts offshore. Coastal upwelling in the Benguela Current region is caused by the predominating southeast trade winds. The coastal winds can be considered as a divergence of the coastal boundary of the trades offshore, resulting from the extensive diurnal pressure changes over the continent (Hart and Currie [60 H]).

Maximum mean southeast trades prevail in spring and summer, from September to March. During these seasons coastal upwelling is, in general, also maximum (Shell [68 S 1]; Andrews and Hutching [80 A]), whereby the maximum seems to shift southward from spring towards summer.

In the south westerly winds prevail during winter, because of the seasonal northward displacement of the west wind belt. No upwelling is found during this season in this area. Offshore the southeast trade winds cause a relatively steady drift current towards NW to W (Defant [36 D 2]). This flow offshore can be considered as the oceanic branch of the Bengueal Current.

Hart and Currie [60 H] denoted the oceanic flow as the southeast trade wind drift and used the name Benguela Current only for the coastal circulation. A seasonal counter current inshore occurs in the south along the coast between Cape Town and Cape Columbine during upwelling season (Andrews and Hutching [80 A]).

The few hydrographic data available over sufficient great depths (cf. Defant [36 D 1, 36 D 2]; Fuglister [60 F]; Hart and Currie [60 H]) indicate the existence of a geostrophic poleward undercurrent along the continental slope. Its core velocity (below 350 m depth at about 20° S) reaches values of $> 15\ \mathrm{cm\ s^{-1}}$. At times, the undercurrent appears to meander onto the shelf (Hagen et al. [81 H]). Hart and Currie termed this subsurface southward flow "compensation current" (Fig. 15). In a schematic picture of some features of the Cape Peninsula upwelling plume Andrews and Hutching [80 A] suggest seasonal southward flow near the bottom over the shelf. The watermass of this flow carries oxygen-depleted water from the Walvis Bay region in the north along the shelf southward (De Decker [70 D]).

The oceanic and coastal branches of the Benguela Current are distinctively different in their water masses. The oceanic surface water is warm (> 18 °C) and more saline than the cool upwelling waters of the coastal Benguela Current. The major constituent of the upwelling water in the Benguela Current region is the South Atlantic Central Water. This water mass ascends, in general, from depths between 200···300 m into the surface layer along the coast.

In the south the waters of the Agulhas Current intermittantly penetrate into the Benguela upwelling region (Darbyshire [63 D]; Shell [68 S 1]).

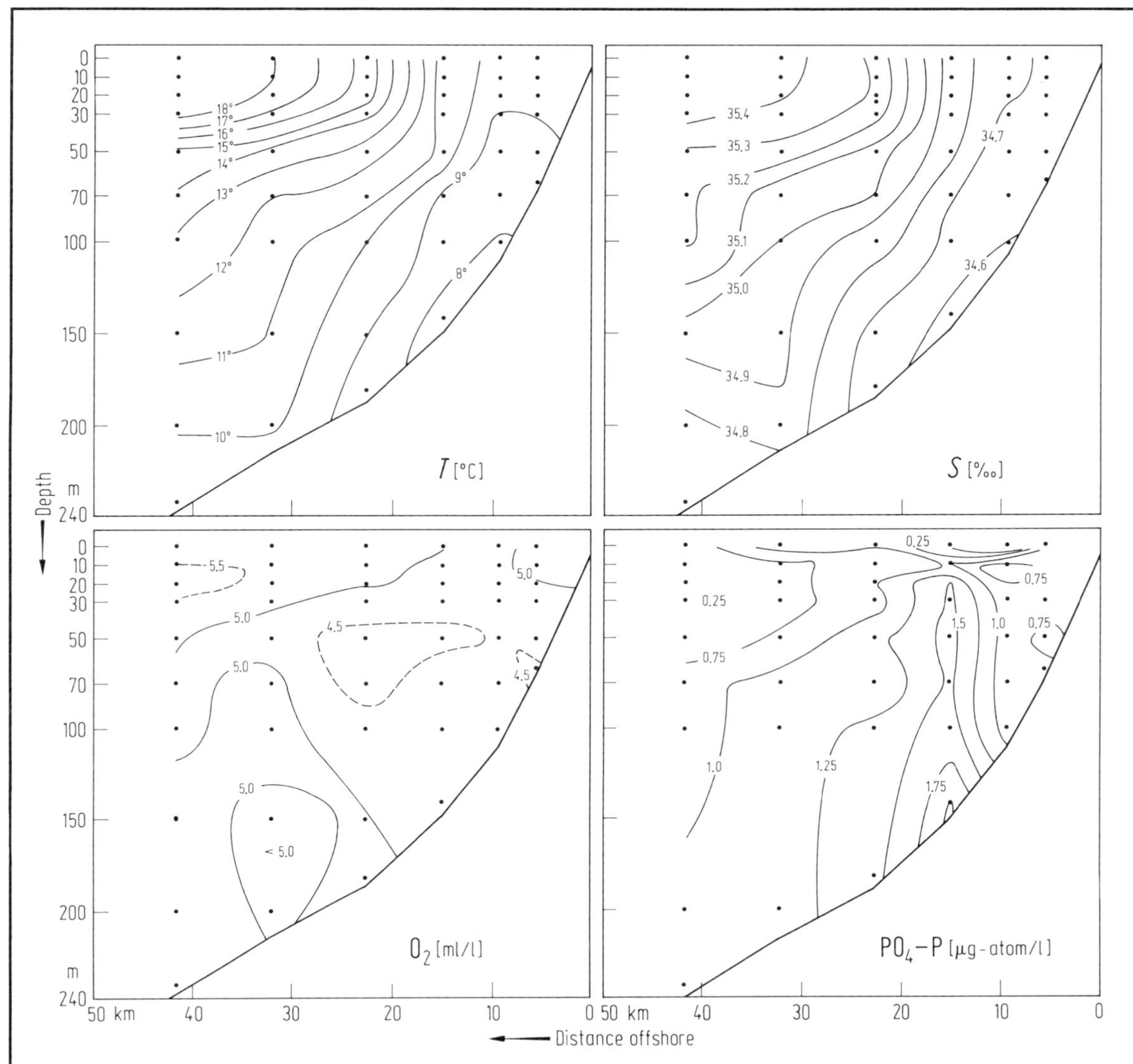

Fig. 14. Example of the vertical distribution of temperature, salinity, oxygen, and phosphate in the Southern Benguela upwelling region (from Andrews and Hutching [80 A]).

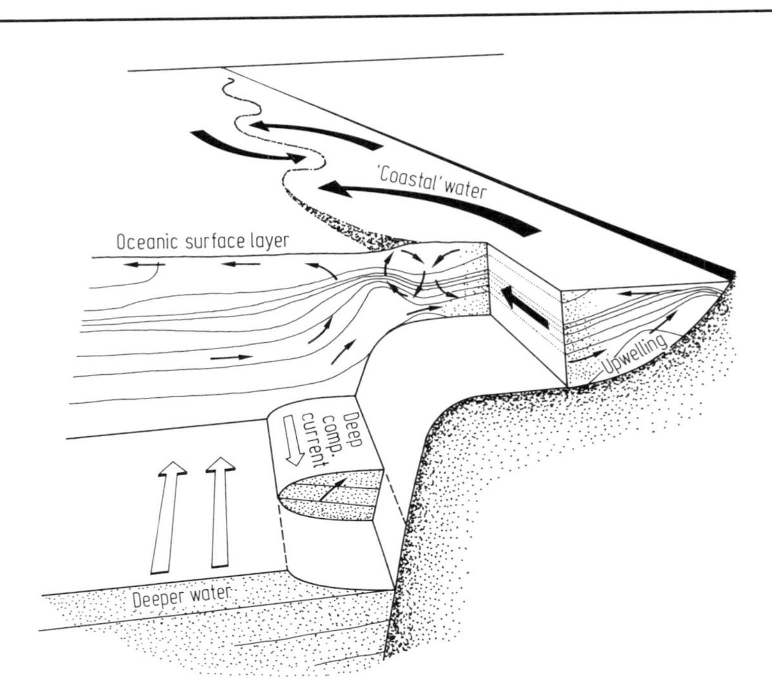

Fig. 15. An idealized picture of the principal horizontal and vertical water movements as derived by Hart and Curry [60 H] from observations in the Benguela Current region. The isotherms are represented by the thin lines on the vertical planes normal to the coast. The curved shaded line along the sea surface indicate the boundary between the upwelled and oceanic surface waters (Comp. Current = Compensation Current).

7.3.4 Guinea Current System

Every summer for a period from early July through September upwelling occurs along the coast between Nigeria and Ivory Coast with maximum intensity off Ghana. The coldest water occurs along the coast between Ivory Coast and central Ghana, where the coastline is nearly parallel to the prevailing Southwest winds (monsoon) indicating a typical wind-driven upwelling situation. But there is poor correlation between the upwelling intensity and the coastal wind. (Houghton [76 H 2]).

The local flow is dominated by the jet-like Guinea Current, which is relatively strong in summer during upwelling with velocities of 1⋯2 knots at the sea surface. Below the Guinea Current is a westward flowing undercurrent (Houghton [76 H 2]), one of the roots of the undercurrent further north off Northwest Africa. The undercurrent below the Guinea Current persists throughout the year and during local upwelling its core moves up onto the shelf.

According to the model simulations by Adamec and O'Brien [78 A] an increase of the trades in the western Atlantic excites an equatorially trapped Kelvin wave which propagates eastward along the equator. At the eastern ocean margin the Kelvin wave propagates poleward as a coastal Kelvin wave and produces the upwelling event along the coast in the Gulf of Guinea.

7.3.5 Canary Current System

According to the seasonal variations of the northeast trade winds, upwelling seasons occur along the northwest African coast (Wooster et al. [76 W]; Speth et al. [78 S]). South of about 20° N upwelling arises during winter and spring. Between 20° N and 25° N upwelling occurs throughout the year with maximum intensity during spring and autumn (Fig. 16). The duration of the upwelling seasons are increasing from 1 month at about 10° N (February) to 6 months at about 15° N (November to May), and to 12 months between 20° N and 25° N (Schemainda et al. [75 S]). North of this region the duration of the upwelling season is about 3 months (July to September).

In summer (July/August) the southern boundary of the trades off the coast lies near 21° N. South of this latitude northerly winds still occur during this season over the coastal waters. But they are usually weak and have an onshore tendency (from NNW to WNW), whereas the trade winds still dominate far offshore. Between Cape Juby (28° N) and Cape Verde (15° N) the very dry and warm Harmattan winds occasionally occur, especially during winter. The Harmattan blows offshore from NE to E. Most of the time the boundary between the maritime trades and the Harmattan runs parallel to the coast over the continent. The boundary is marked by a strong thermal gradient, additionally enhanced by the influence of the cool upwelling water.

In summer, when the southern boundary of the trades has its northernmost position, the coastal northward flowing counter current with its warm surface waters reaches up to the north to about Cape Blanc (21° N). Because of the well-developed trades further north the warm surface water cannot advance beyond these latitudes for an extended period. At this time, the northward flowing coastal counter current forms the inshore limb of a large cyclonic gyre, whose diffuse offshore limb is the southward flowing Canary Current (Mittelstaedt [82 M]). Upwelling has ceased south of Cape Blanc. In late autumn the counter current is gradually replaced again by southward flow associated with upwelling due to the seasonally increasing trades south of 21° N.

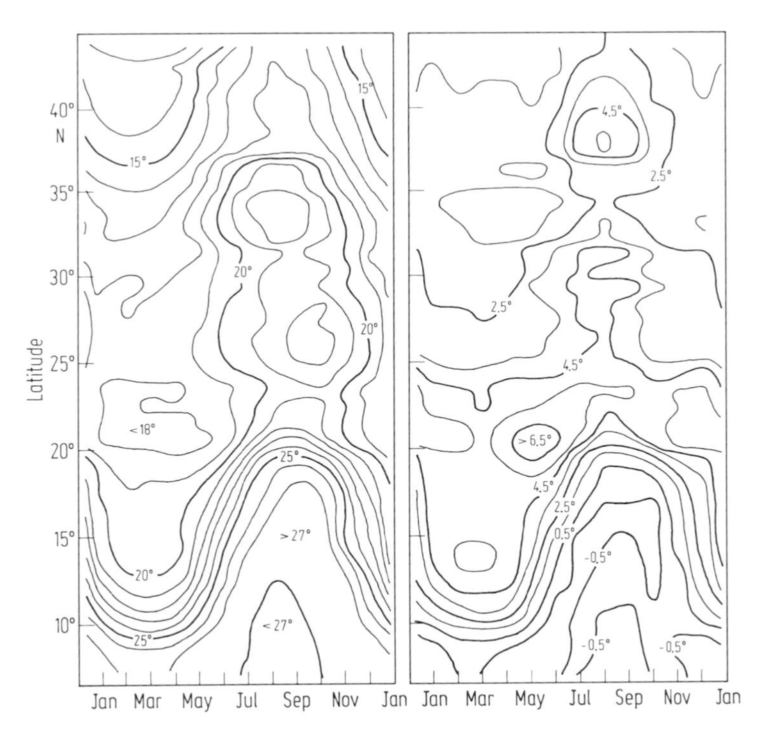

Fig. 16. Monthly variations of the average temperature (left) and the temperature anomalies (right) at the sea surface close to the coast in the eastern north Atlantic (from Wooster et al. [76 W]). Temperature anomalies [°C] represent the difference between near coastal and mid-ocean temperature. Positive values: near coastal sea surface temperature is colder than mid-ocean sea surface temperature.

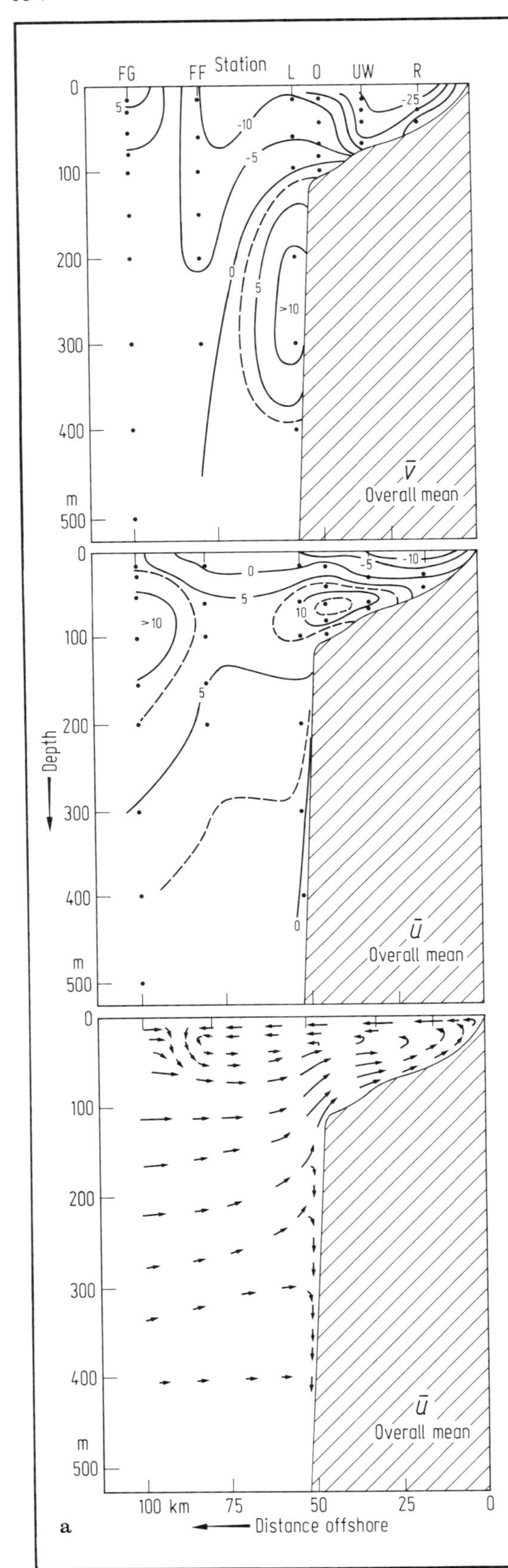

Fig. 17a. Mean coastal flow [$cm\,s^{-1}$] in the Canary Current region during upwelling season. $\bar{v}$: alongshore; $\bar{u}$: onshore; negative numbers: equatorward and offshore current components, respectively; positive numbers: poleward and onshore current components, respectively. Current components at 21°40′ N during February–April 1974 (from Mittelstaedt et al. [75 M, 80 M]).

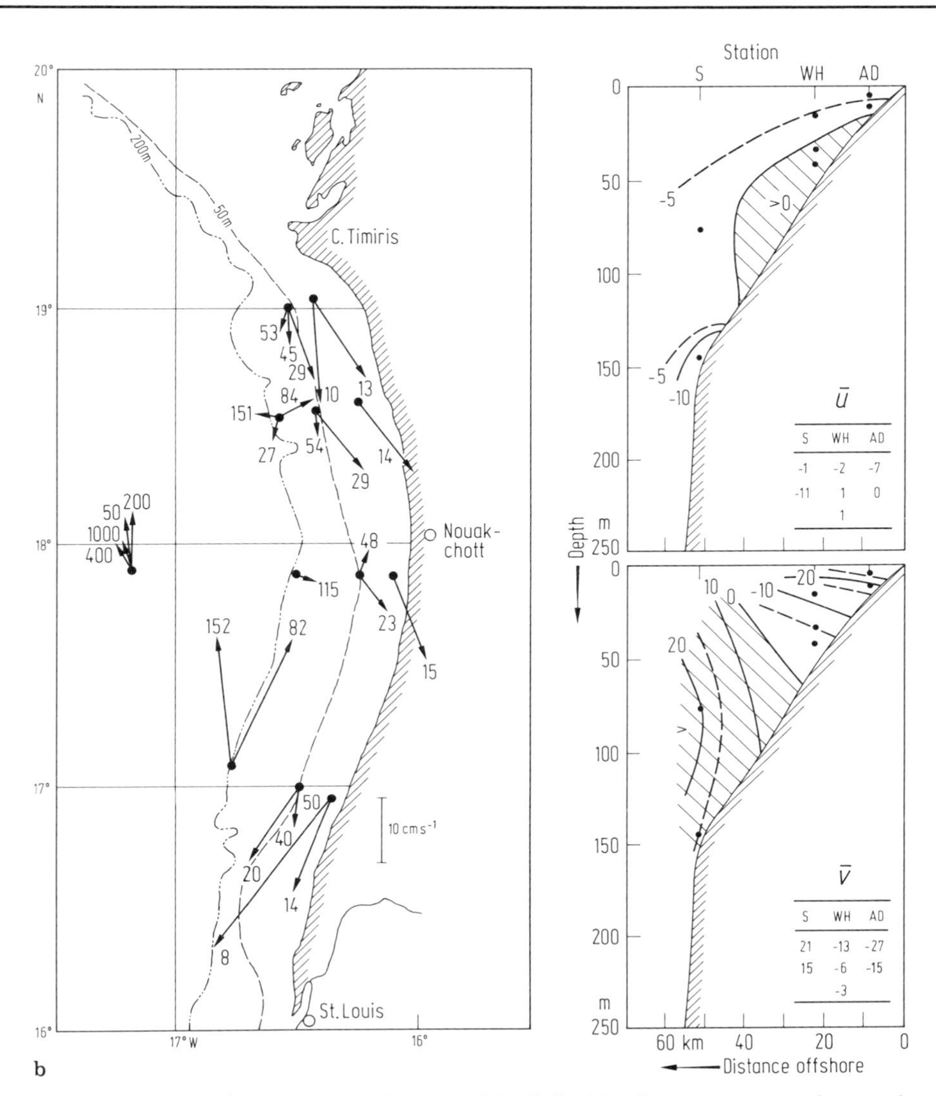

Fig. 17b. Current vectors during January–February 1977 (left). Numbers at vectors denote the depths of observation [m]. Current components at about 17° N during January–February 1977 (right). Numbers on the right of the section, below the station initials S, WH, AD denote the actual values of $\bar{u}$ and $\bar{v}$ in [cm s^{-1}] for depths indicated by solid circles (from Mittelstaedt and Hamann [81 M 2].

The mean flow on the shelf during upwelling season is essentially driven by the local winds. The temporal variations of the coastal winds have typically periods of 5···10 days. Within this time scale occur the local upwelling pulsations (upwelling events) forced by the local winds.

The entire upwelling circulation system extends not beyond the continental slope. The width of the inshore boundary zone with the actual upwelling into the surface layer is still narrower (10···20 km).

The thickness of the surface Ekman layer with offshore flow ranges from less than 20 m inshore to more than 60 m offshore. Below the surface Ekman layer the flow on the shelf is onshore being maximum in the near-bottom layer.

Direct measurements of the vertical velocities by means of freely floating neutrally-buoyant current meters yielded values of $10^{-2}\cdots10^{-1}$ cm s^{-1} (Shaffer [76 S]).

The poleward flowing undercurrent tends to be a rather narrow flow (30···60 km wide), nestling against the continental slope. Its core with mean maximum speeds concentrates at depths of 100···200 m south of Cape Blanc and of 200···400 m north of it (Fig. 17).

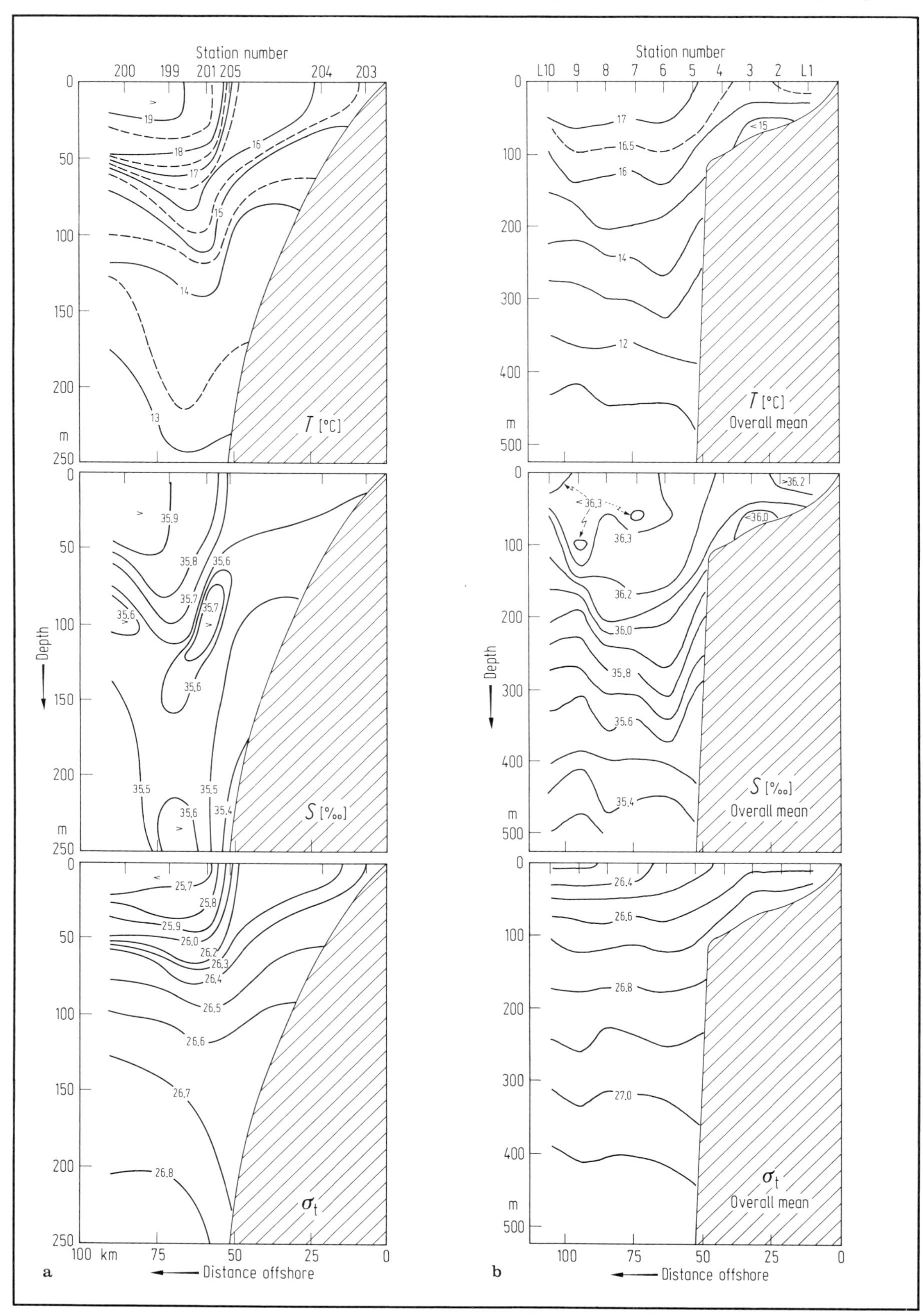
Station number
200
199
201
205
204
203
19
18
17
16
15
14
13
T [°C]
35.9
35.8
35.6
35.7
35.7
35.6
35.6
35.5
35.5
35.6
35.4
S [‰]
25.7
25.8
25.9
26.0
26.2
26.3
26.4
26.5
26.6
26.7
26.8
σ_t
Depth
m
100 km
75
50
25
0
Distance offshore
a
Station number
L10
9
8
7
6
5
4
3
2
L1
17
16.5
16
<15
14
12
T [°C]
Overall mean
>36.2
<36.3
36.3
<36.0
36.2
36.0
35.8
35.6
35.4
S [‰]
Overall mean
26.4
26.6
26.8
27.0
σ_t
Overall mean
Depth
m
100
75
50
25
0
Distance offshore
b

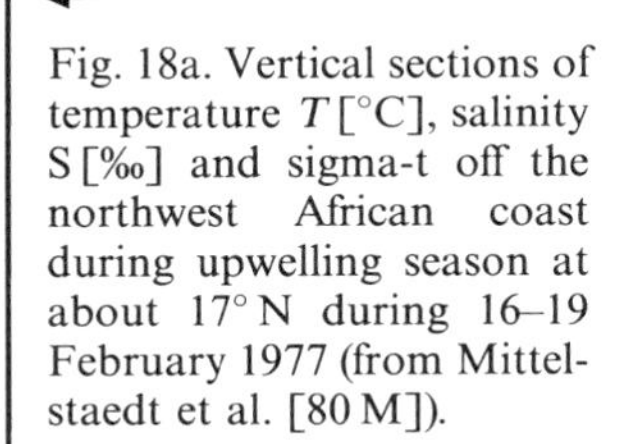

Fig. 18a. Vertical sections of temperature T [°C], salinity S [‰] and sigma-t off the northwest African coast during upwelling season at about 17° N during 16–19 February 1977 (from Mittelstaedt et al. [80 M]).

Fig. 18b. Like Fig. 18a, but at 21°41′ N during February –April 1974. Average sections from repeated surveys (from Barton et al. [75 B]).

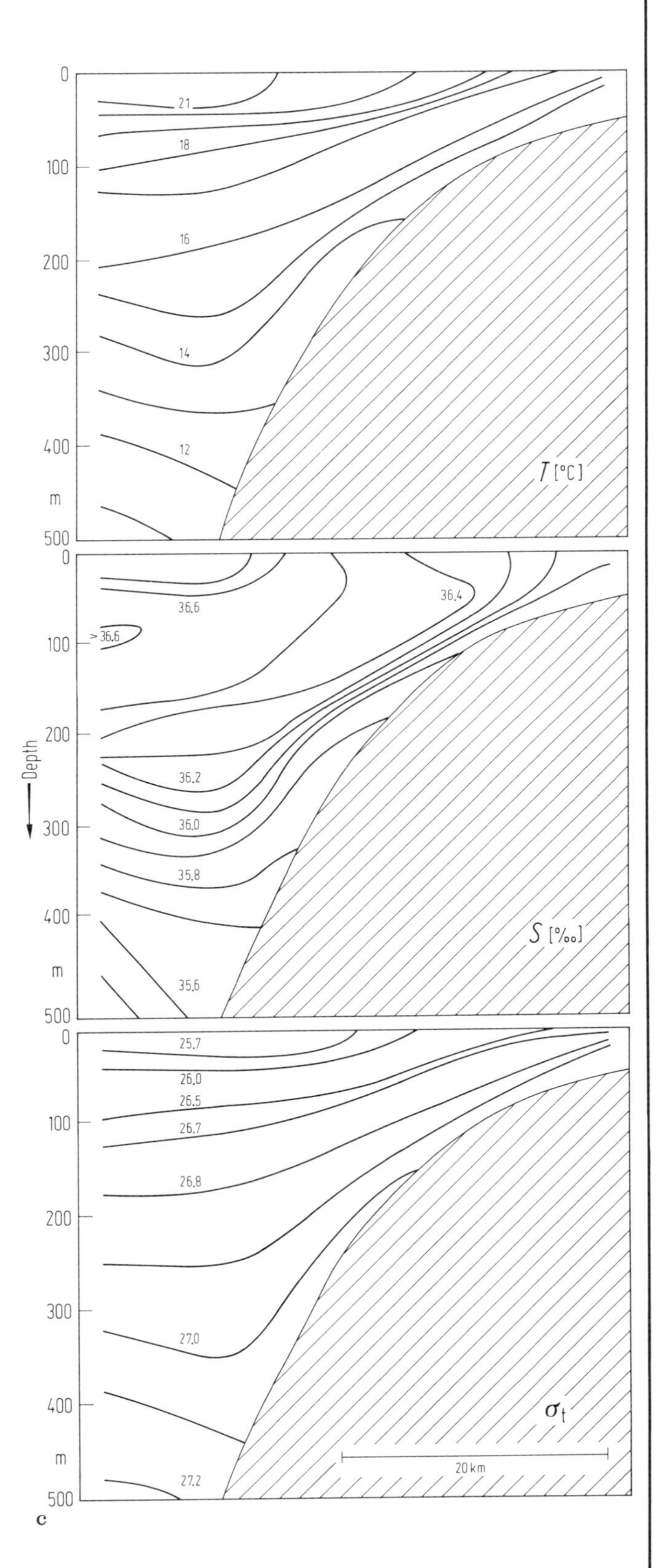

Fig. 18c. Like Fig. 18a, but off Cape Bojador (at about 26° N) on 13 August 1972 (from Hughes and Barton [74 H 2]).

Part of the Northwest African undercurrent originates from the Gulf of Guinea. Another part comes from the Equatorial Counter Current region of the open Atlantic (Defant [36 D 1, 41 D]; Mittelstaedt [76 M 1]; Voiturieux and Herbland [82 V]).

Most of the subsurface waters that compensate the upwelled water inshore come from depths between 100···300 m. So the undercurrent is an important source for the water masses upwelling along the coast, at latitudes south of about 24° N, where it has not yet submerged below the maximum upwelling depth. Due to bottom friction part of the water carried by the undercurrent sinks down the slope within a thin bottom layer of approximately 20 m thickness. Between Cape Bojador (26° N) and Cape Verde (15° N) a northward countercurrent has been observed offshore also in the surface layer during upwelling season (Mittelstaedt [76 M 1, 78 M]; Tomczak jr. and Hughes [80 T]).

The Canary Current upwelling region can be divided into a northern and a southern region. South of Cape Blanc a major constituent of the upwelling waters (salinity: 35,6···35,9‰) is the less saline South Atlantic Central Water (SACW). North of Cape Blanc the upwelling waters (salinity: 36.1···36.4‰) contain a significant amount of the saltier North Atlantic Central Water (NACW). The two upwelling regions are separated by a distinct transition zone (Fraga [74 F]; Hughes and Barton [74 H 1]; Tomczak jr. [78 T, 81 T]) with respect to their water masses and nutrients. The SACW contains more phosphate and silicate than the NACW (Weichart [74 W]; Gardener [77 G]). Within the transition zone intense mixing takes place between both water masses (Tomczak jr. and Hughes [80 T]).

The water with appreciable proportions of SACW comes with the undercurrent towards north, where it ascends along the coast in the southern upwelling region. In the northern region the undercurrent flows at greater depths below the layer from where the upwelling water comes. So the upwelling waters originate from the north in the northern region and from the south in the southern region.

In the south the undercurrent extends from the slope onshore over the outer shelf during upwelling season (Mittelstaedt et al. [75 M]; Mittelstaedt and Hamann [81 M 2]).

The upwelling induced temperature range of the shelf surface water is, approximately, everywhere the same (15···17 °C) in the various areas along the Northwest African Coast.

Offshore from the shelf edge surface temperature is in general higher by 1···2 °C than inshore (Fig. 18). The transition between the cool upwelling water inshore and the warmer mixed oceanic surface water offshore can be very abrupt in forms of frontal zones (Fig. 19). Depending upon its intensity, the horizontal density gradient associated with these frontal zones strengthens the barotropic southward flow by an additional baroclinic current component towards the same direction.

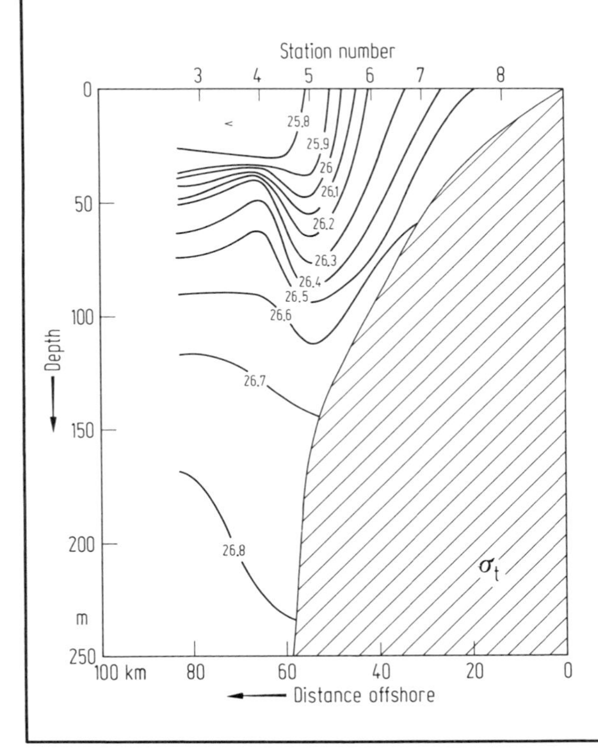

Fig. 19. Example of an upwelling front (from Mittelstaedt et al. [80 M]). The observations have been made on 23 January 1977 during AUFTRIEB '77, four weeks earlier than the section shown in Fig. 18a.

With the poleward flowing coastal countercurrent advancing towards north during summer, tropical warm surface water (> 25 °C) spreads towards higher latitudes until it meets the southern boundary of NACW near Cape Blanc in August (Fig. 20). During this time no noticable upwelling occurs south of Cape Blanc and consecutively no SACW reaches the surface layer in the southern upwelling region. As opposed to this, maximum upwelling occurs in the north, where cool and relatively saline NACW ascends to the surface along the coasts.

With the seasonal reversal of the coastal currents towards south the tropical surface water also retires equatorward and stays in late winter at about 10° N (see Fig. 20), when upwelling has its southernmost extension.

A regional pecularity is the Banc d'Arguin between Cape Blanc and Cape Timiris. On the flat bank (water depth < 10 m) very warm and saline water (salinity up to 39.5‰; Maigret [72 M]), forms by intensive evaporation due to considerable insolation and dry winds blowing from the Sahara towards the sea. Because of its high density the bank water submerges beneath the cool upwelling water and sinks down the continental slope off the Banc d'Arguin, where there is practically no "classical" shelf (Shaffer [76 S]; Peters [76 P]). The bank water sinks down to depths of about 400 m.

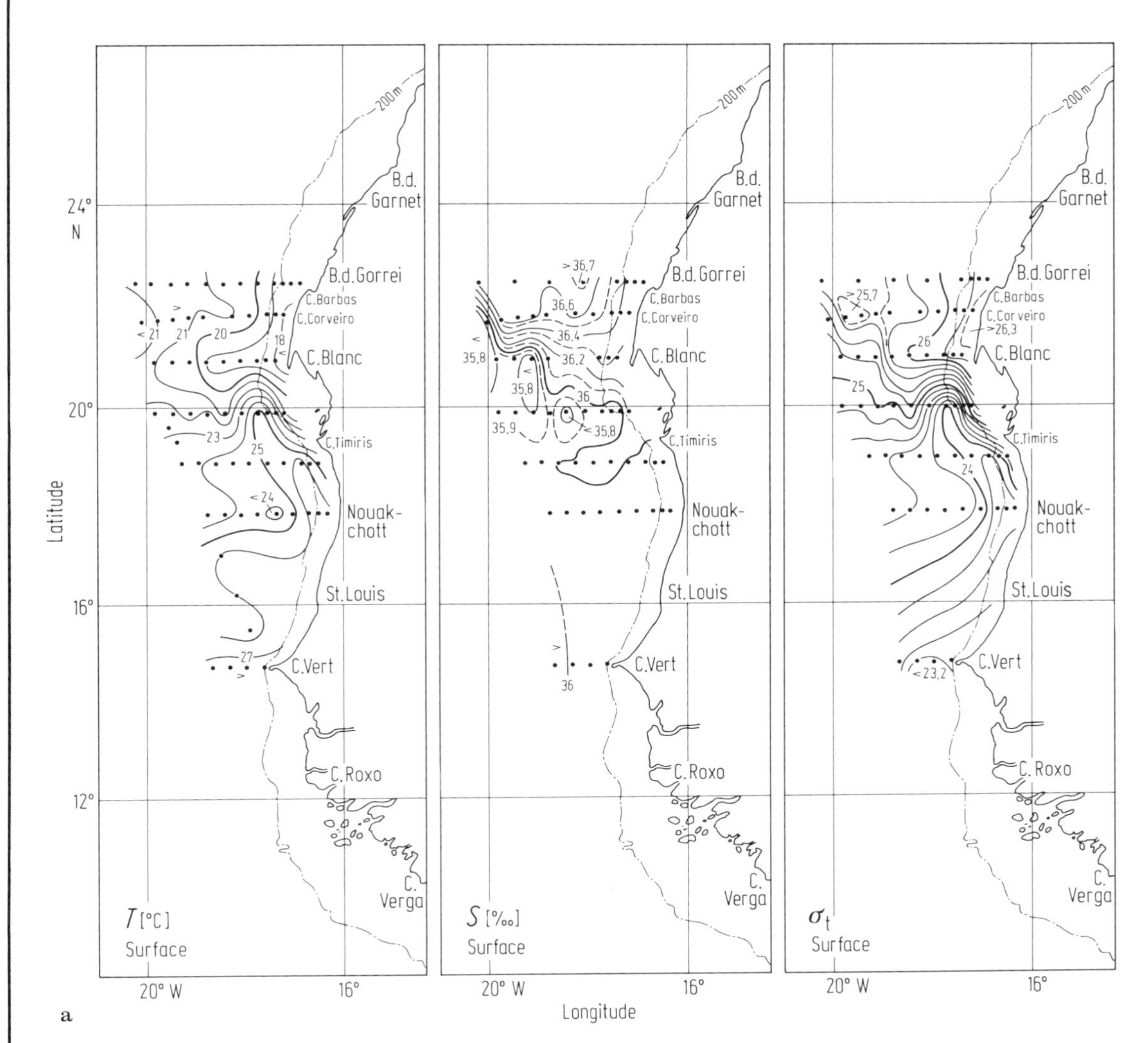

Fig. 20a. Temperature T [°C], salinity S [‰], and sigma-t at the sea surface during summer 1973 (from Schemainda et al. [75 S]).

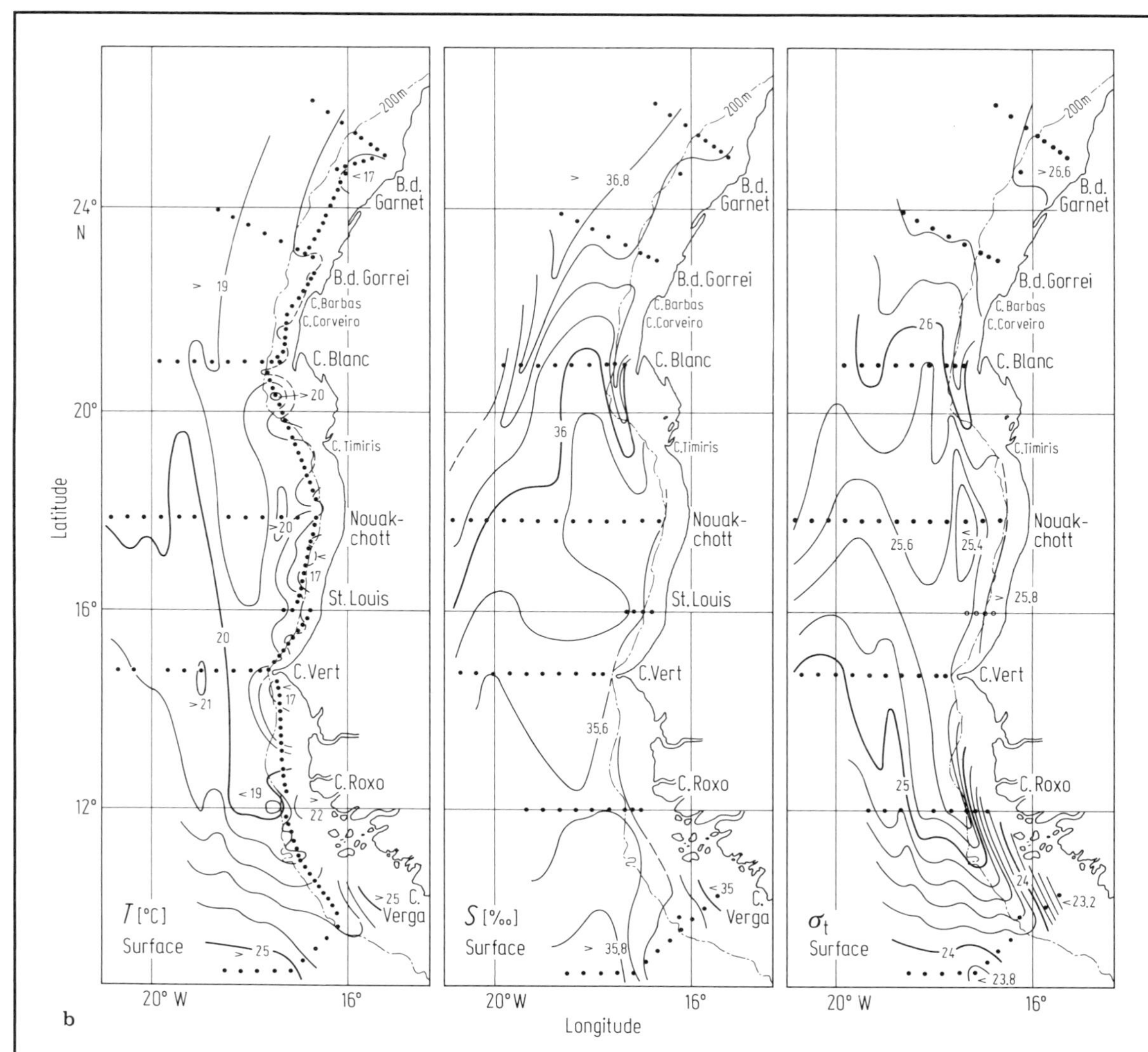

Fig. 20b. Like Fig. 20a, but during winter 1973.

7.3.6 Portugal Current System

Upwelling favorable winds from northerly directions ("Portuguese trades") prevail from June to September. At this time the southward flowing Portugal Current is well developed. Coincidentally wind-driven upwelling occurs along the west coast and south coast of Portugal (Fiúza [82 F 1]).

The upwelling water seems to ascend from depths of 60···120 m. The water mass characteristic of the upwelling water is similar to that one off the Northwest African coast north of Cape Blanc (Fiúza [82 F 2]). This water mass, defined as Eastern North Atlantic Water (Fiúza and Halpern [82 F 3]), originates from a region south of the Azores.

7.3.7 Somali Current System

Upwelling off the coast of Somalia occurs during summer, when the Somali current flows towards northeast during the southwest monsoon. This wind dominates from May through September and is maximum in July/August with mean wind forces of about 6···7 Beaufort (24···30 knots) off Somalia. East of the island Socotra gales of ≧8 Beaufort (>37 knots) have a frequency of 40% during July and further south off Somalia this frequency is still about 20% during the same month (Dt. Hydrogr. Inst. [66 D]). At this time the northeastward flowing Somali Current is strongly developed and reaches mean velocities of 2···3 knots at the surface.

From October through March blows the northeast monsoon. This winter monsoon is, in general, less strong than the summer monsoon, even at the time of its greatest intensity and steadiness during December/January.

With the onset of the northeast monsoon the Somali Current completely reverses, flowing now towards southwest along the coast of Somalia. No upwelling occurs during this season.

The seasonal reversal to the northeastward flowing Somali Current takes place at the onset of the southwest monsoon (April/May). Observations suggest that during spring and early summer the Somali Current is not a continuous feature extending along the Somalia coast, but consists of two large anticyclonic gyres, which are aligned parallel to the coast between the equatorial region and about 10° N (Bruce [73 B 2, 79 B]; Evans and Brown [81 E]). Most intense upwelling is encountered off the coast between 9° N and 11° N, where the northeastward flowing Somali Current leaves the coast and turns into the southwest Monsoon Current of the Arabian Sea (Fig. 21).

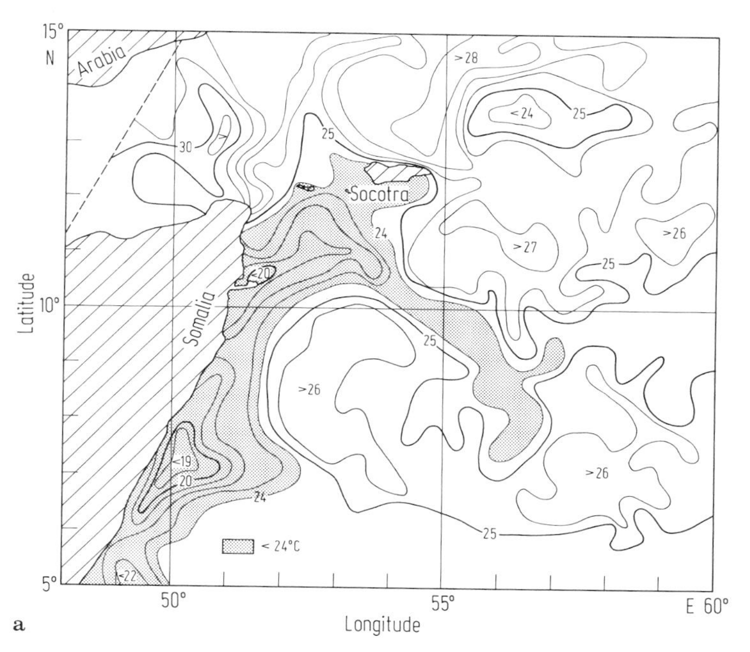

Fig. 21a. Radiation temperature [°C] of the sea surface in the western Arabian Sea on July 3, 1966 (from Szekielda [70 S]). The evaluation is based on radiation measurements of the weather satellite Nimbus 2. Hatched area has surface temperatures below 24 °C.

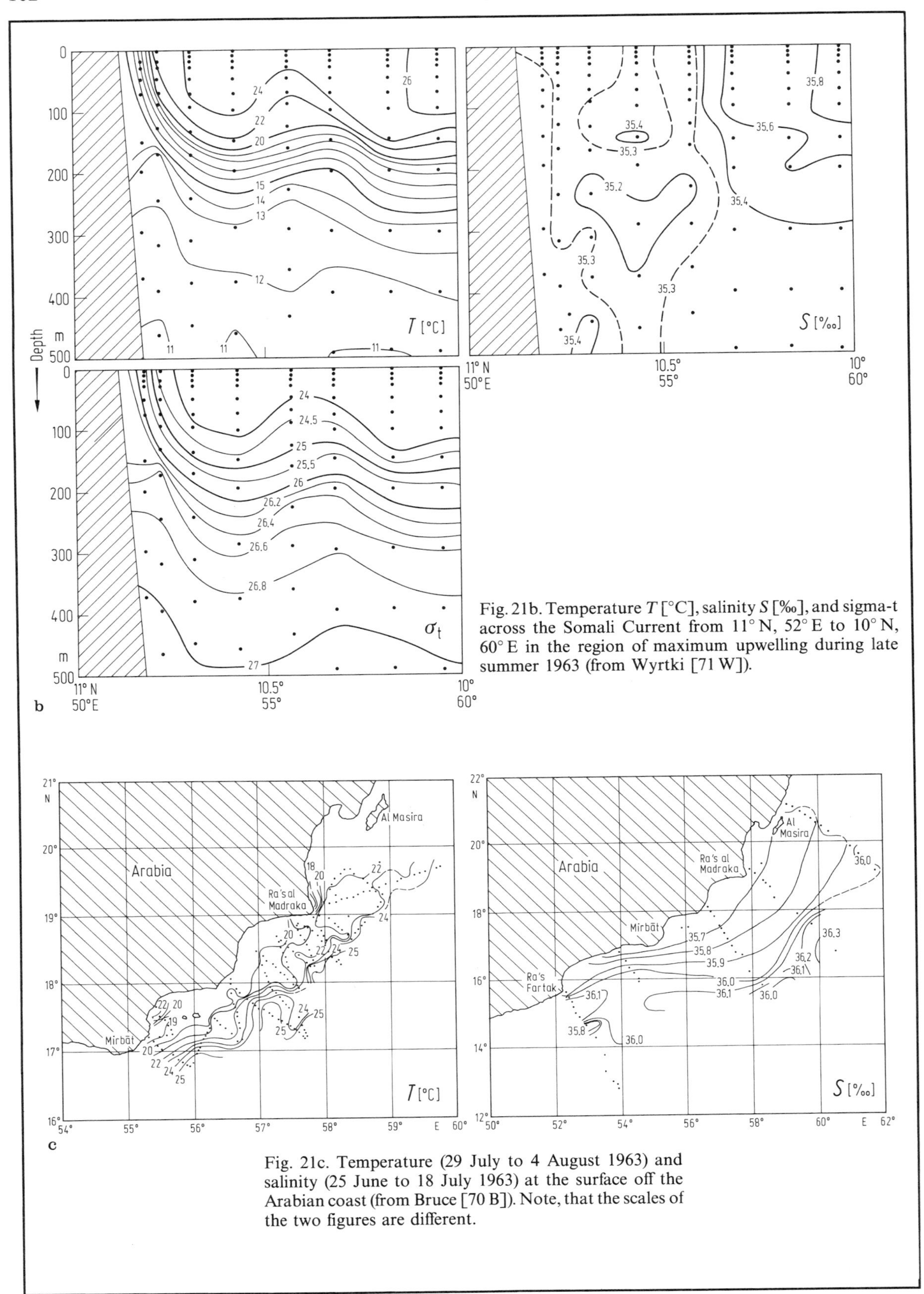

Fig. 21b. Temperature T [°C], salinity S [‰], and sigma-t across the Somali Current from 11° N, 52° E to 10° N, 60° E in the region of maximum upwelling during late summer 1963 (from Wyrtki [71 W]).

Fig. 21c. Temperature (29 July to 4 August 1963) and salinity (25 June to 18 July 1963) at the surface off the Arabian coast (from Bruce [70 B]). Note, that the scales of the two figures are different.

The minimum surface temperature observed during upwelling, there, is lower than 14 °C (Warren et al. [66 W]). The upwelled water comes from depths less than 200···300 m.

Besides the effect of the winds there are evidently further conditions contributing to the coastal upwelling off Somalia:

1. The banking effect due to baroclinic adjustment to the strong Somali Current.
2. The pronounced offshore flow in the north at about 10° N and at the northern flank of the southern gyre.

Recent investigations in the southern upwelling area suggest an southwestward flowing undercurrent below the northeastward flowing Somali Current at depths of about 150···350 m along the continental slope (Leetma et al. [81 L]; Schott [83 S]; Schott and Quadfasel [81 S 1]).

7.3.8 Arabian Sea Upwelling

In general, the strength and persistence of the winds during the southwest monsoon in the northwestern Indian Ocean are higher than are found in any other mid- or low-latitude region. The strong monsoon (maximum in July/August) causes an appreciable offshore Ekman transport at the surface connected with intense coastal upwelling (Smith and Bottero [77 S]).

According to the surface maps of temperature and salinity presented by Bruce [74 B] the coastal upwelling waters are more saline and warmer than the fresh upwelling water in the Somali Current region. But the temperature gradients at the surface of 5···7 °C indicate strong local upwelling processes (see also Fig. 21c). Strong upwelling occurs not only over the shelf but also in the open Arabian Sea. This is a result of the strong steady winds, nearly parallel to the coast, which increase in intensity with distance offshore resulting in a divergence in the surface Ekman layer over an area extending some 1000 km along the coast and 400 km offshore.

Estimates of the vertical velocities yield a magnitude of 10^{-3} cm s^{-1} for both the coastal and the open sea upwelling in the Arabian Sea (Smith and Bottero [77 S]). There exists a subsurface outflow of the relatively saline Persian Gulf water, whose core (salinity-maximum) is at depths less than 270 m along the southeast Arabian coast (Wyrtki [71 W]).

7.4 References for 7

05 E Ekman, V.W.: Ark. Mat. Astron. Fys. **12** (1905) 1.

36 B Böhnecke, G. (1936): Temperatur, Salzgehalt und Dichte an der Oberfläche des Atlantischen Ozeans, Wiss. Ergebn. Dt. Atlant. Exped. "Meteor", 1925–1927, **V**, Atlas.

36 D 1 Defant, A.: Die Troposphäre des Atlantischen Ozeans, Wiss. Ergebn. Dt. Atlant. Exped. "Meteor" 1925–1927, **VI**, **1**, (1936) 289.

36 D 2 Defant, A.: Das Kaltwasserauftriebsgebiet vor der Küste Südwestafrikas, Länderkdl. Forsch., Festschrift N. Krebs, **52** (1936) 52.

41 D Defant, A.: Die absolute Topographie des physikalischen Meeresniveaus und der Druckflächen im Atlantischen Ozean, Wiss. Ergebn. Dt. Atlant. Exped. "Meteor" 1925–1927, **6** (2) (1941) 191.

60 F Fuglister, F.C.: Atlantic Ocean Atlas, Woods Hole Oceanogr. Inst. **1960**, 209 pp.

60 H Hart, T.J., Currie, R.I.: Discovery Rep., Cambridge: Cambridge University Press, Vol. **XXXI** (1960) p. 123–298.

61 D Defant, A.: Physical Oceanography. Oxford, New York, Toronto, Sydney, Paris, Frankfurt: Pergamon Press Vol. **I** (1961) 729 pp.
62 R Reid, J.L., jr.: J. Mar. Res. **68** (16) (1962) 4819.
63 D Darbyshire, M.: Deep Sea Res. **10** (1963) 623.
66 D Deutsches Hydrographisches Institut: Handbuch der Süd- und Ostküste Afrikas, 1. Teil, Nr. 2045, Hamburg: Dt. Hydrogr. Inst. **1966**.
66 N Neumann, G., Pierson, W.J., jr.: Principles of Physical Oceanography. Englewood Cliffs, N.J., USA: Prentice-Hall Inc. **1966**, 545 pp.
66 P Pavlova, Y.V.: Oceanology **6** (1966) 806.
66 W Warren, B., Stommel, H., Swallow, J.C.: Deep Sea Res. **13** (1966) 825.
67 I Ingraham, W.J.: Fish. Bull. **66** (2) (1967) 223.
67 Y Yoshida, K.: Jpn. J. Geophys. **4** (2) (1967) 1.
68 M Mooers, C.N.K., Smith, R.L.: J. Geophys. Res. **73** (1968) 549.
68 S 1 Shell, I.L.: Dtsch. Hydrogr. Z. **21** (1968) 109.
68 S 2 Smith, R.L.: Oceanogr. Mar. Biol., Ann. Rev. **6** (1968) 11.
70 D De Decker, A.H.B.: Notes on an oxygen-depleted subsurface current off the west coast of South Africa, Investigational Report, Division Sea Fish., South Africa **84** (1970) 24 pp.
70 S Szekielda, K.H.: The development of upwelling along the Somali coast as detected with the Nimbus 2 and Nimbus 3 Satellites, Goddard Space Flight Center: Greenbelt Maryland X-651-70-419 **1970**, 52 pp.
70 W Wooster, W.S., Jones, J.H.: J. Mar. Res. **28** (2) (1970) 235.
71 G Gordon, A.L., in: Research in the Antarctic (Quam, L.W., ed.), Washington: AAAS, Publ. No. **93** (1971) 609.
71 W Wyrtki, K.: Oceanographic Atlas of the International Indian Ocean Expedition. Washington, D.C.: NFS **1971**, 531 pp.
72 M Maigret, J.: Campagne expérimentale de pêche des Sardinelles et autre espéces pélagiques, Juillet 1970–Oct. 1971. Ed.: Société Centrale pour l'Equipement du Territoire-International **1972**.
73 B 1 Bang, N.D.: Tellus **25** (1973) 256.
73 B 2 Bruce, J.G.: Deep Sea Res. **20** (9) (1973) 837.
73 C Cutching, D.L., Smith, R.L.: J. Phys. Oceanogr. **3** (1) (1973) 73.
73 H Halpern, D.: Tethys **6** (1973) 363.
74 B Bruce, J.G.: J. Mar. Res. **32** (1974) 419.
74 F Fraga, F.: Tethys **6** (2) (1974) 5.
74 H 1 Hughes, P., Barton, E.D.: Deep Sea Res. **21** (1974) 611.
74 H 2 Hughes, P., Barton, E.D.: Tethys **6** (1974) 43.
74 H 3 Huyer, A.: Observations of the Coastal Upwelling Region off Oregon during 1972, ph.D. thesis, Oregon State University **1974**, 149 pp.
74 H 4 Huyer, A., Smith, R.L., Pillsbury, R.D.: Tethys **6** (1974) 391.
74 Q Quinn, W.H.: J. Appl. Meteorol. **13** (1974) 825.
74 T Tomczak, M., jr., Käse, M.R.: J. Mar. Res. **32** (3) (1974) 365.
74 W Weichart, G.: Meteor Forschungsergeb. Reihe A **14** (1974) 33.
75 B Barton, E.D., Pillsbury, R.D., Smith, R.L.: Vertical sections of temperature, salinity and sigma-t from R/V GILLIS data and low-pass filtered measurements of wind and currents. A compendium of physical observations from JOINT-I, Data Report, References 75-17, Oregon State University **1975**, 60 pp.
75 D Dietrich, G., Kalle, K., Krauss, W., Siedler, G.: Allgemeine Meereskunde, Berlin-Stuttgart: Gebr. Borntraeger **1975**, 592 pp.
75 H Huyer, A., Pillsbury, R.D., Smith, R.L.: Limnol. Oceanogr. **20** (1975) 90.
75 M Mittelstaedt, E., Pillsbury, R.D., Smith, R.L.: Dtsch. Hydrogr. Z. **28** (1975) 145.
75 S Schemainda, R., Nehring, D., Schulz, S.: Geod. Geophys. Veröffentl. IV **16** (1975) 85 pp.
75 W Wyrtki, K.: J. Phys. Oceanogr. **5** (1975) 572.
76 A 1 Allan, J.S.: J. Phys. Oceanogr. **6** (4) (1976) 426.
76 A 2 Allan, J.S.: J. Phys. Oceanogr. **6** (6) (1976) 864.
76 H 1 Halpern, D.: Deep Sea Res. **23** (1976) 495.
76 H 2 Houghton, R.W.: J. Phys. Oceanogr. **6** (1976) 909.
76 H 3 Huyer, A.: J. Mar. Res. **34** (4) (1976) 531.
76 H 4 Huyer, A., Smith, R.L.: Trans. Am. Geophys. Union **57** (1976) 263 pp.
76 K 1 Kundu, P.K.: J. Phys. Oceanogr. **6** (2) (1976) 238.

76 K 2 Kundu, P.K., Allan, J.S.: J. Phys. Oceanogr. **6** (2) (1976) 181.
76 M 1 Mittelstaedt, E.: Dtsch. Hydrogr. Z. **29** (1976) 97.
76 M 2 Mooers, C.N.K., Collins, C.A., Smith, R.L.: J. Phys. Oceanogr. **6** (1976) 3.
76 P Peters, H.: Meteor Forschungsergeb. Reihe A **18** (1976) 78.
76 S Shaffer, G.: Meteor Forschungsergeb. Reihe A **17** (1976) 21.
76 W Wooster, W.S., Bakun, A., McLain, D.R.: J. Mar. Res. **34** (1976) 131.
77 G Gardner, D., in: A voyage of Discovery – George Deacon 70th Anniversary Volume (Angel, M., ed.), Suppl. to Deep Sea Res. **1977**, 305–326.
77 J Johnson, D.R.: Deep Sea Res. **24** (1977) 171.
77 N Nelson, C.S.: Wind stress and wind stress curl over the California Current, NOAA Techn. Rep. NMFS SSRF-714, US Dept. Comm. **1977**, 89 pp.
77 S Smith, R.L., Bottero, J.S., in: A voyage of Discovery – George Deacon 70th Anniversary Volume (Angel, M., ed.), Suppl. to Deep Sea Res. **1977**, 291–304.
78 A Adamec, D., O'Brien, J.J.: J. Phys. Oceanogr. **8** (6) (1978) 1050.
78 B Bryden, H.L.: Estuarine Coastal Mar. Sci. **7** (1978) 311.
78 H Halpern, D., Smith, R.L., Reed, R.K.: J. Geophys. Res. **83** (C 3) (1978) 1366.
78 S Speth, P., Detlefsen, H., Sierts, H.W.: Dtsch. Hydrogr. Z. **31** (1978) 95.
78 T Tomczak, M., jr.: Ann. Hydrogr. Ser. **5** (6) (1978) 5.
78 Z Zuta, S., Rivera, T., Bustamente, A.: Hydrologic Aspects of the Main Upwelling Areas off Peru, in: Upwelling Ecosystems (Boje, R., Tomczak, M., jr., eds.), Berlin, Heidelberg, New York: Springer **1978**, p. 235–256.
79 B Bruce, J.G.: J. Geophys. Res. **84** (C 12) (1979) 7742.
79 H 1 Hickey, B.M.: Progr. Oceanogr. **8** (4) (1979) 191.
79 H 2 Huyer, A., Gilbert, W.E.: Vertical sections and mesoscale maps of temperature, salinity and sigma-t off the coast of Peru, July to October 1976 and March to May 1977, Data Report 79, Ref. 79-16, Oregon State University **1979**, 162 pp.
79 M Martell, C.M., Allan, J.S.: J. Phys. Oceanogr. **9** (4) (1979) 696.
79 S Silva, S.N., Neshyba, S.: Deep Sea Res. **26** A (1979) 1387.
80 A Andrews, W.R.H., Hutching, L.: Progr. Oceanogr. **9** (1980) 1.
80 B 1 Barton, E.D., Argote, M.L.: J. Mar. Res. **38** (4) (1980) 631.
80 B 2 Brink, K.H., Halpern, D., Smith, R.L.: J. Geophys. Res. **85**, No. C 7 (1980) 4036.
80 B 3 Brockmann, Ch., Fahrbach, E., Huyer, A., Smith, R.L.: Deep Sea Res. **27** A (1980) 847.
80 G Guillén, G.O.: The Peru current system. Physical aspects, Proc. Workshop Phenomenon "El Niño", Paris: Unesco **1980**, p. 185–216.
80 H 1 Heburn, G.W.: A numerical model of coastal upwelling off Peru – Including mixed layer dynamics, ph.D. thesis, Dept. Meteorol., Tallahassee, Florida: Florida State University **1980**, 123 pp.
80 H 2 Huyer, A.: J. Phys. Oceanogr. **10** (11) (1980) 1756.
80 M Mittelstaedt, E., Weichart, G., Meier-Fritsch, H., Lüthje, H., Hamann, I.: Zur Hydrographie der Gewässer entlang der Küste von Mauretanien. Meteor-Fahrt Nr. 44, Deutsches Hydrographisches Institut Hamburg: Meereskdl. Beob. u. Ergebn. Nr. **51** (1980) 276 pp.
80 O O'Brien, J.J., Busalacchi, A., Kindle, J.: Ocean Models of El Niño, Resource Management and Environmental Uncertainty, New York: Wiley & Sons, Inc. **1980**, p. 159–212.
80 R Robles, F., Alarcón, E., Ulloa, A.: Water masses in the northern Chilean zone and their variations in a cold period (1967) and warm periods (1969, 1971–1973), Proc. Workshop Phenomenon "El Niño", Paris: Unesco **1980**, p. 83–174.
80 T Tomczak, M., jr., Hughes, P.: Meteor Forschungsergeb. Reihe A **21** (1980) 1.
80 Z Zuta, S., Enfield, D., Valdivia, J.: Physical aspects of the 1972–1973 "El Niño" phenomenon, Proc. Workshop Phenomenon "El Niño", Paris: Unesco **1980**, p. 11–81.
81 E Evans, R.H., Brown, O.B.: Deep Sea Res. **28** A (1981) 521.
81 H Hagen, E., Schemainda, R., Michelsen, N., Postel, L., Schulz, S.: Geod. Geophys. Veröffentl. IV **36** (1981) 99.
81 L Leetma, A., Quadfasel, D.R., Wilson, D.: J. Phys. Oceanogr. **12** (1981) 1325.
81 M 1 McCreary, J.P.: Philos. Trans. R. Soc. London **302** (1981) 385.
81 M 2 Mittelstaedt, E., Hamann, I.: Dtsch. Hydrogr. Z. **34** (1981) 81.
81 M 3 Molinari, R.L., Swallow, J.C., Bruce, J.G., Brown, O.B., Evans, R.H.: J. Phys. Oceanogr. **13** (1981) 1398.
81 S 1 Schott, F., Quadfasel, D.R.: J. Phys. Oceanogr. **12** (1981) 1358.
81 T Tomczak, M., jr.: Oceanologica Acta **4** (2) (1981) 161.

81 W Wyrtki, K.: J. Phys. Oceanogr. **11** (1981) 1205.

82 F 1 Fiúza, A.F.G.: Oceanologica Acta **5** (1) (1982) 31.

82 F 2 Fiúza, A.F.G.: The Portuguese Coastal Upwelling System, in: Proceedings of the seminar on "Present Problems of Oceanography in Portugal", Lisbon 19–20 Nov. 1980, Lisboa: Junta Nacional de Investigacão Cientifica e Tecnológica **1982**.

82 F 3 Fiúza, A.F.G., Halpern, D.: Rapp. P.V. Reun. Cons. Int. Explor. Mer **180** (1982) 58.

82 M Mittelstaedt, E.: Rapp. P.V. Reun. Cons. Int. Explor. Mer **180** (1982) 50.

82 V Voiturieux, B., Herbland, A.: Rapp. P.V. Reun. Cons. Int. Explor. Mer **180** (1982) 114.

83 S Schott, F.: Progr. Oceanogr. **12** (1983) 357.

8 Ice in the ocean

For properties of glacial ice, see LB, NS, vol. V/1b, ch. 8.

8.0 List of symbols

a	empirical coefficient depending upon the rate of sea ice growth, eq. (19), in [$cm^{-1/2}$]
a_0	plate spacing, i.e. the distance between centers of successive brine layers measured parallel to the optical c-axis, Fig. 9
b	empirical coefficient denoting the ratio of the salinity of pack ice at the end of the winter ice growth to the salinity of brine, $b=0.13$, eq. (19)
b_0	center to center spacing of brine cells as measured in the b-direction, Fig. 9
C_l	velocity of longitudinal sound waves
C_t	velocity of transverse sound waves
c	specific heat capacity of sea ice, in [$kJ(kg\cdot K)^{-1}$]
c_b	specific heat capacity of brine, in [$kJ(kg\cdot K)^{-1}$]
c_{fwi}	specific heat capacity of freshwater ice, in [$kJ(kg\cdot K)^{-1}$]
c_s	specific heat capacity of salts, in [$kJ(kg\cdot K)^{-1}$]
E	Young's modulus, in [GPa]
f	ratio of the thermal conductivity of air bubbles to the thermal conductivity of pure ice, $f=\lambda_a/\lambda_p$
G	shear modulus, in [GPa]
g	length of brine cell measured parallel to the growth direction, Fig. 9
g_0	distance between adjacent ice bridges separating brine cells in the growth direction, Fig. 9
H	thickness of sea ice, in [cm]
H_j	thickness of sea ice in part j of the area
$\tilde{H}$	mean thickness of sea ice
K	bulk modulus, in [GPa]
k	stress concentration coefficient
L	latent heat of fusion of sea ice, in [$kJ\,kg^{-1}$]
L_{eff}	effective heat of fusion of sea ice, in [$kJ\,kg^{-1}$]
L_{fwi}	latent heat of fusion of freshwater ice, in [$kJ\,kg^{-1}$]
L_s	latent heat of fusion of salts, in [$kJ\,kg^{-1}$]
l	empirical constant, eq. (21), in [cm^{-1}]
M	total mass of sea ice, in [kg]
M_s	mass of precipitated salts, in [kg]
m	mass of a solid sphere
N	concentration of sea ice, i.e. the amount of sea surface covered by ice as fraction of the area considered
δN_h	concentration of sea ice going into the formation of hummocks and pressure ridges
N_j	concentration of sea ice in part j of the area
n	porosity of sea ice
r_a, r_b	semiaxes of the brine cell cross section in c- and b-direction, respectively, Fig. 9
S	salinity of surface water
S_b	salinity of brine
s	salinity of sea ice
s_{max}	maximum salinity of sea ice by which the ice has still the flexural strength
T	temperature
T_b	temperature of brine
t	time
V	velocity of pack ice
v	specific volume of sea ice, in [$m^3\,kg^{-1}$]
W	total energy of impact
z	height

α	proportionality constant between brine and temperature, $(1+\alpha\Theta)^2\partial S_b/\partial\Theta$
α_k	impact strength of sea ice, in [MPa · m]
β_0	relative spacing of brine cells, $\beta_0 = b_0/a_0$
γ	cubic thermal expansion coefficient of pure ice, in [(°C)$^{-1}$]
δ	ratio of the length of the brine cell to the thickness of the ice bridge between the cells, $\delta = g/g_0$
ε	ratio of the major to the minor semiaxes of the brine cell cross section, $\varepsilon = r_b/r_a$
ε_f	specific energy of fragmentation, in [MPa]
Θ	temperature of sea ice, in [°C]
ϑ	temperature at which sea ice completely melts, in [°C]
λ	thermal conductivity of sea ice, in [W(m · K)$^{-1}$]
λ_a	thermal conductivity of air bubbles, in [W(m · K)$^{-1}$]
λ_b	thermal conductivity of brine, in [W(m · K)$^{-1}$]
λ_{pi}	thermal conductivity of porous ice, in [W(m · K)$^{-1}$]
λ_p	thermal conductivity of pure ice, in [W(m · K)$^{-1}$]
μ	Poisson's ratio
ν_a	air porosity
ν_b	relative brine volume, $\varrho = \varrho_k(1-\nu_b)$
ν_0	relative brine volume at which the ice strength disappears
ϱ	density of a thin plane-parallel ice plate, in [kg m^{-3}]
ϱ_b	density of brine, in [kg m^{-3}]
ϱ_i	actual density of ice, in [kg m^{-3}]
ϱ_k	ice density without cavities, in [kg m^{-3}]
ϱ_{pi}	density of porous ice, in [kg m^{-3}]
ϱ_p	density of pure ice, in [kg m^{-3}]
σ_f	failure strength of sea ice, in [MPa]
σ_0	basic strength of sea ice, in [MPa]
$\dot{\sigma}$	stress rate or rate of loading, $\dot{\sigma} = d\sigma/dt$ in [MPa s^{-1}]
ϕ_{lat}	lateral heat flux for melting of sea ice, in [W m^{-2}]
ψ	porosity coefficient, i.e. the ratio of the area filled by brine cells and air bubbles to the total area of the sample cross section

8.1 Structure of sea ice

8.1.1 Density of sea ice

Sea ice is a polyphase system; consequently its density changes in a complex manner. In general the density of sea ice increases with increasing salinity and decreasing temperature. The relationship between the specific volume v of ice, its salinity s, as well as the density ϱ_b and the concentration (salinity) of the brine S_b is expressed by the formula [45 z]:

$$v = \frac{s}{S_b} \cdot \frac{1}{\varrho_b} + \left(1 - \frac{s}{S_b}\right)\frac{1}{\varrho_p}, \tag{1}$$

where

$$\varrho_b \approx 1000 + 800\, S_b,$$

$$\varrho_p = \frac{916.8}{1+\gamma\Theta}.$$

Both ϱ_b and ϱ_p, the density of pure ice, are in [kg m^{-3}]. Θ denotes the temperature, in [°C], and γ the thermal expansion coefficient of pure ice, in [°C^{-1}]. Results for the density of solid sea ice are represented in Table 1.

Table 1. Density of solid sea ice v^{-1} vs. temperature Θ and salinity s [75 d, 77 d].

Θ [°C]	−2	−4	−6	−8	−10	−15	−20	−23
s [10^{-3}]	v^{-1} [m^{-3} kg]							
2	924	922	920	921	921	922	923	923
4	927	925	924	923	923	923	925	925
6	932	928	926	926	926	925	926	926
8	936	932	929	928	928	928	929	929
10	939	935	931	929	929	929	930	930
15	953	944	939	937	935	934	935	935

The density of sea ice changes considerably from season to season. In accordance with [71 P], the density of first-year ice is 860···920 kg m^{-3}, and for multi-year ice 830···900 kg m^{-3}. In summer the density of very weathered sea ice decreases to 560···640 kg m^{-3} as result of the outflow of brine; the density of multi-year ice varies between 620···933 kg m^{-3}. The highest density occurs in the middle layers of the sea ice, the lowest in the uppermost and lowermost layers.

8.1.2 Porosity of sea ice

It is determined by air and salt inclusions occurring in sea ice. The porosity n and the actual density ϱ_i of the ice are linked together by the relationship [75 d, 77 d]:

$$n = \frac{916.8 - \varrho_i}{\varrho_i}, \tag{2}$$

with ϱ_i in [kg m^{-3}]. The change of the porosity of sea ice with salinity s and temperature Θ, calculated in accordance with [45 z], is shown in Table 2.

Table 2. Porosity n of sea ice vs. temperature Θ and salinity s, with replacement of salt cells by air [75 d, 77 d].

Θ [°C]	−2	−4	−6	−8	−10	−15	−20	−23
s [10^{-3}]	n [10^{-2}]							
2	4.9	2.5	1.8	1.4	1.1	0.8	0.6	0.6
4	9.9	5.1	3.6	2.8	2.3	1.7	1.3	1.2
6	14.9	7.7	5.5	4.2	3.4	2.6	1.9	1.8
8	20.0	10.3	7.3	5.7	4.6	3.4	2.6	2.3
10	25.1	12.9	9.2	7.1	5.8	4.3	3.3	2.9
15	38.2	19.5	13.9	10.7	8.7	6.3	5.0	4.4

8.2 Salinity of sea ice

Sea ice contains, besides air inclusions and other impurities, salts which occur not only in a solid state but also in liquid phase with high salt concentration (brine). Most salts in the solid state fill the space between the ice crystals, while the brine is contained in intercrystalline layers, in capillaries, and in the form of closed cells (pockets).

The salts contained in sea water have different crystallization temperatures. The eutectic temperature for the principle salts of sea water in H_2O are given in Table 3. For these binary systems, no liquid phase exists below the respective eutectic temperature. Owing to the fact that practically all chemical elements are contained in sea water, the crystallization process of sea water differs from that of a binary solution and salts may exist in sea water even below the respective eutectic temperature of the salt in pure water. Apparently, the presence of ions of other salts is responsible for the fact that, for example, Na_2SO_4 precipitates from sea water at the temperature of −7.6 °C, and not at −3.5 °C, as would be the case with a pure solution of this salt. KCl begins to precipitate at −34.2 °C, $MgCl_2$ at −43.2 °C, and $CaCl_2$ at −54 °C. In some laboratory experiments, traces of brine had still been detected after the samples had been cooled down to a temperature of −80 °C [65 p].

The mass determined by several authors of the precipitated salts at different temperatures are given in Table 4.

Table 3. Eutectic temperature of salts in H_2O [75 d, 77 d].

Salt	$CaCO_3$	Na_2CO_3	K_2SO_4	Na_2SO_4	$MgSO_4$
Eutectic temperature [°C]	− 1.9	− 2.1	− 2.9	− 3.5	− 3.9
Salt	KCl	$CaSO_4$	NaCl	$MgCl_2$	$CaCl_2$
Eutectic temperature [°C]	−11.1	−17	−21.1	−33.6	−55

Table 4. Mass of crystalline salts (g per kg of ice) for ice with salinity $s=10^{-2}$ of various temperatures [75 d, 77 d].

Author	T [°C]	−7.6	−9.5	−10.6	−12.3	−17.0	−22.6
Nazintsev		0.049	0.533	0.704	0.849	1.05	1.90
Tsurikov		0.05	0.54	0.68	0.82	0.98	1.80
Savel'ev		0.06	0.54	0.69	0.81	1.53	9.2
	T [°C]	−24.2	−26.0	−28.0	−30.8	−34.2	−35.5
Nazintsev		5.39	6.74	7.28	7.69	7.98	8.07
Tsurikov		5.28	6.61	7.12	7.50	7.89	
Savel'ev		9.06	9.33	9.27	9.30	9.47	9.47

To obtain the mass of crystalline salts for ice of a different salinity it is only necessary to multiply the mass in Table 4 by the ratio of this salinity to 10^{-2}.

Brine concentration S_b vs. temperature is represented in Fig. 1, which is constructed on the basis of empirical data of several authors: brine salinity increases with decreasing temperature; below the crystallization temperature of NaCl the increase in brine concentration is small because the main mass of salts is already in the solid state.

The mass of the brine in sea ice at different temperatures is represented in Fig. 2. Moreover, the quantity of brine depends upon the migration rate of brine through ice. In accordance with ideas to date, this is essentially determined by the following factors: the temperature gradient in the ice, which gives the direction to higher temperature of brine cell migration; the gravitation which assists the downward drainage of the brine; the hydrostatic pressure squeezing out the brine from the cells. The main mass of the brine flows out of the ice during the period of ice formation, when the ice is still comparatively thin and its temperature is comparatively high, as well as in summer.

The salinity of ice depends very much upon the air temperature and the wind speed during the formation of the ice cover. At lower temperatures, the growth rate of ice crystals is greater than the growth rate at higher temperatures. As result of their larger specific surfaces at low temperatures, they retain a larger quantity of brine. Wind and swell mix the crystals chaotically, so that only a smaller quantity of brine can discharge as compared to an ordered, particularly vertical, orientation of the crystals. Therefore, low air temperatures and high wind speed facilitate the formation of ice of increased salinity. The salinity of ice is particularly high when sea water freezes with large quantities of snow on the water surface. The snow retains a large quantity of sea water because of the web-like gossamer structure of the flakes. In Fig. 3 this effect is illustrated by comparison of salinity profiles of young ice which originated under different weather conditions near the coast of the Labrador Peninsula. The upper 7.5 cm thick layer shown in Fig. 3a was formed from a mixture of snow flakes and ice crystals; its salinity in the initial period was only half the salinity of the water from which it was formed.

The salinity of ice decreases very rapidly when ice begins to melt. As a result of the rise in temperature, the mass of the brine increases because the ice on the walls of cells and capillaries melts. Different cells merge together forming capillaries of continuously increasing diameter. Thereby, the migration of brine is facilitated and a rapid decrease of the salinity is attained.

Fig. 1. Brine concentration S_b during freezing of water. Data of Gitterman, Nelson and Thompson, and Ringer, respectively [64 N].

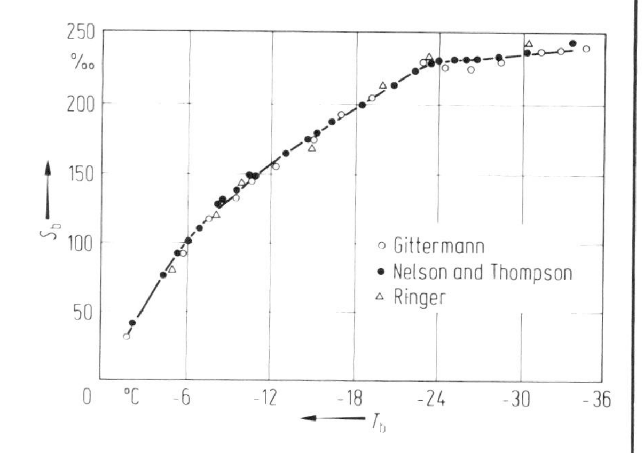

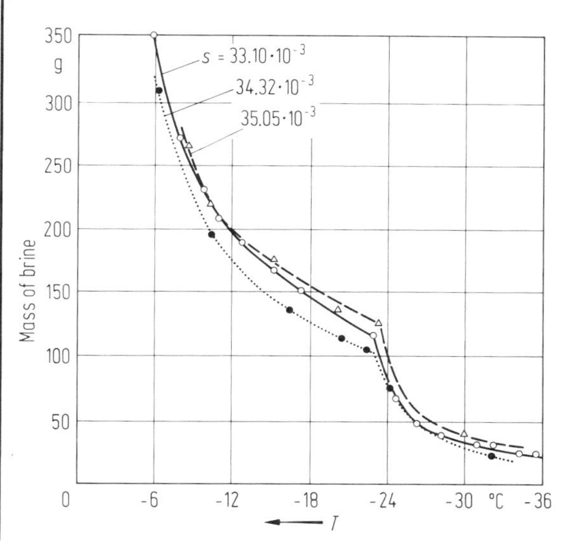

Fig. 2. Mass of brine in 1 kg of sea ice vs. temperature for different salinities s of the ice.
Open circles: $s = 33.10 \cdot 10^{-3}$ (Gitterman's data).
Solid circles: $s = 34.32 \cdot 10^{-3}$ (Assur's data).
Triangles: $s = 35.05 \cdot 10^{-3}$ (Ringer's data) [64 N].

Fig. 3. Salinity profile of young sea ice grown under different weather conditions near Labrador Peninsula. Profiles are shown for consecutive days. The ice was formed (a) in a period of wind-induced swell and (b) under calm conditions [58 W 2]. The dashed line in Fig. 3a separates top and bottom layer. In the top layer the ice consists of a mixture of snow and ice crystals, while the bottom layer is formed of congelation ice.

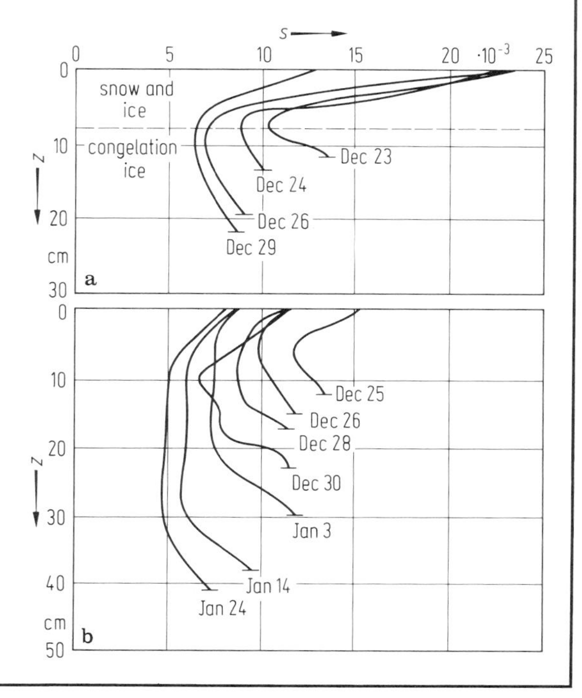

8.3 Thermophysical properties of sea ice

8.3.1 Specific heat capacity of sea ice

As sea ice consists of several components (important are: fresh water ice, brine, air, and salts in solid state), its specific heat capacity is determined from the specific heat capacities and the masses of the components. With every change in temperature each component changes its volume: however, the ratio between solid and liquid phases of sea ice also changes, so that it is necessary to know the amount of salt which solidifies when freezing or which goes into solution when melting, as well as the concentration of brine.

The definitive equation for the specific heat capacity c of sea ice is [75 d, 77 d]:

$$c = c_{\mathrm{fwi}}\left(1 - \frac{s}{S_{\mathrm{b}}} + \frac{M_{\mathrm{s}}}{MS_{\mathrm{b}}}(1 - S_{\mathrm{b}})\right) + c_{\mathrm{b}}\left(\frac{s}{S_{\mathrm{b}}} - \frac{M_{\mathrm{s}}}{MS_{\mathrm{b}}}\right)$$

$$+ c_{\mathrm{s}}\frac{M_{\mathrm{s}}}{M} - L_{\mathrm{fwi}}\frac{s}{S_{\mathrm{b}}^2}\left(1 - \frac{M_{\mathrm{s}}}{Ms}\right)\frac{\mathrm{d}S_{\mathrm{b}}}{\mathrm{d}\Theta}$$

$$- L_{\mathrm{s}}\frac{1}{M}\left(\frac{1}{S_{\mathrm{b}}} - 1\right)\frac{\mathrm{d}M_{\mathrm{s}}}{\mathrm{d}\Theta}, \qquad (3)$$

with c_{fwi}, c_{b}, and c_{s} denoting the specific heat capacity of, respectively, fresh water ice, brine, and salts, while L_{fwi} and L_{s} are the latent heat of fusion of fresh water ice and salts, respectively. M is the total mass of sea ice, M_{s} the mass of precipitated salts.

The following empirical formulae can be used for the specific heat capacities [75 d, 77 d]:

$$c_{\mathrm{fwi}} = (2.12 + 0.0078\Theta)\,\mathrm{kJ\,kg^{-1}\,K^{-1}}, \quad \Theta \text{ in } [^{\circ}\mathrm{C}]$$

$$c_{\mathrm{b}} = (4.1868 - 4.55\,S_{\mathrm{b}})\,\mathrm{kJ\,kg^{-1}\,K^{-1}}$$

$$c_{\mathrm{s}} = 0.8\,\mathrm{kJ\,kg^{-1}\,K^{-1}}. \qquad (4)$$

The data concerning the phase composition of the ice can be taken from Figs. 1 and 2 and Table 4. The specific heat capacity of sea ice as determined from the definitive equation is given in Table 5.

Table 5. Specific heat capacity c of sea ice vs. temperature Θ and salinity s of sea ice, calculated from the definitive equation (3) [75 d, 77 d].

$s\,[10^{-3}]$	2	5	10	15	20
$\Theta\,[^{\circ}\mathrm{C}]$	$c\,[\mathrm{kJ\,kg^{-1}\,K^{-1}}]$				
− 5.6	3.25	5.03	7.97	10.87	13.88
−10.6	2.31	2.65	3.24	3.83	4.41
−15.0	2.14	2.36	2.73	3.10	3.46

8.3.2 Latent heat of fusion of sea ice

Unlike fresh water ice, sea ice does not melt at a specified temperature. Instead, it melts in a whole temperature range. The effective heat L_{eff} which is required in the melting of a unit mass of sea ice of temperature Θ depends upon the heat necessary to melt the pure ice contained in sea ice and the heat necessary for increasing the temperature of the ice and the brine up to the temperature ϑ at which the sea ice completely melts.

The equation for the calculation of the effective heat of fusion of a unit mass of sea ice of temperature Θ and salinity s is [75 d, 77 d]:

$$L_{\mathrm{eff}} = [c_{\mathrm{fwi}}(1 - s) + c_{\mathrm{b}}s]\,(\Theta - \vartheta) + (c_{\mathrm{b}} - c_{\mathrm{fwi}})\frac{s}{\alpha}\ln\frac{\Theta}{\vartheta} - L_{\mathrm{fwi}}\left(1 - \frac{s}{S_{\mathrm{b}}}\right)$$

with α the proportionality constant between S_{b} and temperature, i.e. $(1 + \alpha\Theta)^2\partial S_{\mathrm{b}}/\partial\Theta$.

In Table 6 the influence of the migration of brine during fusion under natural conditions upon the actual heat consumption is not taken into account.

Table 6. Heat L_{eff} required for complete fusion of sea ice of temperature Θ and salinity s [75 d, 77 d].

$s\,[10^{-3}]$	0	1	2	4	6	8
Θ [°C]	L_{eff} [kJ kg^{-1}]					
−0.5	335	300	264	194	124	53
−1.0	336	318	301	266	230	195
−2.0	338	329	320	302	284	264
−3.0	340	334	328	316	303	291

8.3.3 Thermal conductivity of sea ice

The thermal conductivity of sea ice depends upon the thermal conductivities of its components and their arrangement. Their respective ratio changes during the course of time by the action of external factors. The thermal conductivity λ_b of brine is about a quarter of the thermal conductivity λ_p of pure ice; the molecular thermal conductivity λ_a of air bubbles is smaller by two orders of magnitude than λ_p. Therefore, with an increase in the salinity and porosity of the ice, the thermal conductivity decreases.

In accordance with [64 N]

$$\lambda_p = 2.22(1-0.0159\Theta)\ \mathrm{W\,m^{-1}\,K^{-1}}. \tag{5}$$

The thermal conductivity λ_{pi} of porous ice is [63 S]:

$$\lambda_{pi} = \lambda_p \frac{1+0.5f-v_a(1-f)}{1+0.5f+0.5(1-f)v_a}, \tag{6}$$

with $f=\lambda_a/\lambda_p$ and air porosity v_a, see Fig. 4.

The thermal conductivity λ of sea ice with density ϱ_{pi} is [75 d, 77 d]:

$$\lambda = \lambda_{pi} - (\lambda_{pi}-\lambda_b)\left[1+\frac{\varrho_b(S_b-s)}{\varrho_{pi}s(1-S_b)}\right]^{-1}. \tag{7}$$

From this follows that the thermal conductivity of sea ice decreases with increasing salinity s and increases with growing brine concentration S_b, i.e. with decrease in temperature. This correlation is shown in Fig. 5, which also contains the data gained from experimental determinations of the thermal conductivity of laboratory specimens of ice of varying salinity.

On the basis of measurements of λ, which were undertaken in several layers of an ice sheet, the thermal conductivity in the top layers of sea ice is somewhat reduced due to the increased porosity. Comparatively smaller values of λ were ascertained in the lower layers of sea ice also, because there the temperature is higher and the brine content larger, see Fig. 6.

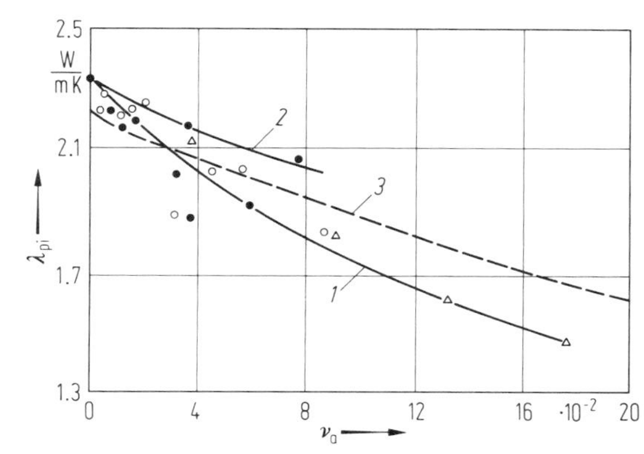

Fig. 4. Thermal conductivity λ_{pi} of porous ice vs. air porosity for ice with (*1*) fine pores and (*2*) large pores, and (*3*) as calculated from eq. (6) for a molecular thermal conductivity of air bubbles $\lambda_a = 0.03\ \mathrm{W\,m^{-1}\,K^{-1}}$. The experimental data was obtained by the electrical analogy method (solid circles), the regular regime method (open circles), by V.V. Shuleikin's determinations (open triangles), and by Abel's formula (solid triangles) [75 d, 77 d].

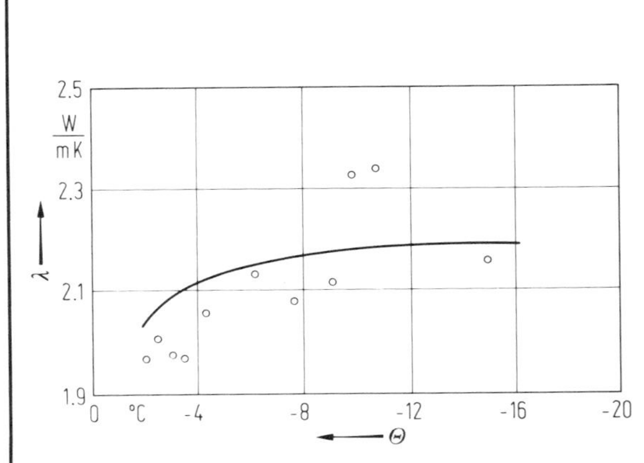

Fig. 5. Thermal conductivity λ of sea ice vs. temperature. Dots indicate experimental data obtained from laboratory specimens of varying salinity s. The solid line represents λ as calculated from eq. (7) for $s=4.7\cdot10^{-3}$ [75 d, 77 d].

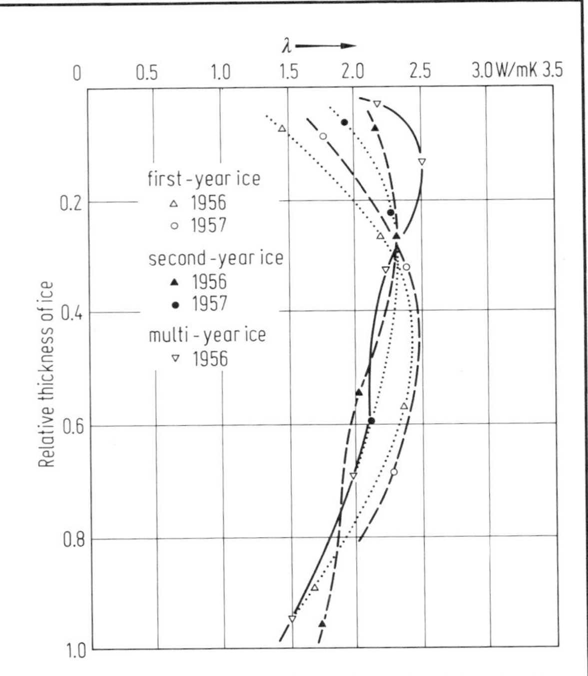

Fig. 6. Vertical profile of thermal conductivity λ of ice sheets. Profiles are depicted for measurements in 1956 on, respectively, first-year, second-year, and multi-year ice, as well as measurements in 1957 on first-year and second-year ice, respectively [64 N].

8.4 Elastic properties and deformations of sea ice

An ice cover, during the whole period of its existence, is subject to actions of different forces. These can be evoked by ice movements, compressions, and other factors. Thereby, a state of tension occurs in the ice cover, which results in crystal structure-changing deformations in the ice.

The elastic part of the deformation is determined by the ratio of the specific deformation energy to the internal energy of the crystal lattice. As the deformation energy per unit mass of the ice, even for decaying ice, is about an order of magnitude smaller than the internal energy of the crystal lattice, the elastic deformations are small. They are related, through Hooke's Law, with the respective stress components. For isotropic bodies this relation contains two elastic constants – the shear modulus G and the bulk modulus K. In place of these constants, the modulus of normal elasticity E (Young's modulus) and Poisson's ratio μ are often used.

The investigation of elastic deformations of sea ice is complicated by the fact that ice in the relevant temperature range represents a multicomponent, multiphase system. The solid component of sea ice is, more or less, pure fresh water ice that forms a continuous structure. Elastic deformations of this structure determine the elastic behaviour of ice as a whole. Therefore, the elastic constants of ice depend only little upon the temperature and salinity of the ice. The liquid phase contained in sea ice (brine) distorts the geometrical structure of ice and leads to a change of its elastic properties, which is the larger the more extensive the distortions are. Such large distortions occur during the quick melting and decay of ice. That is why, under these conditions, the elastic properties decrease.

The elasticity of the ice cover is determined by the elastic constants of each individual ice crystal, by the size and orientation of the crystals, and – to a certain extent – by the impurities and inclusions present. Owing to its fine-grained structure, sea ice can be viewed as an isotropic body. In any case, it is statistically isotropic in the plane of freezing.

A sea ice cover presents a certain anisotropy of its elastic properties in the vertical direction, which is associated with the vertical temperature gradients present and the irregular vertical structure of the ice cover.

By means of seismic investigations, one gains mean integrated values of the elastic properties of ice. For this, the sea ice cover is viewed as a thin plane-parallel plate that is known to be transversally isotropic. Then the elastic constants can be determined according to the following formulae by measuring the propagation velocities of longitudinal and transverse sound waves, C_l and C_t, respectively:

$$C_l = \sqrt{\frac{E}{\varrho(1-\mu^2)}}, \quad C_t = \sqrt{\frac{E}{2\varrho(1+\mu)}}$$
$$E = \varrho C_l^2(1-\mu^2); \quad G = \varrho C_t^2; \quad \mu = 1 - 2\left(\frac{C_t}{C_l}\right)^2. \tag{8}$$

For local measurements of the elastic properties of sea ice, ultrasonic methods are used. They are carried out in situ and under laboratory conditions upon small samples sawn out of the ice cover.

The values of the elastic constants of sea ice gained with the aid of seismic and acoustic methods harmonize fairly well [54 O, 57 L, 58 B, 58 L, 62 L, 71 S].

On the ice island *T*-3, Crary obtained the values: $E = 7$ GPa; $G = 2.6$ GPa; $\mu = 0.325$. With the aid of seismic methods, the following values [54 O] were gained in the arctic ice fields: $E = 6 \cdots 8.5$ GPa; $G = 2.1 \cdots 2.2$ GPa; $\mu = 0.32 \cdots 0.37$.

Table 7 gives the values of the elastic constants of the first-year ice and multi-year ice determined with the aid of ultrasonic methods.

Table 7. Experimental values of elastic constants for different layers of, respectively, first-year and multi-year ice [58 B 1]. Layer depth is increasing from left to right.

	First-year ice				Multi-year ice				
Layer [cm]	0···20	20···130	130···190	average	0···90	90···200	200···215	210···300	300···400
Density [g cm^{-3}]	0.897	0.91	0.913		0.9	0.911	0.9	0.913	0.913
E	2.70	7.50	4.08	5.95	5.85	8.95	5.30	7.96	4.56
G	1.02	2.86	1.54	2.25	2.20	3.36	2.04	2.96	1.69
E/G	2.66	2.62	2.65	2.66	2.66	2.66	2.65	2.70	2.70
K/E	1.02	1.14	1.08	1.02	1.02	1.02	1.08	0.90	0.90
μ	0.33	0.31	0.32	0.33	0.33	0.33	0.32	0.35	0.35

The value of Poisson's ratio μ may be viewed as being almost independent of salinity and temperature. The following values can also be given: $\mu = 0.295$ and $\mu = 0.29$ [65 p].

The elastic constants depend little upon the temperature and salinity of the ice, as the following empirical formulae show:

$$E = 8.76 - 0.021\Theta - 0.00017\Theta^2 \text{ [48 B]}$$
$$E = 10.00 - 0.0351 v_b \text{ [64 P]} \tag{9}$$
$$E = 32.3\varrho - 20.7 + 0.14\Theta - 0.59e^{-0.0129\Theta} \text{ [58 T]}$$

with E in [GPa], Θ in [°C], and density ϱ in [kg m^{-3}]. Brine content v_b is related to ϱ and the density ϱ_k of ice without cavities by

$$\varrho = \varrho_k(1 - v_b).$$

Fig. 7 shows the change of Young's modulus E with the brine content of the ice. The solid line shows the decrease of E for sea ice with increasing brine content. The dashed line describes the decrease of E when the brine flows out of the cells and the salinity of the ice, and thereby the density, decreases.

The changing of E for a 5 m thick cover of multi-year ice during the course of a whole year is about 15% (Table 8).

Fig. 8 gives the observational results in first-year ice [75 d, 77 d]. A strong decrease in E was observed only in the period of ice decay.

Koslowski

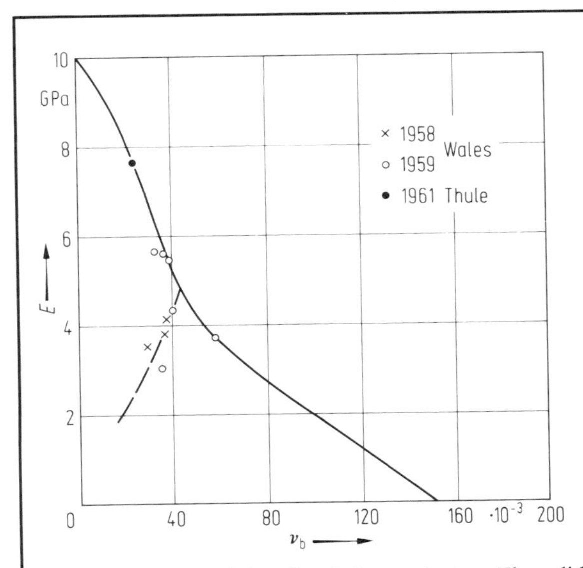

Fig. 7. Young's modulus E vs. brine content v_b. The solid line gives the change in E for ice with increasing brine content v_b, the dashed line for ice with brine flowing out and s decreasing. Observations are from Wales, Alaska, in the years 1958 and 1959, and from Thule, Greenland, in 1961 [63 B].

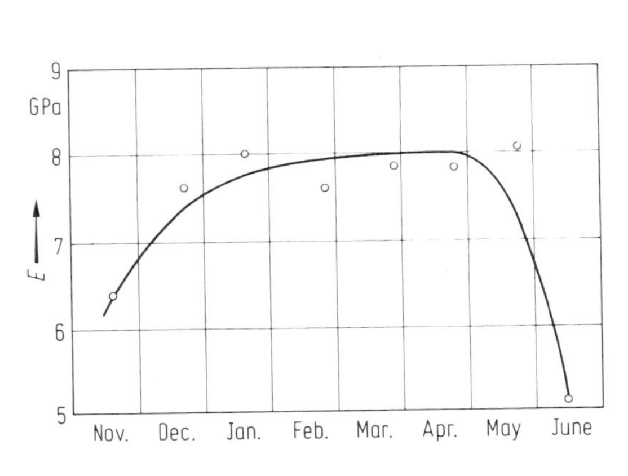

Fig. 8. Change of Young's modulus E of first-year ice in annual cycle of observations [75 d, 77 d].

Table 8. Elastic properties of multi-year ice of 5 m thickness during the course of a year [71 S].

Date of measurement	June 19	Aug. 23	Oct. 6	Oct. 18	Nov. 20	Feb. 18
E [GPa]	9.0	7.7	7.7	8.3	8.5	9.0
μ	0.44	0.36	0.37	0.30	0.43	0.44

8.5 Strength of sea ice

8.5.1 Structural model of strength of sea ice

In cold sea ice, the brine cells are small and the spaces between them are large. The surface of fracture occurring because of tensile and shear stresses runs mainly through pure ice, so that here the strength is only somewhat less than that of fresh water ice. With higher temperatures, the sea ice contains a large volume of brine. The surface of fracture then runs almost exclusively through the brine cells. The strength in this case for fracture is almost zero. Hence, the failure strength σ_f of sea ice changes from a maximum with a brine volume $v_b = 0$ to the value of zero when a certain critical brine volume is reached.

In general, σ_f is described by

$$\sigma_f = \sigma_0(1 - \psi) . \tag{10}$$

σ_0 can be considered the basic strength of sea ice, i.e. the strength of fictitious material that contains no brine, but still possesses the sea ice substructure and fails as a result of the same mechanism that causes failure in natural sea ice.

In the model structure of sea ice, Fig. 9, developed by Anderson and Weekes [58 A 1] and supplemented by Assur [58 A 2], ψ was determined as function of temperature and salinity.

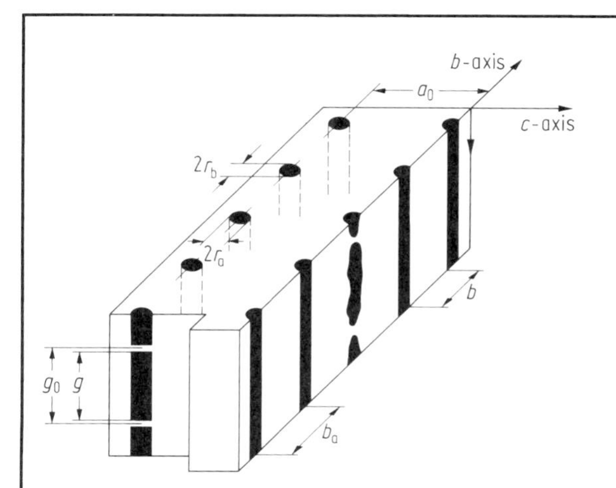

Fig. 9. Model structure of sea ice developed by Anderson and Weekes [58 A 1] and supplemented by Assur [58 A 2], defining parameters a_0, b, b_0, g, g_0, and r_a, r_b, respectively [75 d, 77 d].

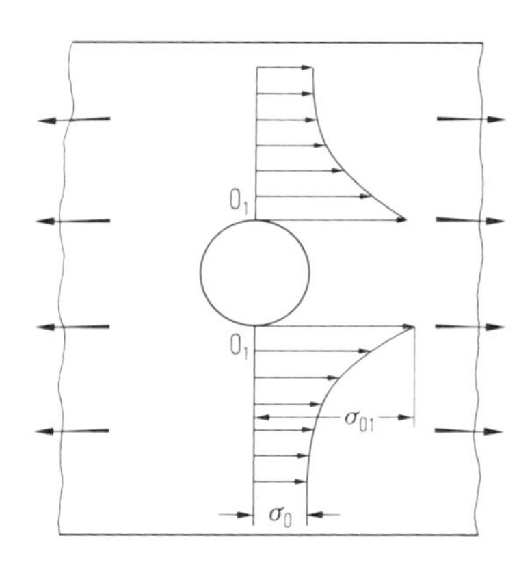

Fig. 10. Stress concentration near spherical brine cell, showing stress σ_0 far from the cell and increased stress σ_{01} at points O_1 of the cell [75 d, 77 d].

In the model, it was assumed that the volume of the air bubbles is small and the brine is concentrated in cells with elliptical cross sections which are arranged in more-or-less regular rows. The solid component consists of almost pure ice which is divided into parallel plates by the rows of brine cells. The optical c-axes are directed perpendicularly to these plates.

In the direction of the c-axes, the tensile strength is a minimum because of the decrease in the area of the cross section of the solid component. In the direction of the b-axis, the shear strength is at the minimum. If one considers the elastic deformation of the ice, then, in accordance with the theory of the stressed state of an elastic body with holes, the stress concentration at the ends of the cells must be taken into account for the determination of σ_0 (Fig. 10).

Theoretically, at the points O_1 in Fig. 10 the stresses abruptly increase: $\sigma_{01}=k\sigma_0$; the coefficient of stress concentration k depends upon the shape of the holes, it varies from 3 to 6 for elliptically to circularly shaped sections.

The plasticity of ice, however, weakens the stress concentration at the holes, likewise the phase transitions which are caused by a change of the chemical potential in the stress field. Thus, in the theoretical model of sea ice it is possible to disregard the stress concentration and to assume that the value σ_0 is equal to the tensile strength of a compact mass of the pure ice [65 p].

In the above equation, both ψ and the brine content v_b can be expressed by geometrical terms of the model (Fig. 9):

$$\sigma_f=\sigma_0\left(1-2\sqrt{\frac{\delta\varepsilon v_b}{\pi\beta_0}}\right). \tag{11}$$

δ denotes the ratio of brine cell length to the mean separation of adjacent ice bridges, $\delta=g/g_0$. ε is the eccentricity of the elliptical cross section of the brine cells, while β_0 gives their relative spacing.

For a brine cell which has the shape of a circular cylinder:

$$\sigma_f=\sigma_0\left(1-2\sqrt{\frac{v_b}{\pi\beta_0}}\right). \tag{12}$$

Accordingly, the strength of sea ice depends linearly upon $v_b^{1/2}$.

The critical value of the brine content at which the ice strength disappears is

$$v_0=\frac{\pi\beta_0}{4\delta\varepsilon}. \tag{13}$$

On the basis of the observations of Frankenstein, the strength of sea ice can be described by a model in which σ_f depends linearly upon both $v_b^{1/2}$ and v_b [72 W]:

$$\sigma_f = \sigma_0 \left(1 - 2\sqrt{\frac{v_b}{v_0}} + \frac{v_b}{\sqrt{v_0}}\right). \tag{14}$$

The value of v_0 ($v_0 < 1$) must be determined experimentally. The strength of sea ice depends upon the brine content as well as the geometrical parameters of its structure. These quantities are affected by the conditions of ice formation and growth of the ice thickness. Between the rate of ice growth $\dot{H}$ and the geometrical parameter a_0, which, according to Fig. 9, specifies the distance between successive brine layers in c-direction, there exists the relationship [64 A, 72 A]:

$$a_0\sqrt{\dot{H}} = \text{const.} \tag{15}$$

The ice growth rate can be determined from the measured values of the vertical temperature gradient in the ice cover and the thermal properties of the ice.

8.5.2 Tensile and flexural strength of sea ice

The principal criteria for the evaluation of the strength of the ice cover are the stresses which by stretching and flexing cause the ice to break.

The determination of the tensile strength takes place according to the following two methods: The ice samples are stretched on a testing machine to the point of fracture, or hollow ice cylinders are compressed in the direction of their diameter. Under such compression, tensile stresses occur at points A, Fig. 11, which cause the fracture. This method is very simple and suitable for the investigation of large quantities of ice samples.

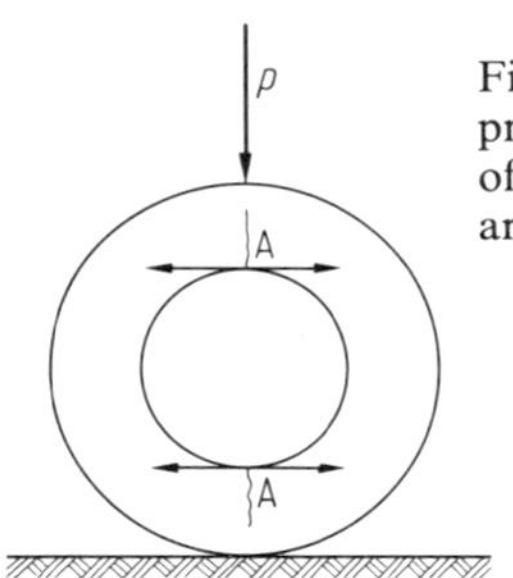

Fig. 11. Hollow ice cylinder under compression, with pressure p applied in direction of a diameter. At points A of the inner circumference, tensile stresses, indicated by arrows, occur causing fracture (wavy lines) [75 d, 77 d].

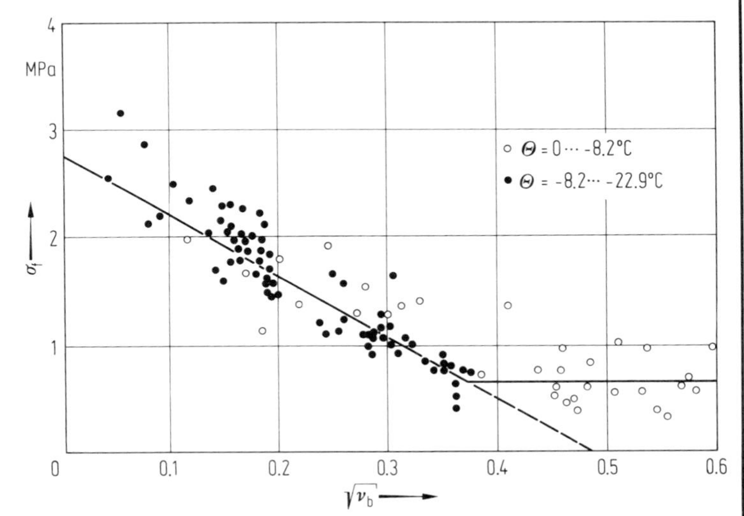

Fig. 12. Ring tensile strength σ_f of sea ice vs. square root of brine content v_b, according to Frankenstein's data, for temperature Θ ranging from 0 to $-8.2\,°C$ and from $-8.2\,°C$ to $-22.9\,°C$. Each circle represents about nine tensile strength tests. The solid line gives an empirical fit to the data, while the dashed line extrapolates the low v_b behavior to higher values of brine content [68 F].

As concerns investigations made upon hollow ice cylinders, extensive data material is available [56 B, 58 A 2, 62 W]. The most complete experimental data were obtained by Frankenstein, who investigated more than 1400 samples, Fig. 12. In the figure, each point represents circa nine tests. For ice with a brine content ν_b of less than 0.16, one obtains the following linear relationship between σ_f, in [MPa], and $\sqrt{\nu_b}$:

$$\sigma_f = 2.796\left(1 - \sqrt{\frac{\nu_b}{0.234}}\right). \tag{16}$$

This equation agrees well with the results gained by other authors [62 W, 64 L].

For a higher salinity of ice, the tensile strength remains constant, the mean value is 0.66 MPa.

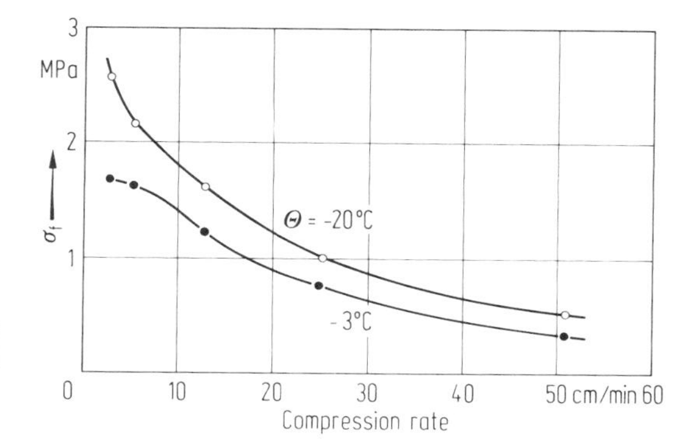

Fig. 13. Tensile strength σ_f vs. compression rate for cylindrical specimens of temperature $\Theta = -3\,°C$ and $\Theta = -20\,°C$ [75 d, 77 d].

The tensile strength of the sea ice determined by this method depends substantially upon the deformation rate of the samples (Fig. 13). This is explained by the fact that with decreasing deformation rate the coefficient of the stress concentration in the hollow cylinders of the ice samples will be smaller, following the plastic relief [72 W].

Direct tensile tests were carried out on artificial saline ice beams which had very similar properties to those of sea ice [67 D]. The ice samples had a cross section of 13 cm^2 and the deformation rate was 1.2 cm/min. The tensile strength depended considerably upon the direction of action of the tensile force relative to the optical axes of the ice crystals. Its value for stretching parallel to the optical axes was almost 2.6 times larger than that for the transverse direction.

The flexural strength of sea ice is determined experimentally by tests on small ice samples (ring specimens) or on floating ice beams and cantilevers sawn out of the sea ice cover. The flexural strength of small samples is considerably greater than that of the ice cover. A clear relationship between σ_f and ν_b was not determined [55 P, 56 B, 59 B, 67 A, 67 T]. The value of the flexural strength depends greatly upon the direction of the optical axes. The flexural strength of vertically cut small beams (tension is parallel to the crystal axes when the ice bends) is significantly greater than the strength of horizontally cut beams (tension is normal to the crystal axes). In the first case, flexural strength values of 2.0···3.0 MPa, and, for the second case, values of 0.5···1.0 MPa were found [59 B].

The influence which the loading rate exerts upon the flexural strength of ice samples is shown in Fig. 14. There, the flexural strength in the mixed zone, Fig. 15, in which not only elastic but also plastic failures occur, has for $\dot{\sigma} \approx 10^{-2}$ MPa/s a minimum. The increase of σ_f for smaller loading rates is explained by the plastic deformation of the ice sample which takes place before the fracture, while the increase of σ_f for larger loading rates is connected with the elastic deformation of the ice samples before the fracture.

Experimental values of the flexural strength of floating ice beams and cantilevers are given in Fig. 16 [56 B, 58 W 1, 63 B]. For a brine content of $\nu_b < 0.12$, the relationship between σ_f, in [MPa], and ν_b has the form

$$\sigma_f = 0.69\left(1 - \sqrt{\frac{\nu_b}{0.202}}\right). \tag{17}$$

For ice with a brine content of $\nu_b > 0.12$, the flexural strength is constant: its value is 0.2 MPa. The flexural strength of sea ice cover is independent of the direction of load action (pull-up, push-down, and pull-sideways) and of the ice thickness. It is constant for a load rate $\dot{\sigma} < 10^{-1}$ MPa/s. With increasing rates, σ_f – owing to the crossover from plastic to elastic deformation – increases proportional to $\ln \dot{\sigma}$.

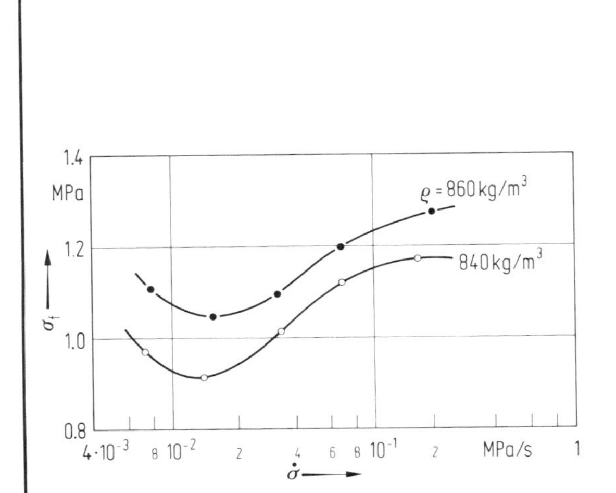

Fig. 14. Flexural strength σ_f of sea ice vs. loading rate $\dot{\sigma}$ for ice of density $\varrho = 860\,\mathrm{kg\,m^{-3}}$ and $\varrho = 840\,\mathrm{kg\,m^{-3}}$ at a temperature of $-9\,°\mathrm{C}$ [67 T].

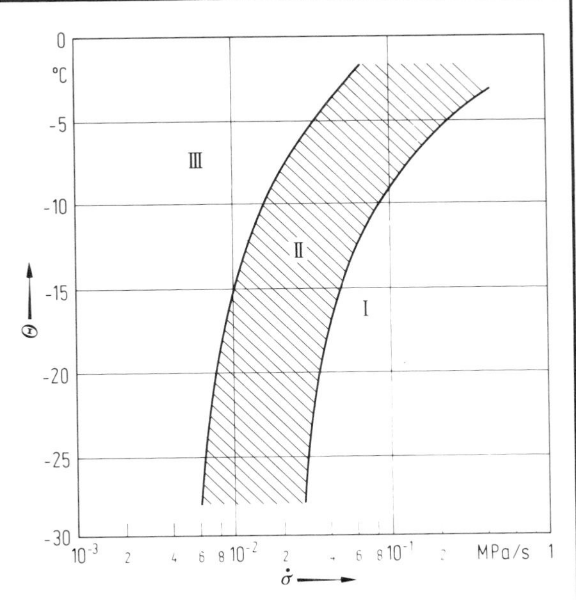

Fig. 15. Approximate boundaries, for ice of salinity $s = (1 \cdots 1.4) \cdot 10^{-3}$ and density $\varrho = 860 \cdots 870\,\mathrm{kg\,m^{-3}}$, of elastic (I), plastic (III), and mixed (II) behavior of sea ice under load as function of temperature Θ and loading rate $\dot{\sigma}$ [67 T].

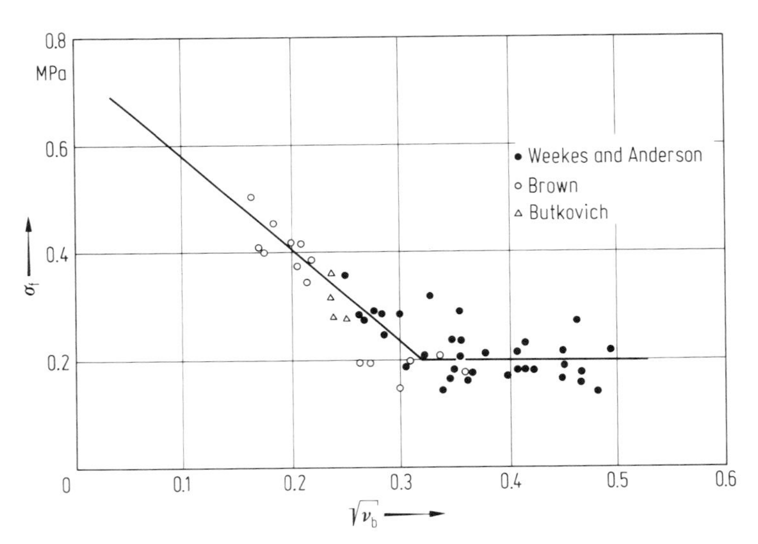

Fig. 16. Flexural strength σ_f of sea ice vs. square root of brine content ν_b in tests on cantilevers afloat. Data are from Weekes and Anderson, Brown, and Butkovich, respectively. The solid line represents a fit to the data [75 d, 77 d].

A prognostical estimation of the flexural strength can be undertaken by means of the empirical relationship [74 R],

$$\sigma_f = \sigma_{0f}\left(1 - \frac{s}{s_{max}}\right), \tag{18}$$

on the basis of the known value of the mean salinity s of the ice. σ_{0f} is the flexural strength of fresh water ice cover at a given temperature, and s_{max} the maximum salinity of the ice at which its flexural strength vanishes, Fig. 17. In the temperature range of practical interest, s_{max} can be taken as constant ($s_{max} \approx 15 \cdot 10^{-3}$).

The salinity s can be determined from the empirical formula [75 d, 77 d],

$$s = S(1-b)\exp(-a\sqrt{H}) + bS, \tag{19}$$

where S is the salinity of the surface water from which the ice is formed, while H denotes ice thickness. The empirical coefficient a depends upon the growth rate of the ice, it varies from $0.35\,\text{cm}^{-1/2}$ at a high rate of ice growth ($\dot{H} = 4$ cm/d) to $0.60\,\text{cm}^{-1/2}$ at a lower rate of ice growth ($\dot{H} < 0.5$ cm/d). When the rate of ice growth is unknown, $a = 0.50\,\text{cm}^{-1/2}$ can be taken. b in eq. (19) is another empirical coefficient, $b = 0.13$.

Fig. 18 represents the change of salinity s with thickness H of young ice for different values of the salinity S of the surface water as calculated from eq. (19) with $a = 0.5\,\text{cm}^{-1/2}$ (solid lines), as well as the observational results for sea ice that originated from frozen water with a salinity S of $34 \cdot 10^{-3}$ [58 A 1], $15 \cdot 10^{-3}$ and $5 \cdot 10^{-3}$ [74 R], respectively.

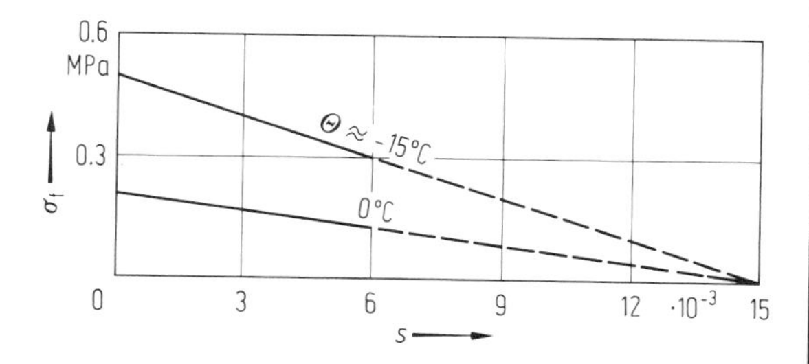

Fig. 17. Flexural strength σ_f of sea ice vs. salinity s for temperatures of approximately 0 °C and −15 °C, respectively (solid lines). The dashed lines extrapolate the low-σ_f behavior to higher values of the salinity [75 d, 77 d].

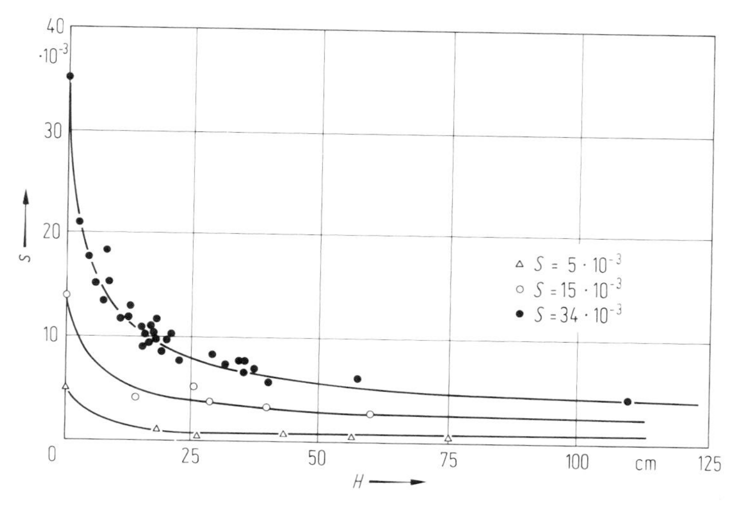

Fig. 18. Empirical relation between salinity s of young sea ice and ice thickness for surface water salinity $S = 5 \cdot 10^{-3}$ and $S = 15 \cdot 10^{-3}$ [74 R], as well as $S = 34 \cdot 10^{-3}$ [58 A 1]. The solid lines are calculated from eq. (19) for $a = 0.5\,\text{cm}^{-1/2}$, $b = 0.13$, and the respective surface water salinity S [74 R].

8.5.3 Compressive and shear strength of sea ice

The experimental determination of the compressive strength of sea ice is performed for cylindrical or cubic samples. The compressive strength depends essentially upon the loading rate of the samples, their salinity, dimensions, and orientation.

Under compression the fractures mainly take place in the directions in which the shear and tensile strengths are near the absolute value. In this case, the possible directions of the extension of cracks are confined to a small angle with the shear planes. For ice of columnar structure the fracture pattern changes considerably if compression takes place either parallel or transverse to the growth direction of the ice. With a compression transverse to the growth direction, i.e. transverse to the long axes of the crystals, the slip planes and, in the case of sea ice, the planes along which the brine pockets are arranged, are orientated in such a fashion that the ice samples easily give way. Plastic deformations lead to the formation of cracks, the sizes of which rapidly increase to the critical value. A rise in temperature increases the plasticity of the ice, thus facilitating the formation of cracks. If the loading acts along the direction of ice growth, the slip planes are vertical to the compression direction, so that the shearing and the formation of cracks is rendered more difficult.

On the basis of the comprehensive data material of Peyton [66 P], the compressive strength of sea ice increases with a change in compression direction from 1.5···8.0 MPa.

The change of the compressive strength as function of salinity is, in general, analogous to the change of the tensile and flexural strength. The dependence of the compressive strength upon loading rate is illustrated in Fig. 19. It is associated with the fact that, with different deformation speeds, the volumes of the plastically deformed materials are diverse. Accordingly, the conditions for the formation of the critical cracks are also diverse.

The shear strength is generally determined by the torsion of ice cylinders when the resulting stress state is close to pure shear. Shear strengths which have been measured normal to the growth direction of the ice (long axes of the crystals) give larger values than those measured parallel to the growth direction.

The following mean values for shear strength are available from a fairly extensive data material: 0.39 MPa for multi-year ice, 0.7 MPa for first-year ice (data without salinity determination) [55 P], 2.1 MPa for a relative brine volume $\sqrt{\nu_b}=0.19$ [56 B].

Further values from extensive data material [67 P] are given in Fig. 20. The decrease of the shear strength with increasing $\sqrt{\nu_b}$ takes place in the same manner as the decrease of the strength in the ring tensile and in-situ cantilever experiments. The values of the shear strength are considerably lower than the tensile strength values gained from experiments on ice cylinders.

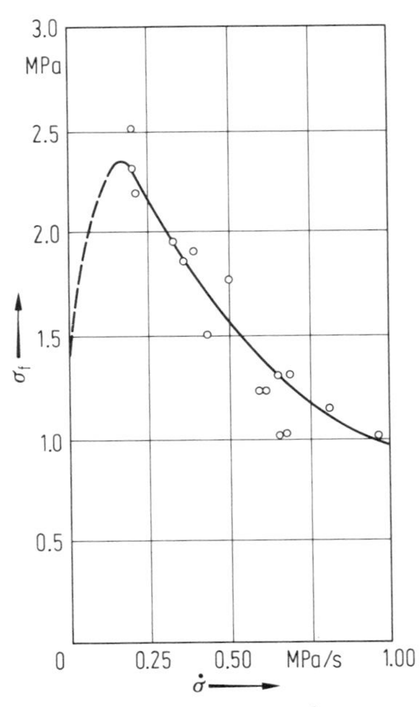

Fig. 19. Relation between compressive strength σ_f of sea ice from Cook Inlet, Alaska, and stress rate $\dot{\sigma}$ [66 P]. The dashed line is conjectural, based upon Arctic Ocean results.

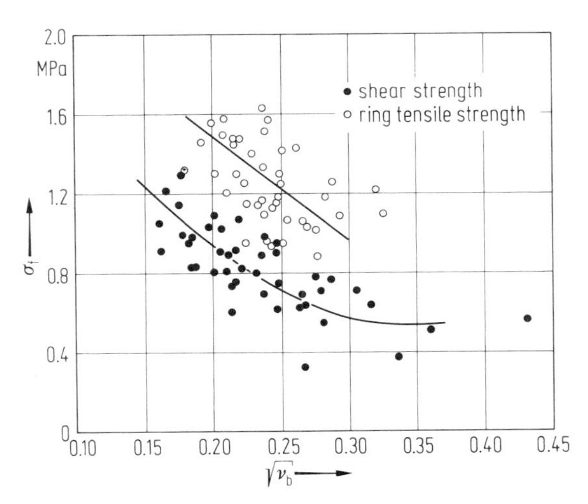

Fig. 20. Shear strength and ring tensile strength of sea ice vs. square root of brine content ν_b [67 P].

8.5.4 Impact strength of sea ice

For the determination of the impact strength of sea ice, two types of experimental procedures are customary. In the first type, the ice samples are crushed on a test machine of the Charpy impact tester type; the impact strength α_k is given, in [MPa m], by the work required for fracture. In the second experimental procedure, weights are allowed to fall upon the ice surface, and the specific energy (work) ε_f of the fracture is determined, in [MPa]. ε_f is ascertained from the total energy of the impact, $W = mgz$, per unit volume of the fractured ice, m denotes the mass of a sphere, z its height, and g the gravity constant. The volume of the fractured ice is determined from the measured diameter of the impression.

In accordance with the Charpy method, the following values were gained: for first-year ice, $\alpha_k = 0.16 \cdots 0.17$ MPa m, whereas for multi-year ice, $\alpha_k \approx 0.40$ MPa m [55 P].

With the aid of a pendulum impact tester, for ice with a salinity s of $(3.8 \cdots 6.1) \cdot 10^{-3}$, the value of $0.35 \cdots 0.62$ MPa m for impact strength was gained [75 d, 77 d].

With the aid of the second method, the following values for the specific energy of the fracture were ascertained: for ice from the northern Caspian Sea, $\varepsilon_f = 0.34 \cdots 3.68$ MPa. The influence of density, temperature, and salinity has not been taken into account. For first-year ice of salinity $s \approx 4 \cdot 10^{-3}$, on the average $\varepsilon_f = 2.7$ MPa [75 d, 77 d].

8.6 Ice cover characteristics of the world ocean

8.6.1 The stages of sea ice development

The following schedule summarizes the stages of sea ice development from the formation of initial ice forms at the beginning of freezing of sea water to the final phase of thermodynamical growth of the ice, in accordance with [70 w].

1. New ice
   Frazil ice: Fine spicules or plates of ice suspended in water.
   Grease ice: A stage of freezing, later than frazil ice, when the crystals have coagulated to form a soupy layer.
   Slush: A viscous mass floating in water after a heavy snowfall.
   Shuga: An accumulation of spongy white ice lumps, a few centimeters across; it is mainly formed from grease ice or slush.
2. Nilas
   Dark nilas: A thin elastic crust of ice, less than 5 cm in thickness; it is very dark in color.
   Light nilas: A thin elastic crust of ice, 5···10 cm in thickness; it is lighter in color than dark nilas.
   Ice rind: A brittle shiny crust of ice formed on a quiet surface by direct freezing or from grease ice, up to 5 cm in thickness.
3. Pancake ice: Predominantly circular pieces of ice, from 30 cm to 3 m in diameter and up to 10 cm in thickness, with raised rims. It may be formed on a slight swell from grease ice, shuga, or slush, or as a result of breaking of ice rind or nilas. Sometimes it also forms in some depth at an interface between water bodies of different physical characteristics, from where it floats to the surface.
4. Young ice
   Grey ice: 10···15 cm thick, less elastic than nilas; it breaks on swell and usually rafts under pressure.
   Grey-white ice: 15···30 cm thick; under pressure it is more likely to ridge than to raft.
5. First-year ice
   Thin first-year ice: 30···70 cm thick.
   Medium first-year ice: 70···120 cm thick.
   Thick first-year ice: 120···200 cm thick.
6. Old ice
   Second-year ice: Old ice which has survived only one summer's melt; it is thicker and less dense than first-year ice. Summer melting produces a regular pattern of numerous small puddles.
   Multi-year ice: Old ice of thickness up to 3 m or more which has survived at least two summer's melt; it is almost salt-free and its color, where bare, is usually blue. Melt pattern consists of large interconnecting irregular puddles and a well developed drainage system.

8.6.2 Extent of ice cover

Sea ice forms in the water of those oceans in which, in the cold season, the heat given off into the atmosphere is greater than the heat transported from the deeper lying layers of the water to the water surface. After the heat storage of the thin surface layer of the water has been completely exhausted by cooling down to the freezing point, further heat loss is compensated for by the heat of crystallization released during ice formation.

From the known seasonal variation in the heat exchange between ocean and atmosphere and a known value of the heat flux between the sea surface and layers lying below it, the limits of the ice distribution can be calculated provided that the ice transport by wind and current is not taken into account. Likewise, the part of the completely melted ice in summer can easily be determined from external heat fluxes when the heat content of the ice during the melting period is disregarded. In this manner, one obtains the largest and the smallest extent of the sea ice, as Fig. 21 shows for the northern hemisphere. The comparison between the calculated and the actual position of the ice limit permits the clear recognition of the fundamental influence of thermal factors upon the extent of the ice cover. Currents and wind do not change the general position of the ice cover very much.

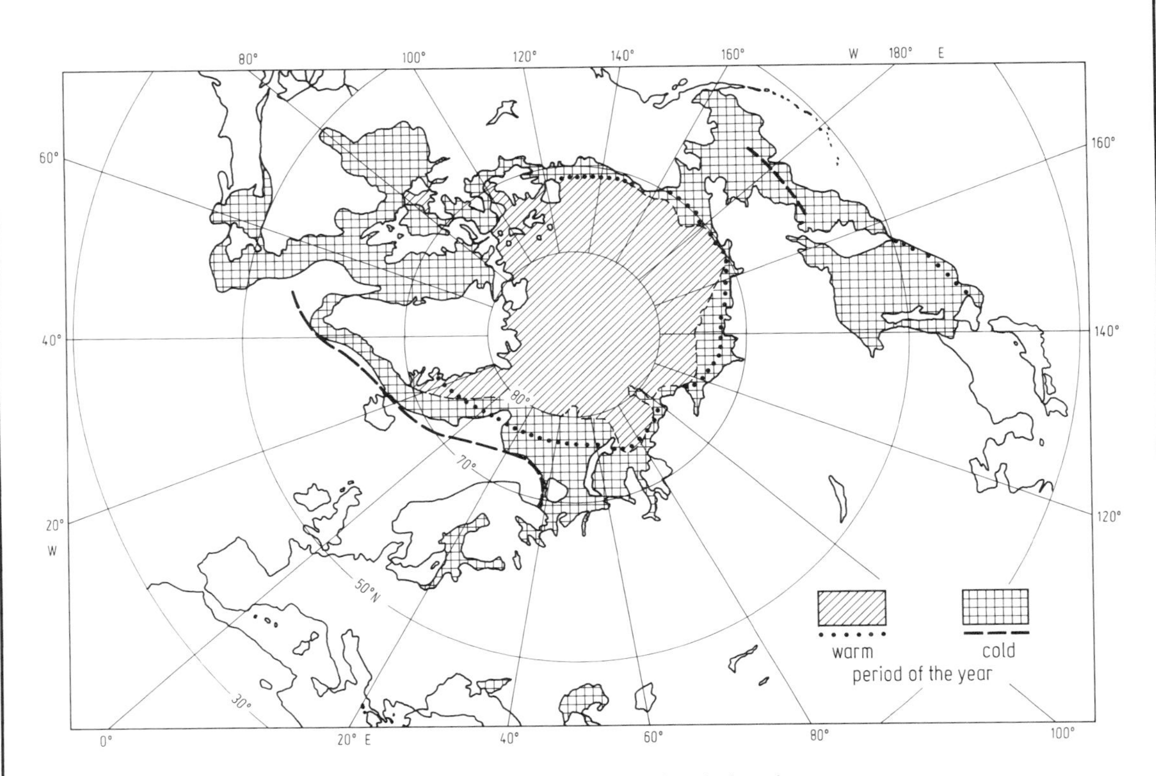

Fig. 21. Extent of ice in the northern hemisphere in warm and cold periods of the year. The computed ice cover boundary in warm and cold periods are depicted by dotted and dashed lines, respectively [66 W].

A mathematical simulation revealed [75 d, 77 d] that an increase in air temperature by 4 °C relative to the long-term average of the months October to March would shift the ice limit 200···250 km towards the North at the time of maximum ice expansion, whereby the arctic sea regions remain ice covered. At the end of summer the effect of this winter anomaly of the air temperature upon the state of the ice cover has fairly subsided. Variations of the air temperature in winter principally influence the quantity of thin ice (ice thickness of up to 1 m) formed in the cold season, which again disappears owing to melting in summer. Multi-year ice, however, does not essentially change its thickness even in a large temperature anomaly.

The mathematical simulation with the same temperature anomaly for the months July and August gave an essential melting of the ice, whereby the multi-year ice would survive in a small circumpolar region. An increase in temperature by 2 °C only would lead to a shift of the actual ice limit by 100···250 km towards the North.

The influence of oceanic heat upon the growth of ice is great. The mathematical simulation [75 d, 77 d] showed that an increase of the heat flux in the winter period from 0···5 W m^{-2} towards the ice would shift the ice limit in the Atlantic sector 500···600 km, in the Pacific sector 600···700 km, towards the North. The same anomaly for the months of July and August gave a northwards directed shift of 160···200 km.

Nazarov [62 n] divided the world ocean into 6 zones with regard to the duration of the ice cover and its origin.

The regions in which the ice cover persists for the whole of the year, although, in the warmer period, the ice concentration decreases, belong to the first zone: the central part of the Arctic Basin, the northern regions of the majority of the seas of the Arctic Ocean and the Amundsen, Bellingshausen, and Weddell Seas. In winter, more ice forms there than ice can melt in summer (multi-year ice). For that reason the ice grows in the course of time until the equilibrium thickness is reached, provided that it is not previously transported into lower latitudes where summer melting exceeds winter growth.

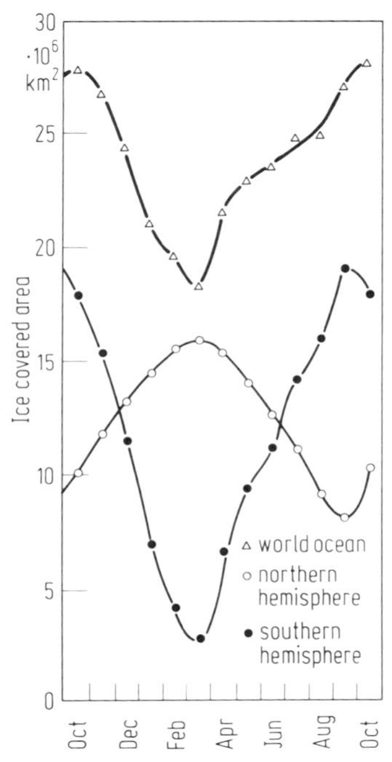

Fig. 22. Seasonal variation of the area covered by sea ice in the world ocean and in the northern and southern hemispheres [78 Z].

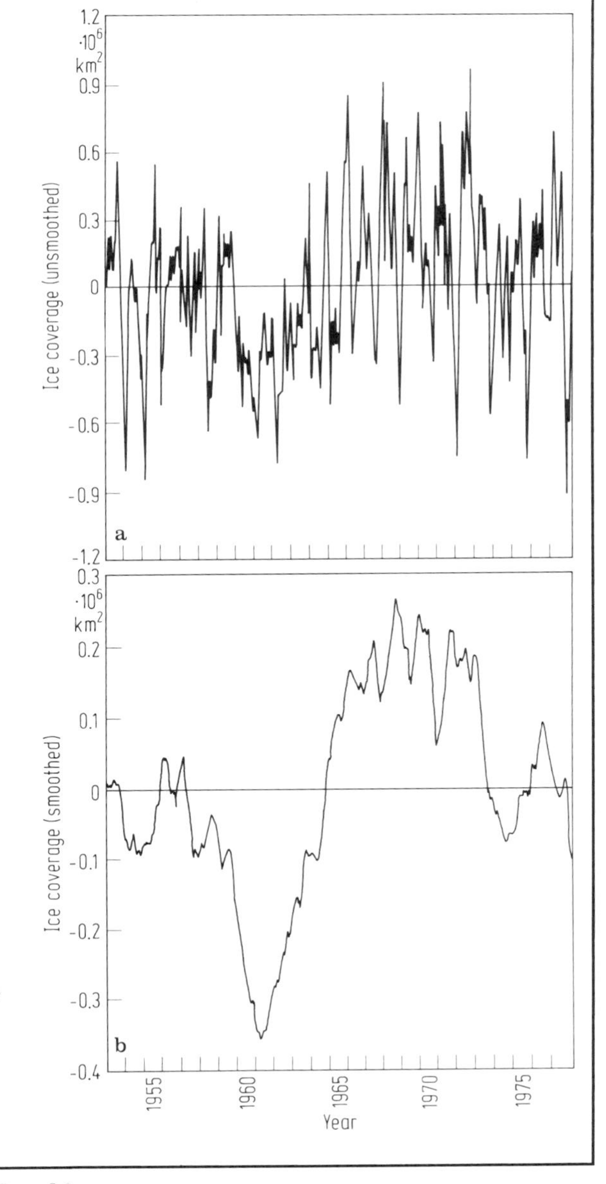

Fig. 23. Monthly deviations from the long-term average of the area covered by Arctic sea ice, showing for the years 1953···1977 (a) unsmoothed values and (b) values averaged over 2-year periods [78 W].

In the second zone the ice cover changes from year to year. The larger part of it melts in summer, but, as a result of ice drift, sea ice can always occur there (e.g., the Kara and Barents Seas).

In the third zone, in each winter, an ice cover is formed which disappears completely in summer: Okhotsk, Japan, White, Baltic, Aral, and Caspian Seas. The duration and the extent of the ice cover can vary considerably from year to year.

In the fourth zone, ice is formed in severe winters only: North Sea, the Sea of Marmara, and the Adriatic Sea.

In the fifth zone, most of the ice which occurs there is transported in from other regions: the Greenland Sea, region of Newfoundland, and a large part of the Southern Ocean including the region of abundant icebergs.

The sixth zone is free of ice.

The total volume of sea ice at the time of maximum development is $25.5 \cdot 10^3\,km^3$ in the northern hemisphere and $30 \cdot 10^3\,km^3$ in the southern hemisphere. In every year an ice volume of $37 \cdot 10^3\,km^3$ melts and is formed again; i.e. $14 \cdot 10^3\,km^3$ in the northern hemisphere, and $23 \cdot 10^3\,km^3$ in the southern hemisphere.

The seasonal change of the ice cover is illustrated in Fig. 22.

The yearly variations of the ice cover can be large, as in the example of Arctic sea ice shown in Fig. 23, which gives, for a period of 25 years, the monthly variations of the ice cover relative to the long-term average.

8.6.3 Influence of drift on the state of ice cover

The ice cover is broken up and displaced by the effect of winds and currents. In winter the open water areas in fractures and leads caused thereby cover themselves with young ice which freezes together with older, thick ice floes. The result of this constant process of break-up and formation of ice breccias is, towards the end of the winter, a horizontal, uneven ice cover composed of a mixture of ice of different thickness, which can react differently to the same thermal and mechanical forcing.

The mean ice thickness $\tilde{H}$ of an ice field can be calculated by the formula [75 d, 77 d]:

$$\tilde{H} = \sum_{j} H_j N_j \qquad (20)$$

if thickness H_j and concentration N_j in part j of the ice field are known. With the convergence of ice, its mean thickness increases – caused by the formation of pressure ridges and hummocks. In this case, $N_j = 1$ is exceeded by an amount which can be determined by the formula [61 G]:

$$\delta N_h = [N(t+\Delta t) - 1] \exp(-lH) \qquad (21)$$

when $(N-1) > 0$. l denotes an empirical constant, while Δt is the time interval for which the change in concentration is considered. δN_h is the amount of ice which passes over by the deformation processes in the formation of pressure ridges and hummocks.

In summer, dynamic factors influence the intensity of the ice melting by changing the area of open water. This influence can be estimated by the formula [70 D]:

$$\frac{\partial N}{\partial t} = -\operatorname{div}(NV) + (N-1)\frac{\phi_{lat}}{L\varrho_i H}. \qquad (22)$$

Therein, the first term on the right-hand side of the equation describes the change of the ice concentration as a result of the dynamic factors, V denotes pack ice velocity, and the second term the reduction of the concentration due to the lateral melting of ice, ϕ_{lat} denotes the lateral heat flux, L the latent heat of fusion, ϱ_i the actual ice density, and H ice thickness.

Table 9 contains the results of a calculation of the melting of ice due to thermal factors (the first line of the table), and the combined effect of lateral melting and divergence (second line) as well as convergence (third line).

Table 9. Change of concentration N of ice under action of thermal and dynamic factors [75 d, 77 d].

	June		July			August			September
N_{therm}	0.89	0.88	0.85	0.82	0.75	0.68	0.53	0.36	0.27
N_{div}	0.85	0.78	0.72	0.60	0.49	0.29	0.09	0	0
N_{conv}	0.95	0.98	0.98	0.94	0.90	0.78	0.64	0.54	0.51

For this calculation, characteristic values of the heat fluxes and the divergence of the ice drift velocities of the marginal seas of the Arctic Ocean were used as initial data. In all cases an initial value of 0.9 for the ice concentration was assumed for the first ten-day period of June. The results made clear the great difference in the intensity of the melting of ice with convergence and divergence, respectively.

In Antarctica the divergence of ice caused by wind and currents predominates. Therefore, the main mass of sea ice melts in summer. Multi-year ice remains only along some coastal sections and in the regions of convergence; for example, the Weddell Sea.

In winter, the regions of the divergence are areas of intensive ice formation. The rates of ice growth, turbulent heat loss to the atmosphere, and salt rejection to the ocean depend strongly upon ice thickness; in the Central Arctic these rates are up to two orders of magnitude larger over a refreezing lead than over perennial (3···4 m in thickness) ice. Although decreasing rapidly with increasing ice thickness, these rates can still be an order of magnitude larger over 0.5 m ice than over 3 m ice [78 M]. Thermodynamic model calculations in which the distribution of the different ice thicknesses in the Central Arctic were taken into account showed that the intermediate thicknesses (0.2···0.8 m) of young and first-year ice – rather than open leads – exerted the greatest influence upon ice production [82 M].

8.6.4 General description of ice cover of the polar oceans

In recent times, the possibility arose to calculate the state of the ice cover owing to the combined effect of thermal and dynamic processes [80 H, 82 M]. A complete quantitative description of the state, expressed as a function of the morphology of the oceanic basin and the oceanographical and meteorological processes which take place there, cannot as yet be supplied.

Arctic Ocean: The main mass of the ice is in motion. The ice cover consists of the freezing together of ice floes of different thickness and expanse, with a proportion of multi-year ice which increases with the geographical latitude. In the Central Arctic, this is 60···70%. There, the ice production in the months from September to May is 1.30 m, the ice melt from June to August is 0.50 m. The daily growth rate and the melting rate for different ice thicknesses can be taken from Fig. 24.

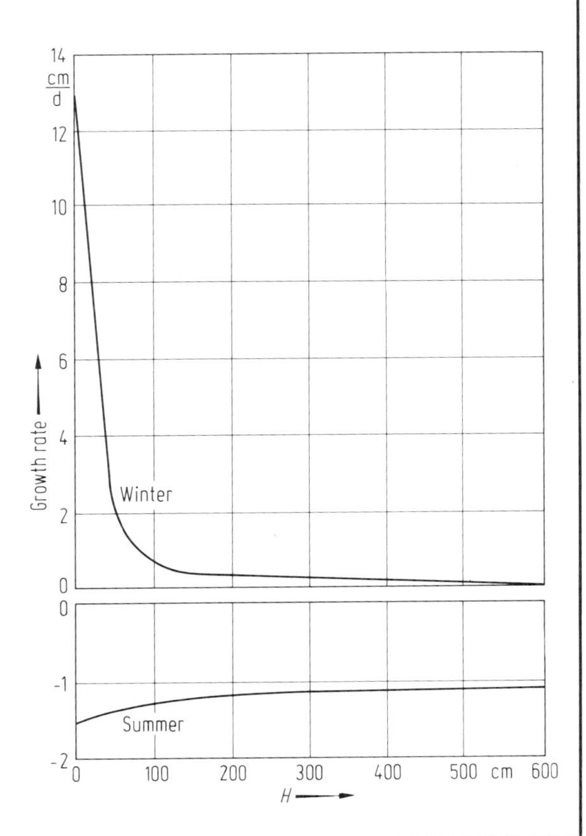

Fig. 24. Typical growth rate of sea ice in the central Arctic vs. ice thickness H for, respectively, winter and summer periods [75 T].

In the Central Arctic, the area of open water in winter comprises not more than 1% of the area of pack ice, and the area of up to 0.8 m thick ice is 8% thereof. In summer, the average value for the area of open water is 10%, and the area of up to 0.8 m thick ice is 4% [82 M].

In regions with large pressure ridges and hummocks the ice thickness can exceed 20···30 m.

Greenland Sea: The amount of ice which is carried away yearly from the Arctic Basin into the Greenland Sea corresponds to an area of circa 900000 km^2. As a result of fluctuations in the meteorological conditions, the interannual variation of the mass of pack ice carried away attains 60% of its mean value.

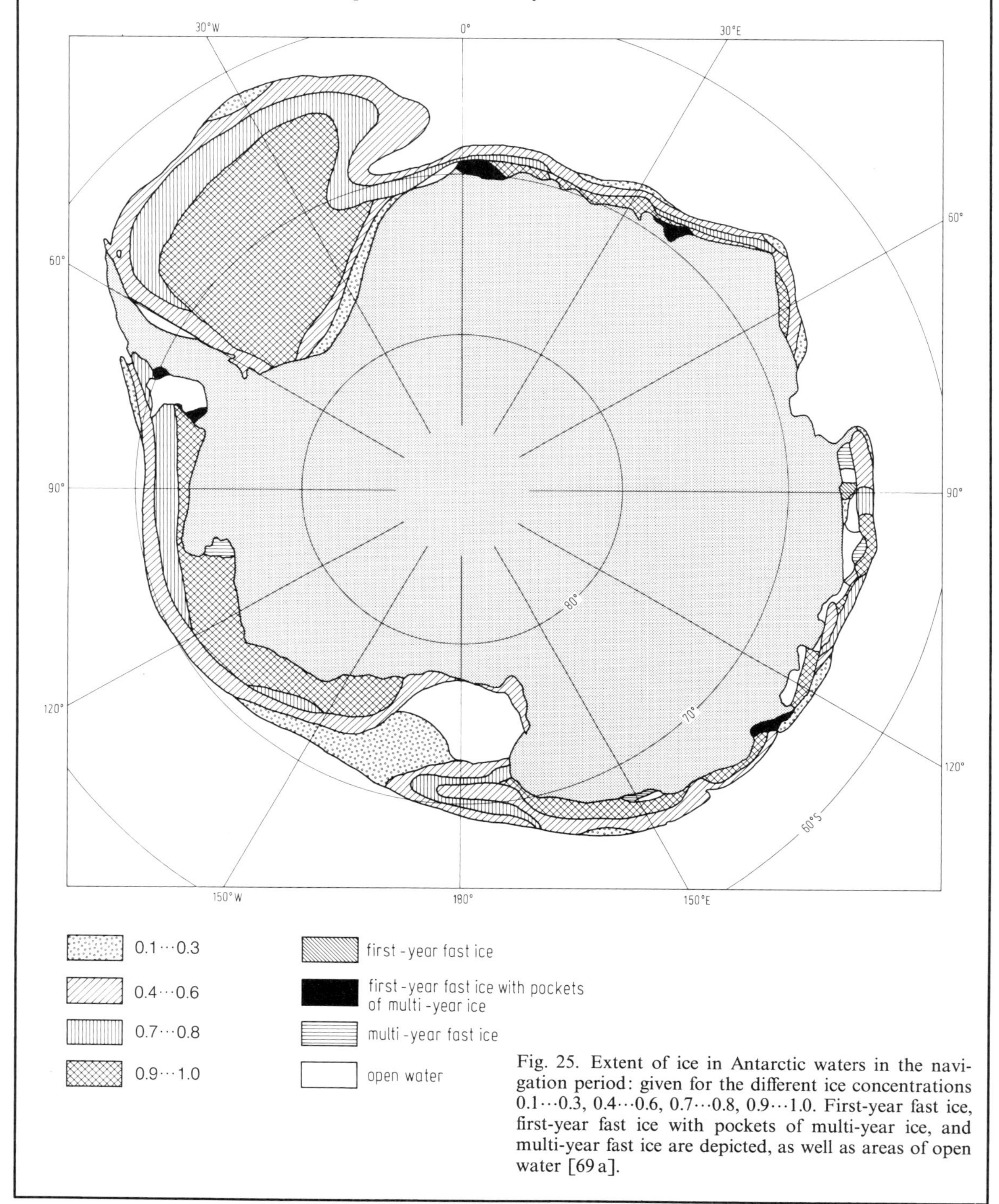

Fig. 25. Extent of ice in Antarctic waters in the navigation period: given for the different ice concentrations 0.1···0.3, 0.4···0.6, 0.7···0.8, 0.9···1.0. First-year fast ice, first-year fast ice with pockets of multi-year ice, and multi-year fast ice are depicted, as well as areas of open water [69 a].

Antarctica (Fig. 25): The main mass of Antarctic sea ice is in motion. It drifts in the immediate vicinity of the continent predominantly towards the West and turns, under the influence of the coastal projections, towards the North, where it then runs into the Antarctic circumpolar current. The diameter of the ice floes decreases towards the North, where – in the main – small ice floes (20···100 m in diameter) and ice cakes (less than 20 m in diameter) occur.

In Antarctic waters the freezing process differs from that of the Arctic Sea in two ways. The first is the formation of slush, caused by the large amount of snow blown away by the wind from the continent and deposited on the water surface in a 30···40 km wide zone.

The second peculiarity of the Antarctic ice cover is the fairly widespread frazil ice, at least, in the coastal waters. Its cause is assumed to be the low density stability of the antarctic waters, which, by rapid cooling in the numerous leads and polynyas, makes possible the supercooling of water – the basic requirement for the formation of frazil ice.

The freezing process begins in the coastal zone in March. From that time onwards, the ice edge advances with a mean rate of 4.2 km per day [62 n] in roughly meridional direction towards the North. It attains its northernmost position, on average, along the 60° S latitude in September-October. Level ice then becomes 1.5···2 m thick. Owing to the relative smoothness of the coastline, the great depths, and the predominance of downslope winds, the Antarctic fast ice is less developed than that of the Arctic. Its average width is only 25···35 km.

8.7 References for 8

8.7.1 General references

45 z Zubov, N.N.: L'dy arktiki (Arctic ice), Moskva, Izd. Glavsevmorputi **1945**.
62 n Nazarov, V.S.: L'dy antarkticheskikh vod (Antarctic Sea Ice), Moskva: Nauka **1962**.
65 p Pounder, E.R.: The physics of ice, Oxford: Pergamon Press **1965**.
69 a Atlas Antarktiki, Leningrad: Gidrometeoizd. Vols. 1 and 2, **1969**.
70 w WMO SEA-ICE Nomenclature, Geneva: WMO **1970** (with supplements).
75 d Doronin, Yu.P., Kheisin, D.E.: Morskoi led, Leningrad: Gidrometeoizd. **1975**.
77 d Doronin, Yu.P., Kheisin, D.E.: Sea Ice, Rotterdam: Balkema **1977**.

8.7.2 Special references

48 B Berdennikov, V.P.: Tr. Gos. Gidrol. Inst. **7 (61)** (1948) 76.
54 O Oliver, J., Crary, A., Cottel, R.: Trans. Am. Geophys. Union **35** (1954) 282.
55 P Petrov, I.G., in: Materialy nabliudenii nauchno-issled. Dreifuiushchei stantsii 1950/51, Leningrad **1955**, T.2, 103.
56 B Butkovich, T.R.: Res. Rep. U.S. Army Snow Ice Permafrost Res. Establ. **20** (1956).
57 L Lin'kov, E.M.: Vestn. Leningr. Univ. Fiz. Khim. No. **16**, P.1 (1957) 57.
58 A 1 Anderson, D.L., Weeks, W.F.: Trans. Am. Geophys. Union **39** (1958) 632.
58 A 2 Assur, A., in: Arctic Sea Ice, Conf. 1958, Washington, N. A. S., N. R. C. **1958**, 106.
58 B Bogorodskii, V.V.: Akust. Zh. **4** (1958) 19.
58 L Lin'kov, E.M.: Vestn. Leningr. Univ. Fiz. Khim. No. **4**, P. 1 (1958) 138.
58 T Tabata, T.: Low Temp. Sci., Sapporo, Ser. A **17** (1958) 147.
58 W 1 Weeks, W.F., Anderson, D.L.: Trans. Am. Geophys. Union **39** (1958) 641.
58 W 2 Weeks, W.F., Lee, O.S.: Arctic **11** (1958) 135.
59 B Butkovich, T.R.: Res. Rep. U.S. Army Snow Ice Permafrost Res. Establ. **54** (1959).
61 G Gudkovich, Z.M., in: Mat. Konf. po Probleme Vzaimodeistvie Atmosfery i Gidrosfery v Sev. Chasti Atlant. Okeana, Leningrad, Gidrometeoizd. **1961**, Vyp. 3–4, 75.
62 L Langleben, M.P.: Canad. J. Phys. **40** (1962) 1.
62 W Weeks, W.F.: J. Glaciol. **4** (1962) 25.

63 B Brown, J., in: Ice and snow processes; properties and applications (Kingery, W.D., ed.), Cambridge, Mass.: The MIT Press **1963**, p. 79.
63 S Schwertfeger, P.: J. Glaciol. **4** (1963) 789.
64 A Assur, A., Weeks, W.F.: Res. Rep. Cold Regions Res. Engng. Lab. **135** (1964).
64 L Langleben, M.P., Pounder, E.R.: J. Glaciol. **5** (1964) 93.
64 N Nazintsev, Yu.L.: Trudy Arkt. Antarkt. Nauchno-Issled. Inst. **267** (1964) 31.
64 P Pounder, E.R., Langleben, M.P.: J. Glaciol. **5** (1964) 99.
66 P Peyton, H.R.: Sea ice strength, College, Alaska, Geophys. Inst., University of Alaska **1966.**
66 W Wittmann, W., Schule, J.J., in: Proc. Symp. on the Arctic Heat Budget and Atmospheric Circulation, Santa Monica, Calif. **1966**, p. 215.
67 A Abele, G., Frankenstein, G.: Techn. Rep. Cold Regions Res. Engng. Lab. **176** (1967).
67 D Dykins, J.E., in: Physics of Snow and Ice, Proc. Int. Conf. on Low Temp. Sci., Sapporo 1966, Vol. **1**, P. 1 (1967), 523.
67 P Paige, R.A., Lee, C.W.: J. Glaciol. **6** (1967) 515.
67 T Tabata, T., in: Physics of Snow and Ice, Proc. Int. Conf. on Low Temp. Sci., Sapporo 1966, Vol. **1**, P. 1 (1967), 481.
68 F Frankenstein, G.: Techn. Rep. Cold Regions Res. Engng. Lab. **172** (1968).
70 D Doronin, Yu.P.: Trudy Arkt. Antarkt. Nauchno-Issled. Inst. **291** (1970) 5.
71 P Perchanskii, I.S.: Trudy Arkt. Antarkt. Nauchno-Issled. Inst. **300** (1971) 4.
71 S Smirnov, V.N.: Trudy Arkt. Antarkt. Nauchno-Issled. Inst. **300** (1971) 56.
72 A Assur, A., in: Ice Symposium, Leningrad **1972**, 1.
72 W Weeks, W.F., Assur, A., in: Fracture, (Liebowitz, H., ed.), New York: Academic Press **7** (1972) 879.
74 R Rivlin, A.Ya.: Probl. Arkt. Antarkt. **45** (1974) 79.
75 T Thorndike, A.S., Rothrock, D.A., Maykut, G.A.: J. Geophys. Res. **80** (1975) 4501.
78 M Maykut, G.A.: J. Geophys. Res. **83** (1978) 3646.
78 W Walsh, J.E., Johnson, C.M.: J. Phys. Oceanogr. **9** (1978) 580.
78 Z Zakharov, V.F., Strokina, L.A.: Sov. Meteorol. Hydrol. **7** (1978) 35.
80 H Hibler, W.D.: Mon. Weather Rev. **108** (1980) 1943.
82 M Maykut, G.A.: J. Geophys. Res. **87** (1982) 7971.

9 Coastal oceanography

9.1 Definitions and spatial extensions

9.1.0 Abbreviations

(Compare also chapter 1 of this volume.)

CCE	Commission on the Coastal Environment (in the IGU)
CNEXO	Centre National pour l'Exploitation des Oceans
CLIMAP	Climate Long Range Investigation, Mapping And Prediction Study
DHI	Deutsches Hydrographisches Institut
EEZ	Exclusive Economic Zone
FAGS	Federation of Astronomical and Geophysical Services
FAO	Food and Agriculture Organization
GARP	Global Atmospheric Research Program (WMO/ICSU)
GC	Geneva Convention
GEBCO	General Bathymetric Chart of the Oceans
IABO	International Association of Biological Oceanography
ICSU	International Council of Scientific Unions
IGN	Institut Géographique National (France)
IGOSS	IOC/WMO Integrated Global Ocean Services System
IGU	International Geographical Union (in UNESCO)
IGY	International Geophysical Year
IHB	International Hydrographic Bureau
IHO	International Hydrographic Organization
IMCO	Intergovernmental Marine Consultative Organization
INQUA	International Quarternaria (Congress or Union)
IOC	Intergovernmental Oceanographic Commission (UNESCO)
JONSWAP	Joint North Sea Wave Project
LANDSAT	Land Satellite of NASA, USA, (NOAA 1983 onward)
MHW	Mean High Water
MLW	Mean Low Water
MSL	Mean Sea Level (msl)
MSS	Multi Spectral Scanner
NAP	Normal Amsterdamsch Peil
NASA	National Aeronautics and Space Administration
NESS	National Environmental Satellite Service
NN	Normal Null (Zero), Map Datum
NOAA	National Oceanographic and Atmospheric Administration
NOS	National Ocean Survey
NWS	National Weather Service, USA
ORSTOM	Office de la Recherche Scientifique et Technique d'Outre Mer
PSMSL	Permanent Service for Mean Sea Level
SCAR	Scientific Committee on Antarctic Research
SCOR	Scientific Committee on Oceanic Research
SEASAT	Sea Satellite of NASA
SHOM	Service Hydrographique et Oceanographique de la Marine
TWS	Tsunami Warning System
UNO	United Nations Organization
UNCLOS	United Nations Conference on the Law of the Sea
USGS	United States Geological Survey
WDC-A	World Data Center-A for Glaciology (Snow and Ice) Colorado
WMO	World Meteorological Organizations

Gierloff-Emden

9.1.1 Definitions

The coastal zone is the area at the margin of the oceans from the continental shelf edge to the coastal plains landwards off the shore. This boundary zone of lithosphere, hydrosphere, and atmosphere is also called a *triple interface*. Table 1, Fig. 1.

Table 1. Classification of coastal terms concerning the triple interface [80 G2].

	Atmosphere ↑	
Litosphere ⟵	Coastal zone	⟶ Hydrosphere
Continent	coast	ocean
Land terrace	cliff coast → ← beach → ← tidal flat	wave-cut platform
Continental	supralittoral → shore ← eulittoral	sublittoral
Subaerial	subaerial → ← (temporarely inundated)	submarine

The sciences of oceanography, geology, geography, and biology have created their own nomenclature of coastal terms [66 B 1, 60 A, 58 G, 59 K, 67 Z, 76 B 1, 80 G 2, 82 S 1, 82 S 2, 85 B, 84 P, 83 K 2, 68 D 2, 68 F, 76 L, 78 E 1, 82 S 3, 83 J, 84 G, 85 D]. Coastal oceanography treated in this chapter is restricted to the marginal zone of the ocean. The sea influences significantly the land and its waters. Fig. 1.

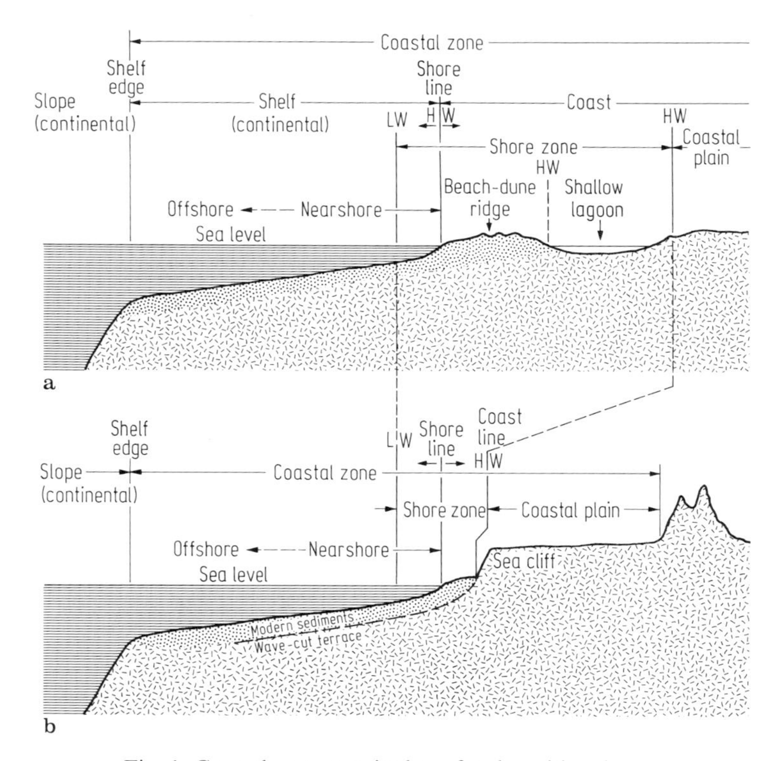

Fig. 1. Coastal zone terminology for depositional coast (a) and cliffed coast (b) [71 I 2]. Model not to scale.

Coastal zone: encompasses the land and sea area bordering the *shoreline* and includes the *shore zone.*

Coast: stripe of land of width up to several nm that extends from the shoreline inland to the first major change in terrain features.

Coastline: 1. technically, the line that forms the boundary between the *coast* and the *shore*;
2. commonly, the line that forms the boundary between the land and the water.

Coastal plain: a plain which borders the sea coast and extends from the *coastline* to the nearest elevated land.

Coastal waters: waters seaward from the low tide shoreline, including nearshore waters (seaward to the low breaker line) and offshore waters (beyond this), and waters of the shallow sea.

Shore: the zone between the water's edge at normal low tide and the landward limit of effective wave action; it comprizes the foreshore, exposed at low tide and submerged at high tide, and the backshore, extending above normal high tide level, but inundated occasionally by exceptionally high tides, or by large waves during storms.

Shoreline: strictly the water's edge migrating to and fro with the tide between high water and low water, i.e. the intersection of a specific plane (level) of water with the shore. For reference level see subsect. 9.1.4.2, Fig. 5, Table 10.

Offshore: 1. in beach terminology the comparatively flat zone of variable width, extending from the breaker zone to the seaward edge of the continental shelf (*shelf edge*).
2. Direction seaward from the shore.

Beach: is an accumulation of loose sediment such as sand, shingle, or boulders, sometimes confined to the backshore but often extending across the foreshore as well.

Barrier: is a bank of beach material which lies offshore, and is exposed at high tide.

Offshore bar: is a similar feature like a barrier, and is submerged for at least part of the tidal cycle.

9.1.2 Units, measurements, and parameters of coastal oceanography

Distances of coastal extensions are given, for horizontal distances, in [m], [km], or in [nm], Table 2, and for vertical distances, in [cm], [m], or in feet [ft] and fathoms [fm], Tables 3a, b.

The length of the coast is size-unit independent if the measurement is made on very large scale charts, i.e. it comes close to the real world. The length of the coast is also size-unit independent if the measurement is made on small-scale charts, i.e. if the size-unit is particularly defined. The measurement of irregular lines can be done by approximation of small circles. An approximation of the length can be done by using the theory of fractals [77 M 1]. The distance of the horizon from the coast or the visibility of the coast from the sea is given in Table 4.

The international *nautical mile* (nm) is equivalent to the average length of a meridian minute, i.e. difference of latitude along a meridian, and corresponds to a mean latitude of $\phi = 45°$ if the earth figure is approximated as a rotational ellipsoid [80 T]. This minute represents the distance of one part in $60 \cdot 360 = 21\,600$ parts of a great circle of the earth.

1 nm ≙ 1852 m (exactly), as defined by IHB, 1954, see Table 2.

Table 2. The nautical mile (nm), different definitions before 1954.

"Admiralty" or nautical mile	British, Hydrograph. Office	1853.232 m	1
Nautical mile	USA, US Coast and Geodetic Survey	1853.248 m	1
Nautical mile, French		1852.276 m	2
Seemeile, deutsch (sm)		1852.276 m	2
International nautical mile (nm)	Intern. Hydrogr. Bureau (1928 calculated, international convention 1954)	1852.000 m	

The nautical mile is the length of one minute of arc on the earth's surface, i.e. 1/21600 of a great circle of the earth, corresponding to 1.8551 km on the equator circle and 1.8519 km on the meridian circle (mean).

1: based on a great circle with mean earth radius
2: based on a "mean meridian minute"

Different numerical results depend also on different reference ellipsoids.

Table 3a. Distances and soundings as they are used on British charts.

	Unit	Abbreviation	Definition	Conversion to metric unit
Distances	nautical mile	nm	1/60th of one degree latitude =6040 ft (on equator) ···6100 ft (89° latitude)	
	nautical mile (mean) Brit. Imperial nm		6080 ft	1853.1840 m [1])
Distances and depths	inch	in		0.0254 m
	foot	ft	12 in	0.3048 m
	fathom	fm	6 ft	1.8288 m
	cable		608 ft	185.3184 m
	nautical mile		10 cables	1853.184 m [1])
	league		3 nautical miles	5559.552 m [2])

[1]) To be distinguished from the International nautical mile [nm]: 1 nm ≙ 1852 m.
[2]) To be distinguished from the league based on the International nautical mile [nm]: 1 league (Int.) = 3 nm = 5556 m.

Table 3b. Nautical measures on different nautical chart issues.

Country	Measures	(Feet)	or (Fathoms)	or (Meters)
Danish	favn	=6.176	1.029	1.882
Dutch	vadem	=5.905	0.984	1.800
Norwegian	favn	=6.176	1.029	1.882
Russian	sakhen	=6.000	1.000	1.829
Spanish	braza	=5.492	0.915	1.675
Swedish	famn	=5.844	0.965	1.765

Depths: Soundings on charts are usually expressed in meters (m).
Velocities: Velocity of currents are given in [cm/s], [m/s], or in [kn = nm/h], see Table 1 in ch. 1.

Table 4. Distance of the horizon from the coast. The table gives the distance, in nautical miles, of the visible horizon for various heights of the eye above sea level. The actual distance varies somewhat as refraction changes.

Height m	Nautical miles nm	Height m	Nautical miles nm	Height m	Nautical miles nm	Height m	Nautical miles nm	Height m	Nautical miles nm	Height m	Nautical miles nm
0.3	1.1	1.8	2.8	3.3	3.8	4.8	4.6	6.4	5.2	7.9	5.8
0.6	1.6	2.1	3.0	3.6	4.0	5.7	4.7	6.7	5.4	8.2	5.9
0.9	2.0	2.5	3.2	3.9	4.1	5.5	4.9	7.0	5.5	8.5	6.1
1.2	2.3	2.7	3.4	4.2	4.3	5.7	5.0	7.3	5.6	8.8	6.2
1.5	2.6	3.0	3.6	4.5	4.4	6.0	5.1	7.6	5.7	4.1	6.3

9.1.3 The length of the coasts (horizontal spatial parameters)

Recording the coast lengths of the world on maps with a scale of 1:1 000 000 leads to a length of 250 000 km and 440 000 km, respectively, depending on the rules of measurements: 50 000 km of these 440 000 km are cliffed coasts [72 H, 77 G 2]. Compare Tables 6, 7.

According to the distribution of land and sea on the earth the majority of the coasts is found on the northern hemisphere. See Tables 6 and 7 [76 I, 53 I, 71 W 1, 76 W 3, 77 M 1, 82 M 1].

Table 5. Dimensions of major topographic features [71 I 1, 71 I 2].

Area of land mass from "National geographic atlas of the world", 1966.

Area of continental shelf and oceans from Sverdrup et al., 1942. Length of the coastline of South America, Australia, Africa, Madagascar, and Great Britain were measured from charts of scale 1:6 000 000. Other coastline lengths were calculated.

Topographic features	Area 10^6 km^2	Length of coastline 10^3 km	Length of shelf break 10^3 km
Continents	138.8	210.0	300.0
Large islands (98 islands > 2500 km^2)	9.3	136.0	149.0
Small islands (25···2500 km^2)	0.8	93.0	
Total land	148.9	439.0	350.0 *)

*) From [74 B].

Table 6. Lengths and development of coasts, classification according to continents [78 L 2]. See also [62 S 1].

Continent	Areas without islands 10^6 km^2	Circumference of a circle with equal parts (V)	Real coastal length (L)	Coastal development ($V:L$)
North America	20.0	15 500 km	75 500 km	1:4.9
Eurasia	50.7	23 950 km	107 800 km	1:4.5
Europe	9.2	10 700 km	37 200 km	1:3.5
Asia	41.5	21 900 km	70 600 km	1:3.2
Australia	7.6	9 700 km	19 500 km	1:2
South America	17.6	14 600 km	28 700 km	1:2
Antarctica, incl. the Shelf Ice	14.0	13 300 km	24 300 km	1:1.88
Antarctica, without the Shelf Ice	13.1	13 200 km	24 700 km	1:1.9
Africa	29.2	18 600 km	30 500 km	1:1.6
Total [1])	139.1	95 650 km	286 300 km	1:3

Coastal lengths after map 1:1 000 000; opposite coasts of islands, which are 10 nm, or less, apart, are not considered. Lengths of highly sinuous coasts with many bays are smoothed out.

[1]) Without the Shelf Ice of Antarctica, and without the subareas Europe, Asia.

Table 7. Lengths of coasts according to states [83 C 3].

Coastline lengths and areal extents of states: As the table shows, the jurisdictional area of some states (small island states in particular) would increase by a very large factor once a 200-mile Exclusive Economic Zone EEZ is established. States without coasts are not named. Source: FAO – UNO. For some states no values are available for the lengths of the coastline.

State	Length of coastline		Sea area within EEZ limits	Total land area	State	Length of coastline		Sea area within EEZ limits	Total land area
	km	nm	10^3 km^2	10^3 km^2		km	nm	10^3 km^2	10^3 km^2
Albania	287	155	12.3	28.7	Ecuador	848	458	1159.0	283.6
Algeria	1104	596	137.2	2381.7	Egypt	2422	1308	173.5	1001.4
Angola	1493	806	506.1	1246.7	El Salvador	304	164	91.9	21.0
Argentina	3926	2120	1164.5	2776.9	Equatorial	341	184	283.2	28.1
Australia	27949	15091	7006.5	7686.9	Ethiopia	1011	546	75.8	1221.9
Bahamas (The)	–	–	759.2	13.9	Fiji	–	–	1134.7	18.3
Bahrain	126	68	5.1	0.6	Finland	1361	735	98.1	337.0
Bangladesh	574	310	76.8	144.0	France	2543	1373	341.2	547.0
Barbados	102	55	167.3	0.4	Gabon	739	399	213.6	267.7
Belgium	63	34	2.7	30.5	Gambia (The)	70	38	19.5	11.3
Benin	–	65	27.1	112.6	Ghana	528	285	218.1	238.5
Brazil	6838	3692	3168.4	8512.0	Greece	3047	1645	505.1	131.9
Bulgaria	248	134	32.9	110.9	Grenada	–	–	n.a.	0.3
Burma	2278	1230	509.5	678.0	Guatemala	330	178	99.1	108.9
Cambodia	389	210	55.6	181.0	Guinea	352	190	71.0	245.9
Cameroon	346	187	15.4	475.4	Guinea-Bissau	–	–	150.5	36.1
Canada	20611	11129	4697.7	9976.1	Guyana	430	232	130.3	215.0
Cape Verde	–	–	789.4	4.0	Haiti	1082	584	160.5	27.8
Chile	5337	2882	2288.2	756.9	Honduras	693	374	200.9	112.1
China	7338	3962	1355.8	9597.0	Iceland	2000	1080	866.9	103.0
Colombia	1893	1022	603.2	1138.9	India	5110	2759	2014.9	3280.5
Comoros	391	211	228.4	2.2	Indonesia	36640	19784	5408.6	1904.3
Congo	196	106	24.7	342.0	Iran	1833	990	155.7	1648.0
Costa Rica	826	446	258.9	50.7	Iraq	19	10	0.7	434.9
Cuba	3235	1747	362.8	114.5	Ireland	1228	663	380.3	70.3
Cyprus	537	290	99.4	9.3	Israel	230	124	23.3	20.7
Denmark	1270	686	68.6	43.1	Italy	4539	2451	552.1	301.2
Djibouti	–	–	6.2	22.0	Ivory Coast	507	274	104.6	322.5
Dominican	602	325	268.8	48.7	Jamaica	519	280	297.6	11.0
East Germany	354	191	9.6	108.2	Japan	8967	4842	3861.1	372.3

Table 7 (continued).

State	Length of coastline		Sea area within EEZ limits	Total land area
	km	nm	10^3 km^2	10^3 km^2
Jordan	28	15	0.7	97.7
Kenya	457	247	118.0	582.6
Kuwait	213	115	12.0	17.8
Lebanon	194	105	22.6	10.4
Liberia	537	290	229.7	111.4
Libya	1 685	910	338.1	1 759.5
Madagascar	3 991	2 155	1 292.0	587.0
Malaysia	3 432	1 853	475.6	329.7
Maldives	–	–	959.1	0.3
Malta	93	50	66.2	0.3
Mauritania	667	360	154.3	1 030.7
Mauritius	161	87	1 183.0	1.9
Mexico	8 978	4 848	2 851.2	1 972.5
Morocco	1 658	895	278.1	446.6
Mozambique	2 504	1 352	562.0	783.0
Netherlands	367	198	84.7	40.8
New Zealand	5 130	2 770	4 833.2	268.7
Nicaragua	824	445	159.8	130.0
Nigeria	769	415	210.9	923.8
North Yemen	452	244	33.9	195.0
Norway	3 056	1 650	2 024.8	386.6
Oman	1 861	1 005	561.7	212.5
Pakistan	1 389	750	318.5	803.9
Panama	1 839	993	306.5	75.6
Papua New Guinea	–	–	n.a.	461.7
Peru	2 330	1 258	786.6	1 285.2
Philippines	12 958	6 997	1 890.7	300.0
Poland	446	241	28.5	312.7
Portugal	1 376	743	1 774.2	92.1
Qatar	378	204	24.0	11.0
Romania	209	113	31.9	237.5
São Tomé and Principe	157	85	128.2	1.0

State	Length of coastline		Sea area within EEZ limits	Total land area
	km	nm	10^3 km^2	10^3 km^2
Saudi Arabia	2 437	1 316	186.2	2 149.7
Senegal	446	241	205.7	196.2
Seychelles	–	–	729.7	0.4
Sierra Leone	406	219	155.7	71.7
Singapore	52	28	0.3	0.6
Somalia	2 956	1 596	782.8	637.7
South Africa	2 708	1 462	1 016.7	1 222.2
South Yemen	–	–	550.3	287.7
Spain	3 774	2 038	1 219.4	770.8
Sri Lanka	1 204	650	517.4	65.6
Sudan	735	387	91.6	2 505.8
Suriname	363	196	101.2	163.3
Sweden	2 517	1 359	155.3	450.0
Syria	152	82	10.3	185.3
Tanzania	1 239	669	223.2	945.1
Thailand	2 406	1 299	324.7	514.0
Togo	48	26	1.0	56.0
Trinidad and Tobago	470	254	76.8	5.1
Tunisia	1 028	555	85.7	163.6
Turkey	3 558	1 921	236.6	780.6
USSR	42 777	23 098	4 490.3	22 402.2
United Arab Emirates	2 421	1 307	59.3	83.6
United Kingdom	5 167	2 790	2 336.5	286.7
USA	21 576	11 650	7 825.0	9 372.0
Uruguay	565	305	119.3	177.5
Venezuela	2 002	1 081	363.8	912.0
Vietnam	2 309	1 247	722.1	332.6
Western Samoa	–	–	96.0	2.8
West Germany	570	308	40.8	248.6
Yugoslavia	789	426	52.5	255.8
Zaire	41	22	1.0	2 345.4

9.1.3.1 The length of the coastline and the shoreline: modern surveying with remote sensing techniques

Detailed surveying: large scale, 1 : 25 000···1 : 100 000, taking features of the scale of 2 m···10 m into account. Local extension based on aerial photography.

General surveying: small scale, 1 : 250 000···1 : 1 000 000 taking features of the scale of 25 m···100 m into account. Global extension based on orbital remote sensing with satellite imagery. See Figs. 2a, b, 3. [79 D 2, 77 G 1, 85 R, 79 I, 77 V 3].

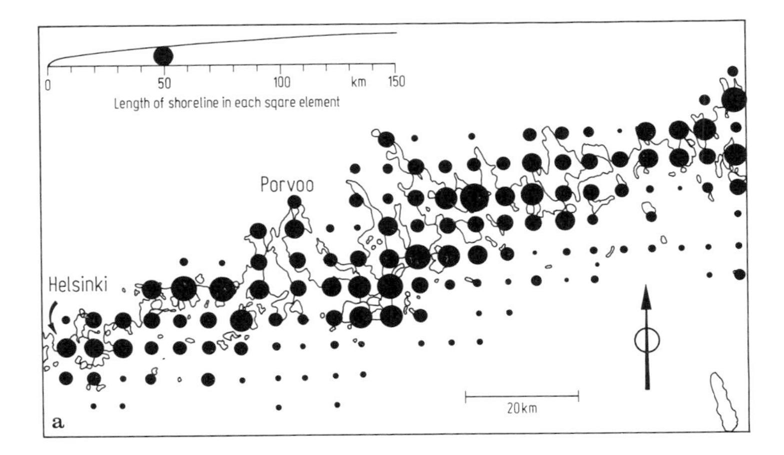

Fig. 2a. Coverage of coast in scale 1 : 20 000; each aerial photograph scene covers 5 km × 5 km. Surveying the length of the shoreline with screen of 5 km × 5 km square elements. The shoreline depth depends on the tidal state. The coastline shows coastal configuration in general. The lengths: shoreline > coastline. Example: coast with archipelago of South Finland [60 G].

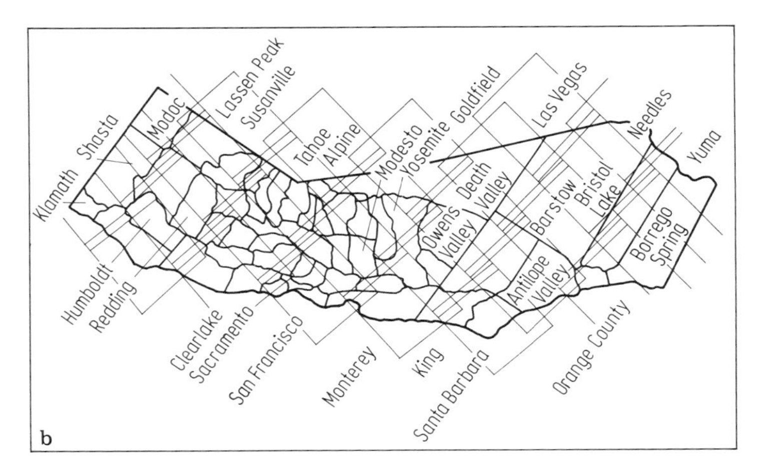

Fig. 2b. Coverage of coasts with LANDSAT; each scene covers 100 nm^2 = 185 km^2. To cover the coast of California (1300 km length) 9 scenes out of 9 orbits are necessary. For one coverage of 250 000 km coasts of the earth's oceans estimated 2500 images would be necessary. Spatial resolution: LANDSAT 1, 2, 3: 80 m × 80 m, 1972···1981; LANDSAT 4, 5: 30 m × 30 m, 1981 onward [82 G 1].

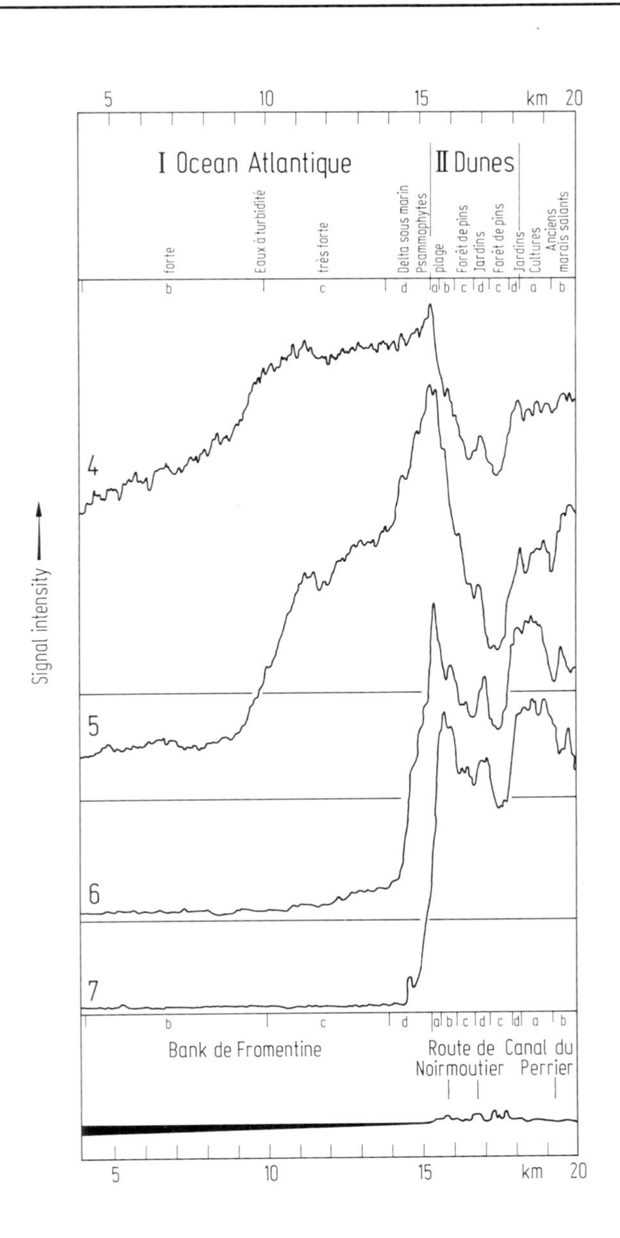

Fig. 3. The coast and the shoreline on orbital remote sensing imagery: spectral reemission, as recorded by LANDSAT-MSS (Multispectral Scanner Imagery). Seen from the space, as on satellite images, the coasts are the most spectacular linear element on the earth's surface [73 V 2].

Profiles through the coastal zone of the Vendée between St. Jean de Monts and Fromentine on the Atlantic Ocean, Bay of Biscay, according to its characteristics on multispectral imagery: The signal intensity of the individual zones, ocean-beach-dunes-forest-cultivated land refers to microdensitometer measurements of MSS Multispectral Scanner of bands 4, 5, 6, and 7 of LANDSAT 1. (Bands 4, 5, and 6: visible light, band 7: reflected infrared). Top: the natural zonation. Bottom: morphologic profile, length in [km], plage = beach with shoreline (boundary I···II). After: M. F. Verger (1973): L'utilisation des donnais par des techniques quantitatives. Bull. Assoc. Géogr. France, Nr. 411···412, S. 723, part of the seaward zone.

9.1.4 Vertical spatial extension and reference datum of the coastal zone

9.1.4.1 The mean sea level (msl)

The msl is defined as the average height of the surface of the sea on the open coast for all stages of the tide over a 19-years period. (This depends on the 18,6-year moon period caused by the rotation of the nodes along the ecliptic). The sea level of the ocean is recorded by means of tide gauges (mareographs) [83 P 2]. The msl is an approximation of the geoid; it is in fact variable in time. See Table 8 and Fig. 4 [66 B 1, 70 S 2, 74 L, 55 P, 60 R, 63 R 2, 83 I 2].

The sea level data: Permanent Service for Mean Sea Level (PSMSL).

The PSMSL has undertaken this survey to determine the present state of sea level monitoring throughout the world (Operational sea level stations, IOC [83 P 2]) by an expanded network of tide gauges and an enhanced system of sea level data exchange including some 100 tide gauges stations of the world. The Permanent Service for Mean Sea Level is a member of the Federation of Astronomical and Geophysical Services (FAGS) established by the International Council of Scientific Unions (ICSU) (Bidston Observatory, Birkenhead, Merseyside).

Table 8a. Variations of the sea level [66 P].

Type	Period *)	Prevalence	Amplitude
Waves	seconds	virtually continuous	up to 20 m
Tsunamis	minutes to hours	occasional	up to 10 m
Tides	12.5 hours	daily *)	up to 20 m
Storm tides	days to years	occasional	one to several m
Annual tides	1 year	yearly	up to 1 m
Long-period changes **)	geologic time	$10^2 \cdots 10^7$ years	up to 200 m

*) Defined as the time interval between successive high waters: not all of these phenomena are strictly periodic [66 P].

**) For long periodic changes caused by eustatic processes compare sect. 9.12.

Table 8b. Mean sea level trends obtained by regression analysis of selected cities [67 R 1, 73 H 3].

Location	Time interval	Mean sea level trend mm/a	Location	Time interval	Mean sea level trend mm/a
New York	1893–1971	$+2.83 \pm 0.12$	Cascais, Portugal	1894–1962	$+1.08 \pm 0.22$
Baltimore	1903–1971	$+3.32 \pm 0.15$	Brest, France	1894–1961	$+2.08 \pm 0.26$
Key West	1913–1971	$+2.07 \pm 0.18$	Aberdeen, Scotland	1874–1962	$+0.78 \pm 0.11$
Galveston	1909–1971	$+5.73 \pm 0.34$	Sheerness, England	1874–1959	$+2.37 \pm 0.17$
Seattle	1899–1971	$+1.89 \pm 0.18$	Den Helder, Netherlands	1874–1962	$+1.45 \pm 0.12$
San Francisco	1898–1971	$+1.95 \pm 0.18$	Vlissingen, Netherlands	1890–1962	$+3.04 \pm 0.19$
San Diego	1906–1971	$+1.95 \pm 0.15$	Kobenhavn, Denmark	1889–1962	$+0.23 \pm 0.16$
Honolulu	1905–1971	$+1.62 \pm 0.21$	Oslo, Norway	1886–1962	-4.11 ± 0.33
Cristobal, C.Z.	1909–1971	$+1.13 \pm 0.15$	Stockholm, Sweden	1889–1962	-3.97 ± 0.19
Trieste, Italy	1905–1962	$+1.40 \pm 0.32$	Helsinki, Finland	1879–1962	-3.15 ± 0.18
Marseilles, France	1894–1958	$+1.69 \pm 0.15$	Swinemünde, Poland	1871–1942	$+0.75 \pm 0.15$

Annual tides are usually small: 0.20···0.30 m in range. There are at least six factors involved (Table 8a):

1. Changing heat content of the water with the season is related to changes in specific volume, and, therefore, in sea level [75 D 1].
2. Changes in salinity due to rainfall, or evaporation, affect specific volume.
3. Changes in the distribution of air pressure over the ocean induce the inverted barometer effect.
4. There are annual and semiannual astronomic tides.
5. The total amount of water in the oceans varies seasonally. In March, there is more water on the continents than in October, largely in the form of northern hemisphere snow [79 G].

Monsoon winds and currents pile up water along coastlines. Compare sect. 9.11.

6. The effect of a 1 mb (millibar) pressure anomaly is to modify msl in height by approximately 10 mm [74 L, 66 F].

The annual variations of the msl are inverse on the two hemispheres, i.e., more exactly, the variations are symmetric with respect to the latitude of 6° N. There is a deviation, however, on the northern hemisphere due to monsoon climates. Fig. 4. [67 R 1, 73 H 2, 85 F 2].

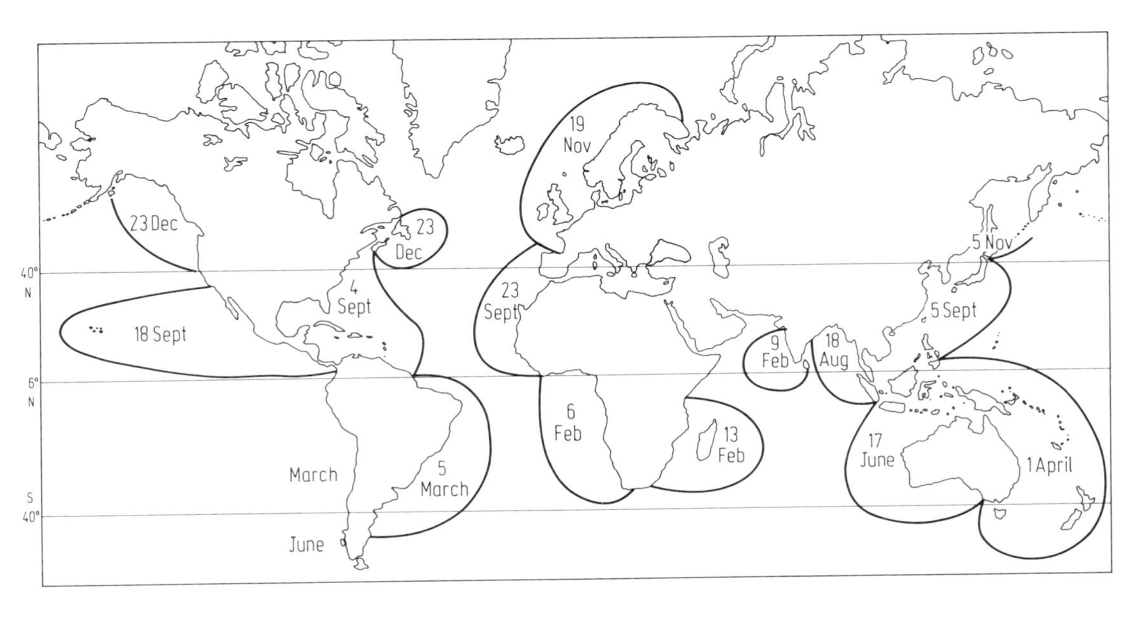

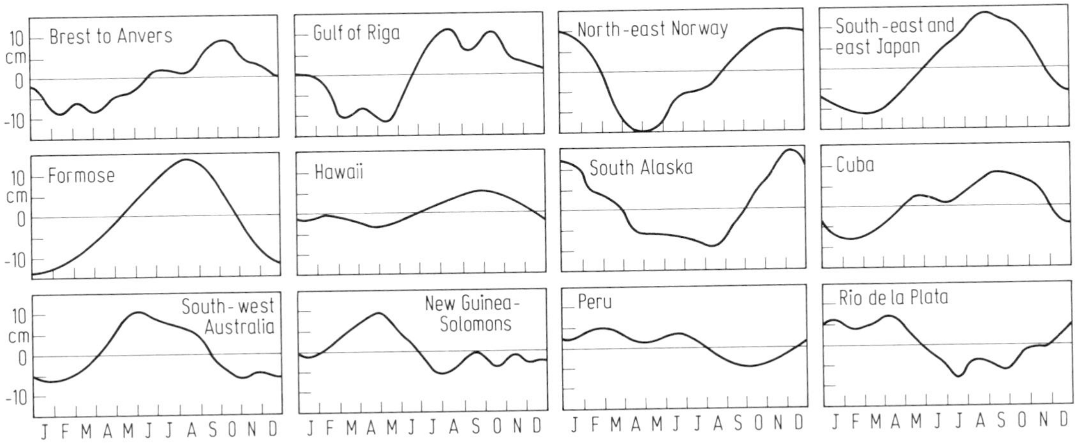

Fig. 4. Annual variations of mean sea level. The map shows the times of mean maxima for groups of stations, according to E. Lisitzin. The curves show annual variations in msl (according to Pattullo et al., 1955) [79 G, 55 L, 55 P].

Gierloff-Emden

9.1.4.2 The reference level for depths on nautical charts (chart datum)

The charted depths on nautical charts are related to the chart datum. Mean low water springs (MLWS) are above, or equal to the chart datum. This differs from country to country. See Tables 9, 11, Figs. 5, 6, 7.

For the German Baltic Sea coast the chart datum is practically identical with the map datum (=NN).

On the German North Sea coast the chart datum (=KN) corresponds to the mean low water spring level (MLWS), "MSpNW" which was derived from numerous gaugings.

Tidal datum planes vary with the type of tides and with the very special regional tide range. Fig. 5. [67 G 1].

Table 9. The chart datum of tidal coasts in different states After [66 S].

State	Chart datum
Fed. Rep. Germany (North Sea)	local MLWS
Sweden (south of 61°42′ N)	local mean low water (MLW)
Norway	approximately the locally lowest low water
The Netherlands	local MLWS (means of several years of the lowest LWS of individual months)
France (north and west coast)	local lowest LW, according to calculations; local exceptions (Brest, Seine-estuary)
Portugal	like France
Spain	like France
Great Britain (UK)	changing from place to place: 0.0 m···0.6 m below the local MLWS
USSR (White Sea)	locally, the lowest possible LW (according to special calculations)

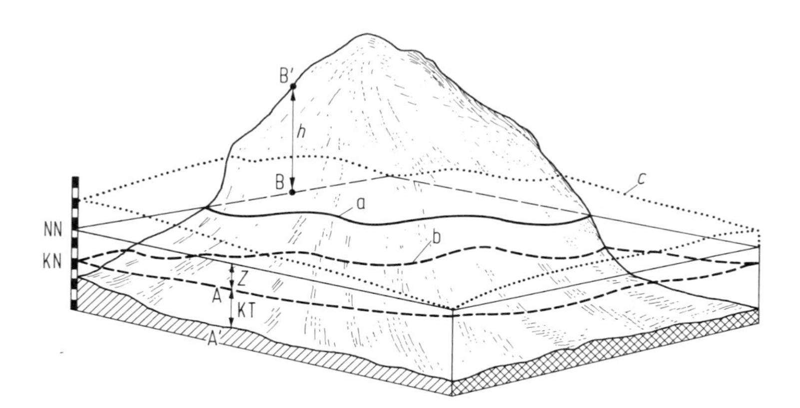

Fig. 5. Chart datum point [44 H]. NN = Normal Null datum level (on maps). KN = datum level (on charts). Z = mean water level (approximately half of mean high water spring). $\overline{AA'}$ = depth related to KN (on charts) = charted depth K_T. a–a = datum on topographic maps. b–b = datum on nautical charts. c–c = approximately high water spring. $\overline{BB'}$ = elevation above datum (NN), on maps the height h of a point = land survey.

9.1.4.3 Coastal datums concerning maps and charts

The sea level datums of different countries are controlled by precision levelling in order to determine either uplift or lowering of the land. These are long-term changes [70 S 2].

In order to determine elevations of the land above a defined sea level (reference level), the definition of the *sea level datum* is necessary. As different countries determined their own national sea level datums, it is necessary to add, or substract, certain values in order to obtain the international reference level. See Tables 9, 10 and 11.

The Soviet Union, Poland, Czechoslovakia, Hungary, Rumania, and Bulgaria uniformly use the zero level of the Kronstadt tide gauge (Leningrad) as reference level for the determination of elevations. In the German Democratic Republic which has also used the Kronstadt level (since 1960), the difference between elevations above NN (≈Amsterdamsch Peil) and elevations above HN (≈Kronstadt tide gauge) amounts to approximately 0.16 m [80 G 2].

Table 10. Map datums of European nations depending on sea level datums [66 S].
N.A.P. = Normal Amsterdamsch Peil, N.N. = Map datum ("Normal Null") of the particular nation.

Nation	National sea level (= zero level of more recent maps)	Point of reference	Starting point H_0 [m]	N.N.-N.A.P. ΔH [m]
Netherlands	N.A.P. = Normal Amsterdamsch Peil (MHW)	O.M. Amsterdam I	1.42_{78}	± 0.00
Belgium	Zéro D.G. = Zéro du dépôt de la guerre = Niveau moyen des basses mers à Ostende (datum main nivellement)	a) Brüssel 1892 IGNMK	100.17_4	-2.33_4
		b) Ostende 1958	6.65_3	-2.31_1
Fed. Rep. Germany	N.N. = Normal Null	DH4-II Berlin. 1912 Normalhöhenpunkt	54.63_8	-0.02_1
France	Z.N. = Zéro Normal	Marseille Rivel Mbc ∅	1.65_{97}	-0.27_3
	N.G.F. = Nivellement General de la France			-0.03_0
Spain	N.Z. = National Zero	Alicante N.P. 1	3.40_{95}	-0.08_5
Portugal	P.N.N. = Portugal Normal Null	Cascais N.P. 1	13.34_0	$+0.12_2$
Italy	Niv. Null 1942	Genova O_M	2.01_{863}	-0.33_2
Switzerland	Niv. Null 1905	Genf 100, R.P.N. = Répère Pierre du Niton	373.60_0	-0.08_2
Denmark	D.N.N. = Dansk Normal Null	1. Aarhus (Kathedrale)		
		2. Erritsø G.I. 1715	22.75_{71}	-0.11_3
Norway	(NGO 1890)	(Oslo 1912)	$(18{,}50_0)$	
	N.N.N. 1954 = Norske Normal Null (21.2 cm below NGO)	Tredge BMD 40 N 55 (Fund. Pkt.)		
		Tredge ID 40 N 53	1.70_6	-0.12_0
Sweden	Normal Null Stockholm	NHP Stockholm 1900	11.80_0	$+0.29_2$
		Stockholm NHP 1950	11.82_0	
Finland	Normal Null 1944	Helsinki Observator.	30.46_{52}	$+0.20_8$
Great Britain	O.D.N. = Ordnance Datum Newlyn			-0.22_6
Austria	Niv. Null			-0.26_7
Yugoslavia				-0.25_3
Greece		Saloniki		-0.07

9.1.4.4 Tidal datums for the USA

The nautical charts of the USA are based on different reference levels with regard to the chart datum:

Along the Atlantic Coast of the USA, where both semidiurnal and diurnal types of tide occur, tidal datum planes are mean high water (MHW) and mean low water (MLW). See Fig. 6.

Along the Pacific Coast, including Alaska and Hawaii, where tides are chiefly of the mixed type, datum planes are mean high water and mean lower low water.

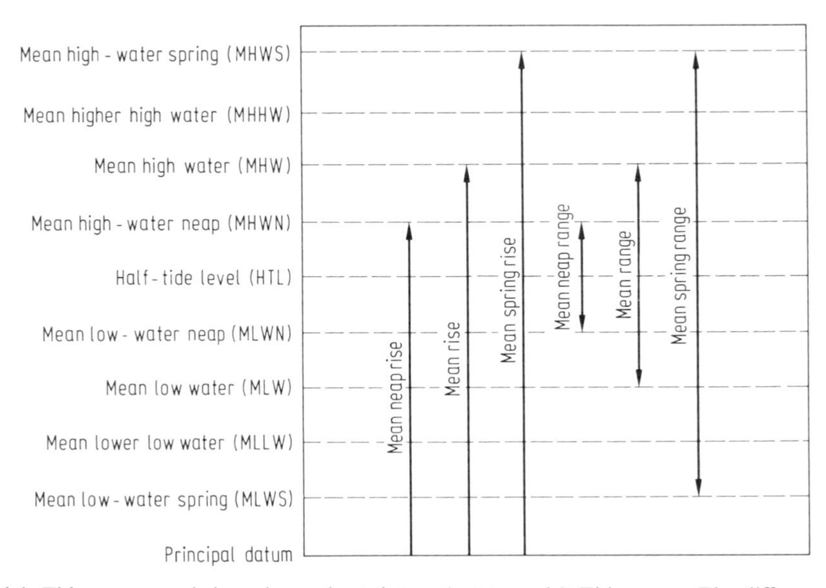

Fig. 6. Model: Tide ranges and rises above chart datum (not to scale). Tide range: The difference in height between consecutive high and low waters. Where the type of tide is diurnal, the mean range is the same as the diurnal range [66 B 1]. Replenished. Principal datum ≙ chart datum. Charted depth: distance from chart datum to the bottom of the sea.

Tidal datum levels useful as reference planes in coastal hydrography. Mean high-water spring: average level of high waters occurring during spring tides. Mean higher high water: average level of higher high waters of each day. Mean high water: average level of all high waters. Mean high-water neap: average level of high waters occurring during neap tides. Half-tide level: level midway between mean high and mean low water. Mean low-water neap: average level of low waters occurring during neap tides. Mean low water: average level of all low waters. Mean lower low water: average level of lower low waters of each day. Mean low-water spring: average level of low waters occurring during spring tides. "Lower low" and "higher high" refer to coasts characterized by semidiurnal tides of unequal amplitude [66 B 1].

Table 11. Abbreviations of tidal datums as reference planes in coastal hydrography [66 B 1].

LW	low water	MHHW	mean higher high water
LHW	lower high water	MHW	mean high water
LHWI	lower high water interval	MHWN	mean high water neaps
LLW	lower low water	MHWS	mean high water springs
LLWI	lower low water interval	MLLW	mean lower low water
MWL	mean water level	MLW	mean low water
MRI	mean rise interval	MLWN	mean low water neaps
MSL	mean sea level	MLWS	mean low water springs
MTL	mean tide level	HTL	half tide level

9.1.4.5 Tidal datums for the Federal Republic of Germany

Official and scientific terms in German language as used in German records and charts: Table 12 and Fig. 7.

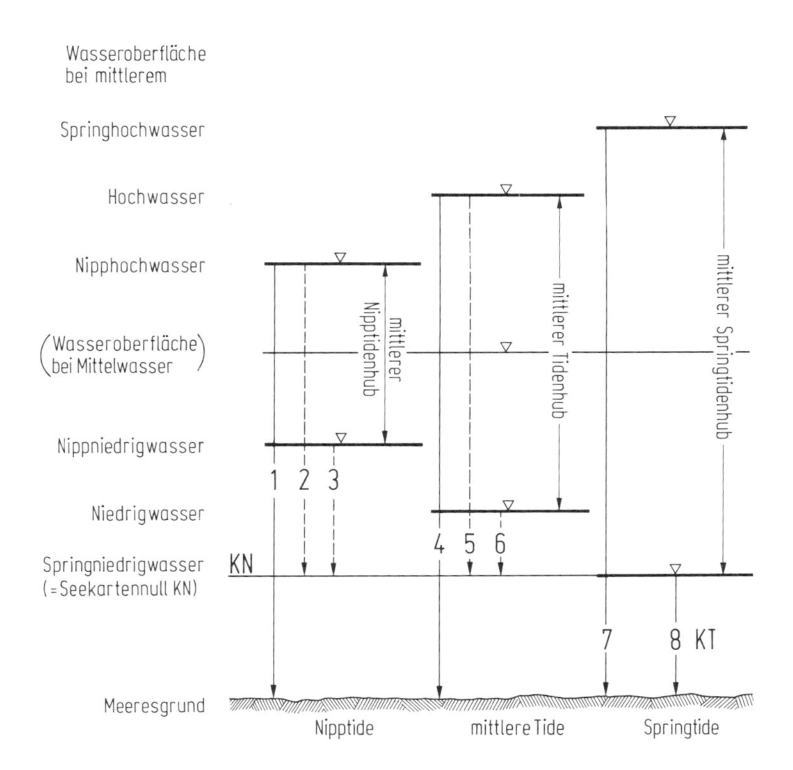

Fig. 7. Nomenclature of tides and tidal elevations, definition of chart datum (KN) [59 S 2]. The tide gauge datum is the reference level of a tide gauge. PN = Pegel Null, Pegel: tide gauge. The tide gauges of the German North Sea coast have a uniform tide gauge datum which is 5 m below the mean sea level ≙ NN (PN = NN – 5 m). See also Fig. 6.

1: Wassertiefe bei mittlerem Nipphochwasser; *2*: mittlere Nipphochwasserhöhe; *3*: mittlere Nippniedrigwasserhöhe; *4*: Wassertiefe bei mittlerem Hochwasser; *5*: mittlere Hochwasserhöhe; *6*: mittlere Niedrigwasserhöhe; *7*: Wassertiefe bei mittlerem Springhochwasser; *8*: Wassertiefe bei mittlerem Springniedrigwasser (= Seekartentiefe KT).

Table 12. Definitions of tidal datums and chart datums of the Federal Republic of Germany [82 D].

A	Amplitude	Mt.	Mitt
F.D.	Falldauer	M.W.	Mittelwasser
Gr.	Greenwich	N.H.W.	Niedrigeres Hochwasser
G.U.	Gezeitenunterschied	N.N.	Normalnull
H	Höhe der Gezeit	N.N.W.	Niedrigeres Niedrigwasser
H.H.W.	Höheres Hochwasser	Np.	Nipp
H.N.W.	Höheres Niedrigwasser	Np.H.W.	Nipphochwasser
H.W.	Hochwasser	Np.N.W.	Nippniedrigwasser
N.W.H.	Hochwasserhöhe	Np.T.H.	Nipptidenhub
H.W.I.	Hochwasser-Intervall	N.W.	Niedrigwasser
H.W.Z.	Hochwasserzeit	N.W.H.	Niedrigwasserhöhe
K.N.	Kartennull	N.W.I.	Niedrigwasser-Intervall
M.E.Z.	Mitteleuropäische Zeit	N.W.Z.	Niedrigwasserzeit
M.G.Z.	Mittlere Greenwich-Zeit	P	Phase
M.H.H.W.	Mittleres höheres Hochwasser	S	Stundenwert
M.H.N.W.	Mittleres höheres Niedrigwasser	S.D.	Steigdauer
M.H.W.	Mittleres Hochwasser	Sp.	Spring
M.N.H.W.	Mittleres niedrigeres Hochwasser	Sp.H.W.	Springhochwasser
M.N.N.W.	Mittleres niedrigeres Niedrigwasser	Sp.N.W.	Springniedrigwasser
M.Np.H.W.	Mittleres Nipphochwasser	Sp.T.H.	Springtidenhub
M.Np.N.W.	Mittleres Nippniedrigwasser	T	Tageswert
M.N.Sp.N.W.	Mittleres niedrigeres Springniedrigwasser	T.F.	Tidenfall
M.N.W.	Mittleres Niedrigwasser	T.H.	Tidenhub
M.O.Z.	Mittlere Ortszeit	T.S.	Tidenstieg
M.Sp.H.W.	Mittleres Springhochwasser	U	P+T+S
M.Sp.N.W.	Mittleres Springniedrigwasser	Z.	Höhe des mittleren Wasserstandes

9.1.5 Spatial extension of coastal zones respecting the Law of the Sea

9.1.5.1 Territorial waters

Definitions of territorial waters:

I. Coastal sea	3-nm belt, territorial waters, sovereignity
II. Contiguous zone	9-nm belt, like territorial waters New coastal sea: 3+9=12 nm
III. Economic zones	e.g. 50 nm, fisheries zone
IV. EEZ: Exclusive economic zone	200 nm (12+188 nm), economic zone
V. High sea	open sea outside the 200-nm-zone.

Table 13. Coastal boundaries: older (GC) and present (UNCLOS) regulations. See Figs. 8, 9, 10, 11 and Table 7.

Geneva Convention 1958 (GC)	Internal waters	Territorial sea (3 nm)	Contiguous zone 9 nm max. 12 nm	High sea	
United Nations Conference on the Law of the Sea 1981 (UNCLOS)	internal waters	territorial sea (12 nm)	contiguous zone 12 nm max. 24 nm	exclusive economic zone 188 nm	high sea

At the present time much of the Law of the Sea must be regarded as being in a transitional phase between the old regime, largely based on the 1958 Geneva Conventions and the new regime, based upon the 1982 UN Convention.

9.1.5.2 The delimitation of the coastal zones according to the Law of the Sea

Along highly sinuous coasts the seaward boundary is determined by means of arcs. Fig. 8.

Along coasts with larger bays whose mouths are not wider than 24 nm (twice the width of the *territorial sea*), the base line is straightened out. The waters on the landward side of this straight base line are regarded as *inland*.

Where islands are situated in front of a bay, and when the distance from these islands to the base line is permanently smaller than 12 nm, the seaward boundary is determined from a straight base line. (Example: Norway). Islands, which are situated more than 12 nm off the coast have their own surrounding *territorial waters* [80 G 2, 78 L 2, 74 A, 76 P, 67 G 1].

Bays are

inland waters if area $\geqq \dfrac{\pi(D/2)^2}{2}$

indentations if area $< \dfrac{\pi(D/2)^2}{2}$

where D denotes diameter of the circle of the bay.

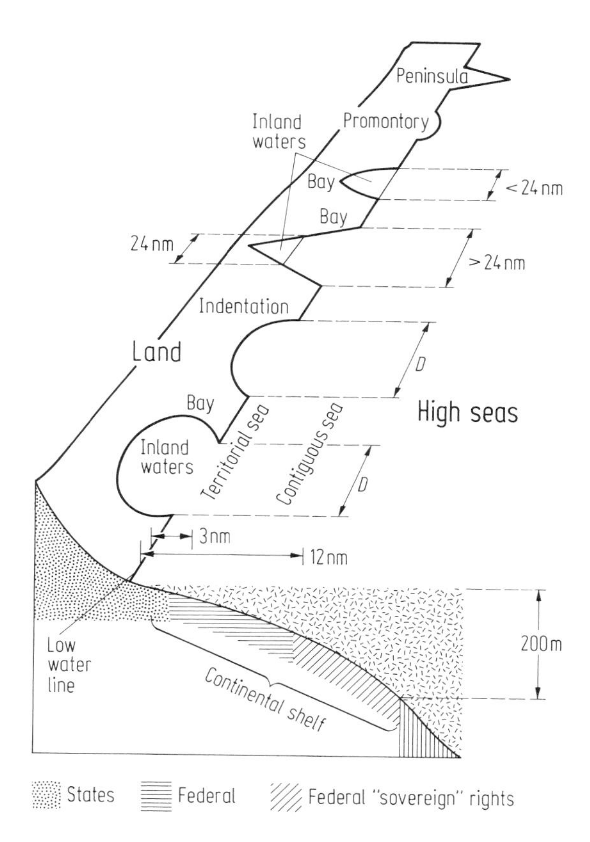

Fig. 8. Generalized application of the principles of seaward boundaries [62 S 1].

9.1.5.3 The spatial extension of coastal zones respecting the Law of the Sea and topographic conditions of the ocean

The EEZ could be extended in cases when the continental shelf extends farther than 200 nm from the coast of an adjacent state. Although the international ratification of this model of an international Law of the Sea as outlined at the end of the last decade by the "United Nations Conference on the Law of the Sea" (UNCLOS) has not been executed, there are some states which claim their adjacent sea in the sense of these designed rules [74 A, 78 L 2]. See Tables 14, 15.

UN "Office of the Law of the Sea Negotiations". A guide to the Law of the Sea, Ref. Pap. 18, United Nations (UN), Dept. of Public Informations: [79 U]. Recommandations of the Intergovernmental Maritime Consultative Organisation (IMCO) are not binding.

The principal physical division of a typical continental margin is shown in Fig. 9. The legal terms do not necessarily bear the same meaning than their physical counterparts. Additional special "formulas" were developed from different countries to fulfill the specific physical conditions concerning width of continental slope and extension and thickness of sediments on continental slope and rise (1982 Convention in accordance with the statutes of the International Court of Justice). See Figs. 9 and 10, Tables 14, 15.

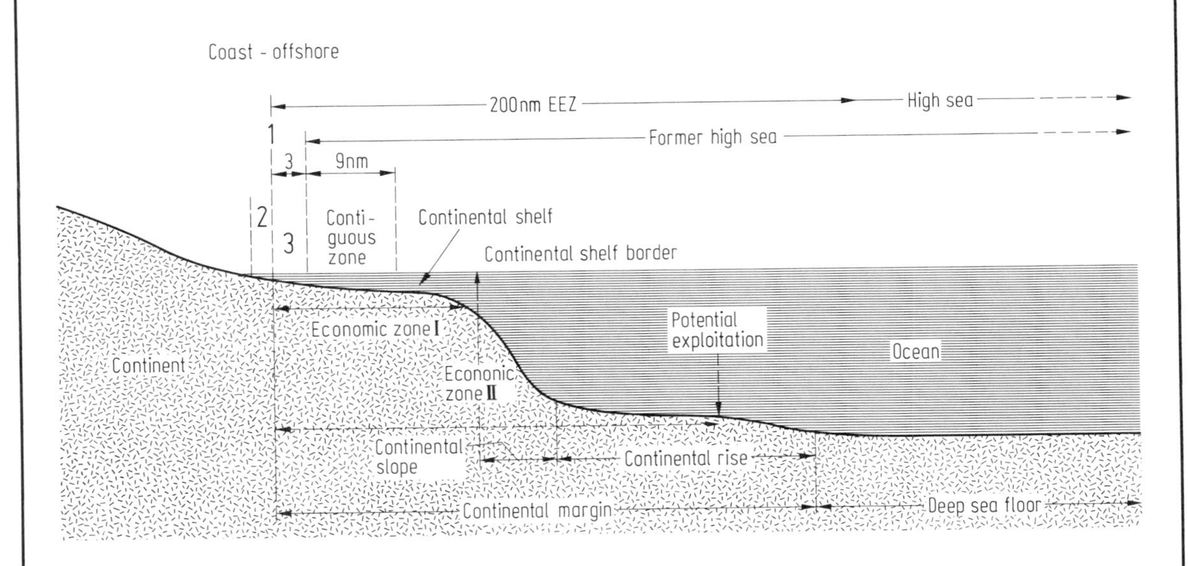

Fig. 9. Legal and economic zones of the sea in relation to the sea floor topography. Model not to scale. The planned 200-nm-EEZ consists of 12 nm territorial waters (3+9 nm), and of 188 nm economic zone. (It may be extended above the continental margin to 350 nm or to 100 nm off 2500 m depth, according to different proposals.) *1*: mean low water (base line); *2*: internal waters; *3*: traditional territorial waters (state borderline at the sea). After different sources. The sea level counts on national datums, compare Tables 9, 10, 11, 12, Figs. 5, 6, 7.

Table 14. Areal extent of shelf and 200-nm-EEZ in relation to ocean areas [78 L 2]. See Fig. 11.

Area	Area	Continental shelf to a depth of 200 m	Extension of the continental platform ("glacis")	200 nm EEZ
	nm^2 (%)	nm^2 (%)	nm^2 (%)	nm^2 (%)
Atlantic Ocean incl. Arctic Ocean	31 040 000 (100%)	4 128 000 (13.29%)	10 709 000 (34.50%)	11 668 000 (37.59%)
Indian Ocean	21 842 000 (100%)	917 000 (4.20%)	3 953 000 (18.10%)	7 064 000 (32.34%)
Pacific Ocean	52 385 000 (100%)	2 986 000 (5.70%)	9 377 000 (17.92%)	19 013 000 (36.29%)
Total	105 267 000 (100%)	8 031 000 (7.63%)	24 039 000 (22.84%)	37 745 000 (35.85%)

Delimitation of coastal and oceanic topographic conditions of the zones of the Law of the Sea:

In about 80% of all global morphological settings the gradients of sea floor areas will be within the limits as given in Table 15. For the remaining 20% of the cases, gradients may lie beyond either one of the two given limits. The change in gradient indicating the edge of the continental shelf, and the beginning of the continental slope occurs at a depth of 150···250 m. As an average the isobath of 200 m is assumed to mark the transition from continental shelf to continental slope. In a similar manner the limits at 2000 m and at 5000 m depth, are average values which, in the physical reality may fluctuate [84 L 1].

In Fig. 10 the under-water provinces are presented according to the Convention on the Law of the Sea, in two hypothetical extreme versions, i.e. with, respectively, a minimum and a maximum gradient based on the values given in Table 15. A fictitious sea floor is shown in between the two extremes. Fig. 10 presents a number of cases distinguished in Article 76 of the Convention on the Law of the Sea.

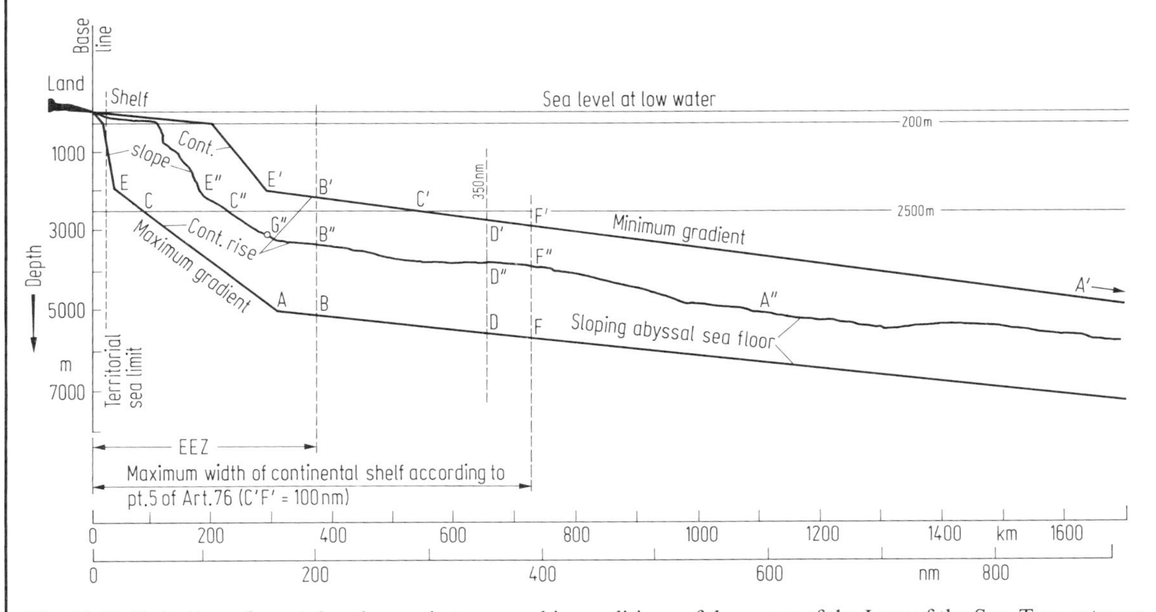

Fig. 10. Delimitation of coastal and oceanic topographic conditions of the zones of the Law of the Sea. Two extreme cases of hypothetical continental margins: 1. with maximum gradient from the continent to the abyssal ocean floor, 2. with minimum gradient (both according to Table 15). Between the two extremes, a fictitious sea floor is shown. Both, the limiting gradients and the limiting depths between shelf, slope and rise are average values which may be surpassed in both directions. Point A' is not shown in the picture as the intersection with the 5000-m-isobath falls outside the type area [84 L 1]. Article 76 UNO, 1981: Delimitations and measurements required to the New convention on the Law of the Sea. Vertical exaggeration 66 times.

Table 15. Sea floor areas with their limiting gradients, and – based on limiting depth assumptions – the minimum and maximum distances of their outer limits from the base line [84 L1].

Feature	Gradient	Depths	Distances of outer limit or isobath from the base line	
			Minimum	Maximum
Continental shelf	1/100···1/1000	0 ··· 200 m	20 km	200 km
Continental slope	1/10 ···1/50	200 ···2000 m	38 km	290 km
Continental rise	1/90 ···1/500	2000···5000 m	308 km	1790 km
Abyssal depth	1/600···hor.	>5000 m	–	–
2500 isobath	– ···–	2500 m	83 km	540 km

9.1.5.4 The 200-nm Economic Exclusive Zone (EEZ)

Territorial waters 3%, EEZ 36%, High Sea 61%. The areal gain of EEZ for a single isolated small island amounts to 400 000 km^2 if calculated with a radius of 200 nm. Fig. 11.

Compare: International boundary study, no. 46. Theoretical areal allocations of seabord to coastal states. The Geographer, Bureau of Intelligence and Research of the Department of the States, USA.

Fig. 11. 200-nm EEZ of the states around continents and islands (not concerning Antarctica). Equal area projection.

1 Nanpo
2 Kuril Islands
3 Aleutian Islands
4 Sable Islands
5 Bermuda Islands
6 Azores
7 Madeira Islands
8 Bonin Islands
9 Mariana Islands
10 W. Caroline Islands
11 E. Caroline Islands
12 Marshall Islands
13 Kiribati
14 Hawaii Islands
15 Palmyra Islands
16 Revilla Gigedo Islands
17 Clipperton Atoll
18 Bahama Islands
19 Guadeloupe Isle
20 Martinique Isle
21 Cocos Islands
22 Galapagos Islands
23 Canary Islands
24 Cape Verde Islands
25 St. Paul Islands
26 Sao Tomé Isle
27 Laccadive Islands
28 Maldive Islands
29 Tuvalu
30 Solomon Islands
31 Fiji Islands
32 Hebrides Islands
33 New Caledonia
34 Phoenix Islands
35 Tahiti Islands
36 West Samoa Islands
37 Tonga Islands
38 Cook Islands
39 Raoul Isle
40 Marquesas Islands
41 Tuamotu Islands
42 Tubuai Islands
43 Easter Islands
44 Sala y Gomez Isle
45 S. Felix Isle
46 Atol das Rocas Isle
47 Noronha
48 Ascension Isle
49 St. Helena
50 Seychelles
51 Comoro Islands
52 Mauritius
53 Reunion Isle
54 Chagos Arch.
55 Christmas Islands
56 Cocos Islands
57 Auckland Islands
58 Chatham Islands
59 Campbell Isle
60 Juan Fernandez Islands
61 Falkland Islands
62 South Georgia
63 South Orkney Islands
64 South Sandwich Islands
65 Tristan da Cunha Isle
66 Gough Isle
67 Bouvet Isle
68 Prince Edward Isle
69 Crozet Islands
70 Kerguelen Islands
71 Heard Isle
72 Amsterdam Isle

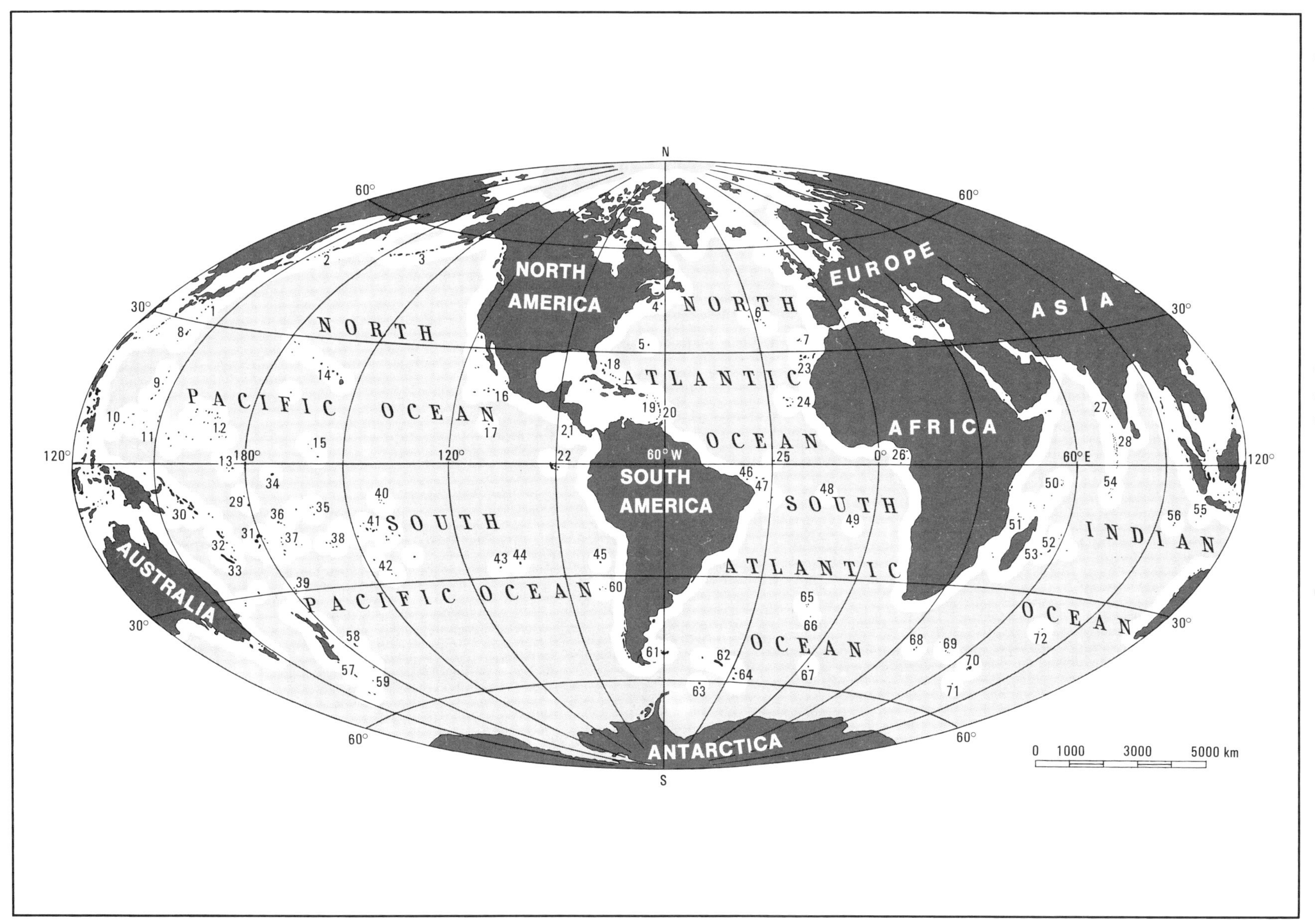
N
NORTH AMERICA
EUROPE
ASIA
AFRICA
SOUTH AMERICA
AUSTRALIA
ANTARCTICA
NORTH PACIFIC OCEAN
SOUTH PACIFIC OCEAN
NORTH ATLANTIC OCEAN
SOUTH ATLANTIC OCEAN
INDIAN OCEAN
60° W
0°
60° E
120°
180°
30°
60°
S
0 1000 3000 5000 km

9.2 Classification of coasts

Coastal classifications are given concerning different aspects, which have been grouped into six main thematic categories:

1. Structural characteristics of the land (oldest [1888 S], reborn with plate tectonic theory [71 I 2]. Table 16.
2. Morphological attributes of the coastal zone.
3. Genetic theories based on processes that have influenced the coast.
4. Dynamic character of the changing geomorphology.
5. Dynamic character of the energy of the marine system.
6. Dynamics of wave characteristics.

For a complete summary, see [82 S 1, 72 V, 76 S 2, 72 K, 58 M 1, 58 M 2, 68 S 3, 77 V 2, 75 H 2, 71 I 2, 60 P, 64 D, 72 D 2].

Table 16. Statistical distribution of first-order features of the world coastal zones [71 I 1].
Data based on measurements of coastline made on charts of 1 : 39 000 000 scale.
The percentages in each row total 100 percent, except for the world which excludes Antarctica.

9.2.1 Coastal classification according to the global tectonics

(compare chapter 1, Topography)

The new approach to coastal classification by Inman and Nordstrom [71 I 2] builds on the structural element of Suess' classification, and is based on global tectonics and sea-floor spreading. Three size scales (1···3) associated with coasts are recognized.

1. Scale associated with moving plates, with linear dimensions of about 1000 km along the coast, 100 km normal to it, and a vertical range of 10 km.
2. Scale brings in erosional and depositional modifications of the first order features, with dimensions of about 100 km, 10 km, and 1 km for the parallel, normal, and vertical direction, respectively. Table 19.
3. Scale is dependent on wave action, type of material and similar variables. Its dimensions are 1···100 km along the coast, 1 km normal to the coast direction, and it includes such features as beach face, berm, and bars. Table 17, Fig. 12.

Table 17. Statistical distribution of second-order features of the world coastal zones [71 I 2].
The percentages in each row total 100 percent.

Table 16.

Continent	Total coast length	Collision coast		Trailing-edge coast						Marginal sea coast	
				neo-		afro-		amero-			
	10^3 km	10^3 km	%	10^3 km	%	10^3 km	%	10^3 km	%	10^3 km	%
Europe-Asia	75.2	9.7	12.9	7.8	10.4	1.7	2.3	31.4	41.7	24.6	32.7
Africa	24.7	1.6	6.5			23.1	93.5				
North America	43.4	11.4	26.3	2.4	5.5			24.2	55.8	5.4	12.4
South America	27.3	9.0	32.9	2.4	8.8			12.5	45.8	3.4	12.5
Antarctica	24.5										
Australia	14.9	2.5	16.8	1.3	8.8			5.7	38.2	5.4	36.2
Large islands (>2500 km^2)	136.1	82.3	60.5			14.5	10.6	39.3	28.9		
Small islands (<2500 km^2)	93.0	60.3	65.0					32.7	35.0		
World	439.7	171.1	39.1	18.9	4.3	29.8	6.8	155.4	35.4	38.8	8.8

Table 17.

Continent	Total coast length	Wave erosion		Wave deposition		River deposition		Wind deposition		Glaciated		Biogenous	
		Length											
	10^3 km	10^3 km	%	10^3 km	%	10^3 km	%	10^3 km	%	10^3 km	%	10^3 km	%
Europe-Asia	75.2	28.9	38.4	15.3	20.3	6.6	8.8	0.5	0.7	23.9	31.8		
Africa	24.7	9.4	38.0	9.6	38.9	0.7	2.8	3.5	14.2			1.5	6.1
North America	43.4	8.2	18.9	7.7	17.7	1.5	3.4	0.1	0.2	23.9	55.0	2.1	4.8
South America	27.3	13.1	48.0	5.7	20.8	3.6	13.2			3.6	13.2	1.3	4.8
Antarctica	24.5									24.5	100.0		
Australia	14.9	5.9	39.7	4.3	28.8	1.4	9.4	1.2	8.0			2.1	14.1
Large islands (>2500 km^2)	136.1	70.9	52.1	8.9	6.5	0.4	0.3			49.6	36.5	6.3	4.6
Small islands (<2500 km^2)	93.0	51.3	55.2		<1					32.7	35.0	9.0	9.8
World	439.1	196.9	44.9	51.4	11.7	14.2	3.2	5.3	1.2	158.0	36.0	13.3	3.0

The term "coastal zone" is used to refer to the first two scales, and the "shore zone" to the third.

The shore zone includes the foreshore, surf zone, and nearshore zone, while the coast zone includes the coastal plain and continental shelf to the shelf edge, often taken conventionally to a depth of 200 m.

In terms of the gross effects of plate tectonics, there appear to be three major classes and several subclasses of coasts as follows:

1. Collision coasts:

a) continental collision coasts, i.e., collision coasts involving the margins of continents where a thick plate collides with a thin plate (e.g., west coasts of the Americas)

b) island arc collision coasts, i.e., collision coasts along island arcs where a thin plate collides with another thin plate (e.g., the Philippines, and the Indonesian and Aleutian island arcs)

2. Trailing-edge coasts:

a) neo-trailing-edge coasts, i.e., new trailing-edge coasts formed near beginning separation centers and rifts (e.g., Red Sea and Gulf of California)

b) afro-trailing-edge coasts, i.e., the opposite coast of the continent is also trailing, so that the potential for terrestrial erosion and deposition at the coast is low (e.g., Atlantic and Indian Oceans coasts of Africa)

c) amero-trailing-edge coasts, i.e., the trailing-edge of a continent with a collision coast, and, therefore, "actively" modified by the depositional products and erosional effects from an extensive area of high interior mountains (e.g., east coasts of the Americas)

3. Marginal sea coasts:

i.e., coasts fronting on marginal seas, and protected from the open ocean by island arcs (e.g., Vietnam, southern China, and Korea).

Table 18. Comparison of the width and type of internal structure of the continental shelves with first-order coastal features [71 I 2].

Percentages based on length of shelf edge, excluding the Antarctic shelf.

	Tectonic class					
	Collision	Trailing-edge			Marginal sea	Percentage of world coast
		neo-	afro-	amero-		
Shelf type:						
Tectonic dam	86.2	1.7			74.2	31.4
Salt dome dam					13.4	1.6
Reef dam	10.5	81.5	4.5		12.4	10.9
Progradational (no dam)	3.3	16.8	95.5	100.0		56.1
Total	100.0	100.0	100.0	100.0	100.0	100.0
Shelf width:						
Narrow (< 50 km)	23.9	6.5	11.2	12.6	3.3	57.5
Wide (> 50 km)	1.4	0.1	3.6	29.1	8.3	42.5
Total	25.3	6.6	14.8	41.7	11.6	100.0

Table 19. Comparison of first and second-order features, based on Tables 15 and 16 [71 I 2]. All figures are percentages.

Second order	First-order						
	Collision	Trailing-edge coast			Marginal sea	Total	Percentage of world coastline
		neo-	afro-	amero-			
Wave erosion	47.9	5.6	3.8	4.9	37.8	100	44.7
Wave deposition	15.5	12.4	21.4	33.3	17.4	100	11.6
River deposition	5.7	6.4	9.9	62.4	15.6	100	3.2
Wind deposition	1.9	18.8	79.3			100	1.2
Glaciated	6.4		7.2	86.4		100	30.7
Biogenous	36.1	21.0	11.3	15.8	15.8	100	3.0
Percentage of world coastline	39.1	4.3	6.8	35.4	8.8		94.4 *)

*) Excluding the Antarctic coastline (5.6%).

Table 20. Comparison of the morphologic classification with the first-order tectonic classification (in %) [71 I 2].

Morphologic class	First-order tectonic class				
	Collision coast	Trailing-edge coast			Marginal sea coast
		neo-	afro-	amero-	
Mountainous coast	97.2	8.0			2.5
Narrow shelf, hilly coast		75.1	14.1		5.6
Narrow shelf, plains coast		15.9	46.2	1.5	
Wide shelf, plains coast			4.0	89.3	3.1
Wide shelf, hilly coast				2.2	77.4
Deltaic coast		1.0	3.4	1.3	5.8
Reef coast			3.0	1.9	5.6
Glaciated coast	2.8		29.3	3.8	
Total	100.0	100.0	100.0	100.0	100.0
Percentage of world coastline *)	39.0	4.6	7.5	35.2	8.1

*) Excluding the Antarctic coastline (5.6%).

9.2.2 Coastal classification according to dynamics of wave characteristics

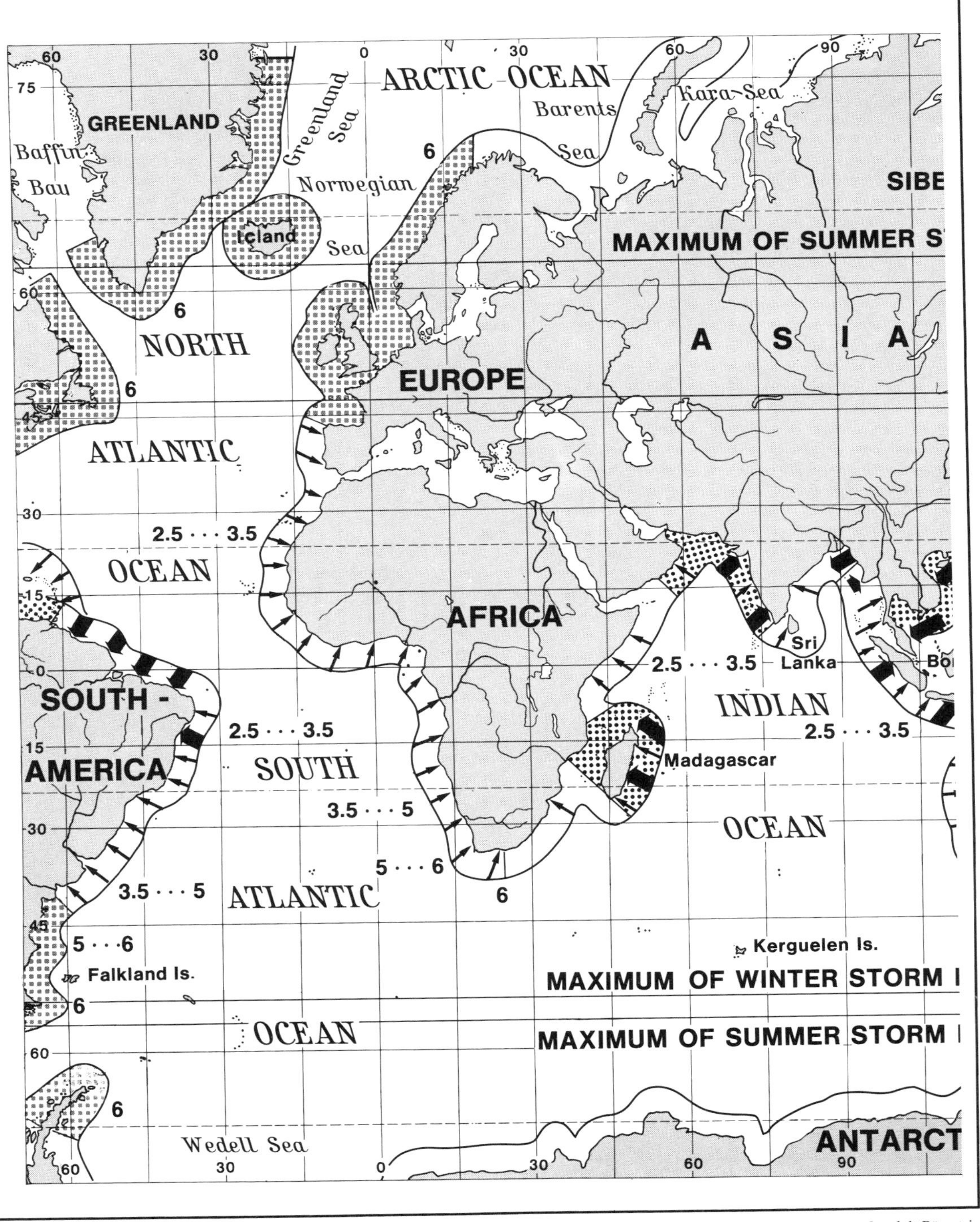

Fig. 12. Classification of coasts based on the dynamic principle: form and strength of prevailing wave systems [73 D]. From summer to winter the zone of the storm wave environment affecting coasts shifts between latitudes 46° N and 62° N on the northern hemisphere, between latitudes 54° S and 66° S on the southern hemisphere. The numbers give the wave height maximum in [m]. Compare [72 D 2, 80 D].

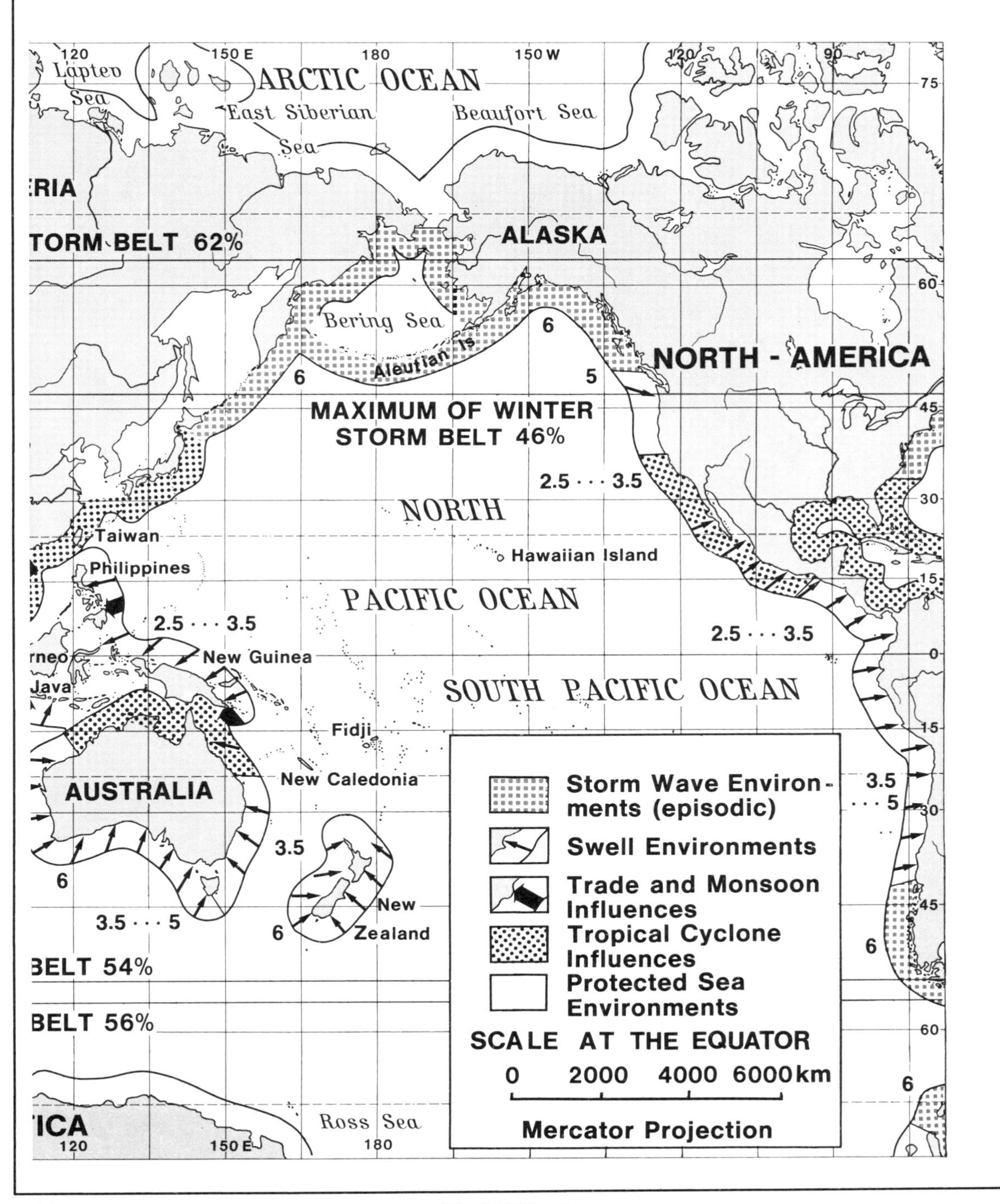

Gierloff-Emden

9.3 The high-energy environment of the coastal zone

The *high-energy environment* can be defined as the environment where waves and currents have sufficient strength to influence geomorphology, or where morphology is sufficiently unstable to be influenced by water movements. Most high-energy environments occur nearshore and on the inner parts of continental shelves. Examples are beaches, tidal inlets and tidal channels, submarine deltas, barrier islands, submarine bottom forms and river mouths [73 I]. See Fig. 13, Tables 21 and 22. Frequently, sand dunes form an essential part of the sedimentary structure, so that wind transport is important [80 G 2, 84 G]. Compare Figs. 84a, b, 85a, b. For energy of man's intervention in coastal zones see Table 23.

9.3.1 The coastal energy budget

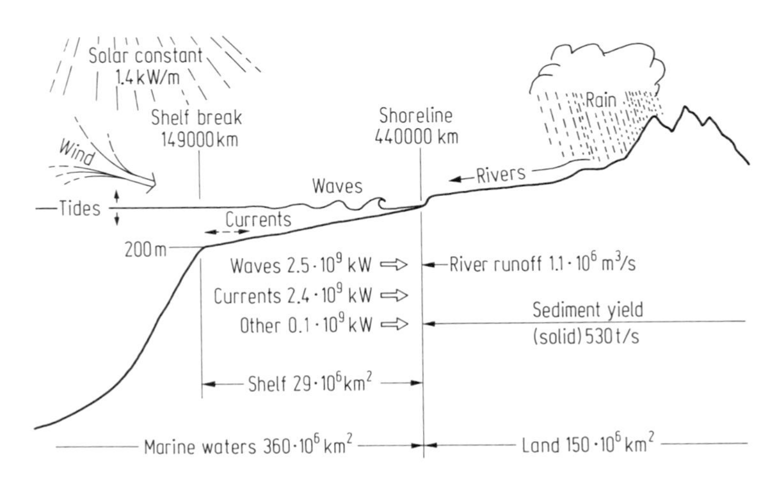

Fig. 13. Budget of energy and land runoff in the coastal zone. Most of the energy comes from the open sea [71 I 1, 73 I].

Table 21. Dissipation of mechanical energy in the coastal zone shallow waters [73 I].

Estimates of the natural rates of dissipation of mechanical energy in the shallow waters of the world.

Source	Dissipation rate
Wind-generated waves breaking against the shoreline	$2.5 \cdot 10^9$ kW
Tidal currents in shallow seas	$2.2 \cdot 10^9$ kW
Large-scale ocean currents in shallow seas (Guiana Current off the northeastern coast of South America, $0.13 \cdot 10^9$ kW; Falkland Current over the Argentine Shelf, $0.03 \cdot 10^9$ kW)	$0.2 \cdot 10^9$ kW
All other sources (wind stress on the beach, $0.01 \cdot 10^9$ kW; internal waves, $0.01 \cdot 10^9$ kW; edge waves; shelf seiche; tsunamis; rivers entering the oceans)	$0.1 \cdot 10^9$ kW
Total	$5.0 \cdot 10^9$ kW

Table 22. Run-off and discharge into the coastal zone waters [73 I].

Estimate of the natural run-off of fresh water and solids from the continents into the coastal waters of the world	
Source	Discharge rate
Discharge of water into the oceans from all rivers	$1.1 \cdot 10^6\ m^3\ s^{-1}$
Flux of dissolved solids	$125\ t\ s^{-1}$
Flux of particulate solids	$530\ t\ s^{-1}$
Average erosion rate of land	$6\ cm/10^3\ a$

Energy flux and budget from man's impact on coastal areas:

Table 23. Energy of man's intervention in coastal areas in terms of the use of energy and the disposal of waste in recent years (1967–1971). The "extrapolated world value" is based on present U.S. standards of use, and a population of $200 \cdot 10^6$ extrapolated to a present world population of $3.5 \cdot 10^9$ [73 I]. Heat units are converted to their mechanical equivalents.

Item, location, source, and amount	U.S.		World
	value per capita	doubling time years	value extrapolated from U.S. standards
Power, United States, 1968			
Total energy consumed, $62.5 \cdot 10^{15}$ British thermal units	10.4 kW	14···20	$36.4 \cdot 10^9$ kW
Electrical energy consumed, $1.33 \cdot 10^{12}$ kWh	0.8 kW	9	$2.7 \cdot 10^9$ kW
Waste heat (coolant) from the generation of electrical energy (1.5 W of coolant per Watt distributed)	1.5 kW	8	$5.3 \cdot 10^9$ kW
Sewage			
Southern California, population: 10^7, 1968, $1.1 \cdot 10^9$ gallons/day	$416\ l\ d^{-1}$		
New York City, population: $8 \cdot 10^6$, 1970, $1.3 \cdot 10^9$ gallons/day	$615\ l\ d^{-1}$		$25 \cdot 10^3\ m^3\ s^{-1}$
Solids from domestic sewers into coastal waters, $500 \cdot 10^3$ tons/year	$15\ kg\ a^{-1}$		
Oil			
Spillage into world oceans		15	$2.4 \cdot 10^6\ t\ a^{-1}$
Solid waste, United States			
Collected per capita in 1967, 5.2 pounds/day	$2.4\ kg\ d^{-1}$	10	$97\ t\ s^{-1}$
Dumped by barge into coastal waters, 1968, $62 \cdot 10^6$ tons/year	$320\ kg\ a^{-1}$	10	$36\ t\ s^{-1}$
Dredge fill, United States, 1971			
Bypass, disposal, and maintenance, $3 \cdot 10^8$ cubic yards	$1.6\ t\ a^{-1}$	4	$172\ t\ s^{-1}$
Mining of sand and gravel, 1970			
California, $23.5 \cdot 10^6$ U.S. tons	$1.1\ t\ a^{-1}$		$122\ t\ s^{-1}$
Great Britain, dredged from the North Sea, $13 \cdot 10^6$ British tons	$260\ kg\ a^{-1}$	5	$29\ t\ s^{-1}$

9.3.2 Time scale of coastal zone processes

Sequences of periodic and non-periodic processes of coastal areas. Figs. 14, 15.

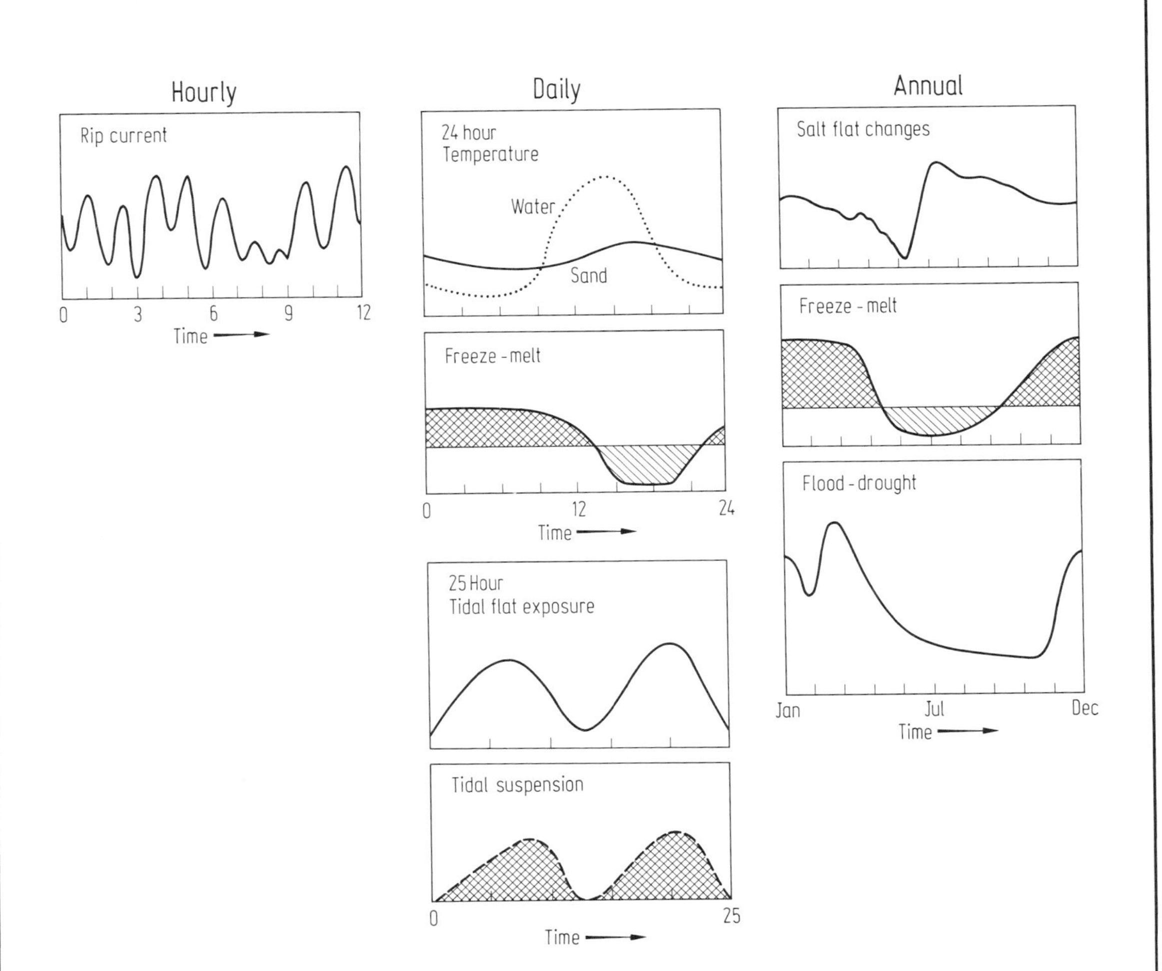

Fig. 14. Examples of some prospective sequential signatures at different time scales: hourly, solar day (24 hours), lunar day (25 hours), and annual. Relative magnitude of reflectance (or emission) is directed along the ordinate, time on the abscissa [72 N 2].

Variability and time-space scale of coastal waters.
Coastal processes: periodic and episodic.

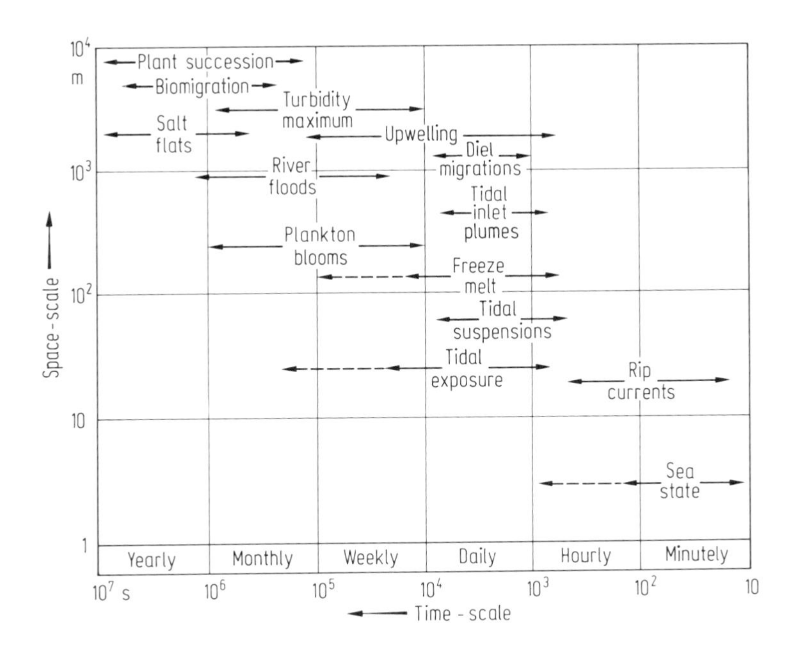

Fig. 15. Time scales of some temporal coastal features at different geometric scales. Full range of geometric scale for each feature is not shown [72 N 2].

9.4 The shore as the interaction zone of sea and land

9.4.1 Shore features of steep coasts and interactions

Cliffs and coastal platforms along steep coasts:

The extension of cliffed coasts of the world amounts to 50000 km.

Definition: The term *cliff* describes breakages along steep coasts [58 G, 64 S, 67 Z, 71 S 1, 71 S 2, 54 V, 67 S 3, 80 W, 84 K 2, 68 S 1, 71 S 3, 71 S 4, 73 S 4, 19 J, 80 P 2]. Fig. 16.

Coastal platforms are unconformable erosional surfaces caused by the breakers; their extension is related to the wave climates, the underlying bedrock (petrovariance), and to tectonic conditions (tectovariance) [77 R 1, 77 R 2]. A special feature of a platform at the coast is the "strandflate" which stretches at and under the level of the low water, and is developed in subpolar zones as there is the Norwegian coast [80 G 2].

Classification of cliffed coasts and coastal platforms [76 B 1]:

1. Cliffed coast with an inter-tidal shore platform: platform with uniform gradient
2. Cliffed coast with a shore platform at about high tidal level: convex slope seaward high water mean
3. Cliffed coast with a shore platform at about low tide level: concave seaward cliff, notch
4. Plunging cliff form with no shore platform: steep submarine slope, continuing the subaerial slope.

9.4.1.1 The mechanism of cliff erosion and platform erosion

Slope control of platform erosion: (compare Figs. 16, 17)
The shore platform lowering is expressed [67 Z] as

$$\frac{dZ}{dt} = \frac{dX}{dt} \cdot \tan\beta$$

where

$$\frac{dZ}{dt} = \text{lowering rate of platform surface}$$

$$\frac{dX}{dt} = \text{rate of cliff recession}$$

$$\tan\beta = \text{platform gradient}.$$

Wave control of cliff erosion:
The cliff erosion by waves is expressed [83 K 3] as

$$X = \phi(f_w, f_r, t)$$

where

$$X = \text{cliff erosion}$$

$$f_w = \text{assaulting force of waves}$$

$$f_r = \text{resisting force of material}$$

$$t = \text{time}$$

and ϕ denoting the functional relationship.
The basic equation of cliff erosion was found to obey the form [77 S 2]

$$\frac{dX}{dt} \propto \ln\frac{f_w}{f_r}, \qquad f_w > f_r,$$

$$\frac{dX}{dt} = 0, \quad \text{otherwise.}$$

Lithological control of cliff erosion:
For the relation of cliff erosion by waves and rock strength, see [77 S 2]. A linear relation should be found between the erosion rate $\frac{dX}{dt}$ and the compressive strength of the material S_c in a semilogarithmic expression:

$$\frac{dX}{dt} \propto (C_1 - \ln S_c)$$

where

$$C_1 = C_1(f_w) = \text{constant}$$

$$S_c = \text{compressive strength in } [\text{kg cm}^{-2}].$$

This relationship has been verified by field data from cliff coasts in Japan [77 S 2].

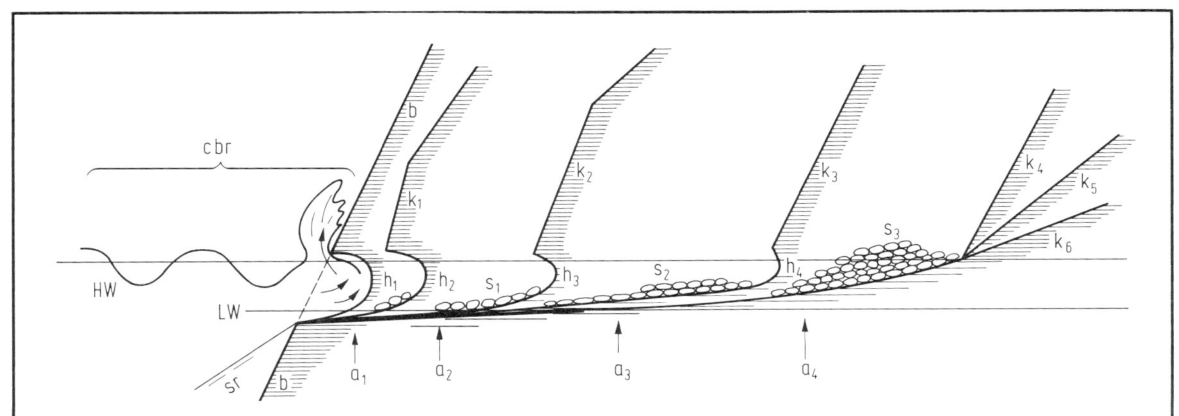

Fig. 16. Beach- and cliff-breakers and morphologic effect on cliffs and platforms [58 M 1]. For classification of cliff notch profiles, see [68 T]. cbr: cliff breakers. sr: submarine sediment ramp. HW: high water. LW: low water. Model not to scale. $a_1 \cdots a_4$: abrasion platforms. bb: initial slope. $h_1 \cdots h_4$: notches from breakers. $k_1 \cdots k_6$: cliffs. $s_1 \cdots s_3$: beach ridges.

9.4.1.2 Cliff erosion rates

A recession distance versus time relationship at a given site along a stretch of coast is schematically given in Fig. 17. Here τ is the length of the time interval under consideration. X/τ is the long-term average erosion rate, and $\Delta X/\Delta t$ the erosion rate in a very short period of time during which erosion actually occurred. The two erosion rates are different: $X/\tau < \Delta X/\Delta t$. Fig. 17, compare Fig. 49.

Erosion rates have been documented or recorded in many countries in the world. The erosion rates listed in Table 24 are average values over the period of 10⋯100 years. Orders of erosion rates summarized on a lithological basis are generally: 10^{-3} m/a for granitic rocks, $10^{-3} \cdots 10^{-2}$ m/a for limestone, 10^{-2} m/a for flysh and shale, $10^{-1} \cdots 10$ m/a for chalk and Tertiary sedimentary rocks, 1⋯10 m/a for Quarternary deposits, and 10 m/a for unconsolidated volcanic ejecta. These results show that lithology strongly controls the erosion rate.

Extremely short-term erosional events caused by storm waves make such average rates as shown in Table 24 unreliable, since episodical erosion rates caused by storms or slumps are reported on the scale of 100 m/d locally. [83 D in 83 K 3].

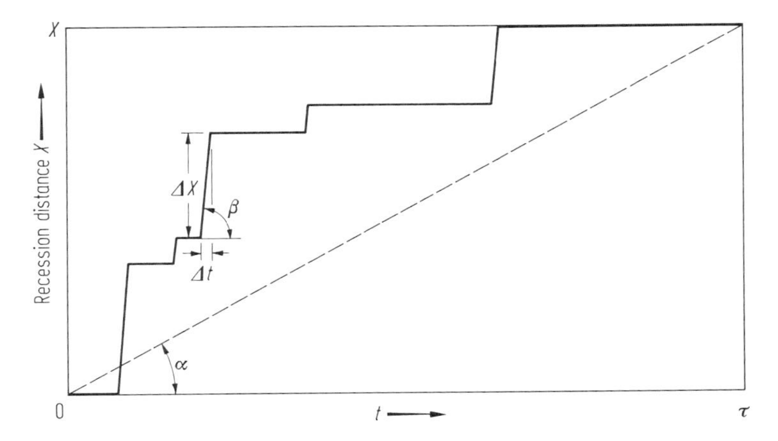

Fig. 17. Schematic representation of temporal change of cliff recession distance. A very short-term erosion rate, $\Delta X/\Delta t$, is generally different from the long-term average value X/τ [83 S 2 in 83 K 4]. Erosion rate: $\tan\beta = \Delta X/\Delta t$. Average erosion rate: $\tan\alpha = X/\tau$.

Gierloff-Emden

Table 24. Coastal cliff erosion rates in different regions and at different rock types. Selected data, after numerous sources [83 S 2, 83 K 3].

Location	Lithology	Erosion rate [m/a]	Interval	Method
Australia				
Point Peron near Perth	limestone	0.0002···0.001	1953···1962	steel pegs
Warrnambool, Victoria	aeolianite	0.014	(130 a)	surveys
Barbados				
Mullins Bay	coral rock	0.002	(4.5 a)	pegs
France				
Ault. Somme	chalk	0.08···0.37	(100 a)	–
Germany				
North of Kiel, Baltic Sea	glacial clay	0.6	1873···1934	–
Heligoland, North Sea	mesozoic sandstone	1	–	–
Cape Arkona, Rugen	chalk	3···4	(100 a)	–
Iceland				
Surtsey Island	lava	25···37	1967···1975	Air-photo-grammetric surveys
Indonesia				
Krakatoa Island	volcanic ash	33	1883···1928	–
Japan				
Haranomachi, Fukushima	pliocene sandstone, mudstone	0.3···0.7	1912···1959	maps
Byobugaura, Chiba	pliocene mudstone	0.73	1884···1969	maps
Toban, Hyogo	pleistocene deposits	1.0···1.5	1893···1955	maps
Southern coast, Atsumi peninsula	pleistocene deposits	0.03···0.6	1888···1959	maps
New Zealand				
Kai-iwi Beach, Wellington	pliocene siltstone	1.5	1876···1893	surveys
Point Kean, Kaikoura	tertiary mudstone	0.24	1942···1974	air photos
Sweden				
Hallshuk, Gotland	silurian limestone	0.018	1899···1955	photos
U.K.				
Fourth Bight, Whitby, Yorkshire	upper lias shale	0.023	1971···1972	micro-erosion meter
Holderness, Humberside	glacial deposits	3.3	–	–
Holderness, Humberside	glacial deposits	0.29···1.75	1852···1952	surveys
Norfolk	glacial deposits	0.9	1880···1967	maps, air photos
Dunwich, Suffolk	glacial deposits	1.6	1589···1753	maps
Seven Sisters, Sussex	chalk	0.51	1873···1962	–
Birling Gap, Seven Sisters,	chalk	0.91	1875···1916	–
Sussex		0.99	1950···1962	surveys
U.S.A.				
Martha's Vineyard, Mass.	glacial deposits	1.7	1846···1886	surveys
North Shore, Long Island, N.Y.	glacial deposits	0.5	(80 a)	charts, air photos
Montara, Calif.	miocene conglomerate	0.26···0.29	1912···1965	maps
La Jolla, Calif.	cretaceous sandstone, eocene shale	0.01···0.2	1940···1979	photos
Sunset Cliffs, San Diego, Calif.	upper cretaceous sandstone	0.012	(75 a)	photos, observations

Table 24 (continued).

Location	Lithology	Erosion rate [m/a]	Interval	Method
U.S.S.R.				
Barents Sea coast	granitic rocks	0.001⋯0.002	–	–
Laptev Sea coast	glacial clay	4⋯6	–	–
Okhotsk Sea coast	volcanic ash	50 (max.)	–	–
Okhotsk Sea coast	quaternary brown loam and clay	40 (max.)	–	–

Very short-term cliff recession distances caused by storm events have been reported: Iceland, Surtsey Is.: 100 m, Japan, Nishinoshima Is.: 120 m, USA, Cape Cod: 1.5 m, USA, Long Island: 12 m.

9.4.2 Shore features of depositional coasts and interactions

9.4.2.1 The beach profile and slope

Dependance of beach gradient on different parameters:

Taking wave energy E, wavelength l, and sand settling velocity v as three variables which include all three quantities, mass, time, and length, and solving dimensionless equations for all the other variables in turn, six dimensionless ratios are arrived at: The beach slope is a function of these dimensionless ratios [53 K]:

$$i = f\left(\frac{h}{l}, \frac{vp}{l}, \frac{v^2 l^3 d}{E}, \frac{l^2 vu}{E}, \frac{s}{l}, \frac{lg}{v^2}\right)$$

i = beach slope, dimensionless
h = wave height in [m]
l = wavelength in [m]
u = viscosity in [kg m^{-1} s^{-1}]
s = sand medium diameter in [m]

p = wave period in [s]
v = sand settling velocity in [m s^{-1}]
d = density in [kg m^{-3}]
E = wave energy in [kg m^2 s^{-2}]
g = gravity acceleration in [m s^{-2}].

Thus it becomes apparent that the beach gradient depends on the wave steepness h/l; the sand size which is itself a function of the settling velocity, is also a significant factor as shown by the second ratio $\frac{vp}{l}$. The wavelength as shown by the ratio $\frac{s}{l}$ is also important [50 S].

The changing of beach slopes in time.

The beach slopes change in time due to episodical, and periodical events (response to 28⋯30 day lunar tide cycle, or yearly climatic cycles): In regions with seasonally changing wave systems typical "summer" and "winter" beaches have developed [59 K]. These phenomena are distinctly present in zones of monsoonal and temperate climates [80 K, 80 N, 78 N 2, 67 S 2, 83 K 4, 71 I 1, 71 W 2, 76 K 2, 76 S 4, 80 O, 84 H 1, 85 D]. See Fig. 20. Compare Figs. 37, 38.

As a modern handbook on this topic, see [84 S 3].

9.4.2.2 Beach type series from dissipative to reflective characteristics

The modal wave height required to favor a particular beach type ranges from >2.5 m for dissipative beaches to <1 m for reflective beaches, with intermediate beaches favored by waves of 1⋯2.5 m in height. The high Southern Hemisphere west coast swell and storm wave environments maintain dissipative conditions all the year round (e.g. Goolwa, Australia). The more moderate and highly variable east coast swell environments of the New South Wales coast favor intermediate modal beach types in the more exposed localities. Reflective beaches can prevail in sheltered embayments in virtually any wave climate.

Modally dissipative beaches are associated with persistant high waves and abundant fine sand to produce the wide, low-gradient beach and surf zone. They are stable beaches possessing low spatial and temporal variability [83 K 3, 83 W]. See Figs. 19 and 20.

9.4.2.3 Modal beach state, temporal variability and environmental conditions: a classification

Over the long term a given beach will tend to exhibit a modal, i.e. most frequently recurrent, state which depends on environment. The beach, in response to the wave climate, consists of a modal beach type determined by the modal wave conditions, and a range of beach morphologies dependent on the range of the wave conditions. Fig. 18.

Association between beach state and the environmental parameter Ω [84 W 4]:

$$\Omega = H_b/(\bar{w}_s T)$$

where
H_b = breaker height
$\bar{w}_s$ = mean sediment fall velocity
T = wave period.

Temporal variability of beach state reflects, in part, the temporal variability and rate of change of Ω, which, in turn depends on deep-water wave climate and nearshore wave modifications.

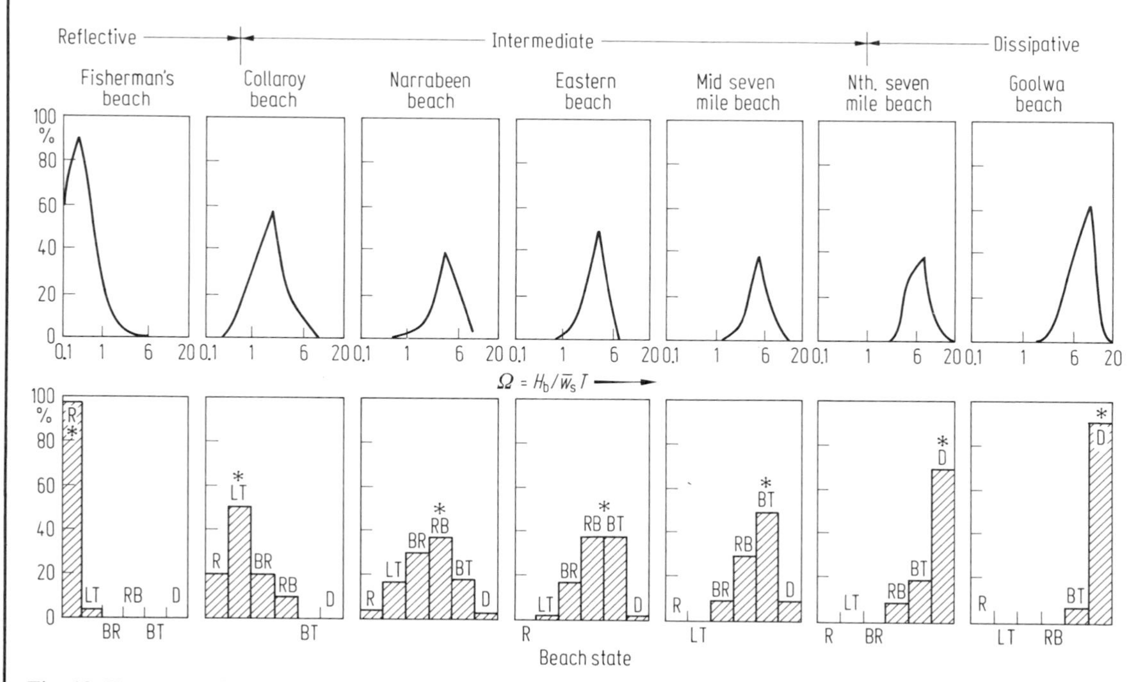

Fig. 18. Frequency distribution of $\Omega = H_b/(\bar{w}_s T)$ and corresponding percentage occurrence of beach states for seven beaches [84 W 4]. R: Reflective; LT: Low tide terrace/ridge-runnel; BR: Transverse bar and rip; *: Modal state; RB: Rhythmic bar and beach; BT: Longshore bar-trough; D: Dissipative.

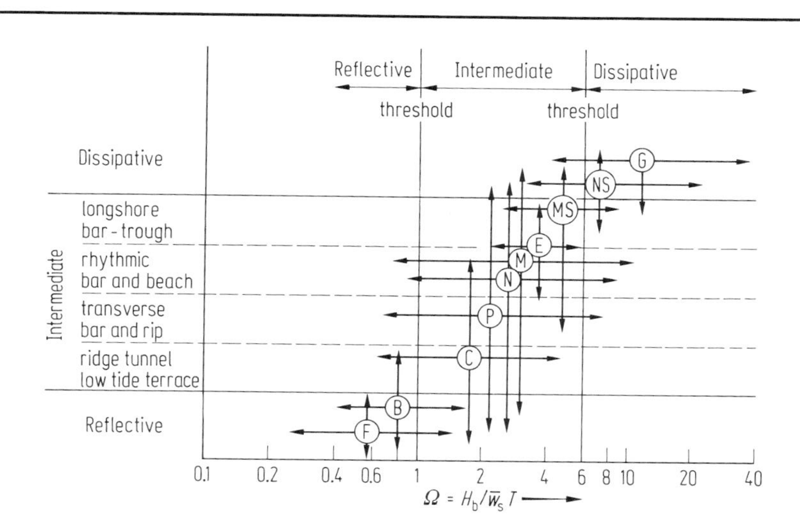

Fig. 19. Modal beach state and temporal variability of beach state as functions of the modal values and temporal variability of $\Omega = H_b/(\bar{w}_s T)$. Horizontal arrows indicate average temporal range of $\Omega = H_b/(\bar{w}_s T)$; vertical arrows indicate average temporal range of beach state [84 W 4]. Circle: modal state, Beaches: G (Goolwa), NS (Nth. Seven Mile), MS (Mid Seven Mile), E (Eastern), M (Moruya), N (Narrabeen), P (Palm), C (Collaroy), B (Bracken), F (Fisherman's).

9.4.2.4 Beach profile mobility in relation to modal beach state and state variability. Classification of a morphodynamic approach

The surf scaling parameter:

Contrary states of beaches are the extreme dissipative and the extreme reflective states. Morphologically these states correspond, respectively, to flat, shallow beaches with relatively large subaqueous sand storage and steep beaches with small subaqueous sand storage.

Morphodynamic distinction by the surf scaling parameter ε [83 K 3]:

$$\varepsilon = a_b \omega^2 / g \tan^2 \beta$$

where:
a_b = breaker amplitude
ω = incident wave angular frequency, $\omega = 2\pi/T$, with T = period
g = gravity acceleration
β = beach/surf zone gradient.

This surf scaling parameter indicates:
complete reflection, when $\varepsilon < 1.0$ (surging breakers occur)
begin of dissipation, when $\varepsilon > 2.5$ (plunging breakers occur)
complete dissipation, when $\varepsilon > 20$ (spilling breakers occur).

The surf zone widens and turbulent dissipation of incident wave energy increases with increasing ε. For breaker types, see Fig. 35.

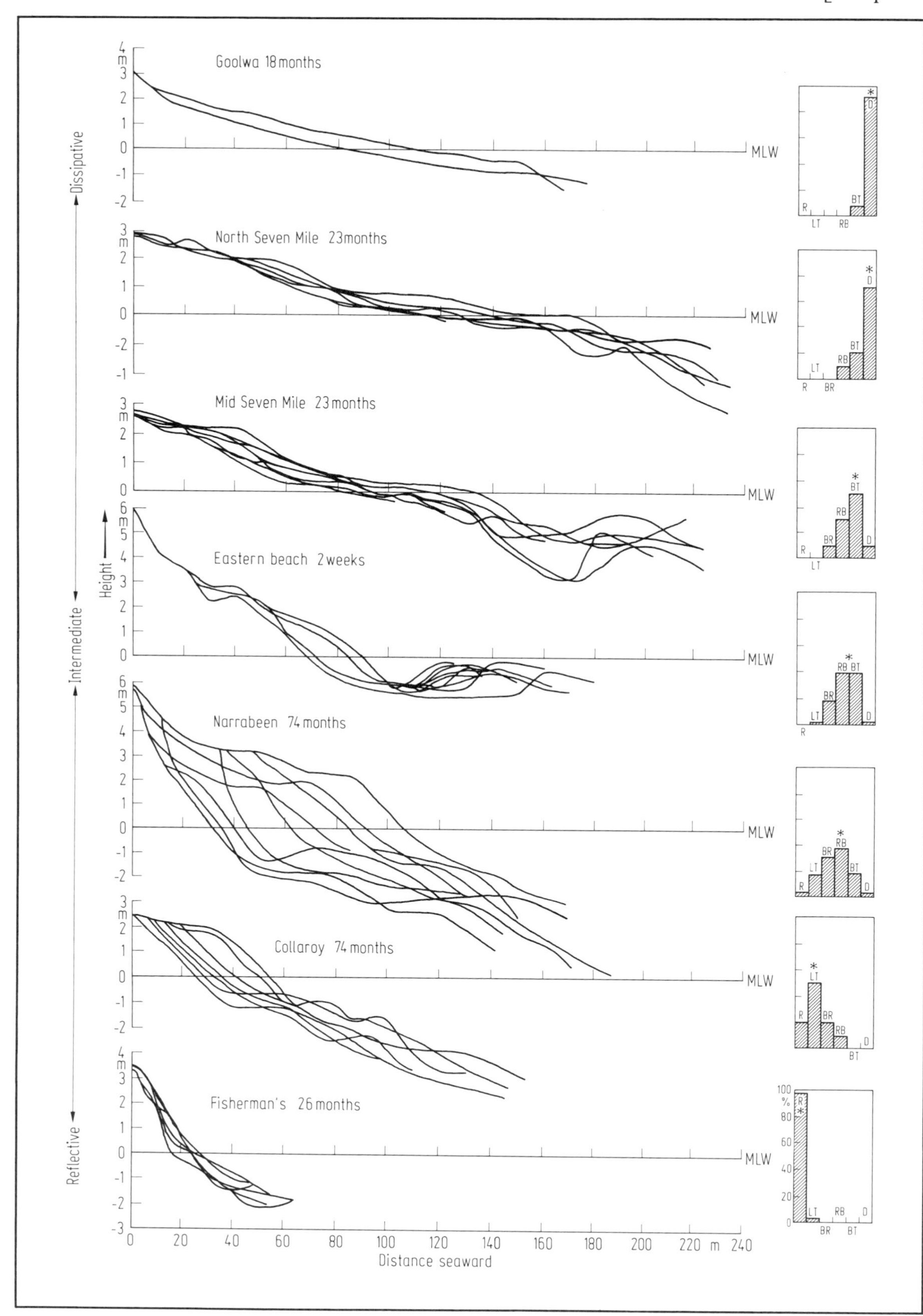
Goolwa 18 months
North Seven Mile 23 months
Mid Seven Mile 23 months
Eastern beach 2 weeks
Narrabeen 74 months
Collaroy 74 months
Fisherman's 26 months
MLW
Dissipative
Intermediate
Reflective
Height
m
Distance seaward
0 20 40 60 80 100 120 140 160 180 200 220 m 240
R LT BR RB BT D
100 % 80 60 40 20 0

←

Fig. 20. Beach profile mobility in relation to modal beach state and state variability. Successive beach and surf zone profiles of different dissipative, intermediate, and reflective beaches. The mobility of these beaches can be considered proportional to the sweep zones occupied by the profiles over prolonged periods [84 W 4]. Studies at beach locations in Australia concerning fine sand beaches. Beach locations and recording times are indicated as well as the respective distribution of beach states. For abbreviations of beach states, see Fig. 18. Asterisks give the modal state.

9.4.2.5 The Bruun rule relating shore erosion and sea level rising. Long time scale beach variation

A model relating shoreline erosion to sea level rise was proposed by Bruun [62 B, 67 S 3], for the case of averaged out seasonal or local variations in the processes involved. The concept is based on a volumetric balance between sediment lost from the shore and sediment deposited in the nearshore zone, defined as the 18 m-depth contour. Rise in sea level causes equilibrium profile to rise and move landward, causing erosion of the shoreface and deposition seaward of a null point. It is assumed that

1. on a submergent shoreline, a shoreward displacement of the beach profile occurs as the upper beach is eroded,
2. the volumes of the material eroded from the upper beach and deposited on the nearshore bottom coincide,
3. the rise of the nearshore bottom due to the material deposited equals the sea level rise, thus maintaining a constant nearshore slope.

The Bruun model suggests that an area with higher bluffs will erode slower than lower bluffs. See Fig. 21. [67 S 3, 78 R 3, 53 B].

Bruun defined the limit of the nearshore zone at ocean beaches as the 18 m-depth contour [62 B].

Besides the Bruun rule there are existing some other concepts of shoreface response to a rise in sea level, which may be more suitable concerning particular regional coastal conditions [81 N 2, 61 F 2].

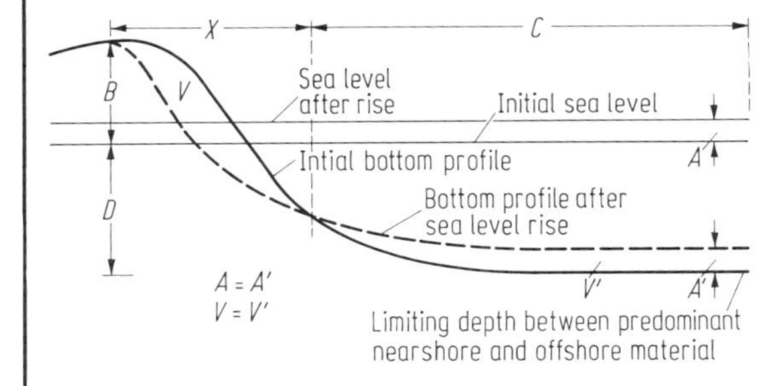

Fig. 21. The Bruun rule relating sea level rise and shore erosion [67 S 2]. $X(B+D)=AC$, where: X = shore retreat, B = shore elevation, D = limiting depth between predominant nearshore and offshore material, A = sea level rise, C = distance to limiting depth from shore, V = volume before sea level rise, V_1 = volume after sea level rise.

Bruun considered the shoreline retreat R that would occur due to an increase in local sea level S, assuming that the beach profile remained the same relative to the actual sea level, that the active profile depth was h_* and the berm height is B. Then the shoreline retreat is $R = LS/(B+h_*)$ or $R = S/\tan\theta$ where $\tan\theta$ is the average beach slope over the active beach width L, i.e. the horizontal recession R is proportional to and an amplified version of the change in sea level S. Thus beaches with mild slopes will retreat more rapidly for a unit of sea level rise than beaches with steep slopes. [83 D quoted].

9.4.2.6 Beaches: hydraulics of sediment transport

The beach is a response object to wave dynamic processes that alter the appearance of the coastline [63 B, 63 S 2, 77 T, 67 Z, 83 W, 63 S 2, 84 G]. Sea bed mechanics [82 L 2, 84 P, 84 S 3, 85 G 1, 79 P 2, 81 N 2, 82 Z]:

Material displacements according to processes and space:

1. Bottom displacement which occurs on the submarine slope under the action of waves that have not yet broken.
2. Shore displacement takes place in the surf zone.

Material displacements according to grain diameter (Fig. 23, Table 28).

1. Silt and fine sand are displaced both by waves and by currents.
2. Coarser sand and gravel are less affected by currents.
3. Shingle and larger material are displaced only by wave action.

The *longshore current* transports beach sediments that have been placed into motion by the waves, and being continuous along extensive stretches of the coast, can potentially move the sediments for many kilometers in the longshore direction. This sand movement is termed the *littoral drift*. See Figs. 22 and 23.

The longshore current is generated by waves approaching the beach under a large oblique angle $\alpha_b > 5 \cdots 10°$. Longshore currents produced by such waves have been shown to be generated by the momentum (or radiation stress) carried by the waves [83 K 3].

The formula derived by Longuet-Higgins [70 L 2] reduces to the simple expression:

$$V_l = 1.19 (g H_b)^{1/2} \sin\alpha_b \cdot \cos\alpha_b ,$$

so that the magnitude of the current V_l can be estimated directly from the measured breaker height H_b and breaker angle α_b (g = gravity acceleration).

According to the handbook "Sea bed mechanics" [84 S 3] there are widely different relationships in comparison for longshore transport and very significant disagreement on the actual values of numerical constants, so that reliable data are very difficult to obtain.

The wave crest can be resolved into two components – one normal to the shore, one parallel to it Fig. 22 shows that the width of the shore-parallel component is equal to the sine of the wave approach angle (since the width AB is set equal to 1 m) Fig. 22 shows that this width of wave crest is "spread" over a length of shoreline equal to BD. Since $BD = 1/\cos\alpha$ the length of wave crest per unit shorelength will be $\sin\alpha \cos\alpha$ (after [84 P]). If the initial unit wave crest can be related to some absolute measure of water discharge – for instance using the equation for mass transport discharge – then the long-shore discharge per unit beach length can be determined, and if the cross-sectional area of the near-shore zone at right-angles to the long-shore flow is known, the velocity of this current may be predicted [84 P, after theory of 63 I].

For long-shore discharge:

$$Q_L = Q_m \cdot \sin\alpha \cos\alpha$$

and for long-shore velocity:

$$U_L = Q_m \cdot \sin\alpha \cos\alpha \tan\beta$$

where:
Q_m = Discharge for unit width of wave
Q_L = long-shore discharge
U_L = long-shore velocity
α = wave approach angle
β = angle of beach face.

Longshore currents generated by an oblique wave-approach vary in magnitude across the width of the nearshore, increasing with distance from the shoreline, reaching a maximum usually just beyond the mid surf positions and decreasing rapidly outside the breaker zone [83 K 3, 80 G 3]. Compare subsect. 9.5.5.2.

The mass of water, the swash, that forms after a wave capsizes is carried by inertia up the beach to a point far beyond the line of the water's edge, and then flows back to meet the following wave.

The swash which has completely lost the wave motion is exposed to the forces of inertia and gravity. When the waves approach the shore obliquely, these forces operate differently. As the swash reaches its limit it slows down, describes a curve, and flows back directly to the sea along the line of steepest gradient.

The swash also affects particles of material which are moved for some distance along the water's edge by each wave and may also migrate up and down the beach [67 Z, 84 C]. See Fig. 23.

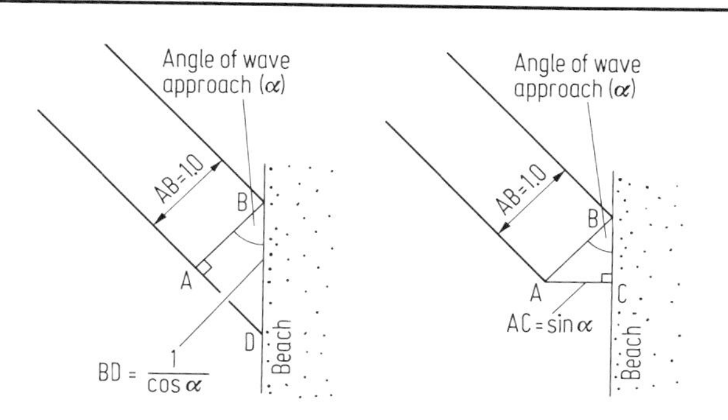

Fig. 22. Geometric conditions of waves approaching the coast [84 P].

Fig. 23 serves to show variations in the distance, measured along the normal, which a particle must be cast up the beach so that its oblique movement carries it a given distance laterally, i.e. the relationship $h/l = \tan\alpha$.

a

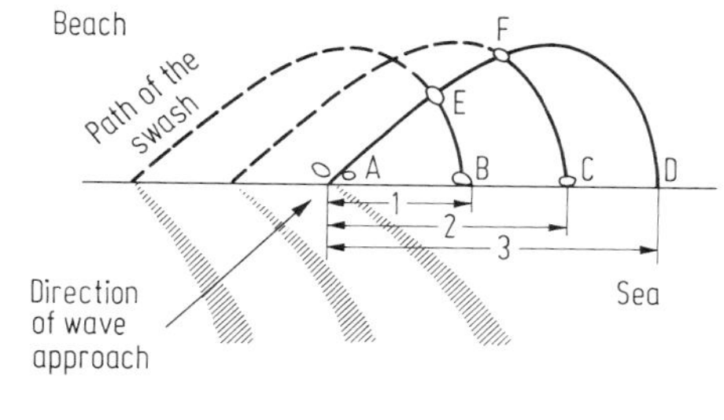

b

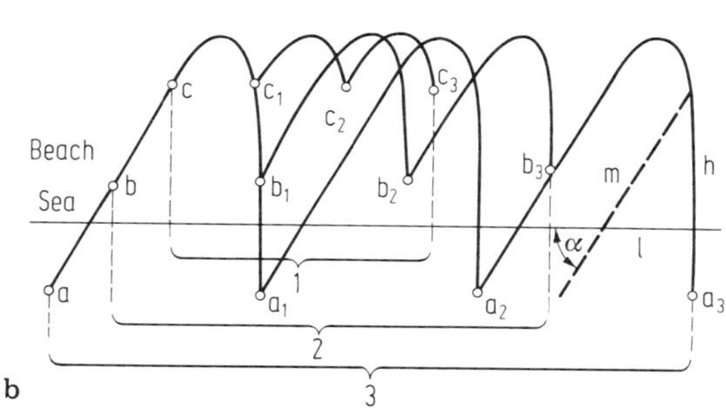

c

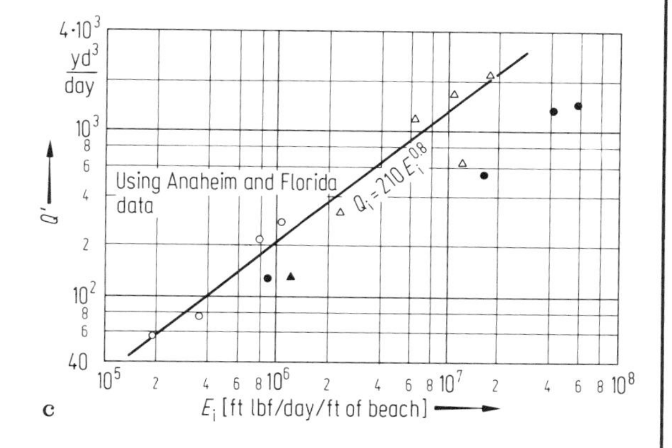

Fig. 23a, b. Diagram of the lateral displacement of material on the beach [67 Z]. (a) Movement of particles of various sizes. AEB trajectory of a large pebble, AFC of shingle, AFD of a gravel particle. (b) Movement of particles of the same size lying at various levels on the beach. As a result of the swash of three waves, particle *a* is displaced to point a_3 by the distance 3. Particle *b* traverses the distance 2, and particle *c* the distance 1. Letter symbols for calculating the coefficient of longitudinal displacement are given on the right-hand side of the diagram. α = angle of pathway of material on the beach, L = length of pathway along the beach, m = pathway of the material, h = height of pathway on the beach.

Fig. 23c. Relation of alongshore drift Q' of sand to alongshore wave energy E_i. From [66 C]. The value of E_i is in turn defined as: $E_i = E_p \sin\theta \cos\theta$, where E_p is the potential energy of the incident wave train and θ is the angle of the wave crest with shore at the point where wave height and length are measured. The potential energy, E_p, is that portion of the total wave energy which moves shoreward as opposed to the kinetic energy. Open circles: data from South Lake Inlet, Fla., open triangles: data from Anaheim Bay, Cal., solid circles: data from predicted wave in North Atlantic, solid triangle: data from B.E.B. wave tank tests. 1 yd³ ≙ 0.765 m³, 1 ft lbf ≙ 1.356 J, 1 ft ≙ 0.3048 m.

Wave dynamics and material transport:

If it is assumed that the estimated total flux of wave energy ($2.5 \cdot 10^9$ kW) is available for transporting sand along the world's coastline, and that the longshore-directed energy flux is equal to one-tenth of the incident energy flux, a rate of immersed weight transport of $1.9 \cdot 10^{11}$ N/s follows which is equivalent to a (dry) mass transport of $3.1 \cdot 10^7$ t/s for the world's beaches, a mass flux $5.8 \cdot 10^4$ greater than that supplied to the coast by all of the erosion of the continents.

The length of path parallel to the coast traversed by particles of material depends on 4 factors:

1. The angle at which waves approach the shore.
2. The size of the material particles.
3. Wave parameters.
4. The angle of slope of the beach.

The amount l of displacement is proportional to the cosine of the angle α at which the wave breaks: $l = m\cos\alpha$. See Fig. 23. It will be assumed that the particle ascends the slope in the direction of wave motion, and falls back along the line of greatest slope, i.e. along the normal to the shore. Since $h = m\sin\alpha$, it follows

$$\frac{h+m}{l} = \frac{1+\sin\alpha}{\cos\alpha}.$$

Quantitative example of beach erosion and transport: Tables 25a, b.

Table 25a. Coefficients of the rate of coastal displacement of material in relation to the angle of wave breaking.

α	$\frac{1+\sin\alpha}{\cos\alpha}$	$\tan\alpha$
85	22.9	11.4
80	11.4	5.6
70	5.9	2.7
60	3.7	1.7
45	2.4	1.0
30	1.7	0.6

Table 25b. Geomorphologic situation of 5 problem areas at Sandy Hook, New York coast. Different sources.

Site	Oceanic front		Head of spit	Bayside lobos	
Breaking wave height	0.82 m	0.58 m	0.21 m	0.18 m	0.10 m
Sediment transport rate 1953–1976	379 000.0 m^3/a	306 000.0 m^3/a	16 000.0 m^3/a	18 000.0 m^3/a	8 000.0 m^3/a
Average width of dry beach (MHW to dune), winter conditions	12.0 m	25.6 m	6.4 m	3.0 m	3.2 m
Rate of shoreline change 1943–1972	−5.3 m/a	1.1 m/a	−0.7 m/a	−3.2 m/a	−3.6 m/a

Examples of the onshore and offshore resultants of material transport within breakers provided threshold wave velocities of 4.5⋯5 m/s between onshore and offshore transport. It can be shown that, for certain wave sizes, each grain size of the sand needs specific beach gradients in order to remain stable, e.g. for sand with a diameter of 1.2 mm a beach with a gradient of 2° can remain stable, while for sand with a diameter of 2 mm a beach with a gradient of 12° is required. See Figs. 24a, b. [82 L 2].

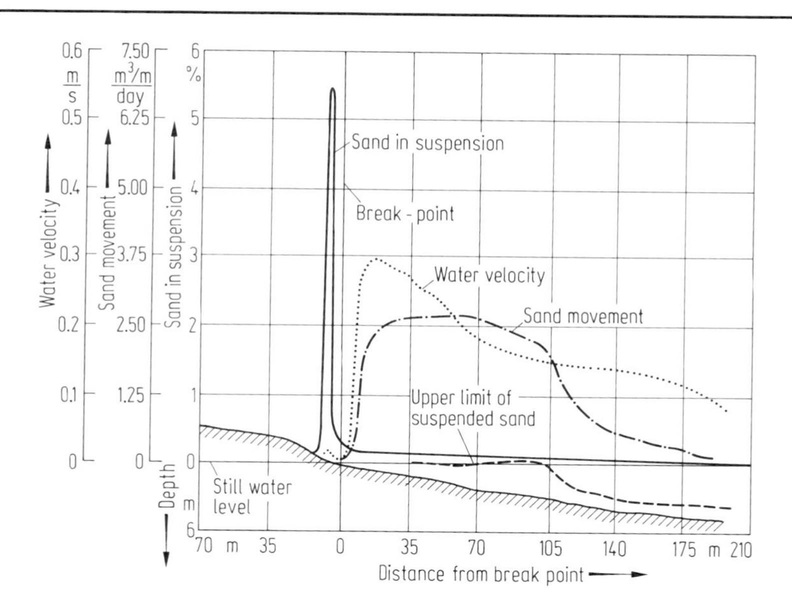

Fig. 24a. Sand in suspension in relation to water velocity, water depth, and position of break point, after measurements from US-Beach Erosion Board at the coast of New Jersey [59 K].

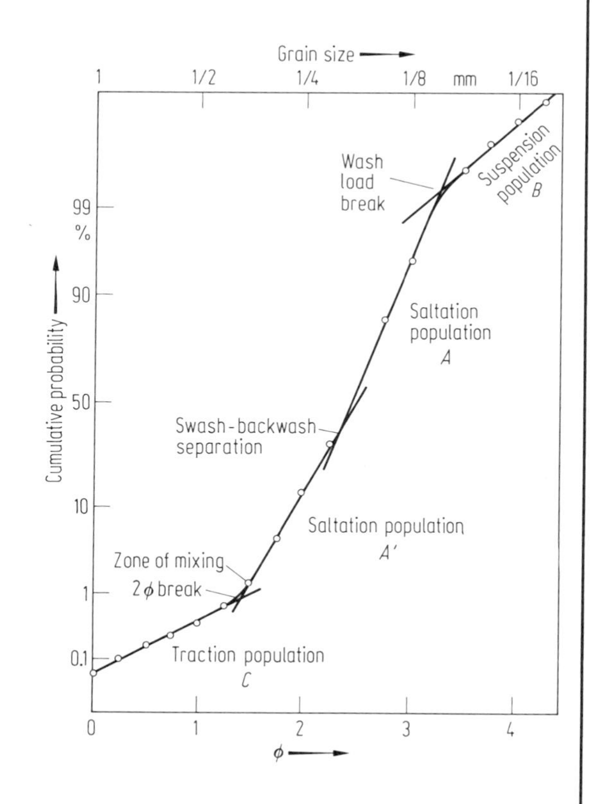

Fig. 24b. Relation of sediment transport dynamics to population and truncation points in a grain size distribution. Example shows four truncated log-normal populations from lower swash zone [75 R 2, after 69 V]. Compare [79 V].

9.4.2.7 Classification and characteristics of beach material

Table 26, Figs. 25, 26.

Table 26. Grain size scale of beach material: classification [76 B1].

Wentworth size class	Particle diameter [mm]	Φ scale *)
Boulders	>256	<−8
Cobbles	64···256	−6···−8
Pebbles	4···64	−2···−6
Granules	2···4	−1···−2
Very coarse sand	1···2	0···−1
Coarse sand	1/2···1	1···0
Medium sand	1/4···1/2	2···1
Fine sand	1/8···1/4	3···2
Very fine sand	1/16···1/8	4···3
Silt	1/256···1/16	8···4
Clay	<1/256	>8

*) The Φ scale is a logarithmic scale of particle diameters d, in [mm], defined by Krumbein (1936) as $\Phi = -\log_2 d$.

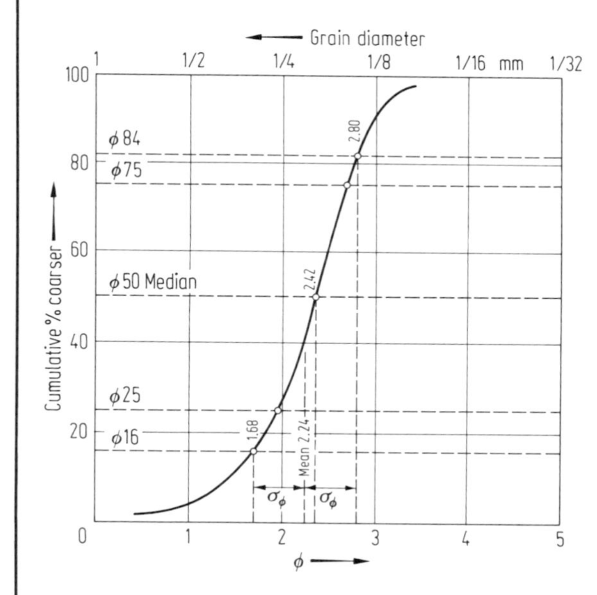

Fig. 25. Graphic representation of a grain-size analysis of a beach sand.

Measured in Φ units, the median diameter, Md_Φ (50th percentile) is 2.42, the 16th percentile (Φ_{16}) 1.68, and the 84th percentile (Φ_{84}) 2.80. Values for the mean (M_Φ), sorting (ϱ_Φ) and skewness (α_Φ) are calculated as follows:

$$\text{Mean} = M_\Phi = \tfrac{1}{2}(\Phi_{16} + \Phi_{84}) = 2.24$$

$$\text{Sorting} = \sigma_\Phi = \tfrac{1}{2}(\Phi_{84} - \Phi_{16}) = 0.56$$

$$\text{Skewness} = \alpha_\Phi = \frac{M_\Phi - Md_\Phi}{\sigma_\Phi} = -0.32$$

(i.e. the mean (2.24) is coarser than median (2.42), the sample is moderately well sorted and the grain-size distribution is strongly negatively-skewed [76 B1].

Table 27. Grain size characteristics of sands of various environments for three truncated populations. See also Fig. 26. After [75 R 2]. C.T.: coarse truncation point, F.T.: fine truncation point.

Environment	Saltation population (A)				Suspension population (B)				Rolling population (C)			
	%	Sorting	Φ		%	Sorting	Mixing A and B	Φ F.T.	%	Sorting	Φ C.T.	Mixing A and C
			C.T.	F.T.								
Fluvial	65···98	fair	−1.5···−1.0	2.75···3.50	2···35	poor	little	>4.5	varies	poor	no limit	little
Natural levee	0···30	fair	2.0···1.0	2.0···3.5	60···100	poor	much	>4.5	0···5			none
Tidal channel	20···80	good	1.5···2.0	1.5···3.5	0···20	poor good	much	3.5···>4.5	0···70	fair good	−0.5···−1.5	average
Tidal inlet	30···65	good	1.25···1.75	2.0···2.5	2···5	fair good	average	3.5···4.0	30···70	fair good	−0.5···no limit	average
Beach	50···99	2 populations excellent	0.5···2.0	3.0···4.25	0···10	fair··· good	little	3.5···>4.5	0···50	fair	−1.0···no limit	average
Plunge zone	20···90	good	1.5···2.5	3.0···4.25	0···2	good	much	3.0···>4.5	10···90	fair poor	no limit	average
Shoal area	30···95	good	2.00···2.75	3.5···>4.5	0···2	poor··· fair	little	3.5···>4.5	5···70	fair poor	0.0···−2.0	much
Wave zone	35···90	good··· excellent	2.00···3.00	3.0···>4.5	5···70	fair poor	much	3.75···>4.5	0···10	poor	0.0···no limit	little
Dune	97···99	excellent	1.0···2.0	3.0···4.0	1···3	fair	average	4.0···>4.5	0···2	poor	1.0···0.0	little
Turbidity current	0···70	fair poor	1.0···2.5	0.0···3.5	30···100	poor	much	>4.5	0···40	fair poor	no limit	much

Characteristics of pebbles and cobbles, the roundness index (Fig. 26):

The material of pebbles and cobbles at marine beaches is rounded by wave action. In shape analysis they are attributed a certain index according to roundness [50 T, 58 G, 73 C 2, 73 C 3, 59 T, 57 G].

Indexes:

$$\text{Roundness} = \frac{2r}{L}, \qquad \text{flatness} = \frac{L+1}{2E}.$$

Since wave action causes a strong roundness, the pebbles at marine beaches tend to a high roundness index of $\frac{2r}{L}$, i.e. approaching 1. With this analysis beach material can be distinguished from other as fluviatil or glacial which is crucial for the determination of paleo-sea levels.

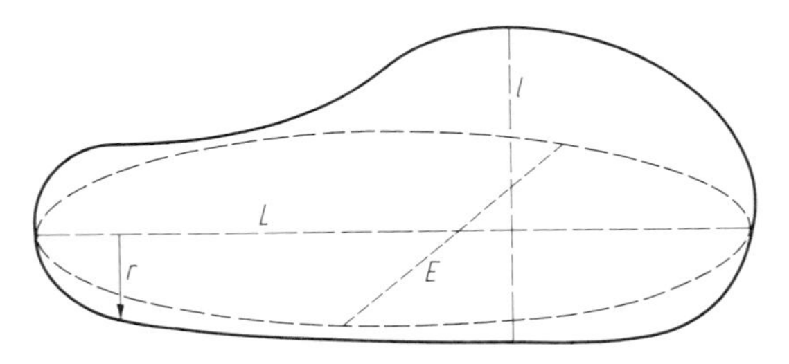

Fig. 26. Roundness index of pebble. L = length, l = width, E = thickness, r = least radius of curvature in the principle plain [58 G].

Table 28. Surface-water suspended sediment concentrations in nearshore muddy coastal waters [77 W 3].

Location	Concentration [100 mg/l]		Source
	maximum	minimum	
Louisiana coast	0.64	0.01	Manheim et al. (1972)
East China Sea	0.70	0.05	Emery et al. (1969)
Venezuela coast	1.00	0.01	Van Andel and Postma (1954)
Gulf of San Miguel	2.00	0.10	Swift and Pirie (1970)
Dutch Wadden Sea	6.20	0.50	Postma (1961)
Gulf of Thailand	9.70	0.01	NEDECO (1965)
Gulf of Po Hai	10.00	1.00	Zenkovitch (1967)
British Guiana coast	26.25	0.05	Delft Hydraulics Laboratory (1962)
Surinam coast	37.49	0.14	this study

9.4.2.8 Current velocity of water and sediment interaction

The tractional force exerted by flowing water is proportional to the square of velocity but the weight of a boulder that can be rolled along is proportional to the sixth power of the velocity, and the diameter to the third power. This relation is obvious: While the weight increases with the third power of the radius, the surface on which the tractional force is exerted, increases with the square of the radius [84 R].

Thus, doubling the velocity causes the pressure per cm^2 to increase 2^2 times, and with a block 2^6 times of the weight the surface is 2^4 times as large. Hence the total force is $2^2 \cdot 2^4 = 64$ times as large, just balancing the increase in weight of $2^6 = 64$.

Relationships between grain diameter and the critical force required for its movement. The critical force approximated by a measurement of velocity taken 1 m above the bed. A more accurate relationship exists between shear velocity and grain diameter. The Shield's coefficient (θ), the ratio between shear stress and grain weight plotted against grain diameters gives a result from depending on both parameters. See [84 P]. Komar and Miller (1973) investigated the problem using experimental data from a wide range of sources. They found that a nondimensional threshold condition similar to the Shield's coefficient (θ) would give good agreement to the experimentally derived data but only if related, not to grain diameter, but to the ratio between grain diameter and orbital diameter of the water particles [84 P quoted]:

$$\frac{\varrho u_m^2}{(\sigma - \varrho) g D} \propto \left(\frac{d_o}{D}\right)^n,$$

where: ϱ = density of water
u_m = maximum orbital velocity
σ = sediment density
d_o = orbital diameter
n = an exponent varying between 0.25 and 0.5
D = grain diameter

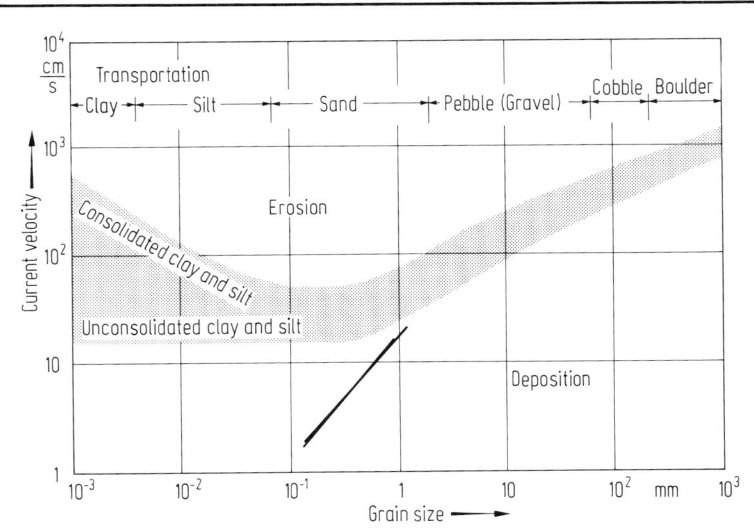

Fig. 27. Diagram relating current velocity somewhat above the bottom to the size of particles of a given class which it erodes, "Hjulstrøm Curve". The graph applies to well-sorted sediment only. These results can only serve as a first approximation [39 H]. For a more detailed presentation of curves of current velocities concerning erosion, transport, and sedimentation, see subsect. 1.5.5.

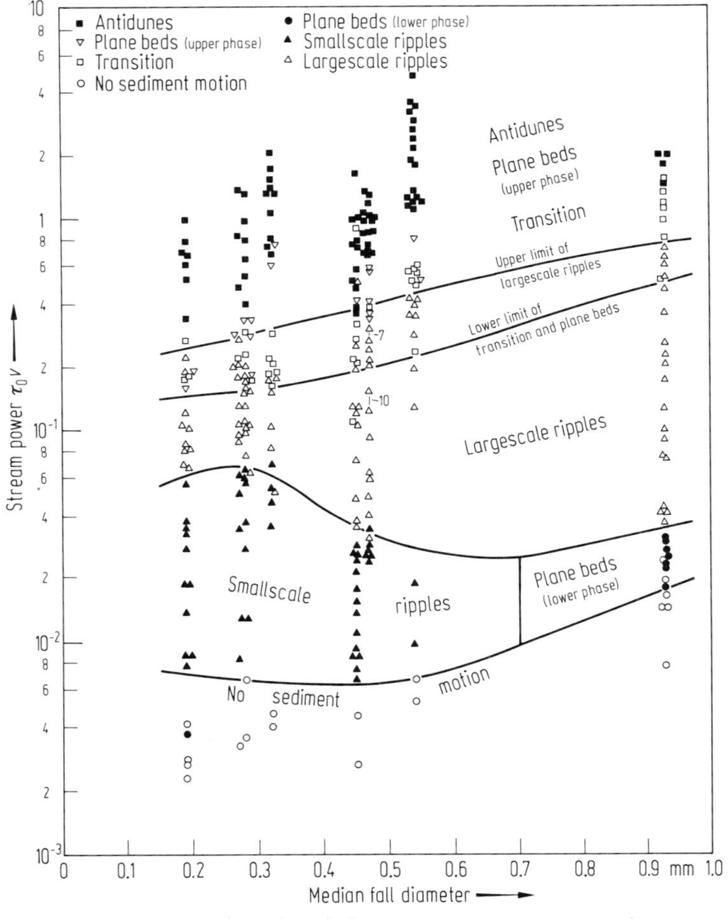

Fig. 28a. Bedform in relation to stream power $\tau_o v$ and grain size [68 A]. Compare [70 A 1].

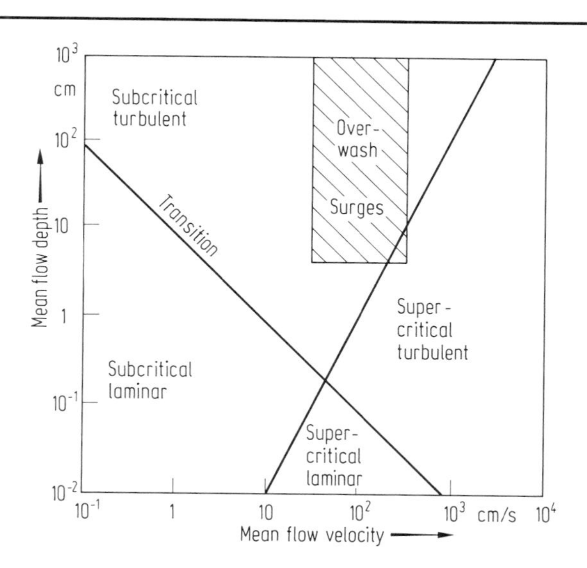

Fig. 28b. Overwash surges. Swell tongues running over barrier beaches. After [79 L 2], generalized. This event happens specifically as induced by storm tides, producing channels. The surges for this overwash fall in both, the subcritical and supercritical turbulent flow regions, indicated by the shaded area. At low velocities or large depths, the flow is described as tranquil or subcritical. At higher velocities or small flow depths, the flow becomes shooting or supercritical. The flow mostly is supercritical for the frontal portion of the surge. The long tail of the overwash bore has much lower velocities as compared to depth, and the flow quickly drops below the critical value to subcritical conditions. Compare [81 L].

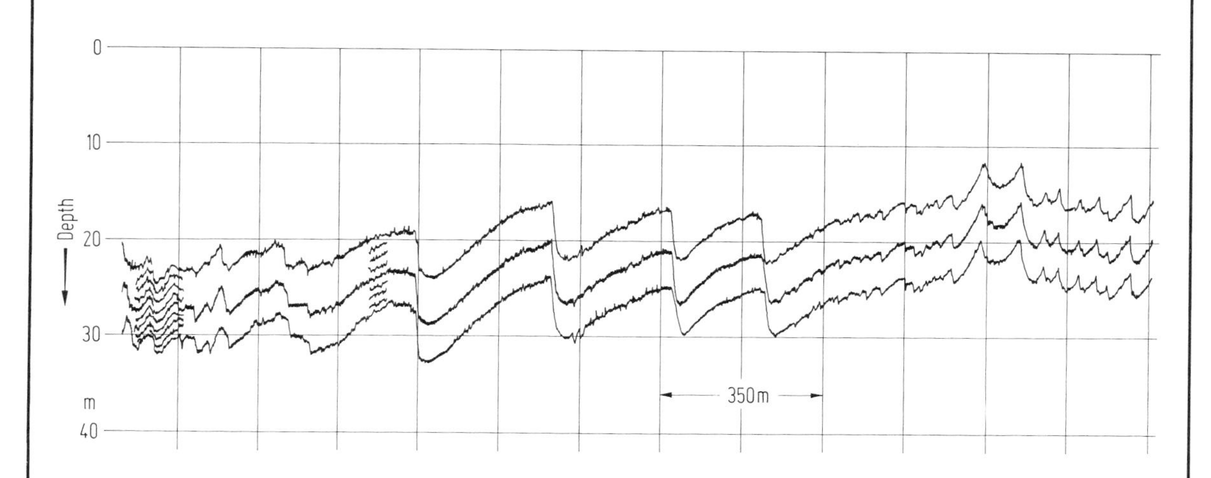

Fig. 28c. Typical example of the succession of giant and mega ripples in the Lister Tief with distinct recognizable running of their crests across the course of the ship [80 U]. Crests of megaripples of lengths of 300 m and heights of 6 m, soundings with 3-beam bands, "fan echo sounder" of the research vessel "Poseidon", see [73 U 2, 80 U].

Table 29. Genetic classification of ripples. After [71 R] and [75 R 2].
L = ripple length, H = ripple height, longitudinal = ripple crest parallel to current direction, transverse = ripple crest at right angles to current direction.

a) Current ripples (transverse)

Name	Nature of crest	Size parameters	Ripple index L/H	Symmetry	Internal structure
Small-current ripples	straight undulatory lingoid rhomboid	$L=4\cdots60$ cm H = up to 6 cm	>5 mostly 8···15	asymmetrical	form-concordant form-discordant Climbing
Megacurrent ripples [1])	straight undulatory lunate lingoid rhomboid	$L=0.6\cdots30$ m $H=0.06\cdots1.5$ m	mostly >15	asymmetrical	form-concordant form-discordant
Giant-current ripples [2])	straight undulatory bifurcating	$L=30\cdots1000$ m (rarely 20···30 m) $H=1.5\cdots15$ m	mostly >30 up to about 100	asymmetrical and symmetrical	known only as form-discordant
Antidunes	straight	$L=0.01\cdots6$ m $H=0.01\cdots0.45$ m		almost symmetrical	form-concordant form-discordant

b) Wave ripples

Name	Nature of crest	Size parameters	Ripple index L/H	Symmetry	Internal structure
Symmetrical wave ripples	straight, partly bifurcating	$L=0.9\cdots200$ cm $H=0.3\cdots22.5$ cm	4···13 mostly 6···7	symmetrical	form-concordant form-discordant climbing
Asymmetrical wave ripples	straight, partly bifurcating	$L=1.5\cdots105$ cm $H=0.3\cdots19.5$ cm	5···16 mostly 6···8	asymmetrical R.S.I. = 1.1···3.8	form-concordant form-discordant climbing

c) Isolated (incomplete) ripples. (Formed on the foreign substratum by sediment paucity)

Name	Nature of crest	Size parameters	Symmetry	Internal structure
Isolated small-current ripples	like small-current ripples	like small-current ripples, but lesser in height	asymmetrical	form-concordant form-discordant
Isolated mega-current ripples	straight curved sichel-shaped	like megacurrent ripples, but lesser in height	asymmetrical	form-concordant form-discordant
Isolated giant-current ripples	like current ripples	similar to giant-current ripples	asymmetrical symmetrical	form-discordant
Isolated wave ripples	straight curved	like wave ripples, but lesser in height	symmetrical asymmetrical	form-concordant form-discordant

Table 29 (continued).

d) Combined current/wave ripples

Name	Nature of crest	Size parameters	Symmetry	Internal structure
Longitudinal current/ wave ripples (direction of wave propagation at right angles to current direction)	straight, unbranched crests parallel to the current direction; also known to occur in mud	$L=2.6\cdots5$ cm	symmetrical and asymmetrical	form-discordant
Transverse current/ wave small ripples (wave direction parallel to current direction)	mostly rounded crests, transverse to current direction		asymmetrical	form-concordant form-discordant

e) Wind ripples

Name	Nature of crest	Size parameters	Ripple index L/H	Symmetry	Internal structure
Wind sand ripples	straight, partly bifurcating	$L=2.5\cdots25$ cm $H=0.5\cdots1.0$ cm	10···70 and more	asymmetrical	laminated sand, rarely few foreset laminae; concentration of coarse sand near the crest
Wind granule ripples	straight, cuspate barchan-like	$L=25$ cm···20 m $H=2.5\cdots60$ cm	12···20	asymmetrical	foreset laminae in opposing directions; on the crest enrichment of granules

[1]) An active megaripple field can be covered by small ripples.
[2]) An active giant ripple field can be covered by megaripples.

9.4.2.9 Spatial and temporal scales of crescentic rhythmic features

The spatial and temporal dimensionalities as well as topographic and process associations of crescentic coastal landforms are the context for standardization of terminology appearing in the literature. These forms include cusplets, cusps, sand waves, and secondary and primary capes [65 R 3, 71 B]. See Tables 30a, b, Figs. 29a, b.

The generating of rhythmic features of the beach may be inclined by edge waves. See subsect. 9.5.3 and 9.5.5, Fig. 44, 45.

Plan geometry of headland-bay beaches:

A "headland-bay" beach is defined as a beach lying in the lee of a headland subjected to a predominant direction of wave attack. Such beaches characteristically have a seaward-concave plan shape resulting from erosion caused by refraction, diffraction, and reflection of waves into the shadow zone behind the headland. Tide-induced currents have no direct effect on the plan shape of headland-bay beaches. Increasing radius of plan curvature with distance from the headland suggested testing the logarithmic spiral $r=\exp(\theta\cot\alpha)$ as an approximation to the shape of headland-bay beaches. See Fig. 30. Changes in shoreline regimen within their area of influence are caused by:

1. interruption of the stream of sediment being carried by shore drifting (the combined process of beach drifting and longshore drifting),

2. dissipation of wave energy by induced turbulence or reflection,
3. redistribution of wave energy by refraction and diffraction of wave trains.

Shore drifting of sediment parallel to the shoreline is a response to waves that break at an angle to the coast line. But where an obstacle, such as a headland, blocks the direct attack of waves against the shoreline, a shadow zone is created in the lee of this obstacle. However, sediment moves through the shadow zone because wave action against this shoreline does not cease.

Table 30a. Crescentic landforms: spatial and temporal scales. Reciprocal action and mutual effect in space and time [74 D 1]. Compare [71 D].

Form characteristic	Cusplet	Cusp	Sand waves	Secondary capes	Primary capes
Spacing	0···3 m	3···30 m	100···3000 m	1···100 km	200 km
Material	fine sand-gravel	sand-boulders	sand	sand	sand-gravel
Topographic association	step	berm, beach face	beach berm-offshore bar system	coastal plains; sufficient sediment	coastal plain deltas
Rhythmicity	yes	yes	yes	often	not always
Motion	fixed	normal to beach	downdrift	probably downdrift	slow downdrift
Temporal	minutes to hours	hours to days	weeks to years	decades	centuries
Suggested processes	swash action on beach face; groove erosion	berm deposition and erosion	wave action, nearshore circulation cells, back eddies of longshore transport currents	kinematic nature of sediment transport, circulation cells	wave action, confluence of coastal currents, back-set eddies, and shoals

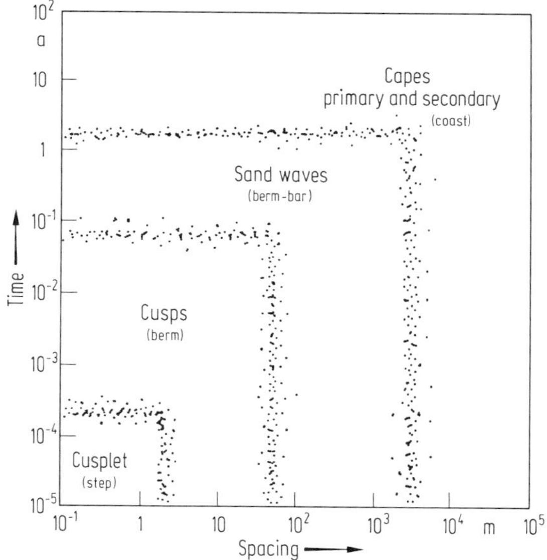

Fig. 29a. Model of space-time scale. Beach morphology and time intervals of development. Life of features, hydrodynamic interface, and substratum in space and time [74 D 1].

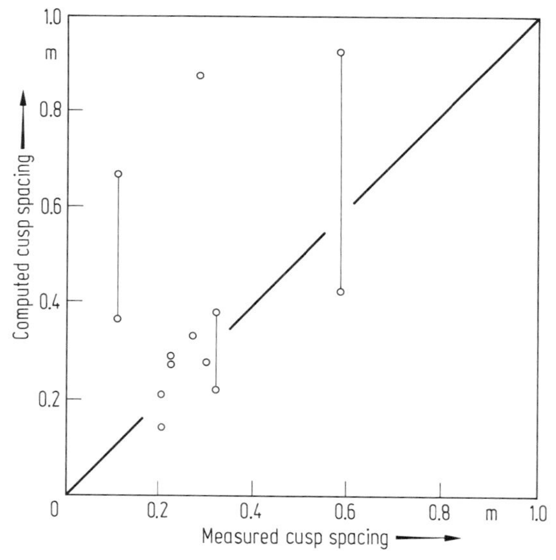

Fig. 29b. Measured cusp spacing vs. computed cusp spacing, based on n = 0 subharmonic edge waves. Data from Komar 1973. Vertical lines connecting data points indicate the range based on computations using horn and trough slopes [77 S 2].

Larger scale beach arcs are common on beaches of different tidal ranges and of different wave climates.

A decadic time scale surveying at the north coast of the island Rügen (Baltic Sea) concerning large scale crescentic beach arcs in the scale of $10^2 \cdots 7 \times 10^2$ m secant length results a typical length distribution with long persistence. [85 G 1].

Table 30b. Large scale beach arcs at the north coast of the island Rügen, Baltic Sea [85 G 1].

Year	Number of beach arcs in 10^2 m							Total number	Average length in 10^2 m
	1	2	3	4	5	6	7		
1957	–	9	14	5	3	1	1	34	3,2
1958	1	10	17	4	4	–	–	36	3,0
1959	1	8	11	8	4	1	–	33	3,3
1960	1	9	13	10	2	–	–	35	3,1
1961	1	9	14	6	2	1	1	34	3,2
1962	1	8	13	9	2	1	–	33	3,3
1963	–	11	10	6	4	2	–	33	3,3
1964	–	7	8	11	4	1	–	31	3,5
1965	–	10	11	6	5	1	–	33	3,3
1966	–	9	9	6	5	–	2	31	3,5
1967	–	9	9	8	5	1	–	32	3,5
1968	–	11	10	7	2	3	–	33	3,5
1957–1968	5	110	139	86	42	12	4	408	3,3

Planimetric geometry of headland-bay beaches.

The logarithmic spiral model:

Because of the decreasing radius of plan curvature that seems characteristically to occur toward the headland (Fig. 30) and because the rate of decrease in radius of curvature appears to be nonlinear, the equiangular (logarithmic) spiral,

$$r = e^{\theta \cot \alpha},$$

is worthy for goodness of fit to the plan shape of headland-bay beaches. r is the length of a radius vector from log-spiral center; α is the angle between a radius vector and tangent to the curve at that point and is a constant for a given log-spiral; see Fig. 30.

The exponential rate-of-increase in radius of curvature from log-spiral center is shown by

$$\frac{dr}{d\theta} = \cot \alpha \, e^{\theta \cot \alpha}.$$

The linear relation between $\ln r$ and θ facilitates use of linear-regression analysis in testing headland-bay beach plan curvature for goodness of fit to a log-spiral. Quoted after [65 Y].

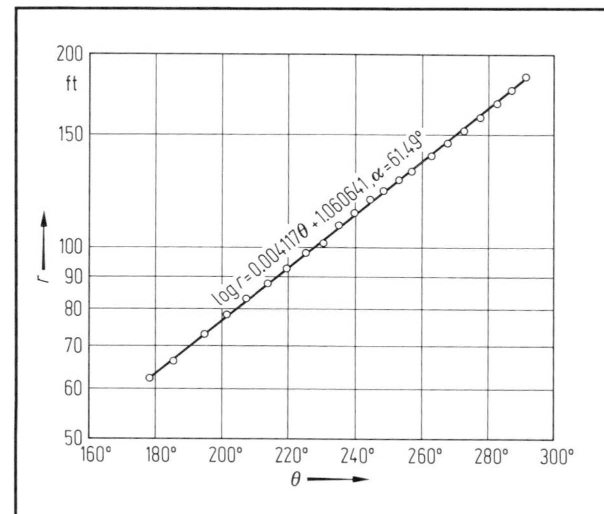

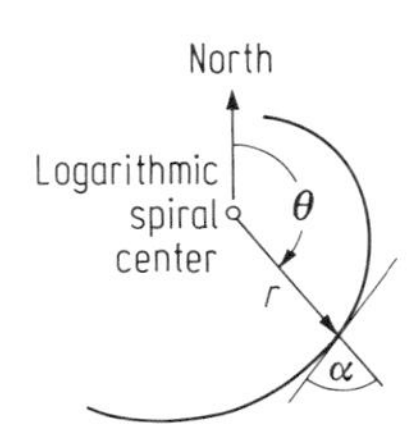

Fig. 30. Logarithmic spiral nomenclature and semi-log plot of r vs. θ for Spiral Beach shoreline curvature. From [65 Y]. 1 ft $\hat{=}$ 0.3048 m. Compare [64 Y].

9.5 Wave effects on coasts

9.5.1 Wave energy on coasts

Wind generated waves of the average height of 1 m transmit ≈ 10 kW/m, i.e. oceanwide (440 000 km coast) they would dissipate $4.4 \cdot 10^9$ kW. Wind generated waves of average height < 1 m oceanwide (in account those shorelines not exposed to the open sea) would dissipate $2.5 \cdot 10^9$ kW at the world's coastlines. [73 I, 67 D, 79 F, 79 S 3, 78 L 1, 85 T, 73 H 3]. See Figs. 13, 31a, b, 32 and Tables 31a, b, 32a, b.

Table 31a. Energy of waves at the coast.

Wave period s	Wave length m	Wave height m	Wave energy $kW\,m^{-1}$
3	14	0.56	0.9
5	39	1.56	12
7	76	3.06	63
9	126	5.06	221
11	189	7.56	603

The discharge of the breaking wave:

Assumed that the total volume of the wave trough moves forward with the breaking wave, then solitary wave theory [Munk, 1949] gives the discharge of a breaking wave as

$$q = \frac{4}{\sqrt{3y^3}} \frac{H_b^2}{T}$$

where

q = the volume transported forward per unit length of wave crest and per unit of time
y = the relative depth
H_b = breaker height
T = wave period.

Fig. 32 represents the wave spectrum in the environment of the trade and storm coast of the Atlantic Ocean, Uruguay [69 D, compare 77 S 1]. See also Figs. 31a, b.

Significant wave: statistical term relating to the one-third highest waves of a given wave group.

Significant wave height (or characteristic wave height: average height of the one-third highest waves of a given wave group. In wave record analysis: average height of the highest one-third of a selected number of waves, determined by dividing the time of record by the significant period.

Significant wave period: arbitrary period generally taken as the period of the one-third highest waves within a given group. In wave record analysis: average period of the most frequently recurring of the larger well-defined waves in the record under study [66 B 1].

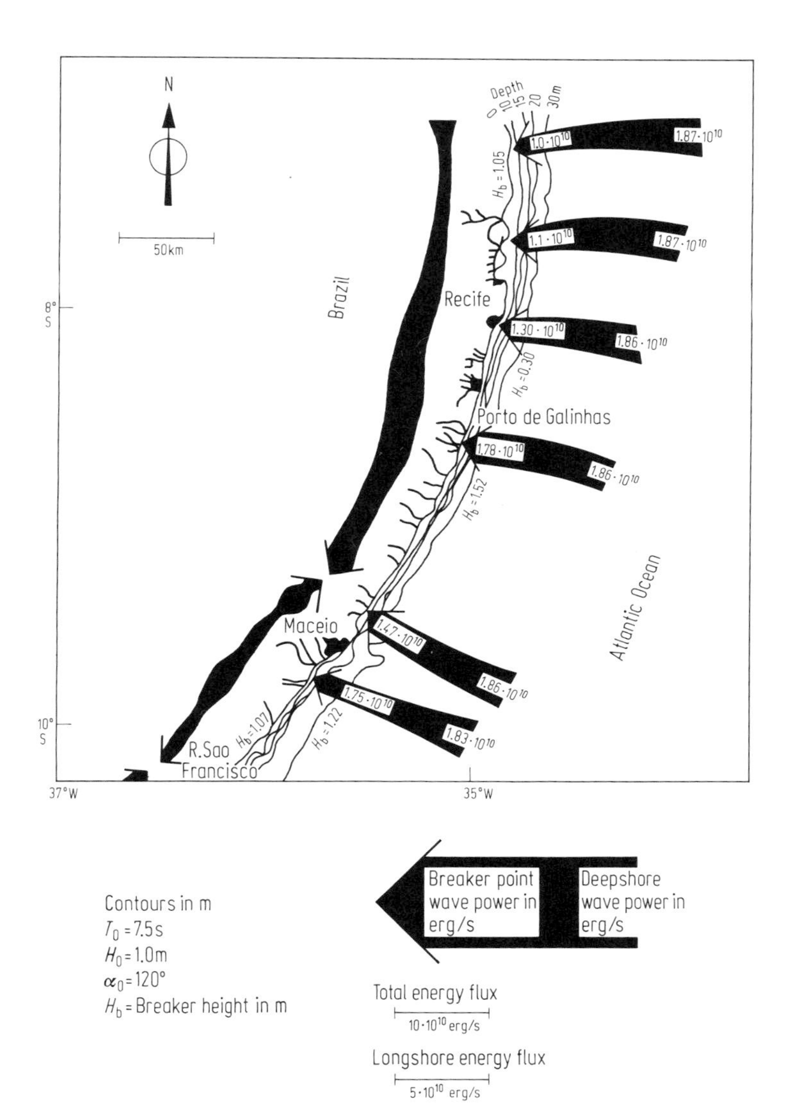

Fig 31a. Spatial variations in the nearshore wave power, longshore power and breaker height along the northeast coast of Brazil. Convergence zones of littoral drift are indicated at Aracaju and Rio São Francisco, and north of Maceio, Brazil. After [77 S 1], compare [80 G 3]. T = wave period, H = wave height, α = angle of wave crest with shore [67 D].

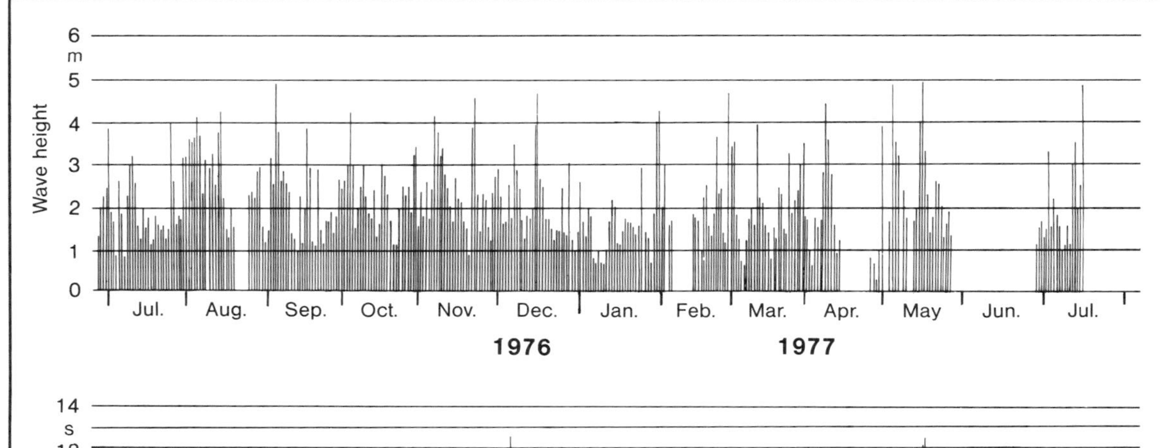

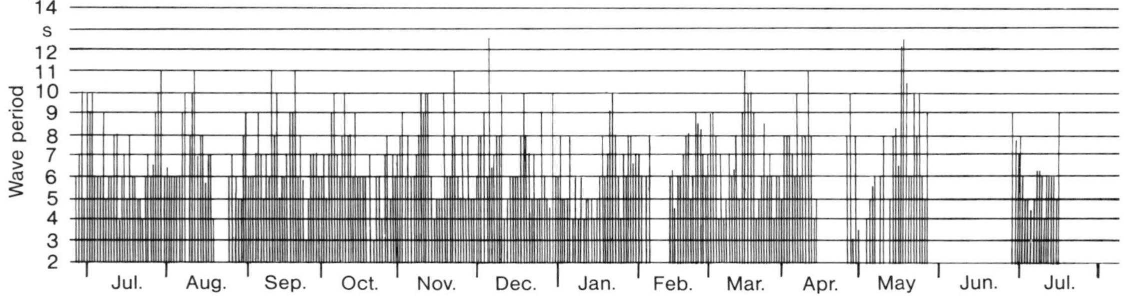

Fig. 31b. Diurnal maximum wave heights on the shore of Uruguay [67 D]. Recorded: 396 days, mean period 7.02 s; maximum height 5.55 m, mean height 2.18 m. Compare Fig. 31a and [77 S 1].

9.5.2 Characteristics of waves at coasts

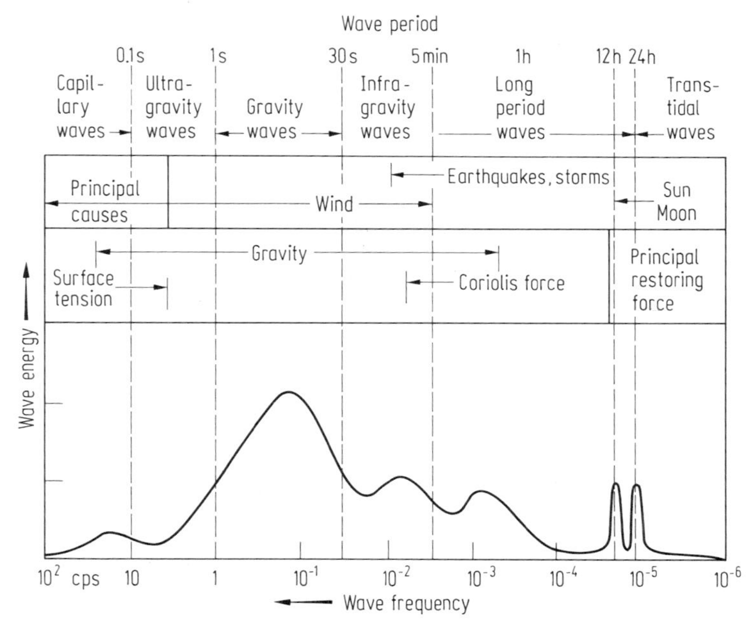

Fig. 32a. Wave classification by wave period. After [65 K].

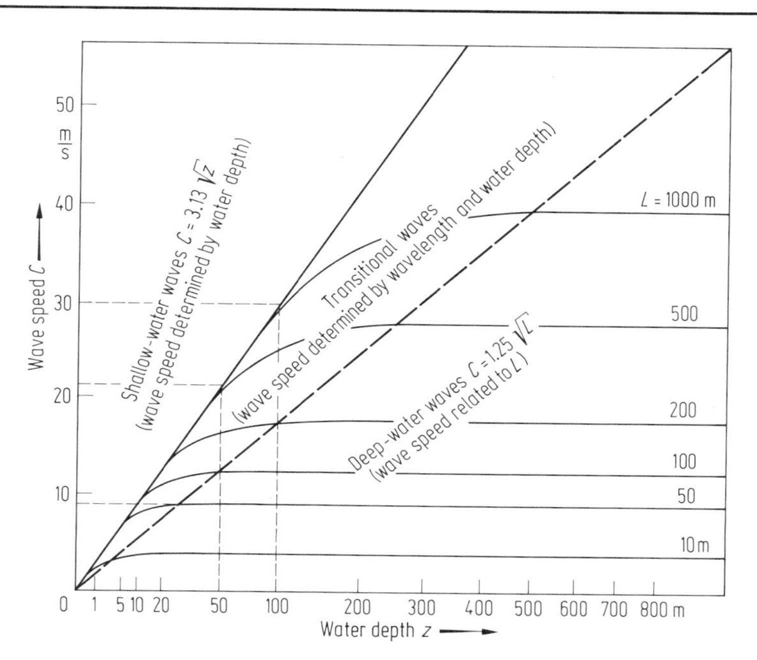

Fig. 32b. Wave speed. To determine the wave speed C of shallow-water waves, move up from the water depth to the line marked shallow-water waves, then over to the wave speed scale (see example for 50 m water depth). To determine the wave speed of deep-water waves, continue the wavelength (L) line to the left until it intersects the wave speed scale (see example for 50 m wavelength). Combine these procedures for transitional waves (lines) (see example for 1000 m wavelength and 100 m water depth). Note that the transitional waves have a lower wave speed than they would have had as deep-water waves. From [85 T]. C in [m/s], z and L in [m].

When the long low swell type of wave gets to a point where the wave height is approximately equal to the water depth, the wave collapses or breaks. In the case of the short steep waves breaking occurs when the wave height is 1/2 of the water depth. The long swell wave usually breaks with a hollow front called a plunging breaker, while the wind wave with a steep front breaks as a spilling type of breaker. If the beach is very steep a surging type of breaker will form with strong uprush and backwash. See Figs. 32a⋯35.

Parameters describing wave characteristics [66 B 1]:

h = water depth
H = wave height
T = wave period
L = wavelength
g = acceleration due to gravity (9.81 m s^{-2})
E = wave energy per unit surface area of wave
f = frequency
C = wave velocity
d = orbital diameter
U_c = velocity of frame of reference
n = wave number
α = angle of wave crest with shore
β = angle of wave inclination of beach slope.

Water particle orbits in shallow water: As the wave enters shallow water so the particle orbits become more elliptical. The velocity of these elliptical movements does not decrease with water depth as in deep water.

Characteristics of orbital paths in

$$\text{deep water: } \frac{d}{L} > \frac{1}{4}$$

$$\text{intermediate water: } \frac{1}{20} < \frac{d}{L} < \frac{1}{4}$$

$$\text{shallow water: } \frac{d}{L} < \frac{1}{20}.$$

After Stokes' wave theory the orbits are not closed so that water transport occurs by the particles.

Terminology for waves at coasts (see also sect. 6.2).

Short crested wave: a wave with the crest length of the same order of magnitude as the wavelength.

Short wave: waves under conditions where the ratio of water depth and wavelength exceeds 0.5, the phase velocity being independent of water depth, but dependent upon wavelength.

Shoaling coefficient: the ratio of the height of a wave to its height in deep water, with the effect of refraction eliminated.

Shoaling effect: the alteration of a wave proceeding from deep water into shallow water [66 B 1].

Surf beat: On beaches facing oceans with an extensive wave-fetch, long swell waves arrive well separated from complicating shorter-frequency waves. Occasionally, however, the waves arriving at the coast consist of two such long-period waves whose frequencies are slightly out of phase. The addition of these two waves creates a series of high waves followed by a series of lower waves, a typical period between the arrival of high waves is two to three minutes. This surf-beat sets up rhythmic vibrations in the near shore which have important morphological implications for the coast. Compare subsects. 9.5.5.3 and 9.4.2.9.

Function	Deep water	Intermediate	Shallow water	Near-breaker
Phase velocity C	$\frac{g}{2\pi} T$	$\left[\frac{g}{k} \tanh kh\right]^{1/2}$	$\left[gh\right]^{1/2}$	$\left[g(h+H)\right]^{1/2}$
Wave length L	$L_d = \frac{g}{2\pi} T^2$	$L_d \left[\tanh k_d h\right]^{1/2}$ ⟶		CT
Wave height H	H_d	$H_d \left[\frac{C_d}{2Cn}\right]^{1/2}$	$\frac{H_d}{\left[4 k_d h\right]^{1/4}}$	$0.32\, H_d \left[\frac{L_d}{H_d}\right]^{1/3}$
Orbital diameter d	Surface $d = H_d$ Bottom $d_0 = 0$	$d_z = H \frac{\cosh kz}{\sinh kh}$	$d_z = \frac{H}{kh}$	$d_0 = \frac{3}{2} H$
Maximum orbital velocity u_m	$\frac{\pi d}{T}$ ⟶		$\frac{1}{2} \frac{H}{h} C$	$u_{m,0} = \frac{1}{3} \frac{H}{h} C$
Limits of application	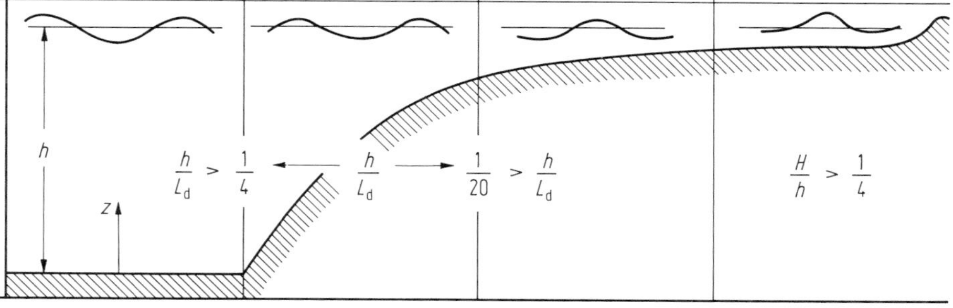 $\frac{h}{L_d} > \frac{1}{4}$	⟵ $\frac{h}{L_d}$ ⟶	$\frac{1}{20} > \frac{h}{L_d}$	$\frac{H}{h} > \frac{1}{4}$

Fig. 33. Wave equations for waves from the sea to the coast and shore [67 S 3]. Compare also [81 K 1].

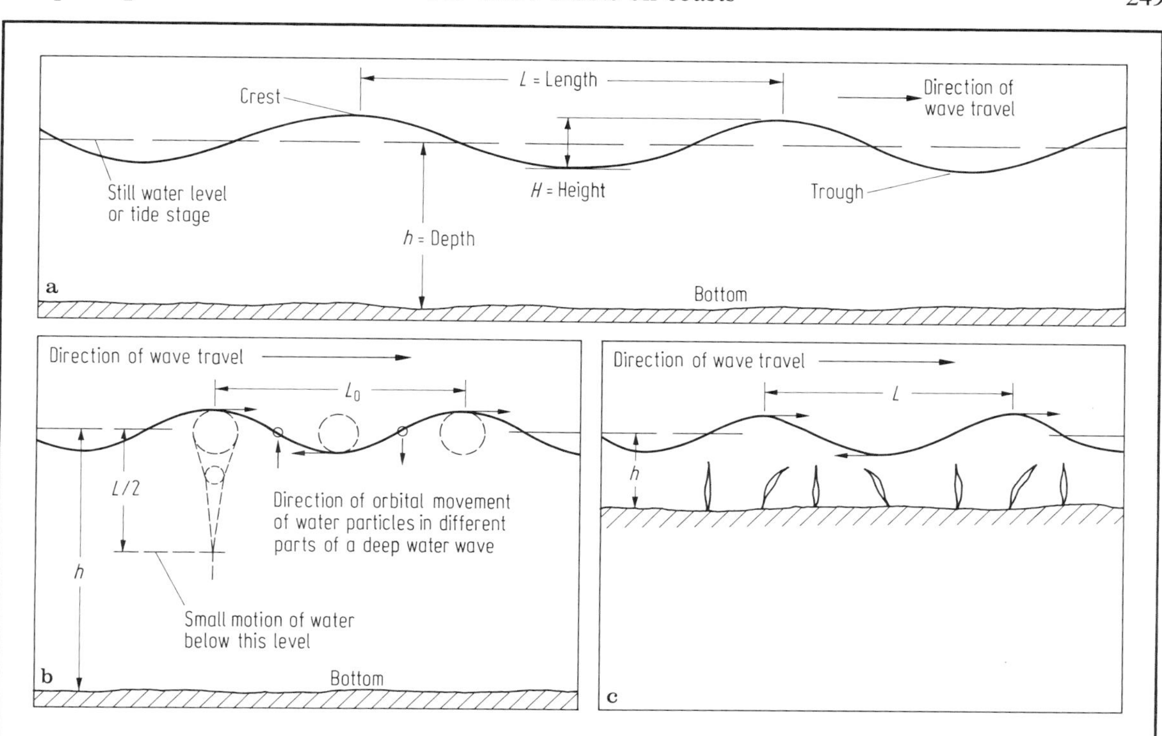

Fig. 34. Terms of waves [53 W]. Model not to scale. a) Definitions of wave characteristics. b) Deep water waves: When the water depth is relatively great, the bottom has no effect on the motion of waves. c) Transitional water waves: When the water depth is relatively small, waves cause the water to move at the bottom. Beach grass shows direction of movement of water particles under various parts of a transitional water wave.

9.5.3 Breakers and surf zone

9.5.3.1 Breaker types

The steepness of the beach has a direct influence on nearshore hydrodynamics, and consequently on beach sand movement. This influence has two major expressions: it affects the transfer of energy from wave motions to long-period motions in the surf zone; and it correlates with run-up patterns in the surf zone.

The surf scaling parameter ε allows the morphodynamic distinction of breaker type and beach. See subsect. 9.4.2.3, Fig. 19.

Terminology for breakers at coasts (Fig. 35, Table 31b):

Breakers, i.e. incident waves breaking on the shore, over a reef, etc., may be roughly classified into three kinds, although the categories may overlap:

spilling breakers occur with shallow beaches and steep waves break gradually over a considerable distance;
surging breakers occur with steep beaches and waves of low steepness tent to curl over and break with a crash;
plunging breakers result from steep beaches and waves of intermediate steepness peak up, but then instead of spilling or plunging they surge up on the beach face;

collapsing breakers are of the transitional type between plunging and surging.

Breaker depth (also called breaking depth): the still water depth at the point where a wave breaks.

Breaker zone: generally used for surf zone.

Reflecting breaker: a wave which is formed by the water mass of a backwash of a former breaker at the coast and which is running perpendicularly off the shore, merging with the next approaching breaker from the sea, steeping up by interference and then breaking to the direction of the sea; all this acting on steep beach and steep underwater beach bottom.

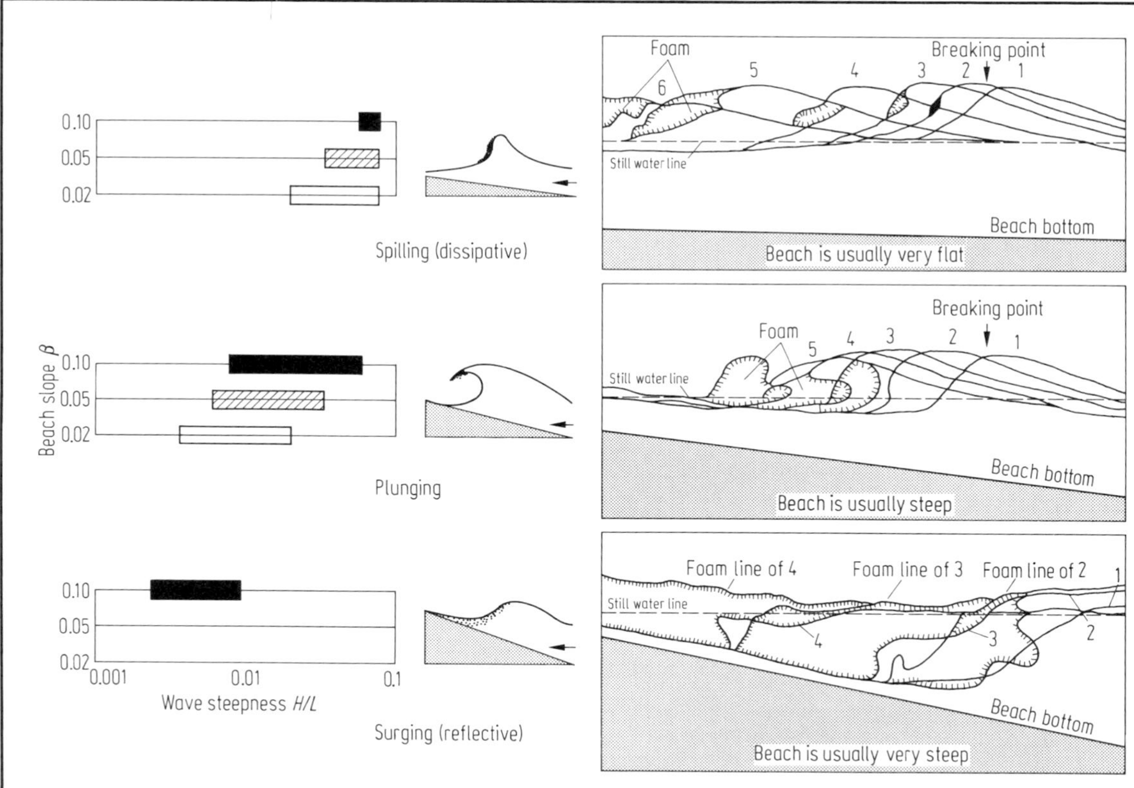

Fig. 35. Diagrams of three types of breakers [53 W]. The three primary types of breaking waves are classified according to beach slope and wave steepness (wave height divided by wavelength, H/L). The breaker type influences the nature of water and sand motions near the surf zone (left Fig). .The sketches consist of a series of profiles of the wave form as it appears before breaking, during breaking and after breaking. The numbers opposite the profile lines indicate the relative time of occurrences (right Fig.) [53 W].

Edge waves: the process of taking energy from incoming surface waves, and pumping it into longer-period fluctuations called edge waves. These edge waves have a longshore periodicity and amplitudes decaying exponentially offshore; the energy is trapped against the shoreline by refraction. The edge wave velocity is about twice the breaker velocity. [52 U, 77 S 2, 74 G, 84 W 4].

Edge wavelengths may coincide with the spacing of rip currents [69 B 2, 71 B].

The key distinction of edge waves is that their direction of progress is purely a longshore with refraction preventing their escape seaward. This tends to concentrate their energy toward the shoreline [83 K 3]. Edge waves are a dominating factor in the inner surf zones, processing and causing a number of nearshore morphologies of the type of longshore rhythmically repeated features like, for example, beach cusps. See Figs. 29a, b, 44, 45. [75 G 5].

The break-point:

There are three explanations normally advanced for wave breaking: [84 P]

1. The increase in wave height and decrease in wavelength in shallow water result in an increase in wave steepness as it progresses shorewards. The Stokes' wave theory predicts that when the angle at the wave crest reaches 120° the wave form becomes unstable and it breaks. Another way of expressing this critical instability is to define the limiting wave steepness at the break point:

$$\frac{H}{L} = 0.147 = \frac{1}{7}.$$

2. The velocity of the orbiting wave particles within the wave also changes as the wave progresses shorewards; as the wave height increases so the diameter and velocity of these orbits also increase. At the same time the velocity of the wave form decreases as the water depth decreases ($C=\sqrt{gd}$), at a critical point the velocity of the water particles becomes greater than the wave form velocity and the water breaks through the wave form.

3. In very shallow water neither Airy nor Stokes' wave theory gives entirely adequate predictions of wave movement. An alternative in these conditions is the solitary wave theory under which waves are envisaged as isolated from others by broad flat troughs. The wave form velocity then depends on the water depth plus the height of the wave at a given moment.
At the wave crest therefore:

$$C=\sqrt{g(d+H)} \quad \text{(i.e. faster)}$$

while at the trough:

$$C=\sqrt{g(d-H)} \quad \text{(i.e. slower)}.$$

Consequently the crests move more quickly than the intervening troughs and waves become asymmetrical with a steep leading edge and a gentle backslope. Eventually this asymmetry leads to instability and the wave breaks.

Each of these three explanations adds to our understanding of the phenomenon of wave breaking although none is sufficient in itself. In all three, however, the important factor is the water depth relative to wave height at the break-point. Thus low waves can run into much shallower water than high waves before breaking and this critical ratio between depth and wave height has been found to be more or less constant. The ratio is generally called γ and its critical value varies from about 0.6···1.2 with a mean of 0.78:

$$\gamma=H/d=0.78 \quad \text{(Galvin 1972)}.$$

The range of γ, although small, is important and has been shown by several authors to be related to the beach slope.

Thus Huntley and Bowen (1975) showed that steep beaches are associated with high values of $\gamma(\gamma=1.2)$ while flatter beaches have lower values ($\gamma=0.6$). Consequently, although on most beaches of average slope and γ values of around 0.78, waves break when the water depth is just less than the wave height, for steep beaches, such as those formed of shingle, the wave would progress into water considerably less deep than this before breaking. Quoted after [84 P].

Table 31b. Comparison of breaker coefficients [84 P]. B_b = breaker coefficient, H_b = wave height at the break point, L = wavelength, g = acceleration due to gravity, s = beach slope, T = wave period, ε = surf scaling factor, a = wave amplitude (height), β = beach slope, P = phase difference = ratio of run-up duration t to wave period T, t = run-up: duration of the onshore water movement caused by a single wave lasting from the moment of breaking until the furthest point that the water moves up-beach.

Author	Theory	Expression	Breaker type transition	
			Surging to plunging	Plunging to spilling
Galvin 1968	Breaker coefficient	$B_b=H_b/gsT^2$	0.003	0.068
Guza and Bowen 1975	Surf scaling factor	$\varepsilon=\dfrac{2\pi a}{gT\tan^2\beta}$	2.5	33
Kemp 1975	Phase difference	$P=t/T$	0.5	1.0

9.5.3.2 The breaker and surf zone

Features of the surf zone:

Existence and width of a surf zone is controlled by beach slope and tidal phase. Steep-sloped beaches rarely possess a surf zone. The breaker zone is followed by the surf zone, which at the shoreline is followed by the swash zone [70 S 4]. See Figs. 36 and 37. [81 W 2].

Interaction of tidal levels, change of beach, and resulting consequences for the course of breakers in the breaker zone. In case of rising tidal level: flatter beach, wider breaker zone, bigger swell of breakers. In case of falling tidal level: steeper beach, narrower breaker zone, smaller swell of breakers. See Fig. 38.

Energy of wave types in the surf zone under tidal condition (see sect. 9.6).

The profile of a beach is shaped by processes including waves and tides. The resulting profile depends on the forces involved in these processes as well as on the time of the activity of the forces, i.e. the time of tidal inundation for the zone of the tidal beach considered.

The concave upward profile of the macrotidal beach depicted in Fig. 38 consists of a subtidal region below the low-tide level, a low-tidal region between spring tide and map low tide levels, a mid-tidal zone between neap low tide and neap high tide levels, and a high-tide zone between neap high tide and spring high tide.

Across the profile the effects of incident waves dominate. Estimates, time-averaged over the lunar half-cycle, of the work performed by waves on the subtidal, low-tidal, and mid-tidal zones indicate that most of the work is due to unbroken shoaling waves rather than surf zone processes. The latter are dominant only in the reflective high tidal zone, which is active for less than 25% of the inundation time. Standing waves, although of minor importance for most of the profile, may be relevant for the high-tidal zone at spring high tide and for the mid-tidal zone at mid high tide, especially at higher infragravity frequencies (0.025···0.02 Hz). In contrast to microtidal beaches, shore-parallel tidal currents over the subtidal and low-tidal zones are less important for macrotidal beaches [83 K 3].

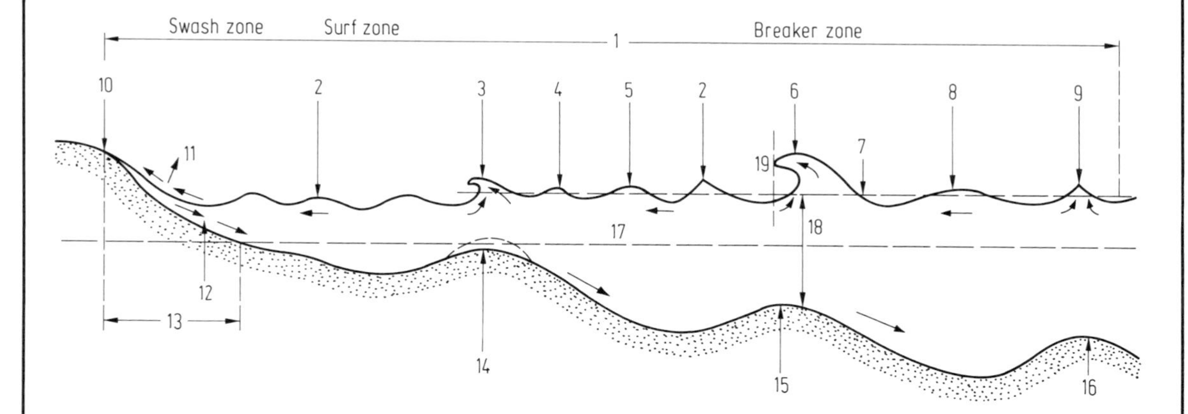

Fig. 36. Schematic diagram of waves in the breaker zone [66 B 1] (not to scale). 1. Surf or breaker zone; 2. Translatory waves; 3. Inner line of breakers; 4. Peaked up wave; 5. Reformed oscillatory wave; 6. Outerline of breakers; 7. Still water level; 8. Waves flatten again; 9. Waves break up but do not break on this bar at high tide; 10. Limited of uprush; 11. Uprush; 12. Backrush; 13. Beach face; 14. Inner bar; 15. Outer bar (inner bar at low tide); 16. Deep bar (outer bar at low tide); 17. Mean lower low water (MLLW); 18. Breaker depth ≅1.3 height; 19. Plunge point. Remark: A reflecting breaker may run off the shore, and may include a breaking wave at 3 or at 6 with breaking-plunging direction off the shore.

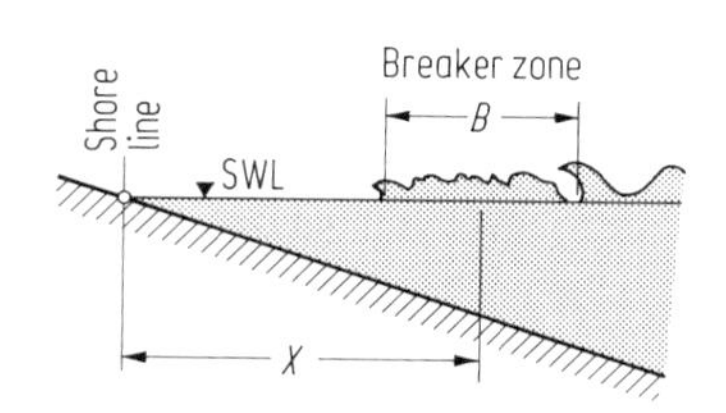

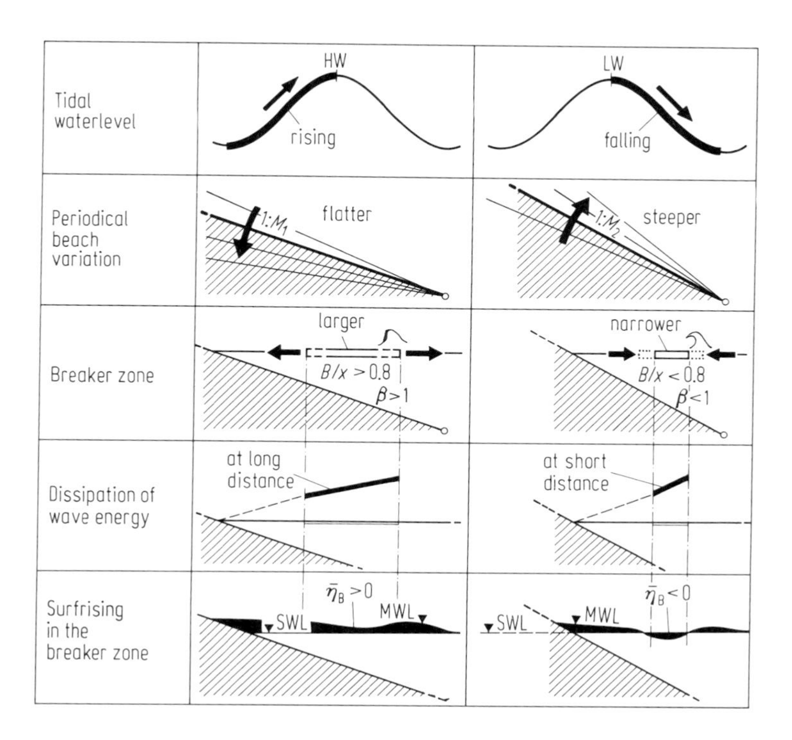

Fig. 37. Model of the relationship of tidal waterlevel, beach variation, and of the thus caused effects concerning the process of surfrising in the breaker zone [79 F in 79 D 1]. Model not to scale. B = breaker zone, x = distance of the breaker zone from the waterlevel at the coast, β = beach slope, η_B = surfrising, SWL = still water level, MWL = mean water level.

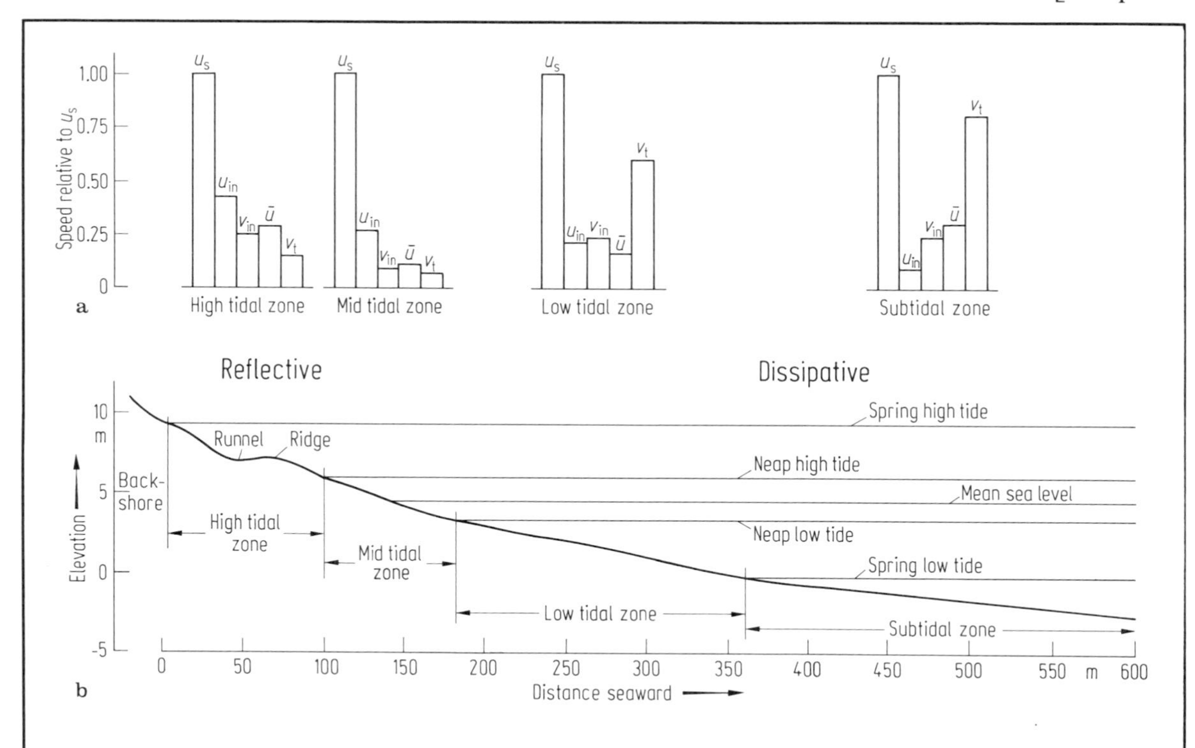

Fig. 38. Wave velocities and surf processes on a macro-tidal beach. Example: Cable Beach, Broome, NW-Australia, Nov. 1980 [83 W in 83 K 3]. a) Process signature, U_s = incident wave velocity, U_{in}, V_{in} = infragravity wave velocity, $\bar{U}$ = mean velocity, V_t = net time average velocity (refers to shore-subtidal parallel currents). (Infragravity waves are those with long frequencies, periods in the order of 100 s). b) Beach profile.

9.5.4 Transformation of waves approaching the shore

9.5.4.1 Velocity, length, height, wave angle

The modifications of waves approaching shallow coastal waters include decrease in phase velocity, wavelength, and wave period. For details, see subsects. 6.2.5.2 and 6.2.5.4. As the ratio of water depth to wavelength gets smaller than 0.5, the wave height first decreases, but increases again for ratios below 0.06. In the same manner the steepness, i.e. the ratio of wave height and wavelength, first decreases, then increases again. In this process the wave crests get narrower and sharper, the troughs wider and flatter, the motion of the water in the wave more and more elliptical. The wave breaks as the shoreward particle velocity in the wave crest exceeds the phase speed of the wave. In this case the wave front collapses sending forth a rush of water (swash) onto the shore [76 B 1].

As a general rule, the decrease in the height of surf due to retraction is negligible for waves that come in at angles smaller than approximately 30°, no matter what their steepness may be offshore. But when the waves are coming in at an angle greater than ≈60°, the decrease in their heights may make landing possible at a place where this would not be so otherwise. Compare [81 K 2, 73 H 3]. See Table 32.

Table 32. Relation of wave angle to shoreline.

Table 32a. The angles which breakers make with a straight shoreline, when all bottom contours are parallel with the beach, for waves of different degrees of steepness in deep water approaching the shoreline at different initial angles. It is assumed that the waves will break where the depth of water is 1.3 times the breaker height [47 B].

Steepness of wave in deep water (length : height)	Angle between wave in deep water and shoreline						
	10°	20°	30°	40°	50°	60°	70°
10 : 1	9	16	23	30	36	41	46
20 : 1	6	10	15	20	23	27	30
40 : 1	5	9	13	17	20	23	25
100 : 1	3	7	10	12	15	16	17

Table 32b. The angles which waves (approaching at different initial angles) make with a straight shoreline in diminishing depths of water (relative to the wave length in deep water) [47 B].

Depth of water in terms of wave length offshore	Initial angle of wave in deep water with shoreline							
	10°	20°	30°	40°	50°	60°	70°	80°
0.500	10°	20°	30°	40°	49°	59°	68°	77°
0.400	10	20	30	39	49	59	68	77
0.300	9	19	29	38	47	56	64	71
0.200	8	17	26	33	42	50	57	61
0.150	8	16	24	31	39	44	50	53
0.100	7	14	21	27	32	37	41	43
0.050	5	10	15	20	23	27	30	31
0.030	4	9	12	16	18	21	24	25
0.025	4	8	11	14	17	20	22	23
0.020	3	7	10	13	15	17	18	20

Table 32c. Percentage decrease in height between deep water and the breaker zone, for waves of different initial degrees of steepness approaching a straight shoreline (with straight and parallel bottom contours) at different angles. It is assumed that the waves break where the depth of water is 1.3 times the breaker heights [47 B].

Steepness of wave in deep water (length : height)	Angle between wave in deep water and shoreline					
	20°	30°	40°	50°	60°	70°
10 : 1	0	3	6	11	18	30
20 : 1	0	5	10	16	26	38
40 : 1	1	6	11	17	27	39
100 : 1	2	6	12	19	28	40

9.5.4.2 Wave refraction and wave diffraction

Wave refraction: Sea floor topography influences the pattern of swell approaching the coast, bending the wave crests until they are parallel to the submarine contours, Fig. 39b. (R) in Fig. 39b is the point of direction turn due to bottom control, where the depth is half the wavelength.

Wave diffraction: Emerged features, such as islands, or reefs awash at low tide, produce complex patterns of wave refraction, and waves that have passed through narrow straits or entrances are modified by spreading out in the water beyond diffraction zone *D* in Fig. 39c.

Wave orthogonals: Orthogonals indicate how the wave energy is distributed, convergence towards a section of coast indicating a concentration of wave energy, whereas divergence of orthogonals indicates a weakening.

Wave refraction coefficient and wave energy: The change in wave heights as waves are refracted is inversely proportional to the square root of the relative spacing of adjacent orthogonals.

A refraction coefficient *R* can be derived as follows:

$$R = \sqrt{\frac{S_0}{S}}$$

where S_0 is the distance between a pair of orthogonals in deep water and S their spacing on arrival at the shoreline. Calculated from a wave refraction diagram, this coefficient is an approximate expression of the relative wave energy on sectors of the coastline. See Figs. 40, 41.

Measurement of wavelength in the refraction zone.

Tables for computation of refraction diagrams are given in [64 W].

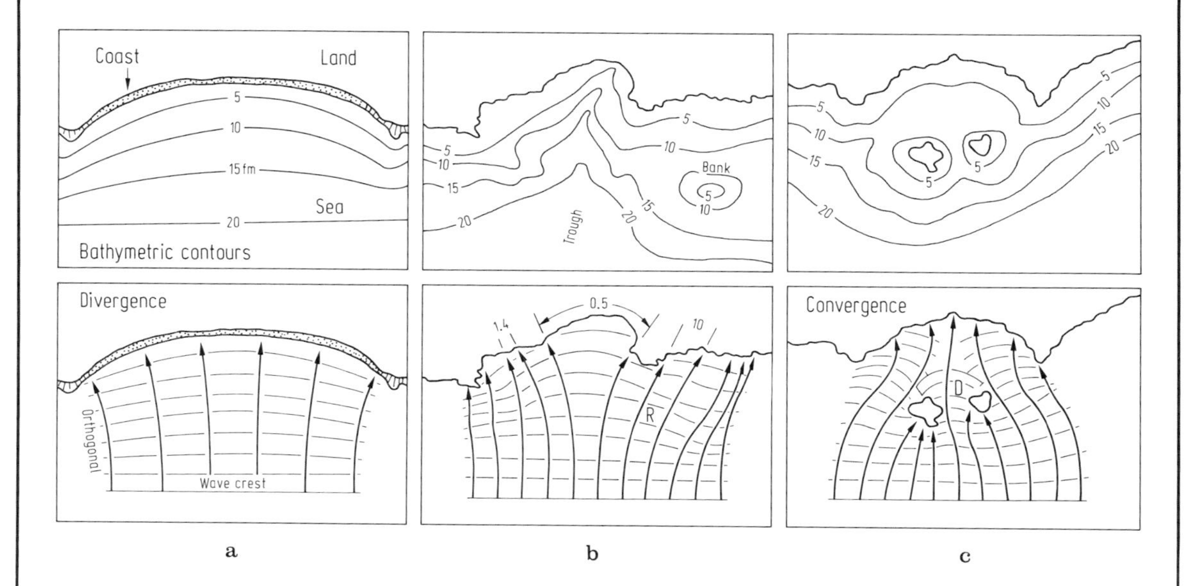

Fig. 39. Wave refraction and wave diffraction on coasts depending on bottom topography [76 B 1]. Model not to scale. a) Wave refraction pattern in an embayment of simple configuration (depths in fathoms). b) Wave refraction pattern in an embayment with a trough offshore and a submerged bank (some refraction coefficients are indicated). R indicates the point the depth is half the wavelength. c) Wave refraction and diffraction patterns formed where a coast is bordered by islands. D denotes diffraction zone.

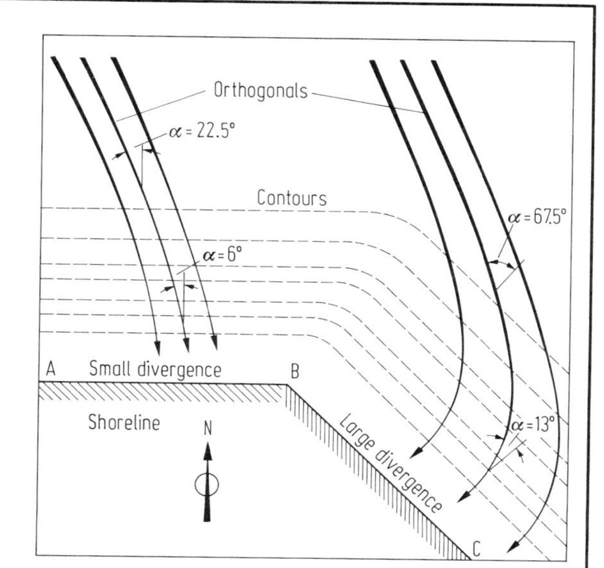

Fig. 40. Refraction of waves along two sections of straight beaches with different exposures [61 D].

The divergence along section A–B is less than along B–C, and the wave height along A–B exceeds therefore the wave height along B–C [63 D 2, 61 D]. The respective refraction factors are 0.97 for the A–B section and 0.63 for B–C.

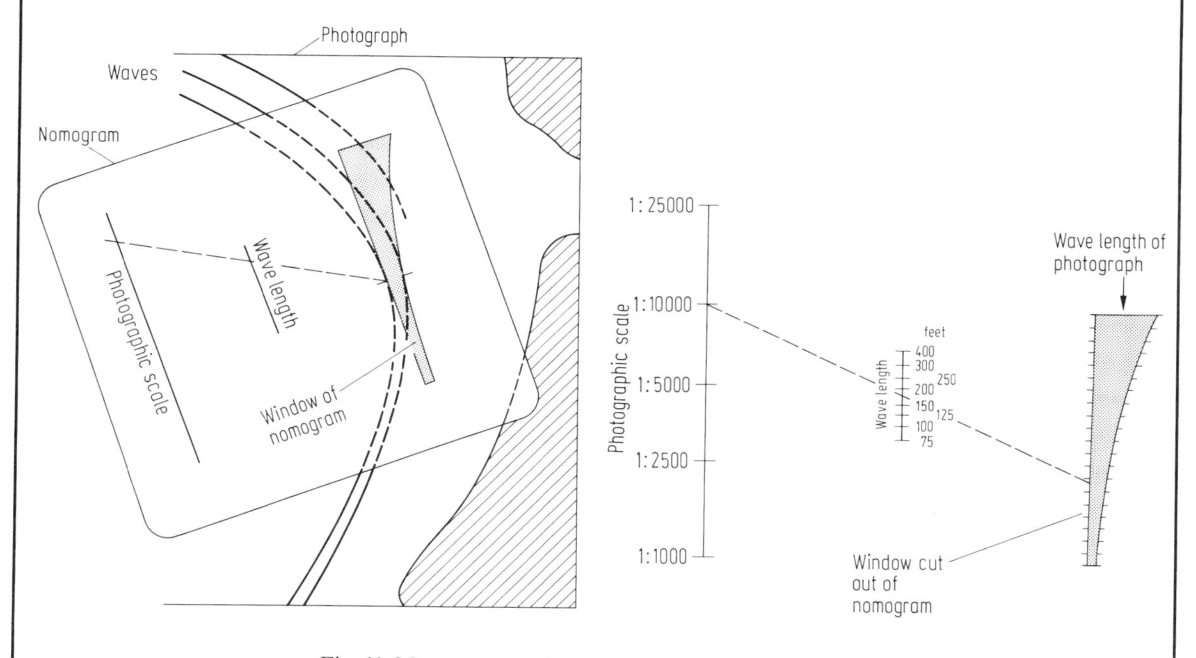

Fig. 41. Measurement of wavelengths in front of the coast and indirect determination of depth in shallow water areas by means of remote sensing methods (aerial photography). The nomogram is fitted to the distance of two consecutive wave crests [68 D 1]. Left: Position of nomogram for the measurement of wavelengths. Right: More detailed presentation of the Dale-nomogram for the measurement of wavelengths. The scales and windows are logarithmic functions.

9.5.5 Wave-generated currents of the shore zone

9.5.5.1 Nearshore circulation cells

Characteristic currents of nearshore circulation are
1. back currents of the bottom,
2. rip currents,
3. breaker currents and gradient currents, *longshore currents*, and
4. coastal drift currents on the seaward side of the breaker zone [63 R 1, 73 I, 63 I].

Pressure fields produced by waves travelling in shallow water change the average water level, reducing it near the breaker zone and increasing it where the waves run up the beach face.

The interaction of surface waves moving toward the beach with edge waves travelling along the shore produces alternate zones of high and low waves that determine the position of rip currents. The pattern that results from this flow takes the form of a horizontal eddy or cell, called the nearshore circulation cell. See Figs. 42 and 43 and Table 33. [71 W 3].

The nearshore circulation system produces a continuous interchange between the waters of the surf zone and the waters of the offshore zone, acting as a distributing mechanism for nutrients and as a dispersing mechanism for land runoff [41 S, 50 S, 84 P].

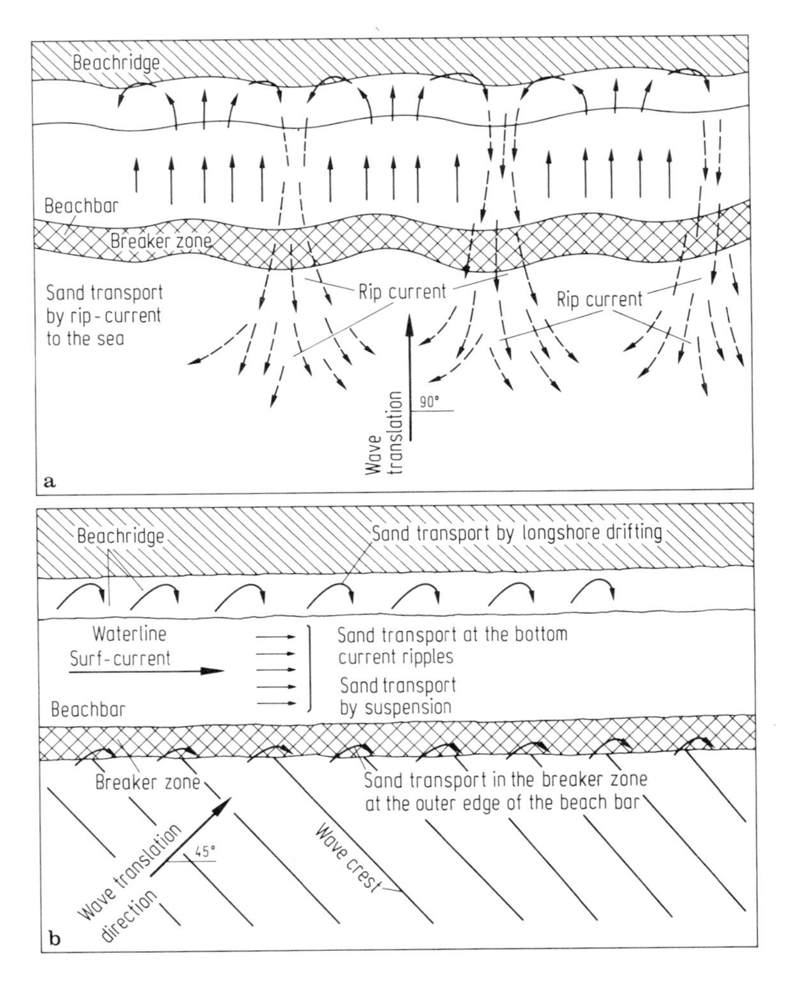

Fig. 42. Nearshore circulation, wave fronts perpendicular (a) and oblique (b) to the coast. After Hensen in [81 K 1]. Model not to scale.

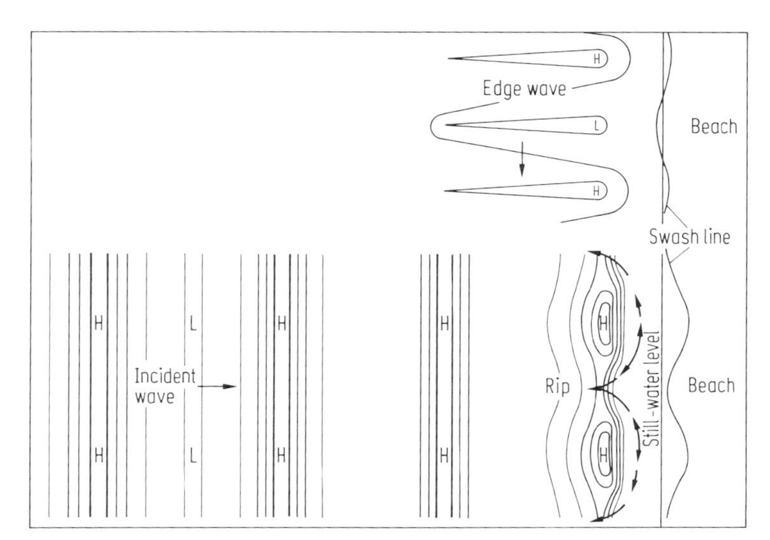

Fig. 43. Formation of rip currents. The interaction of incident waves from deep water with edge waves travelling along the beach producing alternate zones of high (H) and low (L) breakers. Longshore currents flow away from zones of high waves where setup is maximum and converge on points of low waves, causing rip currents to flow seaward. From [71 I 1].

Table 33. Velocity of rip current development depending on short wave period of >5 s [79 G].

Due to wind from the sea $4 \cdots 6\,\mathrm{m\,s^{-1}}$	Close to beach	Middle part	Offshore
Surface	0.1	0.1	0.0
Depth 1 m	0.3	0.3	0.2
Near bottom	0.8	0.6	0.4
Due to wind from the sea $20 \cdots 24\,\mathrm{m\,s^{-1}}$			
Surface	5.5	3.0	2.2
Depth 2 m	7.2	5.8	3.6
Near bottom	10.8	8.6	5.9

Longshore currents (compare subsect. 9.5.1). For extended presentation see [84 P].

In calculating the longshore current generated by incident waves, waves breaking at an angle α with the beach, with a volume discharge q of water in the direction of travel per unit length of wave crest and unit time, are considered. The longshore component of the discharge then amounts to $q \sin\alpha$ per unit length of wave crest, or $q \sin\alpha \cos\alpha$ per unit beach length. The longshore current produced in this way increases linearly with distance along the beach. However, rip currents, with average separation l, transport water away from the beach. Thus the average longshore discharge q_l is obtained by the relation ($q = l^2 T^{-1}$)

$$q_l = \tfrac{1}{2} q l \sin\alpha \cos\alpha = \tfrac{1}{4} q l \sin 2\alpha . \qquad [63\,\mathrm{I}]$$

Assuming that q_l is uniformely distributed over the inshore waters, the average longshore current $\bar{u}_l$ can be calculated as the ratio of q_l and the cross-sectional area of the water inshore from the breaker zone. As the latter quantity is given by $h^2 \cot\beta/2$ where h denotes the depth at which wave breaking occurs and β the average beach slope, $\bar{u}_l$ becomes

$$\bar{u}_l = \frac{ql}{2h^2} \tan\beta \sin 2\alpha . \qquad [63\,\mathrm{I}]$$

If the value of q from solitary wave theory is substituted,

$$\bar{u}_1 = 2\sqrt{\frac{\gamma}{3}}\frac{l}{T}\tan\beta\sin 2\alpha,$$

where T is the wave period and γ is the relative wave height H/h [63 I].

Rip currents

The spacing of the rip currents is equal to the length of the edge wave

$$\lambda = \frac{g \cdot T^2}{2\pi}\sin(2n+1)\beta,$$

where $2n+1$ has integral values up to $(2n+1)\beta < \pi/2$, where β is the slope of the beach. Thus several spacings are possible. [82 B in 82 S 1].

9.5.5.2 Periodic circulation cells induced by waves at the shore and rhythmic features

Rhythmic features are caused by periodic processes as caused by waves, wave-induced currents, and tidal currents. In the case of sand bottom as boundary layer, features of strict geometric forms are produced:

1. Short temporal oscillation ripples (existence from several seconds to hours).
2. Quasi persistent sand bars (existence for years).
3. Short scale (1 tide), or medium scale (1 lunar period), beach cusps. See Figs. 44 and 45.

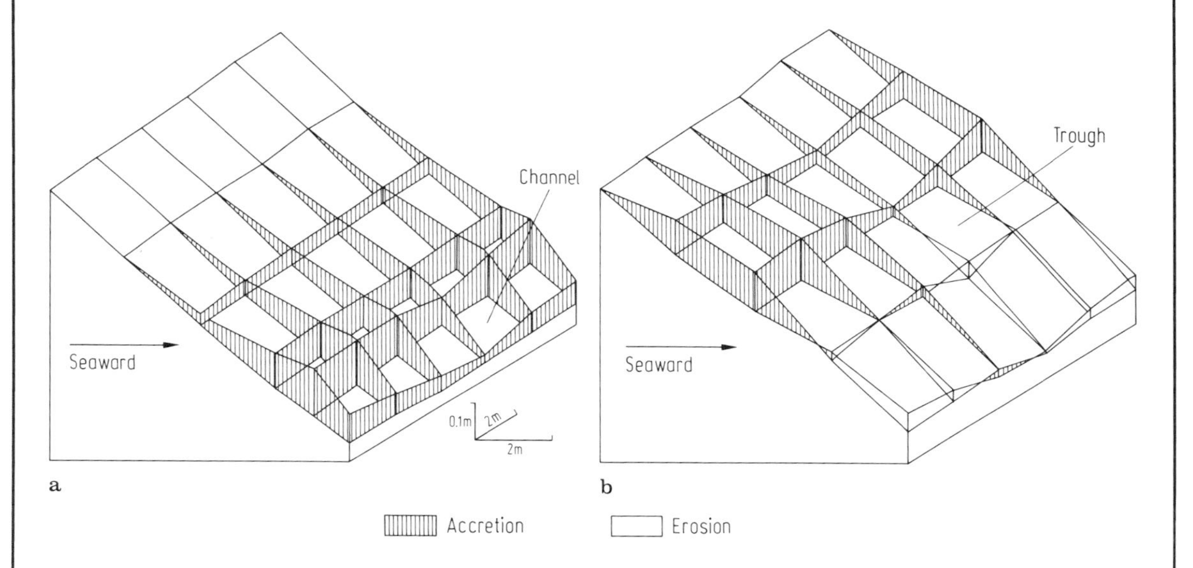

Fig. 44. Block diagrams showing the development of the cuspate form. a) During flood tide a ridge was deposited onto the foreshore with equally spaced channels distributed along its length. The crest of the ridge lays between lines of rods and its position and approximate forms have been added to the diagram. b) The ridge migrated shoreward with the flooding tide. On ebb tide the mouths of channels were progressively widened by swash erosion until adjacent mouths met, effecting a cuspate shape. From [82 G 3 in 82 S 1].

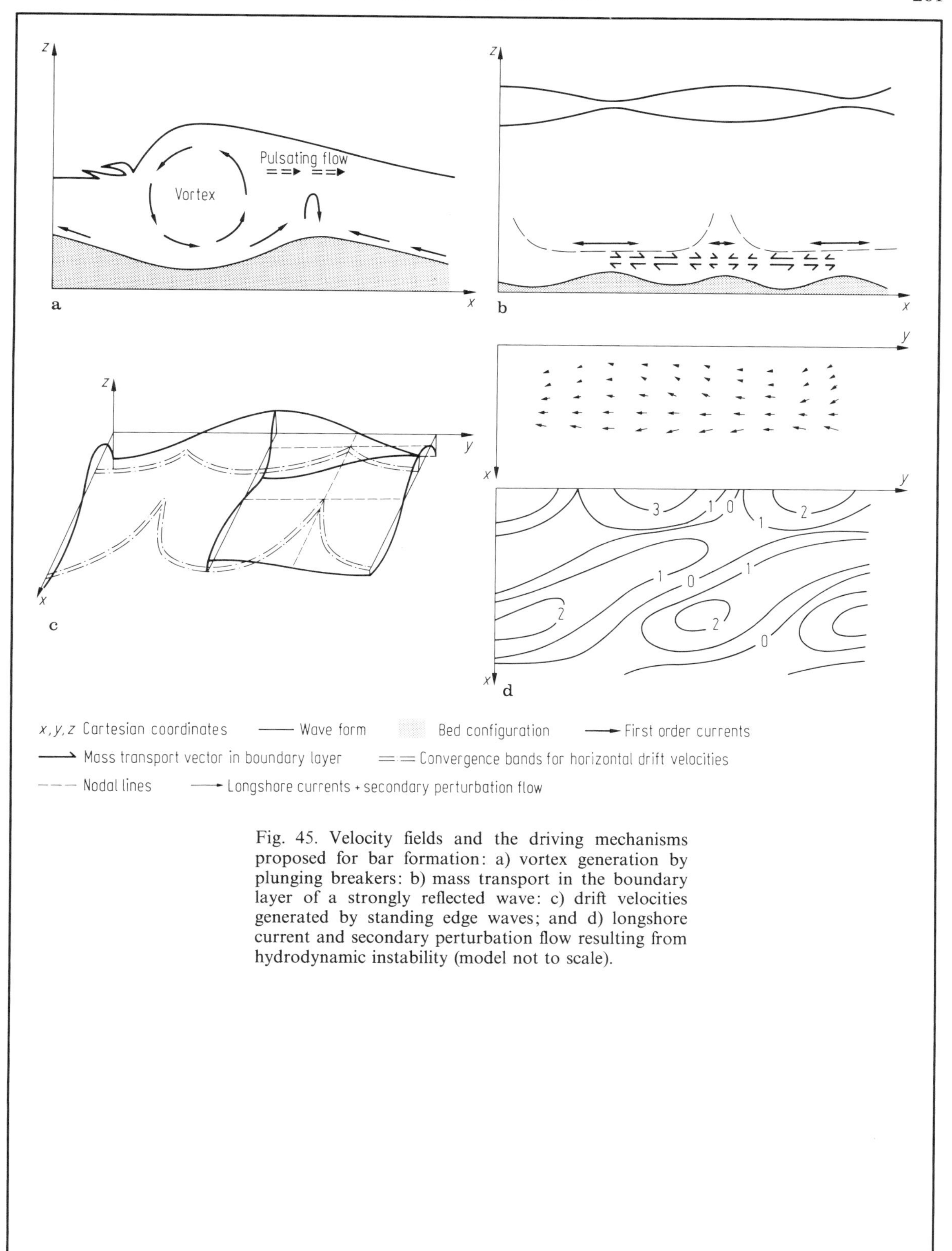

Fig. 45. Velocity fields and the driving mechanisms proposed for bar formation: a) vortex generation by plunging breakers: b) mass transport in the boundary layer of a strongly reflected wave: c) drift velocities generated by standing edge waves; and d) longshore current and secondary perturbation flow resulting from hydrodynamic instability (model not to scale).

9.5.6 Catastrophic waves at coasts: tsunamis

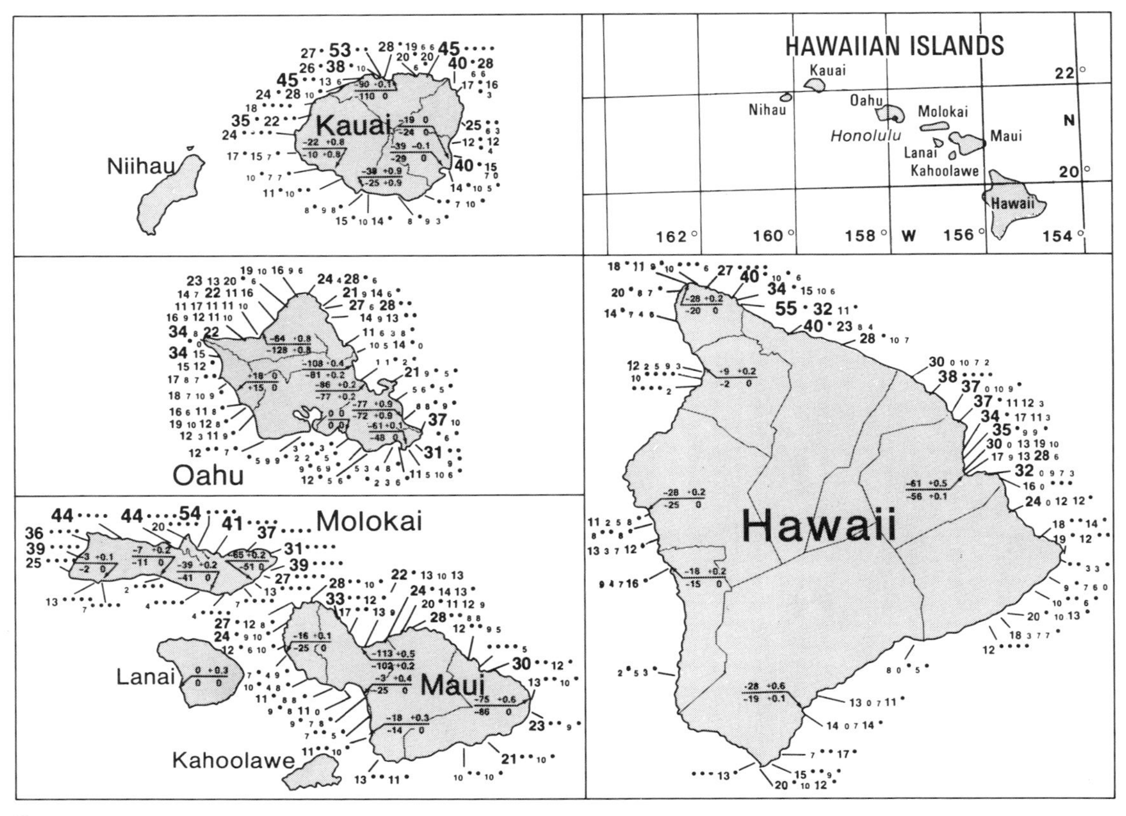

a

Recorded heights of tsunami run-up in feet above mean lower low water datum

(a) (b) (c) (d) (e)
0 0 0 0 0

a 1946 (from Aleutian Islands)
b 1952 (from Kamchatka, U.S.S.R.)
c 1957 (from Aleutian Islands)
d 1960 (from Chile)
e 1964 (from Alaska)

• Not surveyed after tsunami run-up

Tidal differences compared to Honolulu

(a) (b)
0 0
0 0
(c) (d)

a Time of high tide before or after Honolulu (minutes)
b Time of low tide before or after Honolulu (minutes)
c Height of high tide above or below Honolulu (feet or ratio)
d Height of low tide above or below Honolulu (feet or ratio)

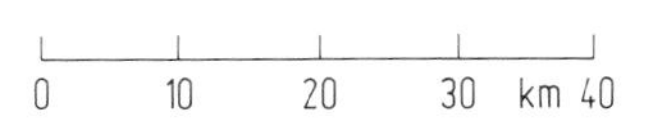

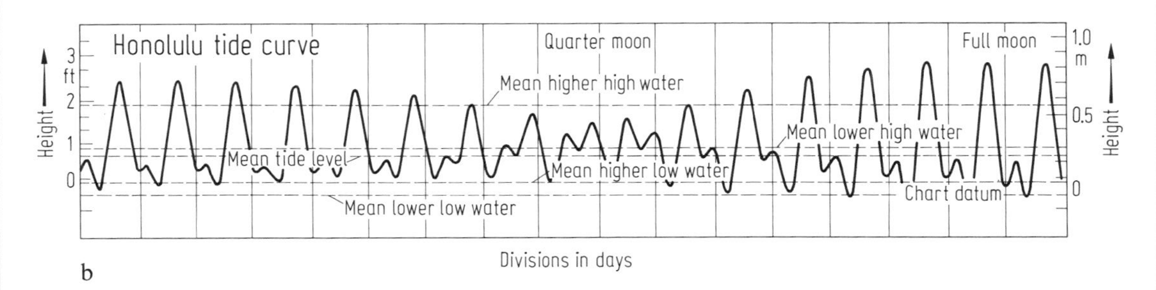

b

Fig. 46. Tsunamis (a) and tides (b) at Hawaii Islands [73 A]. For statistics of run up heights and causes of events of local tsunamis in Hawaii, see [77 C 2].

The tsunami waves are very long and very low. Their period is of the order of 10^3 s; their wavelengths may be as much as 250 km, their height only 0.3···0.6 m in deep water. The slope of the wave front is imperceptible, and ships at sea are unaware of their passage. Because the wavelengths are so long, tsunamis move as shallow-water waves even in the deepest ocean. As such their velocity (C) is controlled by the depth (d): $C=\sqrt{gd}$. Thus, if $g=9.81\ \mathrm{m\,s^{-2}}$, and $d=5000$ m (average depth of the Pacific basin) the velocity of a wave in the deep Pacific is 230 m/s, or 840 km/h. The Pacific Ocean is large so that waves moving at even that speed take considerable time to cross it, and a seismic sea wave warning system has been established to warn coastal inhabitants of approaching tsunamis. Large tsunamis with energy contents as high as $5 \cdot 10^{15}$ J occur only about 5 times per century. The average rate of energy dissipation of a tsunami is about 10^5 kW which is four orders of magnitude less than the total for waves and tides. See Fig. 46. Compare subsect. 1.4.3.4.

The International Tsunami Information Center (ITIC) was established in 1965 by the IOC. The operational center of the Tsunami Warning System (TWS) was established at the Honolulu Observatory.

The International Tsunami Warning System of the Pacific Ocean is based on records from 32 stations at the coast of the Pacific Ocean to compute the tsunami traveltime to the Hawaiian islands, to Honolulu. Recorded tsunami travel times, based on distant-marigraph-data to or from Hawaii, are: California 5^h42^m, Alaska 6^h30^m, Mid-Pacific islands $2^h \cdots 5^h$, Japan $7^h \cdots 8^h$, Chile $13^h \cdots 15^h$. [77 C 2]. For a catalogue of tsunamis see [67 I]. For hydrodynamic modelling of tsunamis see [71 E]. Compare [73 A, 83 I 1, 82 M 3].

9.6 Tide effects on coasts

9.6.1 Tides as coastal phenomena

For astronomical tides, see sect. 6.4.

The tidal phenomena are stronger in coastal areas than on the high oceans. Their strength varies considerably from region to region. [78 R 1, 75 R 2, 85 D, 63 D 1].

Tides as coastal phenomena of vertical and horizontal dimensions: Depending on the coastal configuration, the periodic changes of tidal water levels have vertical or horizontal dimensions: along steep coasts with vertical slopes above and below the water surfaces, a rise and fall of the water level can be observed, i.e. such a steep coast with natural water marks acts like a tide gauge. On shallow coasts with gently sloping beaches above and below the water surface, the tidal phenomenon can be observed as transgression and regression of the shoreline over a certain horizontal distance (apart from the vertical change). Extreme examples of horizontal changes are wide tidal areas, i.e. tidal flats = wadden areas, which are periodically flooded by the tides [82 D, 80 G 2, 45 S, 61 G 2, 78 R 1]. See Figs. 47, 48, 54, 55, 58, 59, Tables 34, 35, 39.

9.6.1.1 Tidal datums

Reference datums that have their origin in the rise and fall of the tides are the most satisfactory of all datums because they possess the advantages of simplicity of definition, accuracy of determination, and certainty of recovery. It is for these reasons that they are used in hydrographic surveying and nautical chart work, and in the demarcation of waterfront boundaries. Figs. 6, 7 and Tables 10···12. Tide gauge records comprize some of the longest and most reliable oceanic time series in existence. For many years mean sea level variations, derived from averaging the data to remove astronomical tide and other relatively short period oscillations of the sea surface, have been collected and stored. A world data center, the Permanent Service for Mean Sea Level (PSMSL), has been in existence at Bidston, UK, since 1933. This center co-operates directly with, and receives, the support of the IOC. International tidal data bank of (IHO) International Hydrographic Organization: [83 P 2].

Since January, 1978, the International Hydrographic Organization (IHO) has in operation a computerized data bank containing tidal harmonic constituent amplitudes and phase lags, as provided by IHO member states from stations around the world. It contains records from 4000 stations over the world's coasts but with a very irregular distribution of density in the different regions. The data base, called the IHO Tidal Constituent Bank, replaces the IHO Special Publication No. 26, which contained the same type of tidal information on single printed sheets.

Requests for data from the Bank should be submitted on an IHB Tidal Data Request Form also obtainable from hydrographic offices or the IHB, B.P. 345 MC-Monaco [78 I 1].

9.6.2 World distribution of tidal types and ranges

Table 34a. Coastal stations with type and range of the tide [33 B].

Station	ϕ	λ	Range [m]
With semidiurnal tides			
Virgo Bay, Spitsbergen	79°43′N	10°52′E	1.07
Melville Island, Canadian Arctic	74 47 N	111 00 W	1.16
Godthaab, Davis Strait	64 12 N	51 44 W	3.81
Ashe Inlet, Hudson Strait	62 33 N	70 35 W	9.15
Vardö, Norway	70 22 N	31 07 E	2.75
Bergen, Norway	60 24 N	5 18 E	1.25
Quebec, Canada	46 49 N	71 11 W	4.45
Fundy Bay, Canada	45 00 N	65 00 W	16.00
St. John, New Brunswick	45 16 N	66 03 W	7.29
Boston, Mass.	42 21 N	71 03 W	3.36
Sandy Hook, N.J.	40 27 N	74 00 W	1.71
Pernambuco, Brazil	8 04 S	34 53 W	2.13
Comodoro Rivadavia, Argentina	45 52 S	67 29 W	5.00
Puerto Gallegos, Argentina	51 37 S	69 00 W	13.91
Cape Town, South Africa	33 54 S	18 25 E	1.40
Duala, Cameroon	4 03 N	9 40 E	1.98
Dakar, Senegal	14 40 N	17 25 W	1.40
Casablanca, Morocco	33 36 N	7 37 W	2.81
Lisbon, Portugal	38 41 N	9 06 W	3.66
St. Malo, France	48 50 N	2 00 W	12.00
Brest, France	48 23 N	4 29 W	5.45
Dover, England	51 07 N	1 19 E	5.49
Liverpool, England	53 24 N	3 00 W	8.85
Heligoland, Germany	54 11 N	7 53 E	2.44
Roter Sand, Weser, Germany	53 51 N	8 05 E	3.97
Dublat, Hooghly River, India	21 38 N	88 06 E	3.41
Rangoon, Burma	16 46 N	96 10 E	5.80
Comoro Islands, Indian Ocean	12 47 S	45 17 E	3.29
Durban, Natal	29 53 S	31 04 E	1.71
Betsy Cove, Kerguelen Island	49 09 S	70 12 E	1.40
Wellington, New Zealand	41 17 S	174 47 E	1.10
Port Hedland, Western Australia	20 22 S	118 36 E	5.18
Apia, Samoa	13 46 S	171 44 E	0.94
Panama (Naos Island), Panama	8 55 N	79 32 W	4.88
Amoy, China	24 23 N	118 10 E	4.72
With diurnal tides			
Vera Cruz, Mexico	19°12′N	96°08′W	0.73
Cristobal, Panama Canal	9 21 N	79 55 W	0.34
Manila, Philippine Islands	14 36 N	120 57 E	1.40
Macassar, Celebes	5 08 S	119 25 E	1.19
Fremantle, Western Australia	32 03 S	115 45 E	0.64
Unalga Bay, Alaska	54 00 N	166 12 W	0.88

Table 34a (continued).

Station	ϕ	λ	Range [m]
With mixed tides			
Baltimore, Md. (Ft. McHenry)	39°16′N	76°35′W	0.43
Nassau, Bahama Islands	25 05 N	77 21 W	1.22
Rio de Janeiro, Brazil	22 54 S	43 10 W	1.28
Buenos Aires, Argentina	34 36 S	58 22 W	0.64
Mar del Plata, Argentina	38 05 S	57 32 W	1.22
Valparaiso, Chile	33 02 S	71 38 W	1.19
Mazatlan, Mexico	23 11 N	106 27 W	1.16
San Diego, Cal.	32 42 N	117 14 W	1.56
Seattle, Wash.	47 37 N	122 20 W	2.81
Sitka, Alaska	57 03 N	135 20 W	3.02
Kodiak Island, Alaska	57 47 N	152 24 W	2.75
Point Barrow, Alaska	71 18 N	156 40 W	0.19
Yokohama, Japan	35 27 N	139 38 E	1.49
Nagasaki, Japan	32 44 N	129 51 E	2.56
Tsushima Sound, Chosen	34 17 N	129 21 E	2.04
Hongkong, China	22 18 N	114 10 E	1.55
Soerabaya, Java	7 06 S	112 42 E	1.49
Singapore, Malaya	1 17 N	103 51 E	2.44
Madras, India	13 05 N	80 17 E	0.94
Colombo, Ceylon	6 56 N	79 51 E	0.61
Bombay, India	18 57 N	72 50 E	3.66
Karachi, Pakistan	24 47 N	66 58 E	2.23
Aden, Arabia	12 47 N	44 59 E	1.49
Reunion Island, Indian Ocean	21 16 S	55 35 E	1.07
Port Darwin, North Australia	12 23 S	130 37 E	5.12
Brisbane, Queensland	27 20 S	153 10 E	1.95
Sydney, New South Wales	33 52 S	151 12 E	1.28
Melbourne, Victoria	37 52 S	144 54 E	0.85
Port Adelaide, South Australia	34 51 S	138 30 E	1.92
Honolulu, Hawaii	21 18 N	157 52 W	0.46

Table 34b. Partial tides. For detailed numeric constants see [75 D 1]. Compare Fig. 47.

Partial tides	Symbol	Speed deg/mean solar h	Period solar h	Coefficient ratio (M2 ≙ 100)
Semi-diurnal components				
Principal lunar	M2	28.98410°	12.42	100
Principal solar	S2	30.00000	12.00	46.6
Diurnal components				
Luni-solar diurnal	K1	15.04107°	23.93	58.4
Principal lunar diurnal	O1	13.94304	25.82	41.5

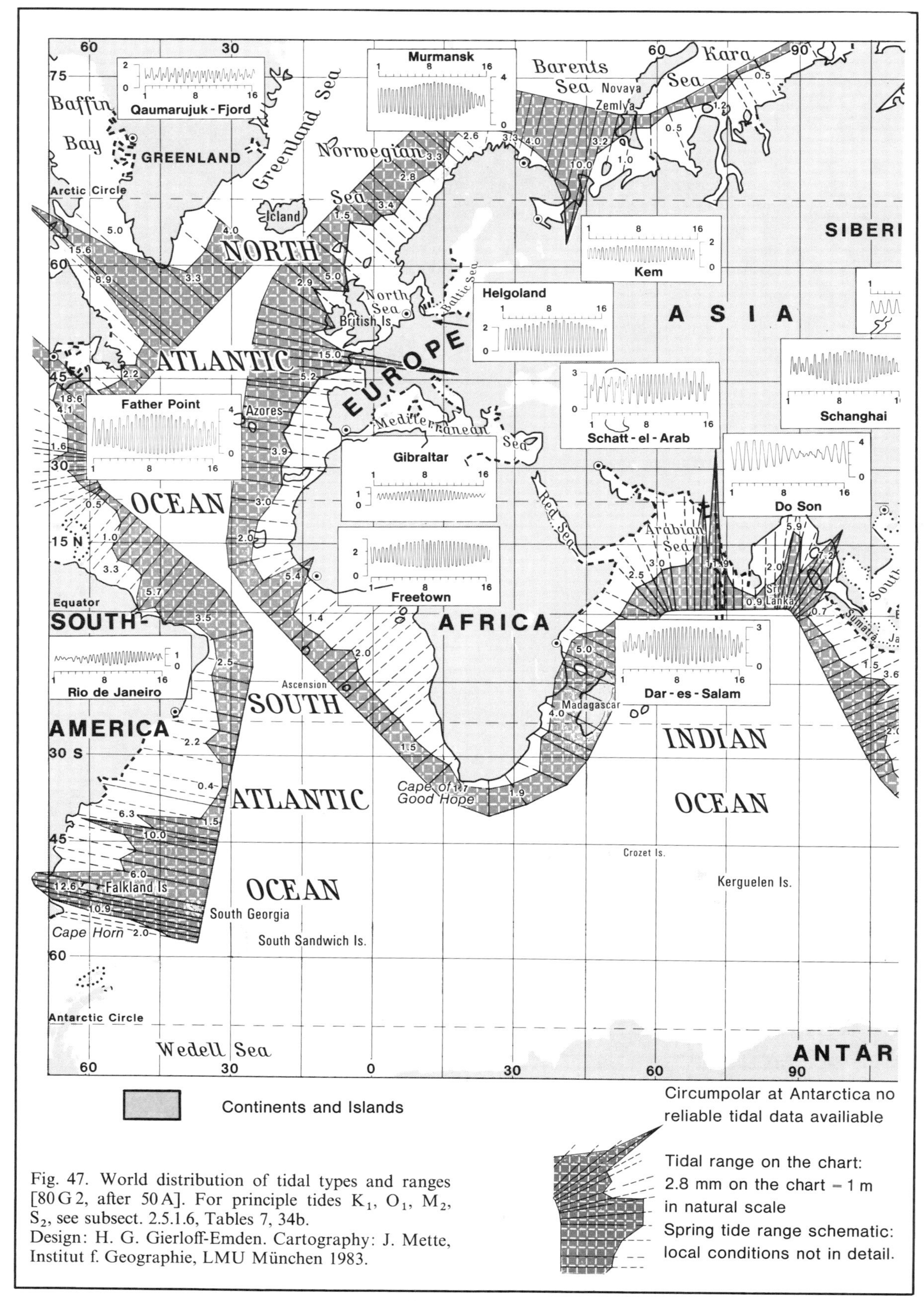

Fig. 47. World distribution of tidal types and ranges [80 G 2, after 50 A]. For principle tides K_1, O_1, M_2, S_2, see subsect. 2.5.1.6, Tables 7, 34b.
Design: H. G. Gierloff-Emden. Cartography: J. Mette, Institut f. Geographie, LMU München 1983.

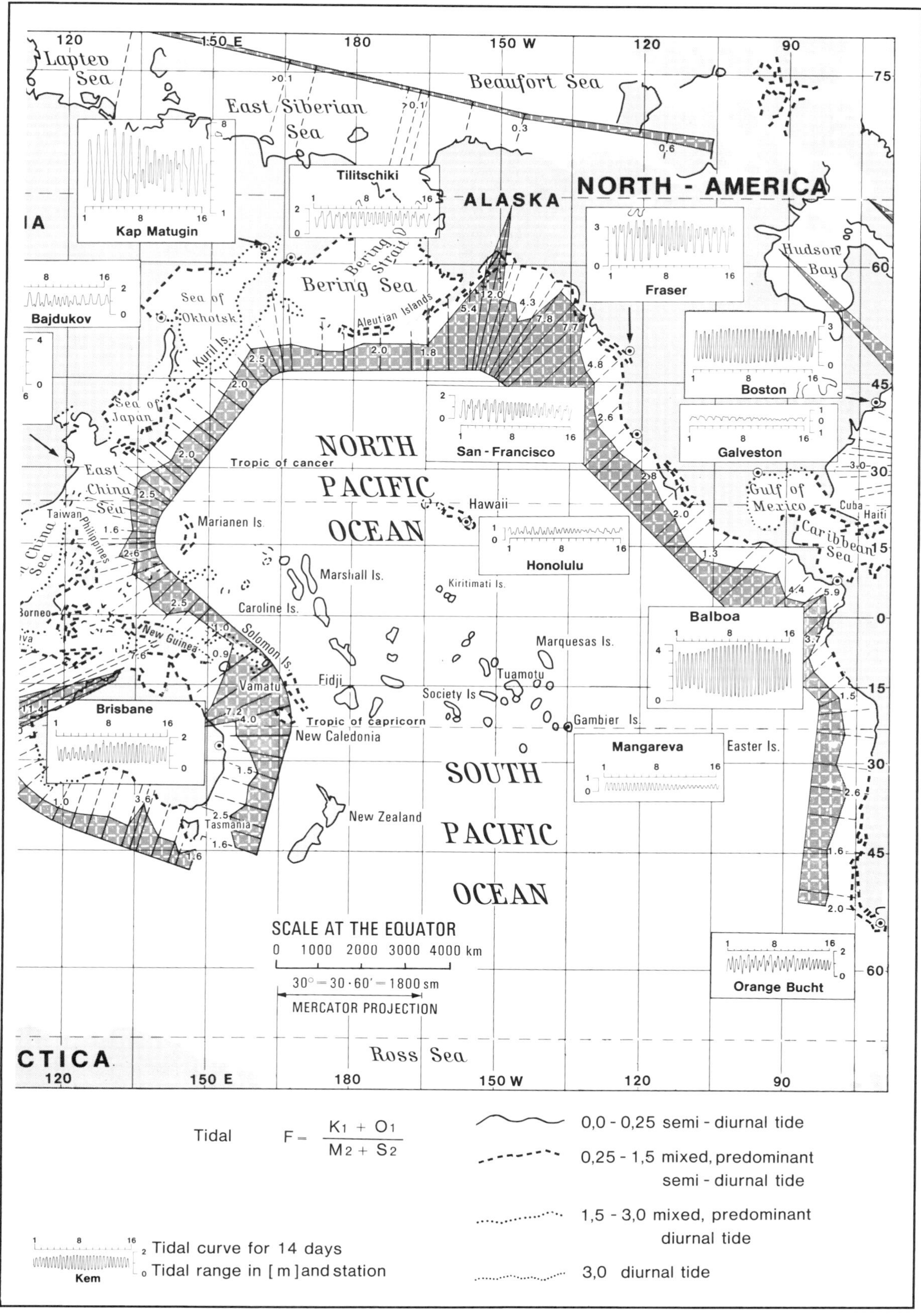

Tidal $F = \frac{K_1 + O_1}{M_2 + S_2}$

0,0 - 0,25 semi - diurnal tide

0,25 - 1,5 mixed, predominant semi - diurnal tide

1,5 - 3,0 mixed, predominant diurnal tide

3,0 diurnal tide

9.6.3 Tidal panorama of a continent: Australia between three oceans

Presentation of the tidal curve of 1/2 month (medium time-scale) as it approaches the coast of a continent simultaneously from the surrounding oceans: Pacific Ocean, Indian Ocean, South-Polar Ocean. (Fig. 48).

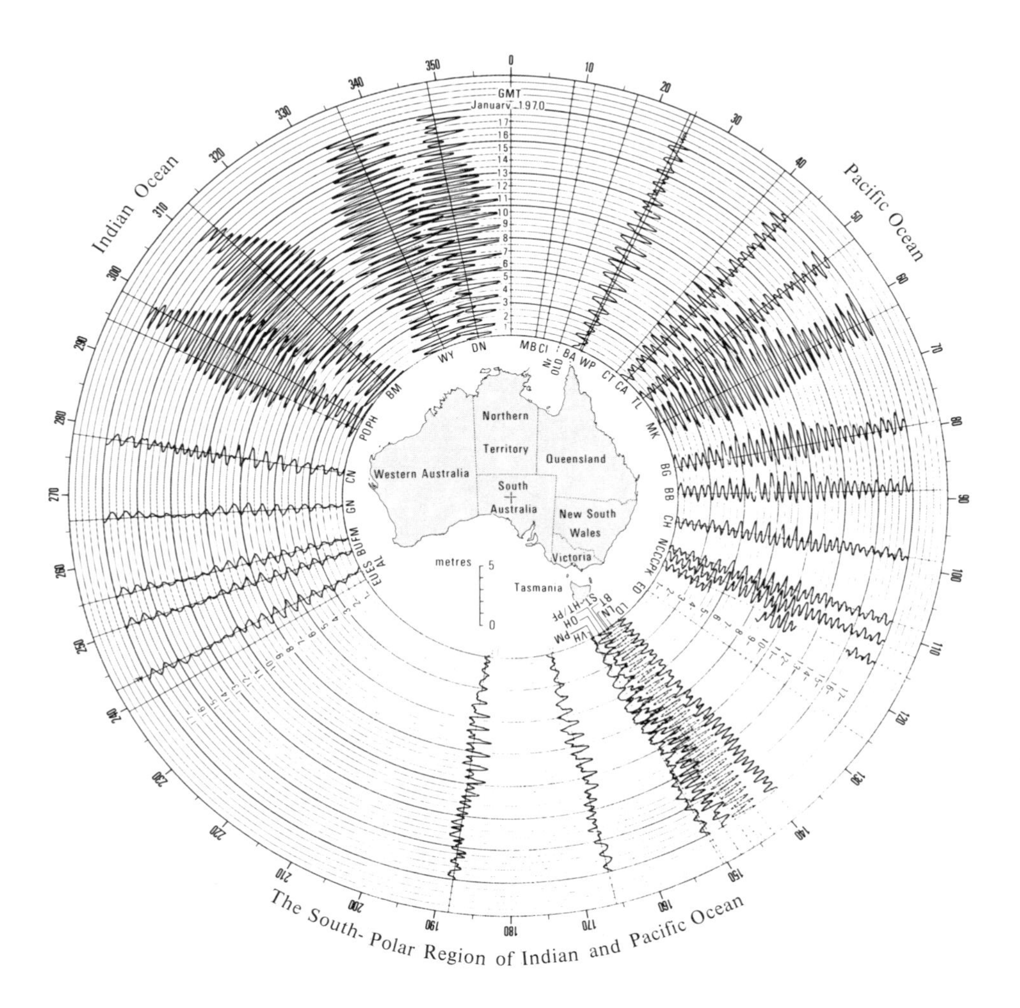

Fig. 48. Tidal panorama of the continent of Australia [76 R 1]. As a typical example, observed tidal ranges for the period 1 to 17 January, 1970 (GMT), a period of a half lunar month tidal cycle with overlapping, are shown for selected stations at which permanent tide gauges are in operation. All heights are referred to mean sea levels computed in most cases from at least three years of records. High tides are plotted in an anticlockwise direction – for example, high tides at Darwin (DN) are drawn to the left of the mean sea level for this station, while at Thevenard (TV) high tides are to the right, unless the plate is inverted. Orientation: Meridional, positive anticlockwise circles equidistant, with demarcate days. Tidal ranges local datum, ranges in [m]. Acronymer at inner circle: Localities of tide gauge. Examples: DN = Darwin, BB = Brisbane, TL = Townsville, BM = Broome, GN = Geralton, FM = Freemantle, AL = Albany, EC = Esperance, TV = Thevenard, NC = New Castle. For the macro-tidal beach of BM (Broome) see Fig. 38.

9.6.4 Tidal ranges in polar regions

The tidal ranges in polar regions are in general less than 1 m. See Table 35 and Figs. 76a, b, c.
Records for the Arctic Ocean are relatively rare but reliable.
Records for the Circum-Antarctic Oceans are not yet present or reliable since extended coasts of the Antarctic continent exist as ice fronts of the ice shelves, which are envolved in tidal range variation as floating bodies.

Table 35. Tidal data at spring tide for points at the North Siberian Shelf [61 D] (according to Sverdrup [27 S]).

Location (old and new name)	Geographical positions		Spring tide		$\frac{S_2}{M_2}$	$\frac{K_1+O_1}{M_2+S_2}$	Age of spring tide
	ϕ	λ	range cm	establish-ment h			d
Point Barrow	71.3° N	156.7° W	15	9.65	0.40	0.46	1.6
Cape Serdze Kamen (Cape Dezhneva)	66.9° N	171.8° W	14	6.73	0.70	0.44	–
Pitlekaj	67.0° N	173.5° W	7	11.71	0.38	0.73	2.3
Ajon Island (Aion)	69.9° N	167.7° E	5	12.33	0.48	0.26	2.0
Bear Island (Medvezhi O.)	70.7° N	162.4° E	3	3.2	0.4	0.6	1.7
Station 3	74.7° N	166.4° E	18	8.2	–	–	–
Station 7	76.5° N	144.0° E	92	5.4	–	–	–
Station 8	76.5° N	141.7° E	210	5.0	–	–	–
Bennett Island	76.7° N	149.1° E	105	6.6	–	–	–
Cape Chelyuskin	77.6° N	105.7° E	34	3.88	0.42	0.40	1.5

The tidal range varies from 210 cm (7 ft) directly north of the New Siberian Islands to only 3 cm (1 in) at Bear Island. The amplitudes of the waves decrease along the co-tidal lines from right to left (taken in their direction of propagation) and also in the direction of propagation of the wave itself. The amplitude drops from 210 cm (7 ft) at the New Siberian Islands to 14 cm (1 ft) near Point Barrow, and in the direction of propagation they drop from 18 cm (1.5 ft) at a point 400 nm off the coast to 5 cm (2 in) near the Aion Island and to 3 cm (1.5 in) near Bear Islands [61 D].
For principle tides K_1, O_1, M_2, S_2, see sect. 6.4 and LB, NS, vol. V/2a, subsect. 2.5.6.1 (Table 7).

9.6.5 Tidal range phenomena at shore areas

See Fig. 49, Table 36.
Tidal range and tidal power plant
The unique large tidal plant in operation (up to 1985) is still the "Rance Work" near St. Malo at the Channel.
Technical data:

tidal range:	max. 13 m
volume basin:	$184 \cdot 10^6$ m^3
area:	22 km^2
dam:	750 m length
construction:	1961···1966
generators:	24
each:	10 MW
annual production of energy:	$544 \cdot 10^6$ k Wh

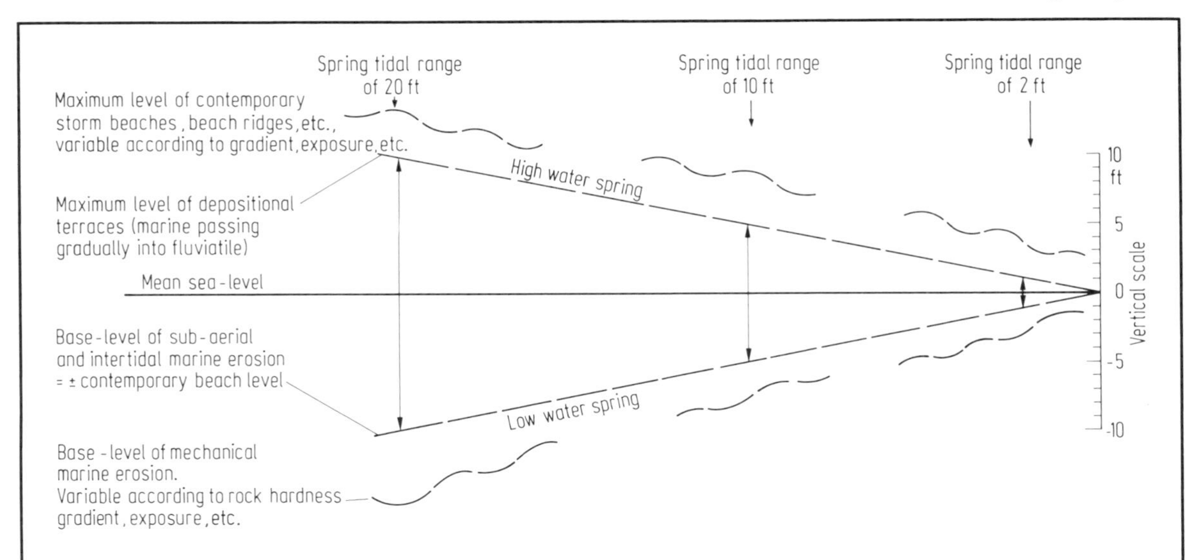

Fig. 49. Tidal range and shore features. Determination of the sea level at any time depends upon recognition of characteristics of intertidal and shore features. Both physiographic and biological zonation of any shore depends upon tidal range and exposure. Base-level of subaerial erosion is mean low water spring tide level: the height of contemporary bench planation, coral reef growth, etc. Maximum height of depositional terraces (marine, estuarine and thalassostatic fluvial) is mean high water spring tide, locally plus a factor associated with floods or storm conditions. Beach ridges build to swash limit, plus a variable capping of eolian character. The greater the tidal range, the greater the limits between these different features, and the greater the margin of error in identifying former sea level stands [61 F 1].

Table 36. Tidal range effect in river estuaries.

River	Country	Tide boundary	Distance from mouth km	Tidal range of mouth (max) m
Elbe	Germany	Geesthacht (barrage)	130	3
Schelde	Belgium	Gent, sluice	90	4.5
Seine	France	Rouen	140	6.5
Thames	England	Teddington (sluice)	110	6.5
Delaware	USA	Trenton Falls	230	1.5
Mississippi	USA	Baton Rouge	400	0.75
St. Lawrence	Canada	Three Rivers	700	4
Amazon	Brazil	Obidos ("mar dulce", $300 \times 600\,km^2$)	850	6
Yangtze Kiang	China	Nanking	300	4.5

Bores

Bore is a phenomenon of interaction between river flow and tide oscillation. The rising tide penetrates into the estuary as a long wave, deformed by a large slope at the advancing front running against the river current. The speed of such a wave is: $C=\sqrt{g \cdot (H+h)}$, where g=gravity, h=river depth, H=wave height. The speed can rise to $>10\,m/s$. [82 M 3].

9.6.6 Coastal seiches

Definition: The term "seiche", supposedly derived from "siccus" (Latin), meaning dry, hence exposed, has had centuries of usage for describing the occasional rise and fall of water at the narrow end of Lake Geneva. See Table 37. A free oscillation in the case of a fluid in an enclosed or semi-enclosed basin is called a "seiche". Where water is the medium, the oscillating system is typified by the shape of the basin and the depth of water within it, while the necessary restoring force is provided by the action of gravity tending always to maintain a horizontal surface of the water [61 D].

Table 37. Coastal seiches in typical gulfs, bays, and harbors; observed modes of oscillation [66 W 2 in 66 F].

Name of gulf, bay or harbor	Country	Observed periods of oscillation (approximate) min					
St. Johns Harbor	Bay of Fundy, Canada	74	42				
Narragansett Bay	Rhode Island, U.S.A.	44	46				
Vermillion Bay	Louisiana, U.S.A.	180	120				
Galveston Bay	Texas, U.S.A.	75					
San Pedro Bay	Los Angeles, California, U.S.A.	55···60	27···30	15	9···11	2···5	
San Francisco Bay	California, U.S.A.	116	47	34···41	24···27	17···19	
Monterey Bay	California, U.S.A.	60···66	36···38	28···32	22···24	16···20	10···15
Hilo Bay	Hawaii, U.S.A.	20···25	10	7			
Guanica	Puerto Rico, Caribbean	45					
Lerwick	Scotland	28···30					
Port of Leixoes	Near Porto, Portugal	20···25	13···15	3···5			
Bay of Naples	Italy	48	17···18				
Gulf of Venice–Gulf of Trieste	Italy	210···240	60	40	10	5	
Euripus (Gulf of Talanta)	Greece, between Is. Euboe and mainland	105	60				
Algiers	Algeria, North Africa	20···26					
Casablanca	Morocco, North Africa	35···40	18···20				
Table Bay, Capetown	South Africa	58···62	38···43	25···30	18···21	14···17	10···11
Algoa Bay, Port Elizabeth	South Africa	69···75	57	42···52	35	20···25	16···17
Tamatave	Madagascar	15	8···10	1···2			
Tuticorin, Gulf of Mannar	India-Ceylon	180					
Bay of Hakodate	Hokkaido, Japan	45···57	21···24				
Bay of Aomori	Honshu, Japan	295	103	23···26			
Bay of Ofunato	Honshu, Japan	41···44	36···39	12···17	5···6		
Bay of Nagasaki	Kyushu, Japan	69···72	54	44···45	40	32···38	22···25
Wellington	New Zealand	28	?	?			
Lyttleton	New Zealand	156	?	?			

9.6.7 Tidal currents

A selected list of tidal currents at the coasts is presented with Table 38.

The vertical profile of tidal currents is explained with Fig. 50.

Measurements of the phases of tide and current in one of the bends of the Patuxent River estuary, Fig. 51, show time relationship: slack of current at extreme tides, maximum strength of both currents at approximately mean water level. This condition seems to be a major requirement for the formation of estuarine meanders, see Figs. 58 and 59. [82 H 2].

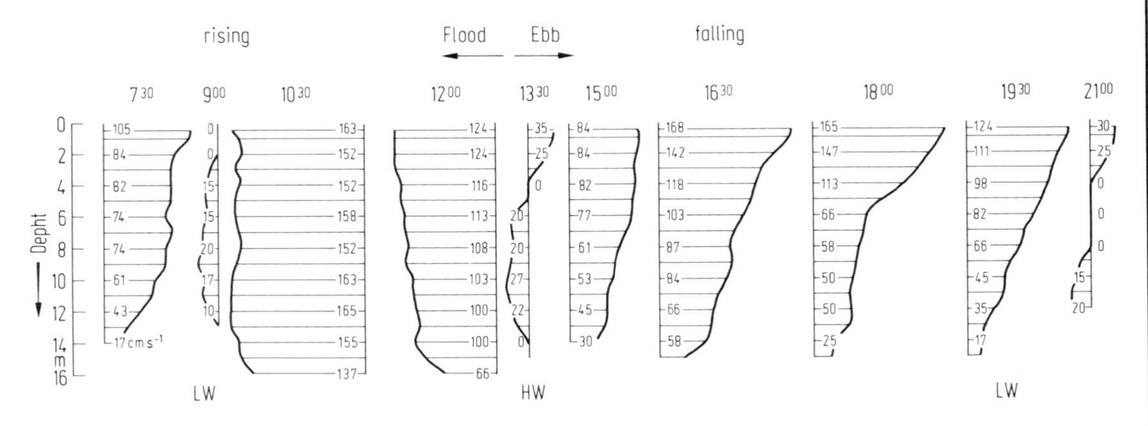

Fig. 50. Current conditions of a water column from the water surface to the sea floor. Shallow tidal area of the North Sea. Observed period: half a day, velocity profiles: velocity in [cm/s]. From [72 L]. From 7.30 h (LW) low water, rising water level, to: 13.30 h (HW) high water, and falling water level, to: 20.00 h (LW) low water, and rising water level. Velocity measurements on May 5, 1967 from the outer Elbe waterway north of Scharhörn. Remark: At times of LW the water column is approximately 2 m lower because of the tidal range [72 L].

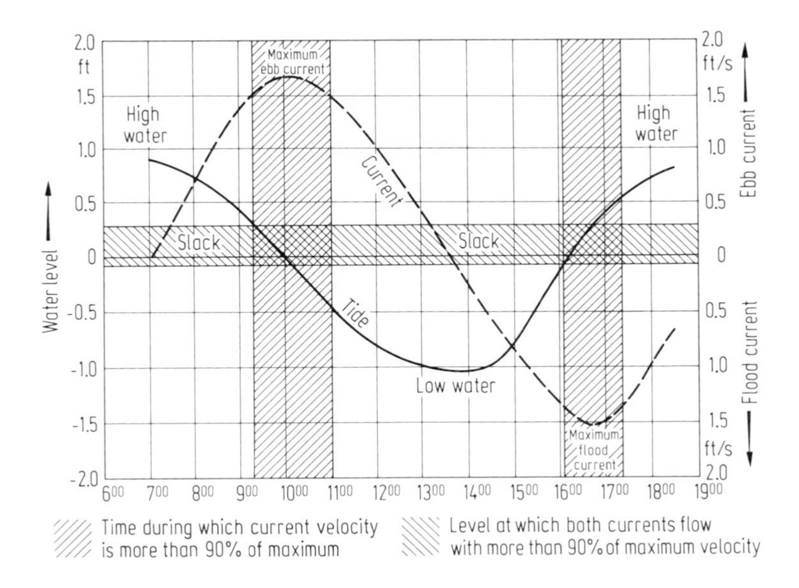

Fig. 51. Tide and current in the Patuxent River at Nottingham, Maryland, USA (April 4, 1957) [60 A].

Table 38a. Tidal currents, maximum velocity in different regions of the coasts [59 S 1].
E: ebb, F: flood.

Continent	Country	Location	Current velocity kn	
Asia	Australia/New Guinea	Torres Strait	8	
	Philippines (Luzon)	San Bernardino Strait	8	E
	Philippines (Mindanao)	Basilan Strait	7	
	Japan	Inlandsea, Naruto	10	F
		Inlandsea, Kurushima	8	F
		West coast Shimonoseki	8	E
North America Pacific coast	Alaska	Aleutian Islands	>10	E
		Cross Sound	> 9	E
		Chatham Strait	>10	E
	Canada	Okisollo Channel	11	E
		Discovery Passage/ Seymour Strait	14	E
		Georgia Strait	11	E
North America Atlantic coast	USA	Fundy Bay	7	E
		Long Island Sound	5.5	E
		East River (New York)	> 5	E
		North Edisto River	> 5.5	E
South America Atlantic coast	Argentine	San José Gulf, Puerto Deseado	8	
		Rio Gallegos, Rio Grande	6	
		Santa Cruz	7	E
		Le Maire Strait	6	
		Falkland Islands	10	
	Chile	Primera Angostura, Magellan Strait	8	
		Canal Gajarao, Magellan Strait	8	
		Angostura Kirne Magellan Strait	10	
Europe			F	E
	FRG	Sylt	2.8	1.9
		Cuxhaven	1.9	5.0
		Outer-Weser	2.9	3.6
	Netherland	Texel	1.4	3.0
	Great Britain	The Downs	3.7	4.0
		Themse mouth	3.0	3.0
		Pentland Firth	> 8	>8
		Orkney Islands	7	7
		Shetland Islands	6···7	6···7
		Solway Firth	4.8	4.0
		Dover Harbour	5.0	2.8
	France	Calais	3.0	3.0
		Roads of St. Malo	6.0	6.0
		Bay of St. Malo (La Rance)	9.0	8.0
	Norway	Saltfjord/Skjerstadfjord	–	16
		Lofotes	> 7	–
		Tromsesund, Skagöysund, Kvalsund, Mageröysund	> 6	–

Table 38b. Tidal current measurements on the North Siberian Shelf. Direction of rotation of current in all cases cum sole [61 D].

Date	Geographical positions		Bottom depth m	Depth of tidal current measurement m	Maximum tidal current			Minimum tidal current	Ratio min. max.
	Latitude ϕ	Longitude λ			cm/s	Against true direction	Phase tidal hour		
8–10 Aug. 1922	71.3° N	175.0° W	76	0···20	16.5	S	9.5	13.0	0.79
20–21 March 1923	74.2° N	169.7° E	50	40	20.0	S 45° W	8.4	18.0	0.90
24–27 Nov. 1923	75.2° N	159.7° E	38	28	9.5	S 55° W	9.1	7.5	0.78
6–9 Feb. 1924	75.2° N	157.7° E	38	28	9.5	S 55° W	8.9	5.0	0.58
17–20 May and 29 May–2 June 1923	74.7° N	166.2° E	56	0···56	3.8	S 55° W	8.2	3.0	0.79
27 Aug.–2 Sept. 1923	76.2° N	164.0° E	64	0···64	6.5	S 15° W	7.5	4.2	0.65
30 June–3 July 1924	76.5° N	144.0° E	35	0···35	16.5	S 10° E	4.8	5.5	0.33
18 July 1924	76.5° N	141.5° E	22	0···22	38.0	S 45° E	3.0	5.0	0.13
1 Aug. 1924	76.6° N	138.5° E	22	0···22	22.5	S 50° E	2.2	13.0	0.58

9.6.8 Coastal zone, tidal process response, models and features

9.6.8.1 Inlets, bedforms, ripples

For inlets, bedforms, ripples, see [61 G 2, 75 G 1, 79 H, 80 K, 60 S, 67 S 1, 65 S, 73 U 1, 73 U 2, 80 U, 84 R, 69 O, 73 S 3, 82 S 4, 82 W 2, 85 D].

A wave dominated barrier inlet is characterized by sand bodies exclusively on the landward side of the inlet gorge. The gorge itself is relatively stable, the ebb tidal delta (outer shoal) is present only as a minor subtidal shoal, frame 1 in Fig. 52, see also Fig. 53.

Mixed energy, high tide range, inlets:

The gross differences in morphology appear to reflect primarily the larger tide range, frames 3 and 4 in Fig. 52.

Inlets with a large ebb-tidal delta with a nearly continuous arc of swash bars along its margin, the "reef-bow": frames 4 and 5 in Fig. 52.

The high wave energy appears to cause rapid swash bar migration contributing to the instability of the seaward end of the main ebb channel.

The inlets are very wide. Typical are multiple channels as effect of Coriolis force on tidal currents, frames 4 and 5 in Fig. 52.

River inlets have estuarien tidal feature, multiple channels, wave action is not of importance, frame 6 in Fig. 52.

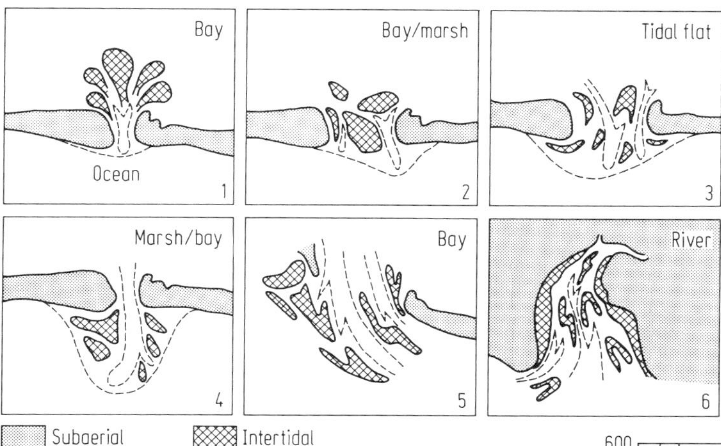

Fig. 52. Tidal inlet morphological models with sand bars. Frames 1 through 6 reflect an increasing role of tidal currents to wave domination in inlet sediment dispersal [78 N 2]. Model not to scale. See also Fig. 72 and Tables 45, 46. Compare [61 G 2, 85 D].

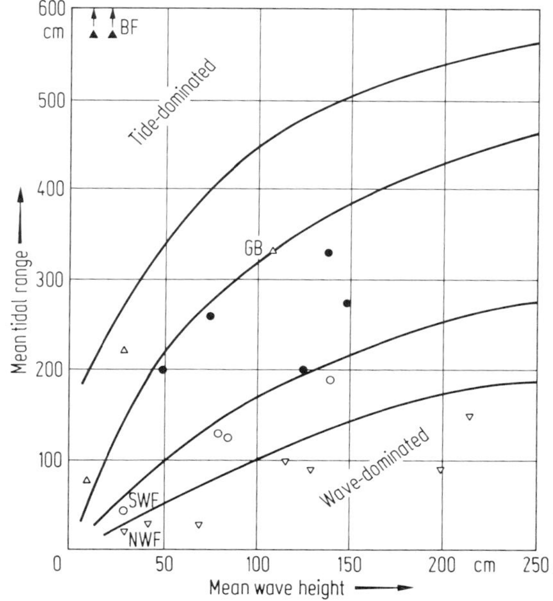

Fig. 53. Inlet morphologic types as functions of mean annual wave height and tide range. 19 barrier island coasts in North America and Western Europe were used to establish this diagram with classification boundaries. Modified [83 K 3], after [78 N 2]. BF = Bay of Fundy, GB = German Bight, SWF = Southwest Florida, NWF = Northwest Florida. Solid triangles: tide-dominated high; open triangles (upward): tide-dominated low; solid circles: mixed energy tide-dominated; open circles: mixed energy wave-dominated; open triangles (downward): wave-dominated.

9.6.8.2 Tidal prism and tidal basin

The incoming flood currents fill the tidal basin with tidal waters, e.g. estuaries, river mouths, bays, wadden basins. The filled volume has the shape of a prism, the tidal prism (Figs. 54, 55, and 56).

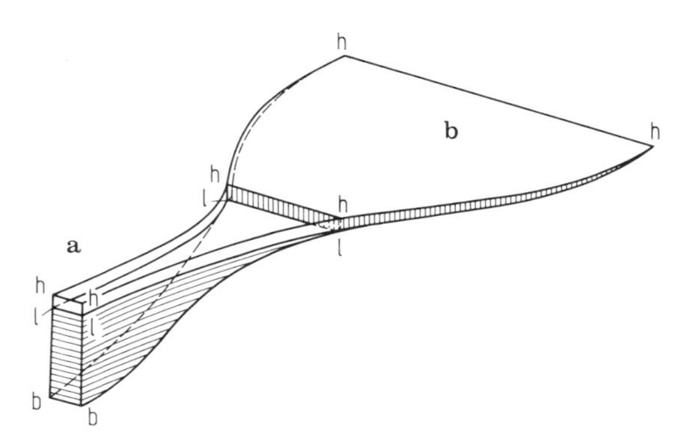

Fig. 54. Tidal prism: volume of the waterbody within the limits hh and ll. A large part of the water of the wadden sea is alternately spread out horizontally (b) and concentrated in narrow deep channels (a), h···h: high tide level, l···l: low tide level, bb: bottom of the tidal channel, ll(middle)···hh(right): bottom of the tidal flat [76 Z 1].

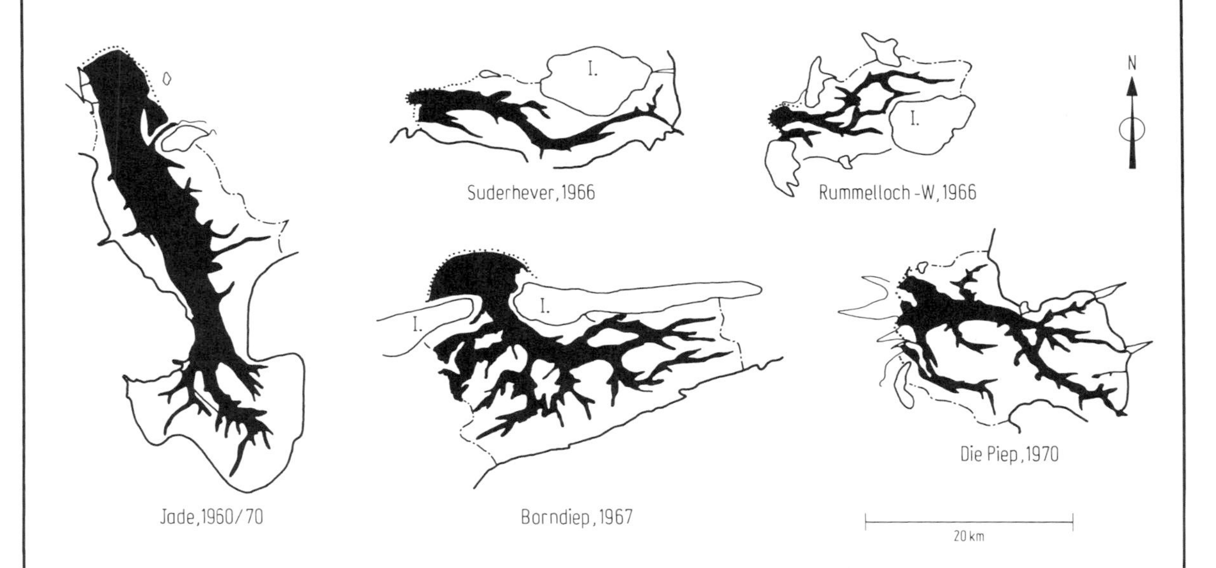

Fig. 55. Tidal basins of different type and size: tidal flats, estuary and tidal channel of the German Bight, North Sea [76 Z 1]. Black: Low water fill, permanently submarine. I. = island.

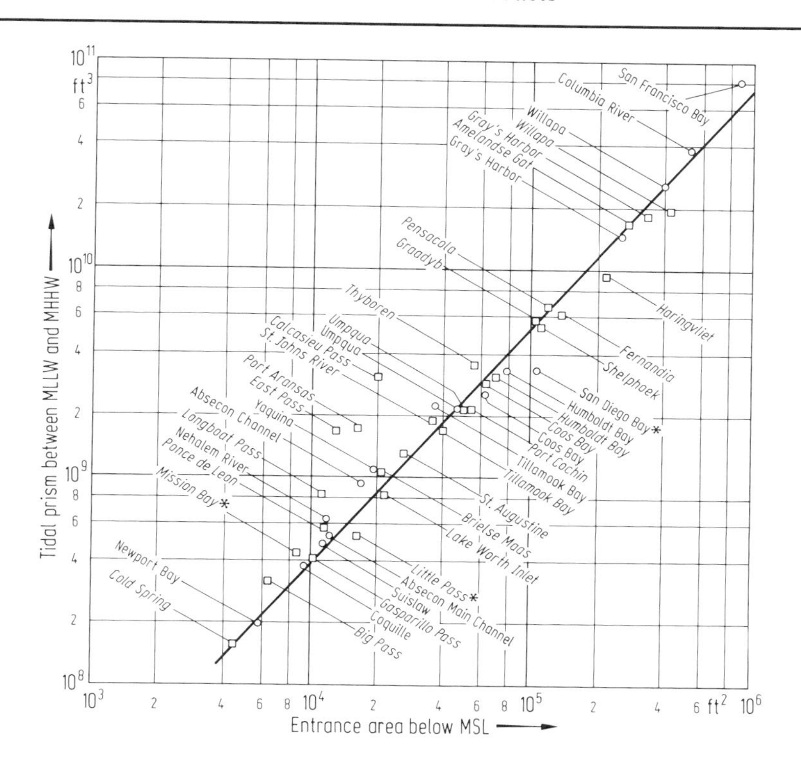

Fig. 56. Relationship between tidal prism and entrance area of different lagoonal bays and harbors. After [64 W]. ○ O'Brien (1931). □ Bruun and Gerritsen (1958). Tidal prism for estuaries with diurnal tides based upon mean tide range. * Entrances dredged.

Definitions [66 B 1]:

Tidal prism: The difference between the mean high water volume and the mean low water volume of an estuary.

Tidal prism method: A theoretical procedure for determining the flushing time of a harbor or estuary. The method assumes that the contaminant is initially distributed uniformly throughout the harbor or estuary, and that during each tide cycle a volume of water and contained contaminant equal to the tidal prism is removed from the harbor and replaced by a new volume of sea water which mixes completely and uniformly with the water present in the estuary at low water. Therefore, the amount of contaminating material removed on each tidal cycle may be expressed as a percentage of the contaminant in the harbor during the previous tidal cycle:

$$\frac{\text{tidal prism volume}}{\text{high water volume of harbor}} \cdot 100 = \text{percent of contaminant removed from harbor.}$$

In order to fill and drain a part of the tidal areas south of the islands of Juist and Norderney, North Sea, $200 \cdot 10^6\ m^3$ of water have to pass the tidal channel between the two islands in the course of one normal tidal period. The water movement which occurs twice per day in the Jadebusen is even larger. This bay of approximately $200\ km^2$ is flooded by ca. $450 \cdot 10^6\ m^3$ of North Sea water, and then drained again during the following ebb tide.

9.6.8.3 Tidal hydrodynamics: tidal meander channeling and classification

See Figs. 57, 58, 59, Table 38c.

Tidal creeks are normally sinuous with parallel or subparallel banks and relatively high depth/width ratios. Tidal creek meanders often differ from riverine meanders in that the bends of the former tend to be angular rather than rounded. Tidal creeks mostly widen progressively to the mouth. [76 K 1, 77 C 1, 78 F, 74 C, 83 P 1, 81 T, 76 C 1, 68 V 1, 75 U].

The tidal creeks develop in plain view as dentritic network in the intertidal flood plain of the wadden and saltmarsh area. The tidal creeks are tributaries to the main channels of the estuarine meanders. Quantitative analysis of channel numbers and channel lengths of tidal creeks. The tidal creeks number increases upward the length of 10···20 m in general 4 times from one order to the next.

Table 38c. The numbers and mean lengths of tidal creeks (Strahler's system) in 6 different network regions of the Chonsu Bay, South Korea [83 P], out of an area of 10 km^2, mean tidal range 3.3 m, mean spring tidal range <6 m.

Region	I		II		III	
Order \ Network	Number	Mean length [m]	Number	Mean length [m]	Number	Mean length [m]
1	1104	45.78	635	35.35	1003	37.24
2	288	54.60	189	59.52	274	60.22
3	69	69.71	50	102.50	63	92.46
4	20	87.08	11	327.27	12	295.83
5	4	81.41	2	537.50	3	816.67
6	1	117.12	1	850.00	1	450.00

Region	IV		V		VI	
Order \ Network	Number	Mean length [m]	Number	Mean length [m]	Number	Mean length [m]
1	813	34.33	683	67.25	572	83.73
2	201	63.37	200	110.64	146	178.35
3	55	110.45	52	168.37	33	280.88
4	6	204.17	8	240.63	5	339.05
5	3	666.67	2	1362.50	2	457.1
6	1	550.00	1	375.00	1	463.00

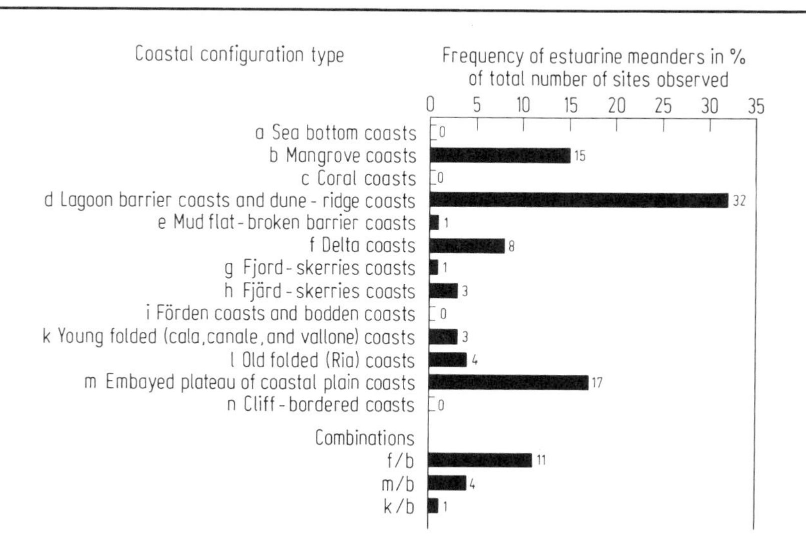

Fig. 57. Relative frequency of estuarine meander sites on different coastal configuration types, after [63 A].

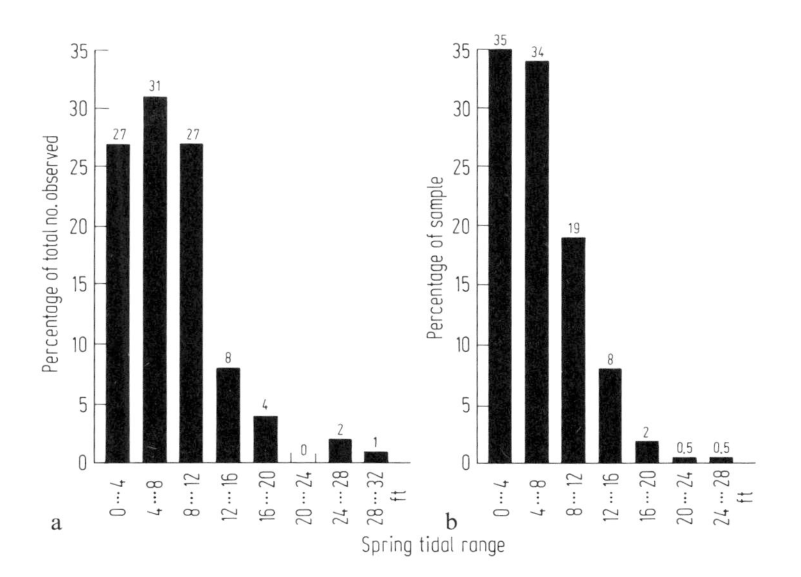

Fig. 58a, b. The frequency of estuarine meanders of the world's coasts in relation to spring tidal ranges. a) Frequency of estuarine meander sites in relation to spring tidal range (total number of observed sites: 358 of the world's coasts). b) Frequency of spring tidal ranges (sample of 400 tidal stations). From: Tide tables, US-Coast and Geodetic Survey [63 A], with the stations spaced equally throughout according to their sequential position in the tide tables.

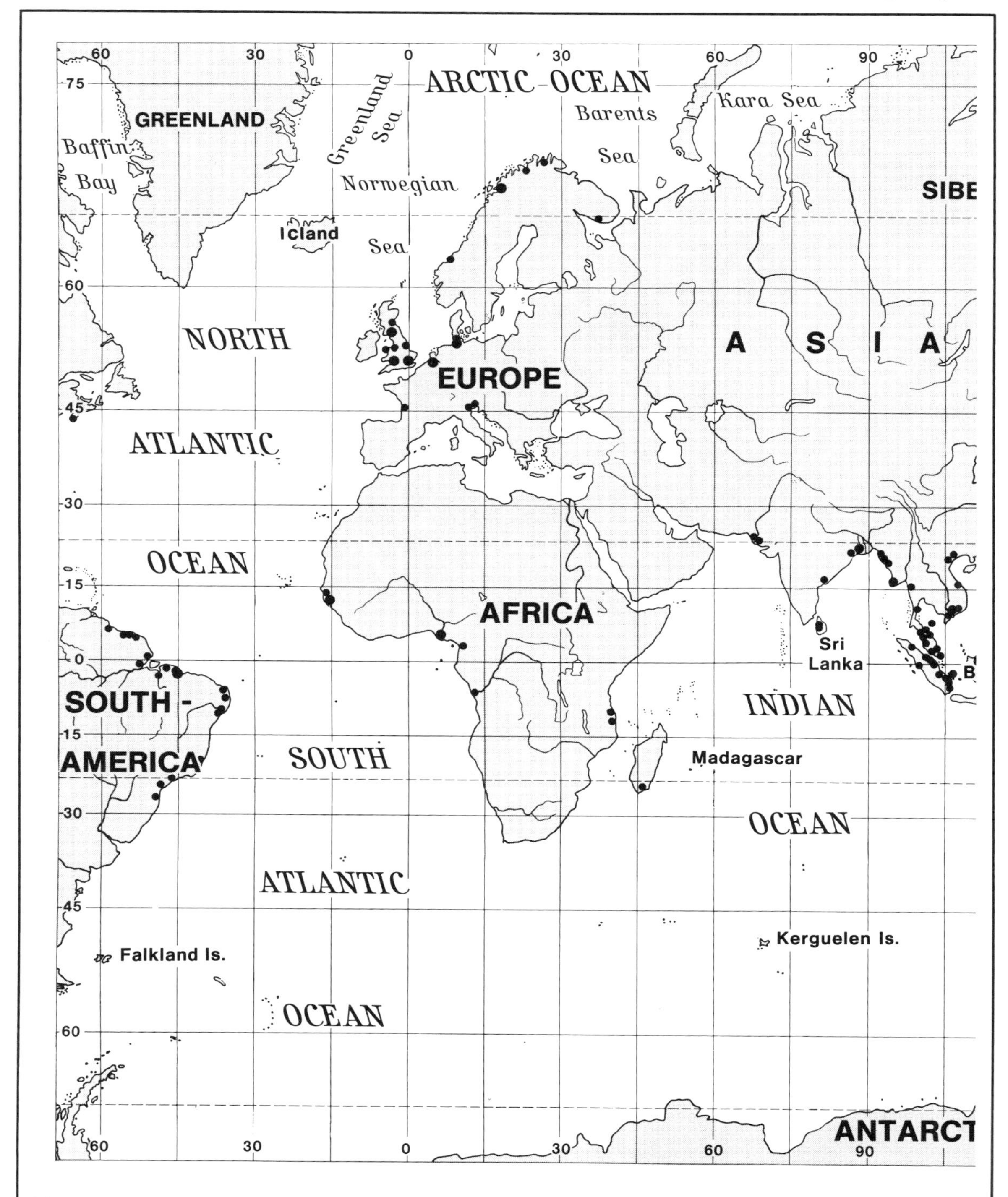

Fig. 59. Distribution of estuarine meanders in tidal marsh regions of the world. Most sites are scattered in the regions: 1. Atlantic Ocean coast of the USA, 2. Pacific and Indian Ocean coast of South Eastern Asia, 3. Atlantic coast of South America, 4. Pacific coast of Australia. Drawn after [63 A].

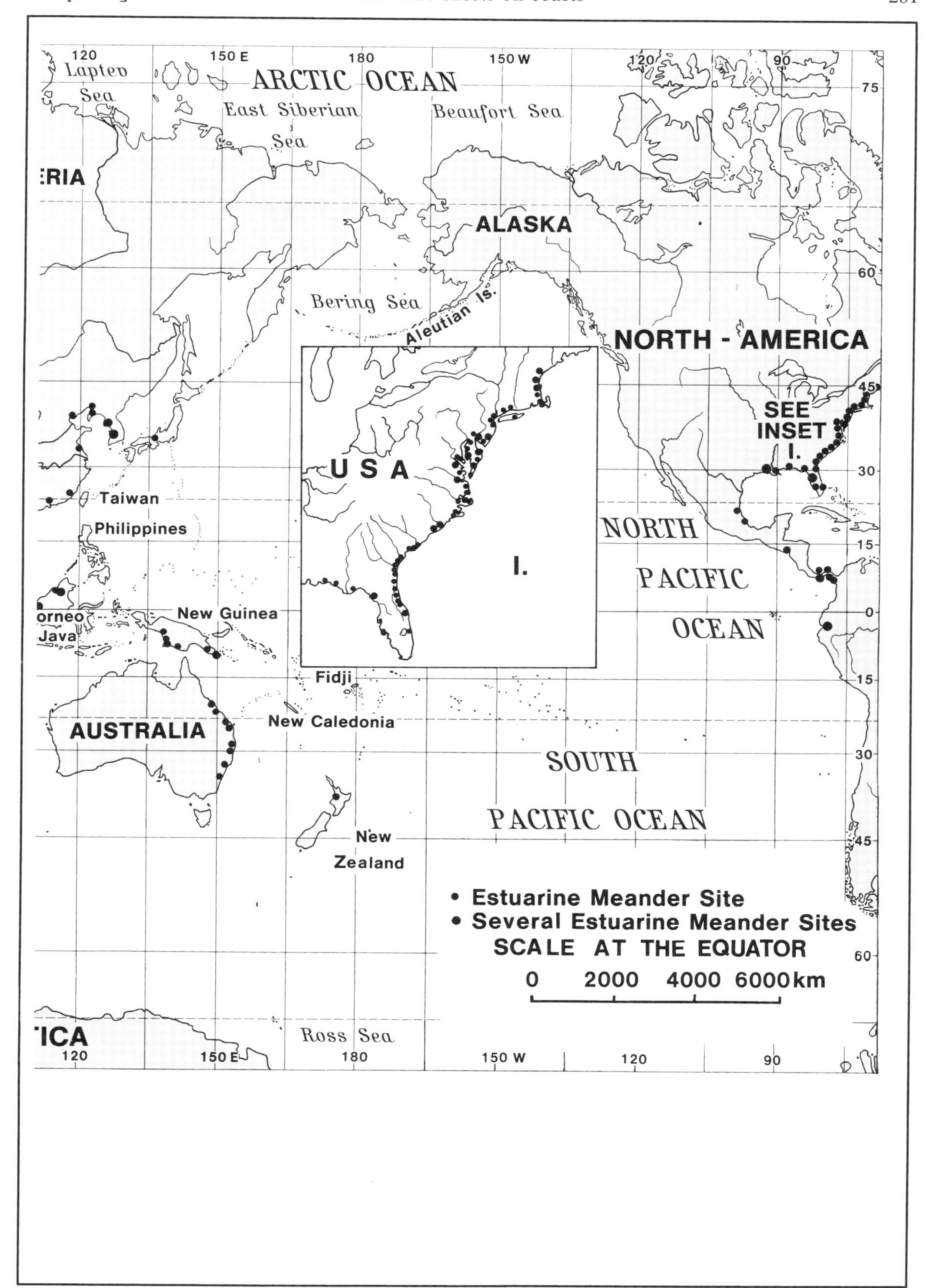

Gierloff-Emden

9.6.8.4 Exposure and sediments of the wadden sea bottom

See Figs. 60a, b, 61, Table 39.

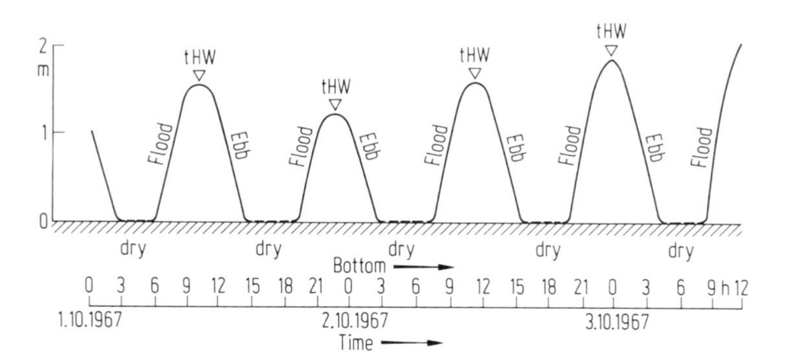

Fig. 60a. Water cover, ebb and tidal curve over wadden, and inundation time and aerial exposition of the wadden areas. Tidal flats exposition time interval due to lunar time versus solar time. Ebb = tLW = tidal low water, flood = tHW = tidal high water [73 H].

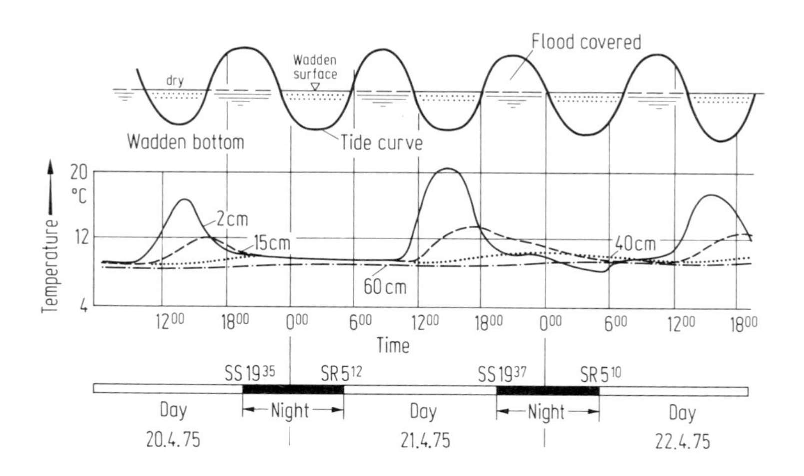

Fig. 60b. Variation of temperature in humid brown sand wadden soil influenced by tide periods, yearly season and depth of data measurement during slightly clear to cloudy weather. SS: sunset, SR: sunrise [84 W 2].

Table 39. Time-scale as interface of satellite data acquisition systems against coastal water and tidal region processes (Gierloff-Emden). [82 G 1]. Compare [80 R 1].

Time scale facing
- Remote sensing system dependent time-scale parameters MSS-Landsat
- Object-natur processes dependent time-scale parameters (coastal environment)

Time-scale	Time-scale	Processes		micro		meso		macro
			Time-scale	seconds	minutes	hours – days	weeks	months (year)
			Processes	remote sensing bounded				
			→ ↓	6–30 s (orbit) =Nadir path.	30 s scanning process data acquisition	1) overlapping 2) fixtime: 9^{00} (orbit)	18 days repetition multi-temporal	4–8 months repetition 1) singularity 2) multitemporal
micro	minutes – seconds	processes	0–15 s illuminat.	hot spot location				
micro	minutes – seconds	processes	8–15 s waves 30–12° s rip current 50–80 s longshore current (unsteady)	tidal water cover. in minutes distances over 1 pixel – 2 p. – different spectral cond. of pixel due to scanning pr.	breaker zone water line kinematic (spectral change)			
meso	days – hours	object	12^{25}–24^{50} h tidal periodic high-w-low-w. bidiurnal-diurnal tideperiodic 24 h + 50 min versus 24 h day – night 1–24 h wind 3–6 h biology			tidal flats: tidal coverage waterline translation (spectral change) coverage exposure =colour of alges (biological clock)		
macro	months – weeks	bounded	tidal range 2 weeks – spring tide nipp tide (tidal expos. time)				tidal coverage waterline spring-tide nipp-tide salt flats evaporation (spectral change)	
macro	months – weeks	bounded	climatic period summer – winter vegetation freezing					vegetation types contrast, discriminating different (spectral change) ice: albedo
macro	months – weeks	bounded	tidal date $18^1/_2$ year					coverage-multitemporal

The detection of objects and of boundaries requires the analysis of the following relationships: 1) sensor versus spectral resolution; 2) object versus spatial resolution; 3) sensor system versus time scale (repetition of satellite); 4) image versus visual resolution.

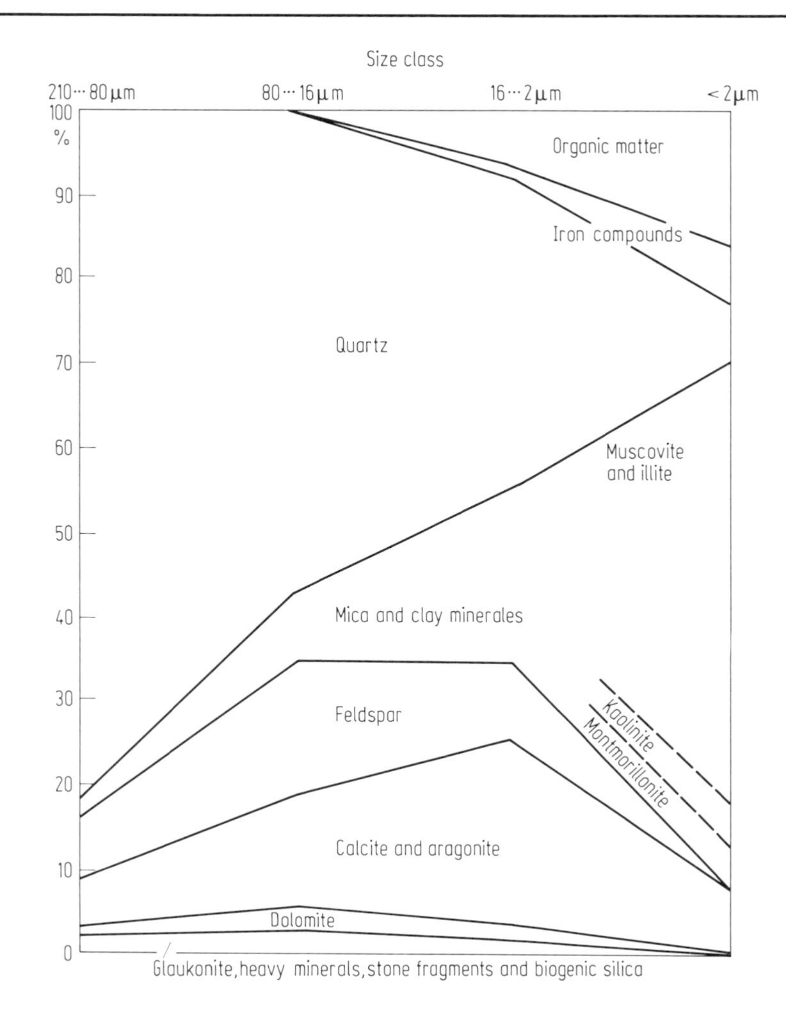

Fig. 61. Average mineralogical composition of wadden sea sediments. Data are presented for four size classes: 210···80 μm, 80···16 μm, 16···2 μm and < 2 μm. After [64 V]. Compare [75 G 1, 75 R 1, 75 R 2].

According to Kubiëna (1953) the wadden sea sediments belong to the marine gyttjas. Below an oxydized layer with a thickness of millimeters in mudflats or of centimeters in sandflats an anaerobic reduced layer occurs. The upper layer shows the brown colour of iron hydroxide whereas the lower layer looks bluish-black due to the presence of reduced iron in the form of $FeS \cdot nH_2O$. Burrows of bottom organisms in the latter layer mostly show tubes with oxydized walls. Below the black layer the pyrite zone with light grey colours (due to FeS_2) occurs. Pyrite is found in sizes between 2 and 25 μm diameter. [64 V].

9.7 Estuaries and lagoons as coastal water bodies

9.7.1 Hydrographic conditions

Definition: "An estuary is a semi-enclosed coastal body of water which has a free connection with the open sea and within which sea water is measurably diluted with fresh water derived from land drainage" [67 P].

"Estuaries, in contrast to lagoons, are characterized by poikilohalinity and the instability of environmental factors" [67 P, 67 G 2, 72 N 2, 78 N 1, 74 P, 52 S, 67 L, 80 B 1, 77 G 1, 77 C 1, 79 D 2, 67 E, 82 G 1, 63 G, 77 K, 72 N 1, 52 P 2, 56 S, 76 W 2, 82 K 2, 77 M 2, 79 L 1, 69 P 1, 69 P 2, 72 N 1, 75 K 2, 76 S 1, 78 O 2, 80 O 2, 81 O, 83 K 2, 85 N]. Tables 40 and 41, Fig. 62.

The mean excursion of the water at the mid-point of the gradient is illustrated in Fig. 62 by the horizontal arrow, and the extent of the salinity fluctuation between high and low tide is shown by the vertical arrow. The latter is derived from excursion and the slope of the gradient.

Table 40. Features and parameters of estuarine fronts.

Feature	Unit	Estuarine		Coastal wind and current induced
		tide and bottom related	tidal wedge	
Location		upper and lower bay	lower bay	coast and shelf
Frontal alignment		parallel to river flow axis	perpendicular to river flow axis	any direction
Transverse velocity	$cm\,s^{-1}$	5···20	10···60	5···40
Transverse movement	km	±0.3	>10	>10
Convergence velocity	$cm\,s^{-1}$	5···40	2···20	2···20
Shear velocity	$cm\,s^{-1}$	5···20	1···5	1···15
Secchi depth ($m_1 \rightarrow m_2$)	m	0.4→1.6	1.0→2.2	3→8
Color change		strong	moderate	moderate/none
Foam, oil and detritus capture		strong	moderate	moderate/weak
Wave refraction and damping		strong	moderate	moderate
Temperature change	°C	0.3	0···2	0···2
Salinity change	‰ S	0.5···3	1···4	0.5···2

Table 41. Salinity of the water in estuaries and the marsh [73 S 3].

30···45% Euryhaline	marine zone:		
18···30% Brachyhaline	sea water	salt marsh	20%
10···18% Polyhaline	gradient zone:		
3···10% Mesohaline	brackish water	brackish marsh	
0.5···3% Oligohaline			0.5%
	fresh water zone:		
0.5‰	fresh water	fluvial marsh	< 0.5%

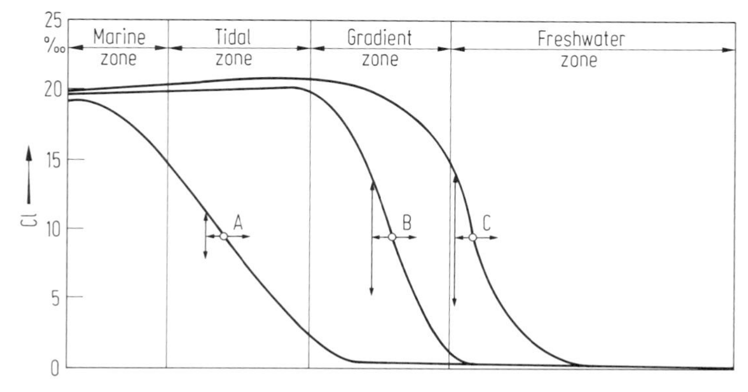

Fig. 62. Zonal distribution of chlorinity Cl in a schematic estuarine system in Australia (after Rochford, 1951). From [83 K 2]. A: Flood gradient; B: Equilibrium gradient; C: Drought gradient.

9.7.2 Hydrographic classification of estuaries

The concentration and the vertical distribution of the salinity differs from estuary to estuary; thus it is possible, to classify estuaries on the basis of their salinity layers. Four main estuarine types can be distinguished [75 D 1, 52S]:

type A completely vertically mixed
type B moderately vertically layered
type C strongly vertically layered
type D salt wedge.

For each of these types, the salinity distribution is schematically represented in Fig. 63, i.e. for the four stations 1, 2, 3, and 4, whose position in the estuary is indicated on the top figure. On the left, the vertical curves of salinity are shown for all stations; on the right, the longitudinal sections of salinity from the surface to the bottom and from the head to the mouth of each estuarine type are shown.

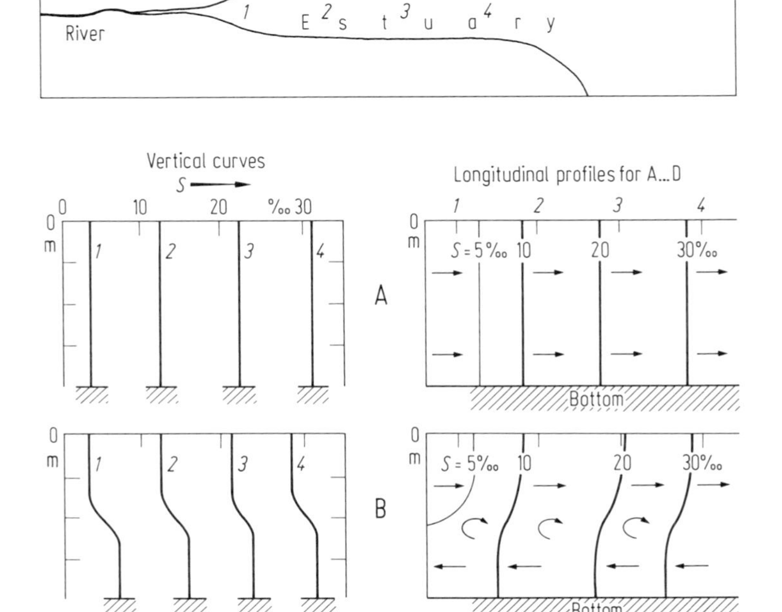

Fig. 63. Schematic cross section of the salinity layers of the four main estuarine types [75 D 1], S = salinity, A: Vertically mixed. B: Weakly layered. C: Strongly layered. D: Salt wedge. Top: Positions of the stations in the estuary. Arrows: Estuarine circulation.

Type A: Unlayered and shallow depths, increase of salinity from head to mouth, this means that, considering the terrestrial discharge, at all depths water is transported seaward, as indicated by the arrows in Fig. 63. The salt concentration migrates landward by horizontal exchange.

Type B: Always or temporarily two layers on the average, outflow in upper and inflow in the lower layer. The vertical mixing, indicated by semi-circles along the transitional layer in Fig. 63, causes the increase of salinity from head to mouth. There is, especially during high water, a distinct salinity difference between surface and bottom, which proves the upstream movement of salt water.

Typ C: Always two layers, usually great depths, fjord type. Increase of salinity in the thin top layer, but almost constant salinity in the thick bottom layer from head to mouth of the estuary. Outflow in the top-layer, inflow in the bottom layer. Vertical mixing inbetween, which affects only the upper part of the bottom layer. All Norwegian fjords, the fjords of Greenland and of the Canadian west coast – the so-called inlets – belong to this type. A special feature of most fjords, are so-called fjord bars, i.e. a ridge crossing the fjord bottom near the mouth. They may exert a considerable influence on the fjord circulation by impeding the typical full circulation, thus permitting only partial circulation [83 F 2].

Type D: Two layers, salt wedge. Examples can be found in estuaries of rivers with low vertical mixing and tidal currents, such as on the Mississippi, but also in long and shallow channels with fresh-water supply at the head.

Ria estuaries:

Rias are defined as river systems partly or wholly flooded by the sea. The degree of drowning depends on the magnitude of the base-level movement and on the altitude of the river's source. The subaerial origin of rias is demonstrated by the occasional existence of incised meanders as on the Aulne at Landevennec in the Rade de Brest. The total length of the coast largely exceeds the length of a smooth line through its headlands. Flooding results from the Flandrian transgression, but also from subsidence, especially for the Pacific islands, as, e.g., in New Caledonia. [58 G, 68 S 2, 76 G, 68 G, 82 G 4].

9.7.3 Flushing time

The *flushing time* is the time required to replace the existing fresh water in the estuary at a rate equal to the river discharge, see subsect. 9.6.8.2 and Table 42. The flushing time $T = Q/R$ where Q is the total amount of river water accumulated in the whole or a section of the estuary and R is the river flow. In New York Bight the flushing time varied between 6.0 and 10.6 days with the river flow [83 K 2]. See also Figs. 54, 55 for tidal prism method.

The fraction of fresh water methods

The mean fractional fresh water concentration over any segment is

$$f = \frac{S_s - S_n}{S_a}$$

where S_s is the salinity of the undiluted sea water and S_n is the mean salinity in a given segment of the estuary. The total volume Q is found by multiplying the fractional fresh-water concentration f by the volume of the estuary segment. Calculating the flushing time of the Mersey Narrows by this method and at a river discharge of 25.7 m^3/s the flushing time was 5.3 days. Larger river flows decreased the flushing time [58 P].

Table 42. Flushing time (T), volume (V), and dilution potential ($1/f$), for various rivers, estuaries and coastal regions After [83 K 2].

Region	T d	V $10^6\ m^3$	$1/f$
Raritan River (New York, U.S.A.)	8	36	2
Houston ship channel (Texas, U.S.A.)	56	67	1
Delaware River (U.S.A.)	28	1 130	2
Boston Harbor (Mass., U.S.A.)	2	67	24
Mersey Estuary (England)	4	147	7
Raritan Bay (New York, U.S.A.)	14	800	12
Passamaquoddy Bay (N.B., Canada)	16	4 700	41
New York Bight (U.S.A.)	8	23 000	63
Bay of Fundy (Canada)	76	880 000	52

9.7.4 Meteorological and climatological stress on estuaries

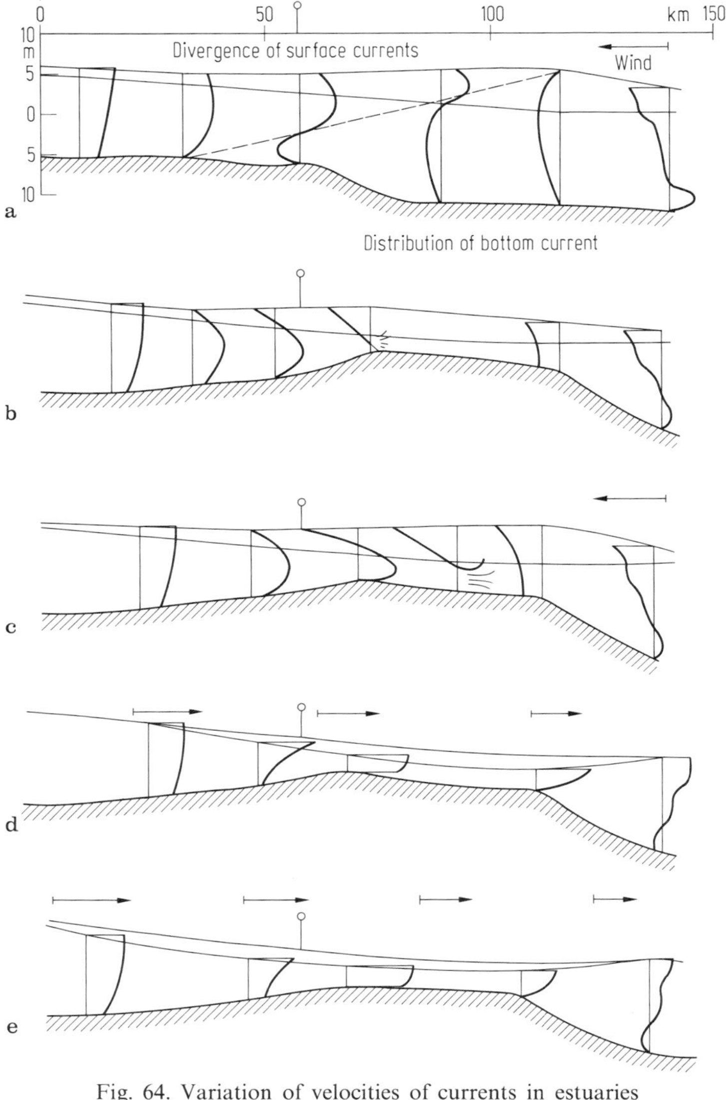

Fig. 64. Variation of velocities of currents in estuaries caused by wind. Vertical profiles [56 S]. a: Wind to the land, deep coastal waters. b: Wind to the land, flat coastal waters, maximum at delta. c: Wind to the land, flat coastal waters, maximum at slope. d: Wind to the sea, with maximum negative deformation of sea level in coastal waters. e: Wind to the sea, with maximum negative deformation of sea level in river mouth.

9.7.5 River estuaries

The river, by definition, enters the estuary at its "head". The head may be defined as that region where the surface of the river first comes to approximately sea level, or the riverward extremity of detectable admixture of salt water. One defines the "inner end" of the estuary as the section above which the volume required to raise the level of the water from low to high water is equal to the volume contributed by the river during a tidal cycle [63 A, 79 E 3, 70 L 1]. Very intensively investigated are the estuaries: Elbe, Gironde, Loire, Schelde, in Europe, and the St. Lawrence [79 E 3] in Canada, and the Cheasapeake Bay system with rivers in USA (Table 36).

Surfaces of spring and neap tide conditions in an estuary, Fig. 66:

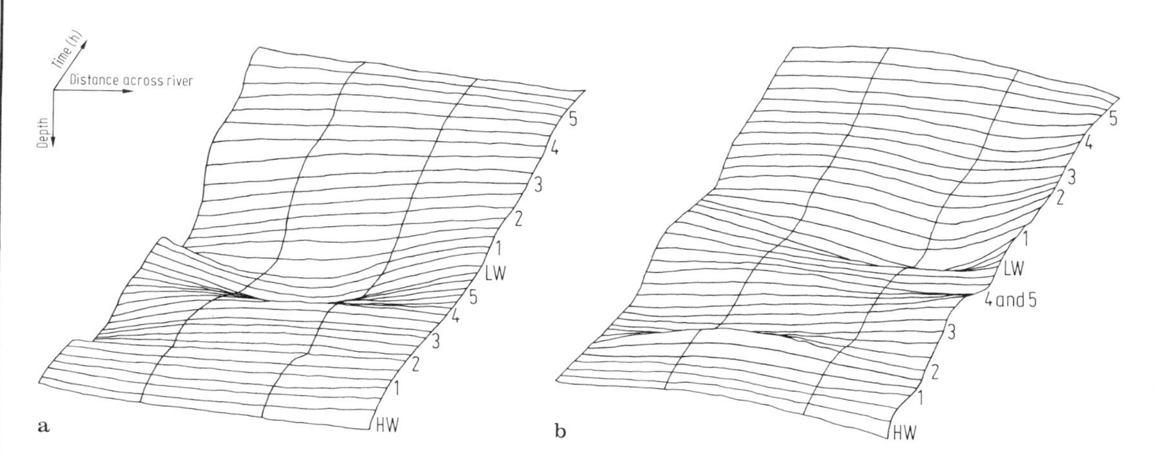

Fig. 66. Isopycnal surfaces passing a section as a function of time for spring and neap tide conditions [79 M 3]. The section is located in the middle estuary of the St. Lawrence River, i.e. 150 km downstream Quebec, 400 km from the mouth of the St. Lawrence Gulf. Parameters of the section: width: 20 km, depth: 10···70 m, mean tidal range: 4 m. The estuarine water surface is deformated by tidal effects (waves) causing large vertical amplitudes, which can cause strong currents. a) Isopycnal surface depth vs. time at section spring tides. b) Isopycnal surface depth vs. time at section neap tides.

9.7.6 Fjord estuaries

Fjords are, according to the hydrologic conditions, a type of estuaries. (Compare type C of the classification Fig. 63).

The terms "fjord" (Danish and Norwegian), "fjard" (Swedish), "Förde" (German), "fjadur" (Icelandic), and "firth" (Scottish) are all linguistic variants of the same word, and refer to coastal inlets of various forms and origins possessing the common characteristic of relatively great length compared with width. Table 43.

General description and oceanographical classification of fjords.

Models of overflow of the sills are presented in [83 F 2].

One of the more frequently quoted classification schemes is that of Hansen and Rattray (1966). It is derived under certain mathematical constraints on the hypothesized salinity and velocity profiles, and on assumptions about vertical and horizontal diffusion, so as to fit their similarity solutions. The scheme is based on two parameters, representing the circulation and the stratification. Different estuarine regimes occur for different values of the parameters. The classification is shown in Fig. 68a. The surface velocity $\bar{U}_s$ is difficult to measure accurately, especially in the presence of surface waves and wind effects; nevertheless, the scheme does serve to distinguish the deep fjord estuaries with relatively thin and thus relatively highly sheared near surface circulation, from the shallow, partially mixed estuary and the arrested salt-wedge. From [83 F 2, quoted].

Fjords are developed on the coasts of British Columbia, southern Alaska, eastern Canada (especially eastern Baffin Bay), Norway, Iceland, Greenland, Spitzbergen, and South Island, New Zealand. They thus occur mostly in higher latitudes (more than 45°), and are usually backed by rugged and glacially dissected highlands. They have the features associated with glacially eroded troughs [66 G 1, 66 G 2, 60 H, 72 D 2, 75 E, 74 R, 79 S 2, 73 S 2, 80 S 2, 70 T, 66 W 2, 73 W 1, 73 W 2, 80 F, 75 H 1, 83 C 1]. See Table 43, Figs. 67···70.

Table 43. Selected fjords of the North Atlantic Ocean. Length and depth below present sea level.

Name	Locality	Length [km]	Depth [m]
North Fjord	Norway	120	640
Hardanger Fjord	Norway	150	870
Sogne Fjord	Norway	180	1308
Upernavik Fjord	Greenland	55	1055
Scoresby Sound – Franz Josef Fjord	Greenland	320	998
Ice Fjord	Spitzbergen	110	700

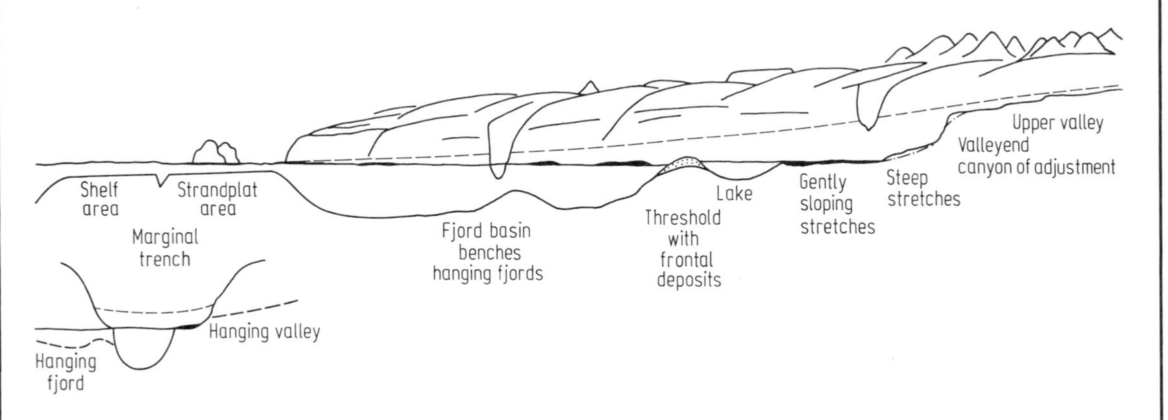

Fig. 67a. Characteristic features of fjords and fjord valleys of Norway, longitudinal profile and cross profile through the fjord. The strandflat, the fjord-side benches, and the gently sloping valley stretch at the head of the fjord are found near the same level which is nearly in accordance with the present sea level. The canyons of adjustment in the steep stretches and in the valley end are indicated by dash-dotted lines. A possible preglacial valley floor is indicated by a dashed line [66 G 1].

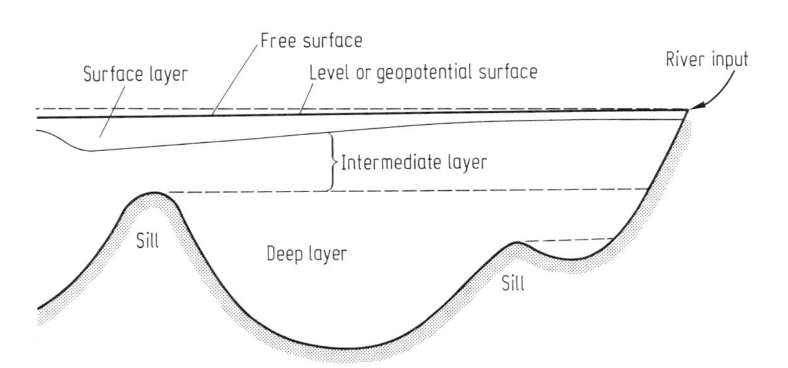

Fig. 67b. Schematic diagram of a fjord, showing sills and deep, intermediate and surface water layers. In reality the oceanography of a fjord is constantly changing, the transient effects often dominating the mean so that a simple representation of this type can be misleading. Moreover the distinction between different layers is seldom as clear-cut as implied in the figure. The slope of the free surface, typically of order 1 cm per 10 km, provides the hydraulic head that drives the brackish layer seaward. Other factors influencing the circulation include wind effects, tides and changes in the outer boundary conditions caused by fluctuations in the coastal density structure. From [83 F 2].

Pacific fjords, water characteristics:

The Pacific fjord regions are in the Coast Mountains of Alaska and British Columbia, the southern Andes of Chile, and the Fjordland mountains of the South Island of New Zealand, all relatively young mountains. The fjords extend along coastline lengths of about 550 km for Alaska, 850 km for British Columbia, 300 km for Vancouver Island, 1500 km for Chile and 230 km for New Zealand. Elevations of the mountains surrounding the fjords are from 2000···4000 m along Alaska/British Columbia, to 2200 m in Vancouver Island, to 3500 m in Chile, and to 2760 m in New Zealand. The fjords themselves are largely the result of earlier glacial gouging and they extend from the coastline to as much as 150 km inland [80 P 1]. See Figs. 68 and 69.

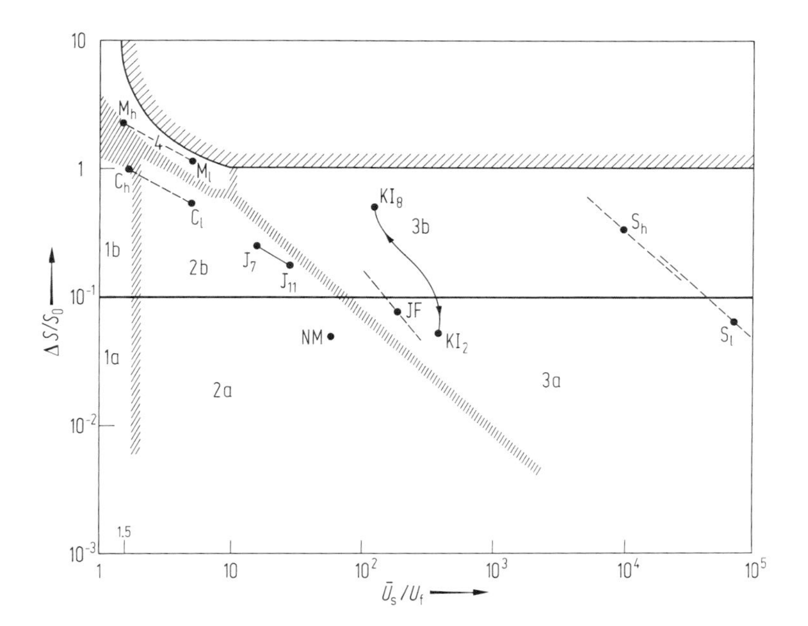

Fig. 68a. Estuarine classification scheme due to Hansen and Rattray (1966). Relative, tidally averaged salinity difference from bottom to surface is plotted against the ratio of tidally averaged net circulation velocity at the surface to the sectionally averaged net river run-off flow velocity. The location of a point on the plot depends on both location in the estuary and time of year. The entries KI_8 and KI_2 are August and February positions (respectively) on the plot for a station about 30 km from the head of Knight Inlet. These represent approximate extreme positions for Knight Inlet, a station would generally lie somewhere between these two points. Other points are plotted for: Mississippi River mouth (M); Columbia River estuary (C), James River estuary (J); Juan de Fuca Strait (JF); narrows of the Mersey estuary (NM) and Silver Bay (S). Subscripts h and l refer to high and low river discharge. Fjords fall into region 3, partially mixed estuaries into 2, salt wedges into 4 and well mixed estuaries into region 1. ΔS: tidally averaged salinity difference from bottom to surface. S_0: Depth and tidally averaged salinity. U_f: Cross-sectionally averaged river flow velocity. $\bar{U}_s$: Circulation velocity measured at the surface [83 F 2], quoted from "The physical oceanography of Fjords." [66 H 2].

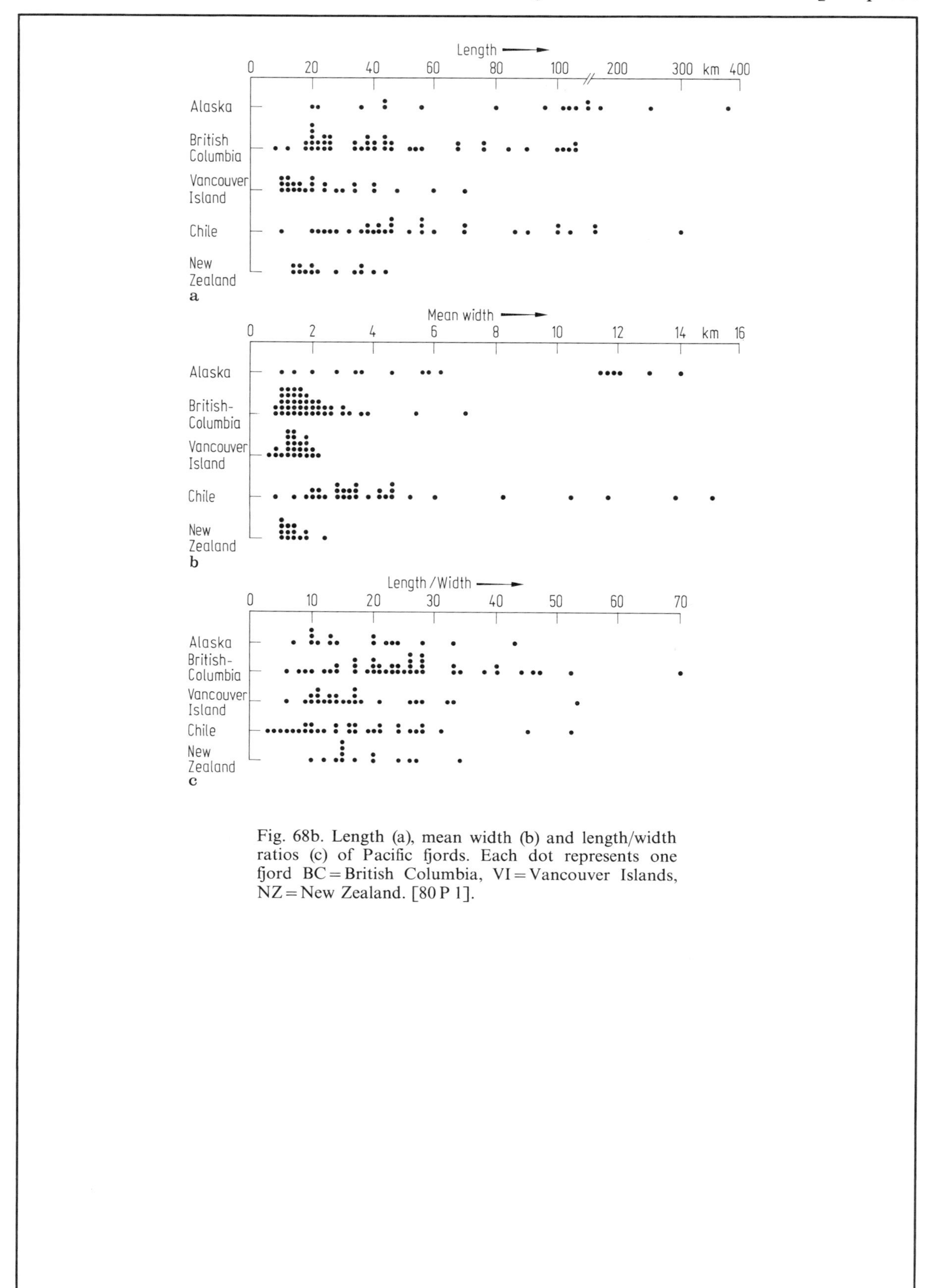

Fig. 68b. Length (a), mean width (b) and length/width ratios (c) of Pacific fjords. Each dot represents one fjord BC = British Columbia, VI = Vancouver Islands, NZ = New Zealand. [80 P 1].

In length, the fjords range from a few kilometers to nearly 400 km and in mean width from 0.6 to 15 km. The dot diagrams of Figs. 68 and 69 show the lengths and widths of individual fjords in the five regions. 90% of the lengths lie in the range 10 to 200 km and 90% of the widths from 1 to 6 km. Length to mean width ratios range from 3 to 70 with 90% being in the 5 to 30 range as shown in Fig. 68 [80 P 1].

In Fig. 69 are shown the individual sill, mean, and maximum depths at lowest normal tides. The effective depths will be greater at higher stages of the tide; typical values for the tidal range are given in a later section on the tides. The mainland American fjords generally have greater sill depths than the Vancouver Island and New Zealand ones, and it is noted that very shallow entrance sills are uncommon. (The definition adopted here for a "very shallow" sill is that it should be shallow enough to hinder inflow of saline water from outside, i.e. that its depth should be equal to or less than the thickness of the low salinity upper layer which is usually less than 10 m) [80 P 1].

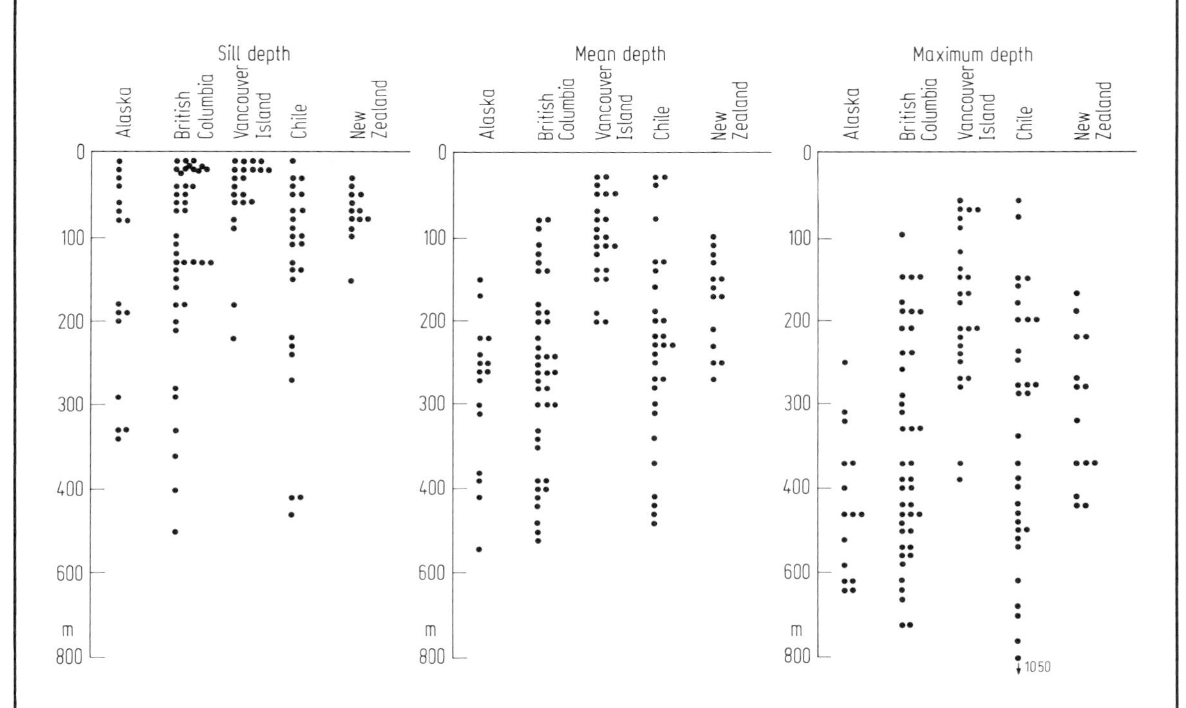

Fig. 69. Sill, mean, and maximum depths of Pacific fjords [80 P 1].

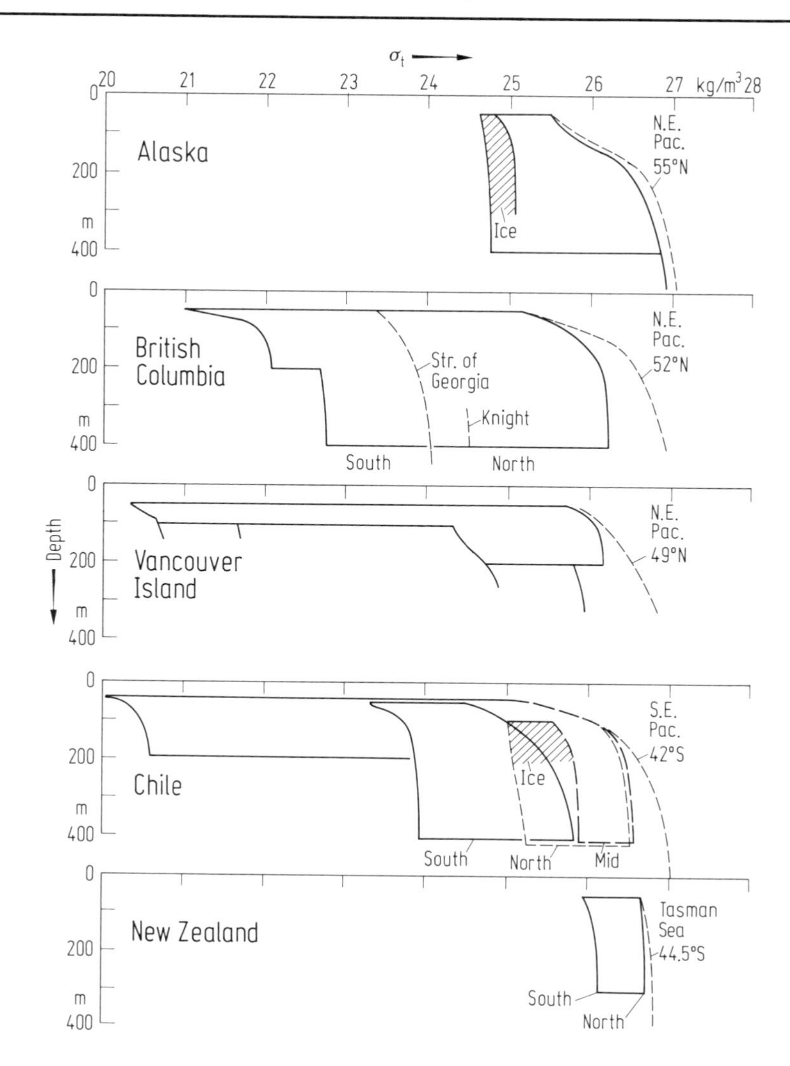

Fig. 70. Envelopes of density, depth profiles for Pacific fjord regions [80 P 1].

9.8 Coastal lagoons

9.8.1 Definition and classification

A coastal *lagoon* is a semi-enclosed body of water connected with the open sea by inlets through a system of barriers, See Fig. 71, Table 44.

Lagoon as a collective term is intended to imply: lagoon (English), Lagune, Haff, Strandsee (German), Noor (Danish), étang, lagune (French), bai (Dutch), estero, laguna (Spanish), laguner (Swedish), liman (Slav) and other [59 G, 61 G 1, 77 G 1, 80 G 2, 81 I 1, 81 I 2, 82 L 1, 82 G 1, 69 C, 70 C, 67 L, 67 P, 82 P, 73 S 1].

Spatial scale of lagoons:
Length 0.1···200 km,
width 0.01···25 km,
depth 1···30 m.

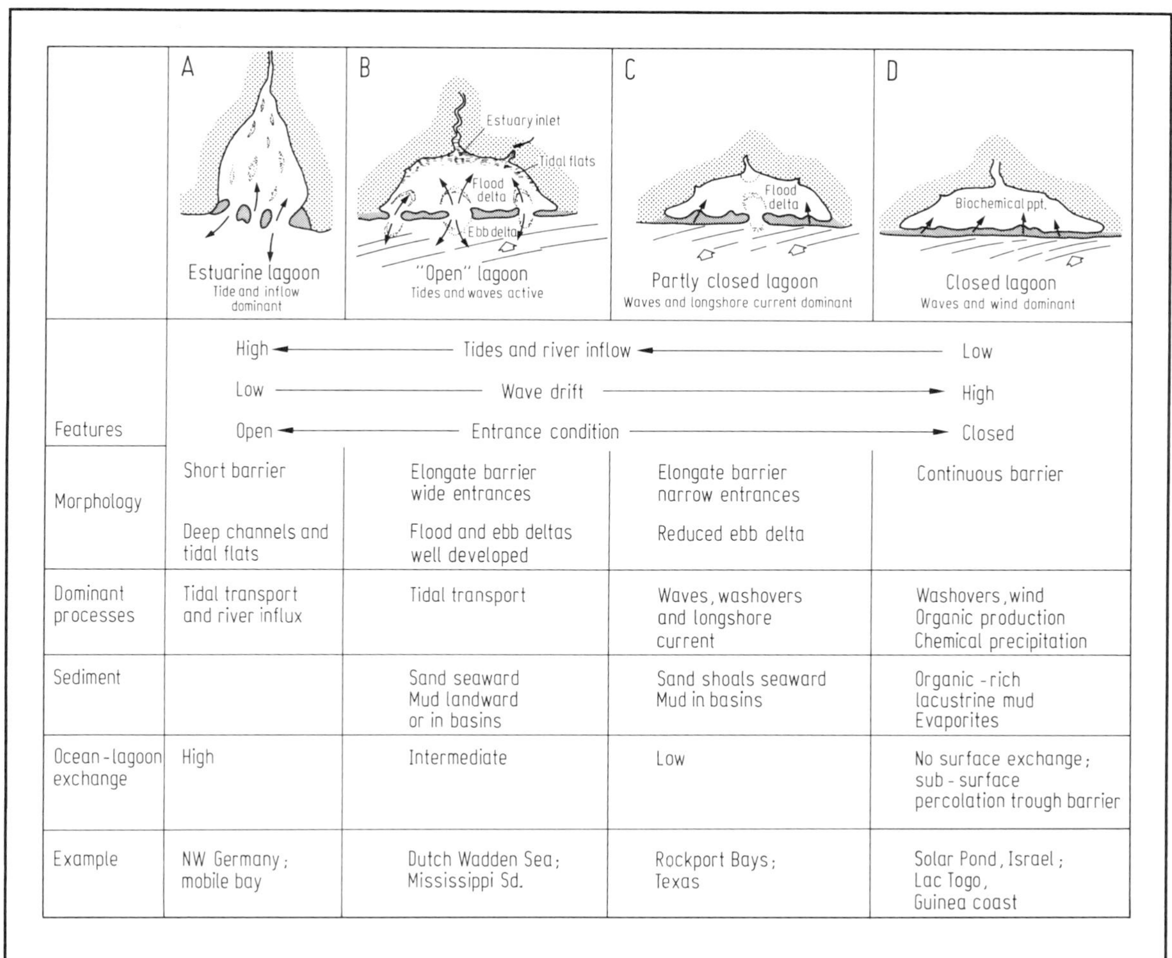

	A	B	C	D
	Estuarine lagoon Tide and inflow dominant	"Open" lagoon Tides and waves active	Partly closed lagoon Waves and longshore current dominant	Closed lagoon Waves and wind dominant
Features	High ← Tides and river inflow ← Low Low — Wave drift → High Open ← Entrance condition → Closed			
Morphology	Short barrier Deep channels and tidal flats	Elongate barrier wide entrances Flood and ebb deltas well developed	Elongate barrier narrow entrances Reduced ebb delta	Continuous barrier
Dominant processes	Tidal transport and river influx	Tidal transport	Waves, washovers and longshore current	Washovers, wind Organic production Chemical precipitation
Sediment		Sand seaward Mud landward or in basins	Sand shoals seaward Mud in basins	Organic-rich lacustrine mud Evaporites
Ocean-lagoon exchange	High	Intermediate	Low	No surface exchange; sub-surface percolation trough barrier
Example	NW Germany; mobile bay	Dutch Wadden Sea; Mississippi Sd.	Rockport Bays; Texas	Solar Pond, Israel; Lac Togo, Guinea coast

Fig. 71. Processes dominant in different lagoon types [81 N 1].

9.8.2 Global zonal classification of lagoon features and processes in different climatic zones

Lagoon shores are affected most by waves where they are exposed to the greatest fetch. Long narrow lagoons parallel to the significant wind direction have the strongest wave-action along the maximum fetch. Waves erode embayments or promontories, and build spits, cusps and cuspate spits. As the spits grow lagoonward, they break up a smooth shore circular basin [61 G 1]. Table 44.

By segmentation, a lagoon adjusts its form to patterns of waves and wind-generated circulation cells. Segmentation can be part of a dynamic equilibrium between forces shaping the basin according to the fetch and the depth of effective wave scour. By fill and scour or segmentation, lagoons can maintain a characteristic ratio of average width to maximum depth. Fig. 72.

Table 44. Comparison of lagoon features and processes in different climatic zones [81 N 1, 81 I 1, 81 I 2].

Feature		High-latitude, polar Alaska [1]) ppt: 23 cm/a T: −35 °C···15 °C Wind: NE, onshore 16···100 km/h Veget.: tundra	Mid-latitude, humid Texas [2]) ppt: 115 cm/a T: 6 °C···35 °C Wind: SE, onshore Veget.: dense	Low-latitude, arid Persian Gulf [3]) ppt: 5 cm/a T: 12 °C···46 °C Wind: NW, onshore Veget: sparse, strong xerophytic	Low-latitude, tropical Guinea coast [4]) ppt: 250 cm/a T: 20 °C···33 °C Wind: SW, oblique to coast Veget.: dense rainforest
Morphology and development	Lagoon	shallow broad basin coast parallel depth: 3 m max.	shallow broad basin coast parallel and transverse depth: 3 m max.	shallow small basin coast parallel, dissected by barrier accretion lobes depth: 2···5 m	shallow narrow basin with elongate arms coast parallel depth: 1.3···3.0 m
	Barrier	narrow gravel spit or low sandy islands moderate longshore migration	sandy islands with ridges 2···3 m high stable, washovers locally	wide sandy islands with 12 m high landward accretion with spillovers	wide sandy islands with multiple beach ridges seaward progradation and longshore migration
	Mainland margins	ice and tundra	marsh, wind-tidal flats, beach, deltaic flats	tidal flats numerous algal flats, wind-tidal flats, beach ridges mangroves locally	mangroves, swamps, marshes; tidal flats locally
	Inlet	wide, numerous open in summer, closed by ice in winter	narrow, intermittent open throughout year, mainly stable	wide, numerous open throughout year, stable prominent tidal channels	narrow, few ephermal barrier breaches
Water	Character	salinity: 2··· 26‰ stratified to well-mixed anoxic under ice	salinity: 1···34‰ well-mixed	salinity: 42···67‰ well-mixed	salinity: <1···30‰ partly-stratified to well-mixed
	Exchange lagoon-sea	moderate in spring and summer	low tidal exchange except high during hurricanes	high tidal exchange	moderate tidal exchange, high during river floods
Sediment sources	Primary secondary	rivers and streams shore erosion and barrier washover	rivers and streams barrier washovers, shore erosion, oyster growth	biological CO_3 production lagoon water via ppt	rivers and streams barrier and shoreface via inlet and washovers

Table 44 (continued).

Feature		High-latitude, polar Alaska [1]) ppt: 23 cm/a T: −35 °C···15 °C Wind: NE, onshore 16···100 km/h Veget.: tundra	Mid-latitude, humid Texas [2]) ppt: 115 cm/a T: 6 °C···35 °C Wind: SE, onshore Veget.: dense	Low-latitude, arid Persian Gulf [3]) ppt: 5 cm/a T: 12 °C···46 °C Wind: NW, onshore Veget.: sparse, strong xerophytic	Low-latitude, tropical Guinea coast [4]) ppt: 250 cm/a T: 20 °C···33 °C Wind: SW, oblique to coast Veget.: dense rainforest
				barrier via onshore wind	
Transport processes	River discharge	low except in spring	moderate river floading	very low inflow	high river inflow and flooding
		overflow by-passes shore			
	Tide	low astronomic range, 0.3 m	low range, 0.1 m	intermediate range, 1.0···3.5 m	low range, 0.3···0.1 m
		high wind tide, 1.3 m strong drift currents	high wind tide, 1.0 m tidal-deltas	moderate wind tides, strong tidal currents, 0.65 m/s	moderate tides
	Wave energy	high in summer	moderate on shoreface, bed resuspension, longshore currents	moderate onshore to barrier and shores	low activity
				bed resuspension and beach ridge formation	bed resuspension on shoals
	Wind energy	strong, but effective only in summer	moderate, eolian transport along barrier and shores	strong eolian transport from barrier and shores	weak, limited influence
	Unique processes	ice shove, gouging local rafting thermal niching river overflow on ice	dredging hurricane tide and surge	chemical ppt of evaporates	rainwash on barrier and flats barrier breaching, seaward
	Seasonality	strong, spring thaw with flooding	moderate, northerly wind winter, southeast, summer	seasonally stable	seasonal low to high rainfall lagoon flooding, flushing through inlets

Gierloff-Emden

Table 44 (continued).

Feature		High-latitude, polar Alaska [1]) ppt: 23 cm/a T: −35 °C···15 °C Wind: NE, onshore 16···100 km/h Veget.: tundra	Mid-latitude, humid Texas [2]) ppt: 115 cm/a T: 6 °C···35 °C Wind: SE, onshore Veget.: dense	Low-latitude, arid Persian Gulf [3]) ppt: 5 cm/a T: 12 °C···46 °C Wind: NW, onshore Veget.: sparse, strong xerophytic	Low-latitude, tropical Guinea coast [4]) ppt: 250 cm/a T: 20 °C···33 °C Wind: SW, oblique to coast Veget.: dense rainforest
Despositional processes		settling of river-borne suspended load	settling of river-borne suspended load	biologic extraction of CO_3	settling of river-borne suspended load
		strudal infilling	rapid accretion at river delta	entrapment by algal mats	entrapment by algae and mangroves
		flocculation probable	entrapment in marshes washovers and inlet deltas	chemical ppt on flats settling of resuspended sediment	flocculation accretion at river delta
			Oyster reef desposition	biodeposition and bioturbation	
			bio-desposition and bioturbation		
Bed sediments	Central lagoon	silt, clay and organic layers	silty with oyster reefs	pelletal and foram mud	organic rich silt and clay or sandy mud
	marginal zones	sand and silt	silt and clay in marsh sand and shall on beaches	shell beach ridges evaporite flats, gypsum, anhydrite pelletal and foram sand shoals Oolite on tidal deltas	shelly sands

[1]) Elson Lagoon (Faos, 1969) and Simpson Lagoon, Alaska (Neidu and Hevett, 1975).
[2]) Aransas and San Antonio Bay (Shepard and Moore, 1960) and Galveston Bay, Texas, USA (Fisher et al., 1972).
[3]) Trucial Coast, Arabia (Evans and Duch, 1969; Purser and Evans, 1973).
[4]) Lagos-Lekki-Lagoon-Complex, Nigeria (Webb, 1957; Allon, 1965a and b) and Dahomey Coast Lagoons (Guilcher, 1959).

9.8.3 Configuration of inlet-bay geometry of lagoons and littoral drift

See Figs. 72a, b, c, d, Tables 45, 46.

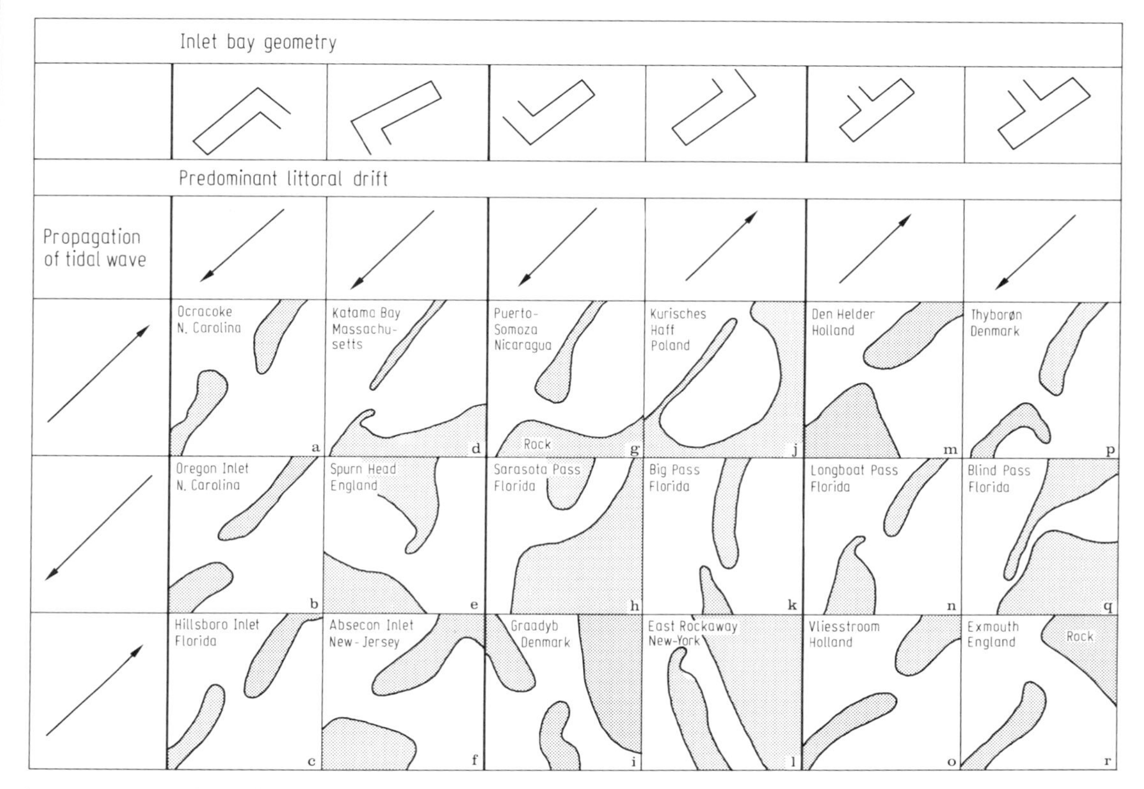

Fig. 72a. Inlet configuration in relation to inlet and bay geometry, propagation of the tidal wave, and the direction of the predominant littoral drift [60 B 2]. Model not to scale. Compare Fig. 47, and [81 B 3].

Table 45. Origin of some tidal inlets [60 B 2].

Ocracoke Inlet, N.C.	break-through
Oregon Inlet, N.C.	break-through
Hillsboro Inlet, Fla.	break-through
Katama Bay, Mass.	formation of barrier across bay
Spurn Head, England	formation of barrier at bay-river mouth
Absecon Inlet, N.J.	break-through
Puerto Somoza, Nicaragua	formation of barrier at river mouth
Sarasota Pass, Fla.	break-through, possibly caused by sinking of land mass or rise of sea level
Graadyb, Denmark	break-through, possibly caused by sinking of land mass
Kurisches Haff, Poland	formation of barrier across bay
Big Pass, Fla.	break-through, probably caused by sinking of land mass or rise of sea level
East Rockaway, N.Y.	formation of barrier
Den Helder, Holland	break-through, probably caused by consolidation of soil and rise of sea level
Longboat Pass, Fla.	break-through, possibly caused by sinking of land mass or rise of sea level
Vliestroom, Holland	probably an old river mouth now enlarged because of rise in sea level
Thyborøn, Denmark	break-through
Blind Pass, Fla.	formation of barrier across inlet
Exmouth, England	formation of barrier across bay

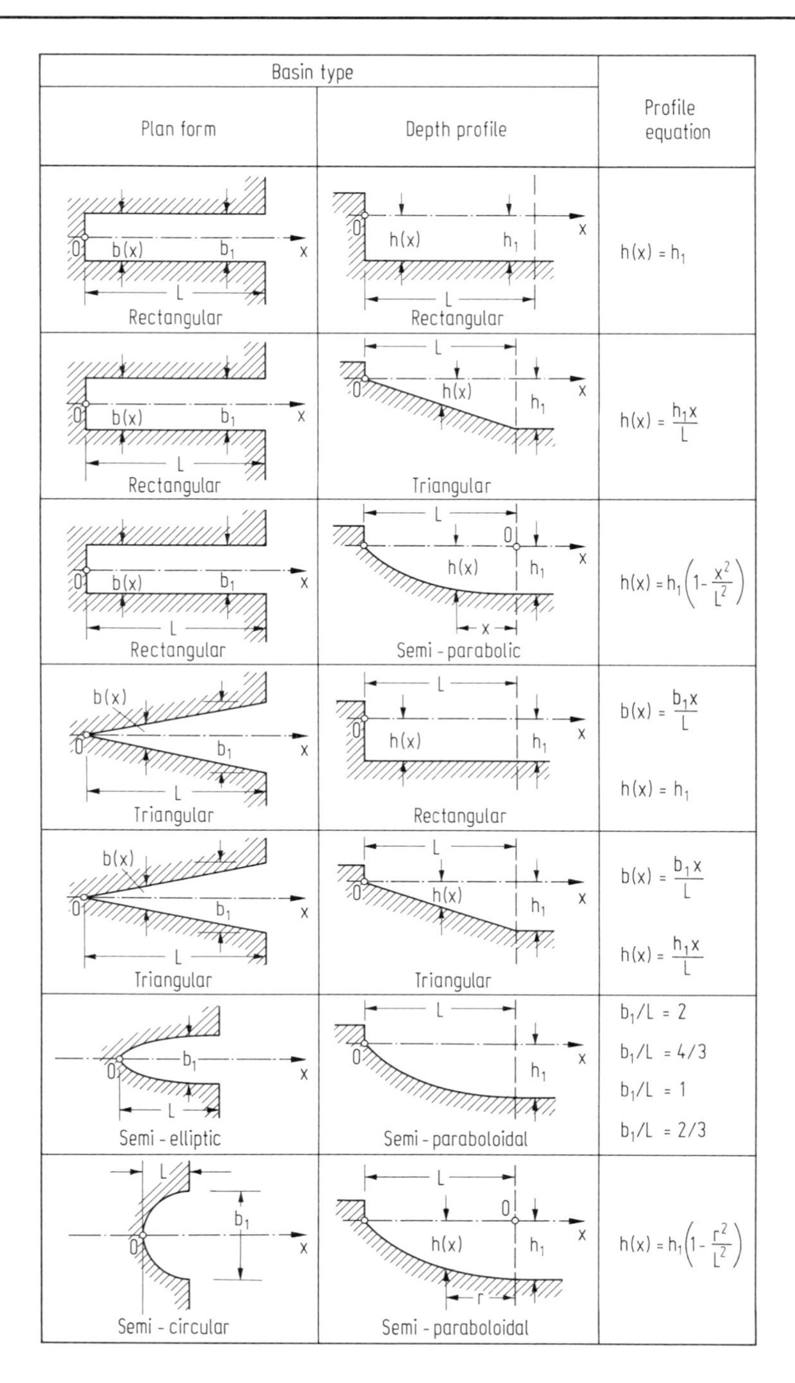

Fig. 72b. Features and parameters of bay topography [66 W 2].

Table 46. Flow and littoral drift characteristics for some inlets of estuaries and bays [60 B 2].

Ω = spring tide, Q_{max} = maximum discharge, M = predominant littoral drift, 1 cu yd = 0.7646 m^3.

Total amount of littoral drift interfering with the inlet may deviate from this value if drift direction is not too predominant and/or the inlet is not improved.

Inlet (kind of improvement)	Ω 10^6 cu yd/ half cycle	Q_{max} cu yd/s	M 10^6 cu yd/a
Amelandse Gat, Holland (bank stabilization on north side)	600	36600	1.0
Aveiro, Portugal (jetties)	150	9000 (increasing)	0.75
Big Pass, Florida (none)	12	700	<0.1
Brielse Maas, Holland before closing	40	2700	1.0
Brouwershaven Gat, Holland (will be closed)	430	30000	1.0
Calcasieu Pass, La. (diurnal) (jetties and dredging)	110	2600	0.1
East Pass, Florida (diurnal) (dredging)	60	1720	0.1
Eyerlandse Gat, Holland (none)	270	19000	1.0
Figueira Da Foz, Portugal (dredging)	20	1200	0.5
Fort Pierce Inlet, Florida (jetties and dredging)	80	3700	0.25
Gasparilla Pass, Florida (none)	15	900	<0.1
Grays Harbor, Washington (jetties and dredging)	700	48000	1.0
Haringvliet, Holland (being closed)	350	25000	1.0
Inlet of Texel, Holland (stabilization of south side)	1400	115000	1.0
Inlet of Vlie, Holland (none)	1400	110000	1.0
Longboat Pass, Florida (none)	30	1400	<0.1
Mission Bay, California before dredging (jetties and dredging)	15	1100	0.1
Oosterschelde, Holland (will be closed)	1400	100000	1.0
Oregon Inlet, N. Carolina (occasional dredging)	80	5100	1.0
Ponce De Leon Inlet, Florida (none)	20	1500	0.5
Port Aransas, Texas (diurnal) (jetties and dredging)	65	1900	0.1
Thyborøn, Denmark (minor dredging)	140	7500	0.9
Westerschelde, Holland (some dredging)	1600	115000	1.0

Process and geometrical configuration of segmented coastal lagoons

Coastal lagoons often are of elliptic or elongated form in plain view, stretching parallel to the coast.

Varying degrees of ellipticity exist in different geographic areas as a function of wind regime, wave climate, and geologic or environmental conditions.

A segmentation of geometric regularity due to development of facing spits and a final feature of elliptical coastal lakes are common, where microtidal ranges are prevailing. [62 C].

The dominant process of wave action in the elongated flat basins depends on direction and fetch of winds. See Fig. 71c.

The spit development will be orientated $\approx 45°$ to the wave direction. The interior shoreline of the lagoons and coastal lakes can be described by an index [61 G 1, 67 Z].

According to Bruun [53 B], in basins of finite dimensions maximum velocity of littoral drift occurs at points on the downwind shores where the angles between wave orthogonals and normals to the shorelines are 50° (Fig. 72). Minimum littoral drift occurs where these angles are 0°, the nodal points, or directly on the downwind sides. Lagoons have large ratios [82 A]. For shorelines of limited length, a cycloid equilibrium outline results. Reversal of winds gives rise to two opposite shorelines of cycloid form, hence a basin of elliptical shape. [61 R, 62 C].

The shoreline development ratio (or index) is a number that relates the measured shoreline length of a given to the shoreline length of a perfectly circular of equal area. The formula for the ratio is

$$D_L = L/(2\sqrt{\pi A})$$

where L is the measured shoreline of a lake and A is the area of the lake. The ratio can be no smaller than 1.0, the ratio for a perfectly circular lake or lagoon, those in oxbow configuration, have large ratios [82 A].

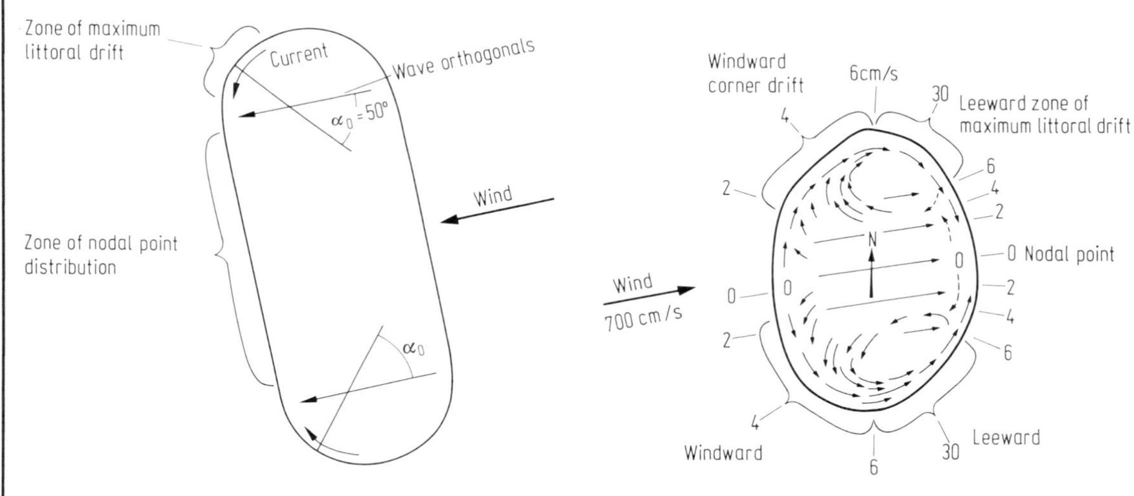

Fig. 72c. Relationship between wave orthogonals and zones of maximum littoral drift in leeward corners according to the prediction of Rex [61 R, 62 C].

Fig. 72d. Wind circulation system and littoral drift of a closed coastal lake. [62 C].

9.8.4 Coral atolls as oceanic lagoons

See Figs. 73, 74a, b

Compare subsect. 9.10.3, Fig. 82

Hydrography of coral atolls (typical example: Bikini-Atoll).

Living coral reefs that encircle volcanic islands have two forms: a barrier reef separated from the island by a lagoon, and a fringing reef attached to the island except at the mouths of rivers. A fringing reef forms an extension of a low-tide, wavecut platform, whereas an island with a barrier reef has in effect a triple coastline: an outer reef coast exposed to the full force of the ocean, particularly boisterous on the windward side; an inner, lagoonal reef coast, and an innermost lagoonal volcanic coastline.

The atoll consists of a coral reef surrounding and protecting a usually tranquil lagoon from a usually restless ocean and supporting small vegetated islets 3···4 m high [1842 D, 15 D, 57 W, 62 W, 78 S].

Reef width and lagoonal depth: equatorial regions provide optimum conditions for coral reef growth and dependent atoll reef widths. Maximum lagoonal depth varies inversely with the average lagoonal radius [57 W, 15 D, 1842 D, 54 E, 62 W].

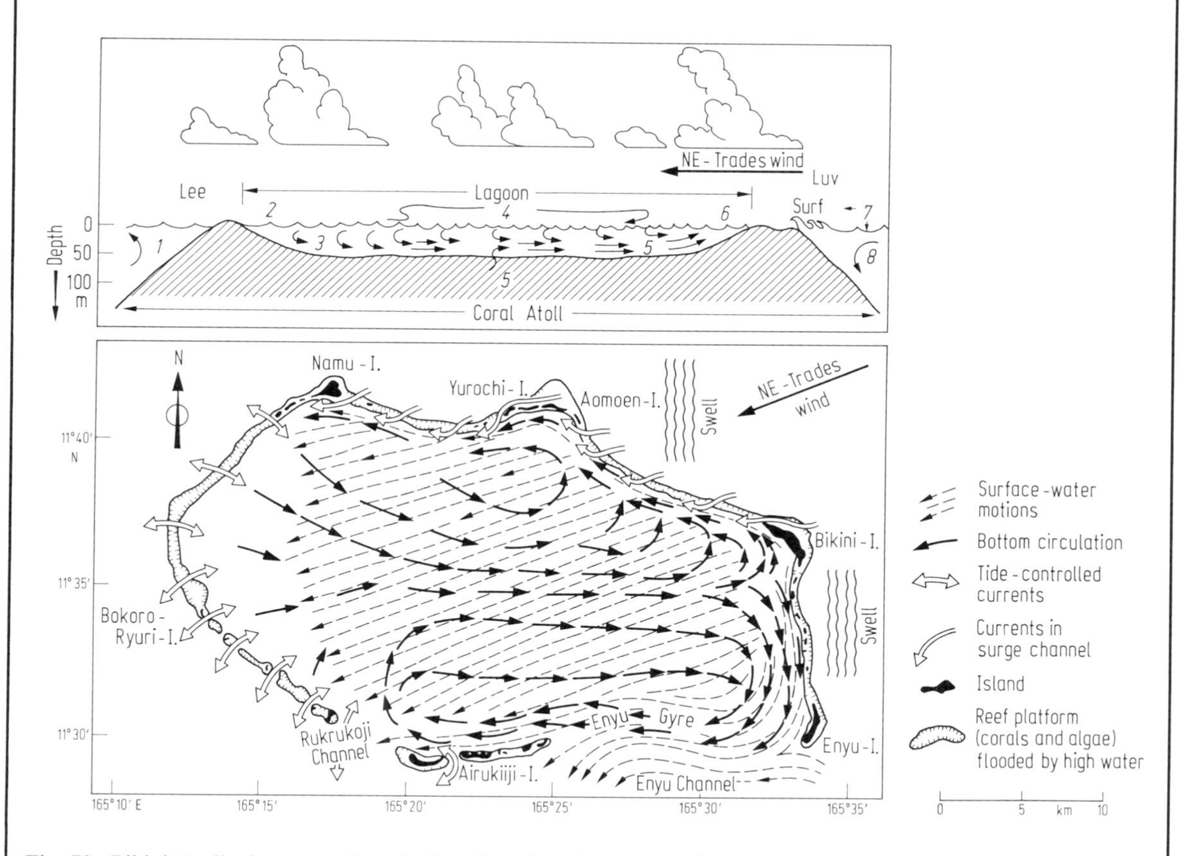

Fig. 73. Bikini Atoll: the atmosphere-hydrosphere interface as coupled model of the environment of a coral atoll with lagoon. The Bikini Atoll has a mean depth of 40 m in the north, in the centre and in the south a depth of 55 m. [After chart 6032 DHI. 54 A, 80 G 2].

Top figure: 1: upwelling current on the exterior lee-slope of the atoll; 2: waves and surf within the lagoon; 3: sinking water; 4: increasing waves towards lee, depending on fetch and thicker layer of surface current; 5: bottom current of thick layer; 6: upwelling, calm water; 7: incoming swell; 8: sinking current on the exterior luff-slope of the atoll.

Spatial parameters of atoll-lagoons (coral-reef type).

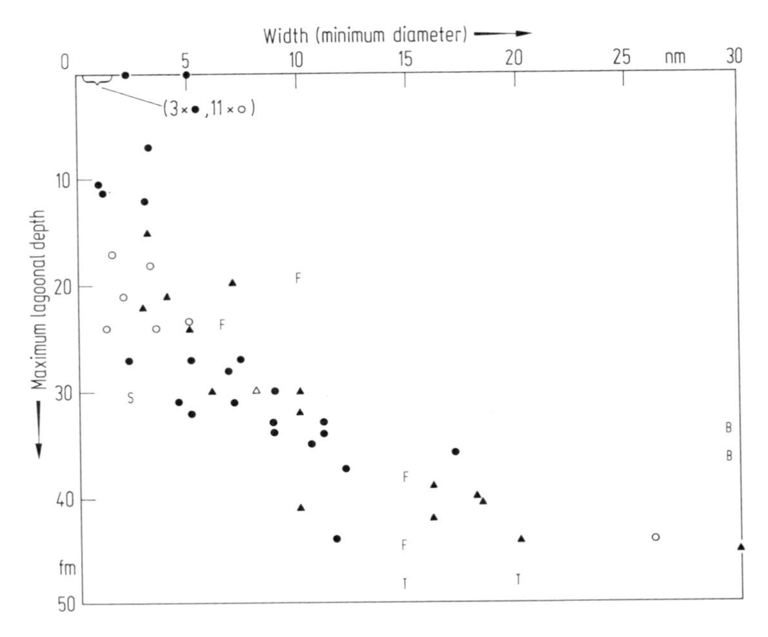

Fig. 74a. Plot of oceanic atolls showing relationship of depth and diameter for atoll lagoons: [54 E]. Location: all Pacific ocean, except the Maldive in Indian Ocean, based on 60 selected atolls. For depths of feet of corals, see [73 B]. Solid circles: Marshall; open circles: Caroline; open triangles: Gilbert and Ellice; F: Fiji; T: Tonga; S: Society; B: Great Barrier; solid triangles: Maldive. Compare [71 S 5].

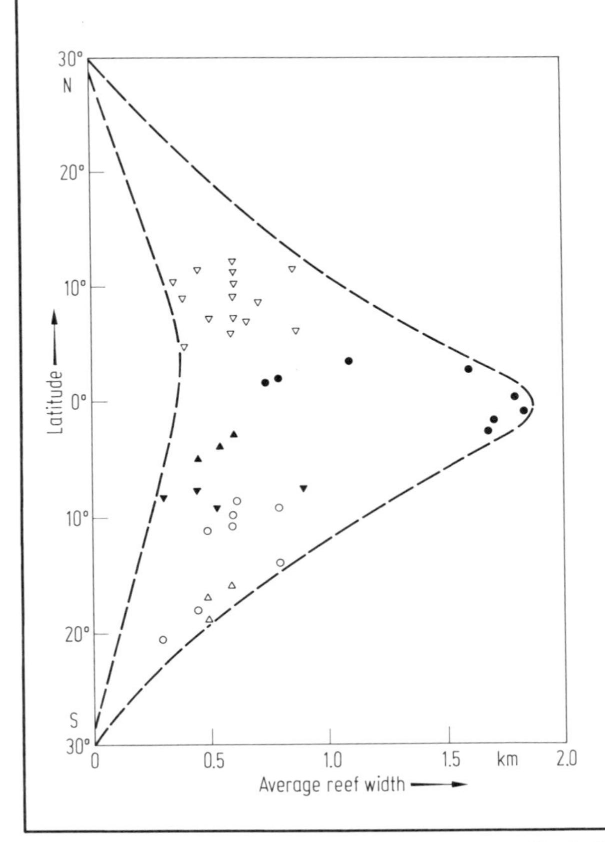

Fig. 74b. Average coral reef width relative to latitude. [After Schofield, J. C. in 82 S 1, p. 624]. Open triangles (downward): Marshall I.; solid circles: Gilbert I.; solid triangles (downward): Ellice I.; open circles: Cook and Tokelau I.; open triangles (upward): Lau I.; solid triangle (upward): Phoenix O. For distribution of coral islands compare subsect. 1.4.3.2, Fig. 57.

9.9 River discharge affecting coastal waters

9.9.1 Terrestrial run-off

According to most recent research in the context of the International Hydrological Decade, NACE 1970, an amount of 924 000 m^3/s was given (=29.1 · 10^3 km^3/a), neglecting the areas of internal discharge (103 · 10^6 km^2). Mainly unknown is still the discharge of the large ground water reservoir. It is estimated to be about 7000 m^3/s, i.e. less that 1% of the surface run-off. For discharge of main rivers, see sect. 1.5, Table 23. [75 B 2, 82 R 1, 81 H 2, 83 C 4].

The terrestrial run-off is unevenly distributed on individual rivers, and these again are unevenly distributed in relation to the oceans. Both facts are clearly shown by Table 47 in which all large rivers with an annual mean discharge of more than 2.0 · 10^3 m^3/s are listed. There are 33 of them. They provide 58% of the terrestrial run-off. Among them the Amazon-river, the most abundant in water, provides 19.5%. Considering the continental watersheds, the Atlantic Ocean has the largest drainage basin, and gains also the largest amount of the terrestrial discharge. The European rivers are of minor importance. Only the run-off of the Danube, Petchora, Dvina, Neva, and the Rhine is larger than 2.0 · 10^3 m^3/s. The Volga with 8.1 · 10^3 m^3/s drains into an inland sea, the Caspian Sea. In the European Mediterranean Sea (without the Black Sea) there is no river since the damming of the Nile in 1965 with a discharge larger than 2.0 · 10^3 m^3/s.

The distribution of the rivers' drainage basins is of importance for the type of sediment. Figs. 75a, b, c.

This relationship and the quantity and quality of the petrographical and chemical river transported material (suspension and dissolved) to the coastal water of the oceans is treated in "River transport to the oceans" [81 H 2] in "The oceanic lithosphere" [81 E].

Table 47. Annual mean terrestrial discharge of rivers larger than 2.0 · 10^3 m^3 s^{-1}, grouped according to oceans [75 D 1].

Atlantic Ocean	10^3 m^3 s^{-1}	Pazific Ocean	10^3 m^3 s^{-1}
Amazon	180.0	Yangtze Kiang	34.0
Kongo	42.0	Mekong	15.9
Orinoko	28.0	Amur	11.0
La Plata *)	24.1	Sikiang	11.0
Mississippi	17.5	Columbia	6.7
Yenisei	17.4	Yukon	4.3
Lena	15.5	Fraser	3.9
Ob	12.5		
St.-Lorenz-River	10.4	**Indian Ocean**	**10^3 m^3 s^{-1}**
Magdalenen	8.0		
MacKenzie	7.5	Brahmaputra	20.0
Danube	6.5	Irawadi	14.0
Niger	5.8	Indus	3.9
Petchora	4.1	Zambezi	2.5
Kolyma	3.8		
Dwina	3.5		
San Francisco	3.3		
Grijalva	3.3		
Newa	2.6		
Pyasina	2.6		
Nelson	2.3		
Rhine	2.2		

*) Paraná and Uruguay.

Gierloff-Emden

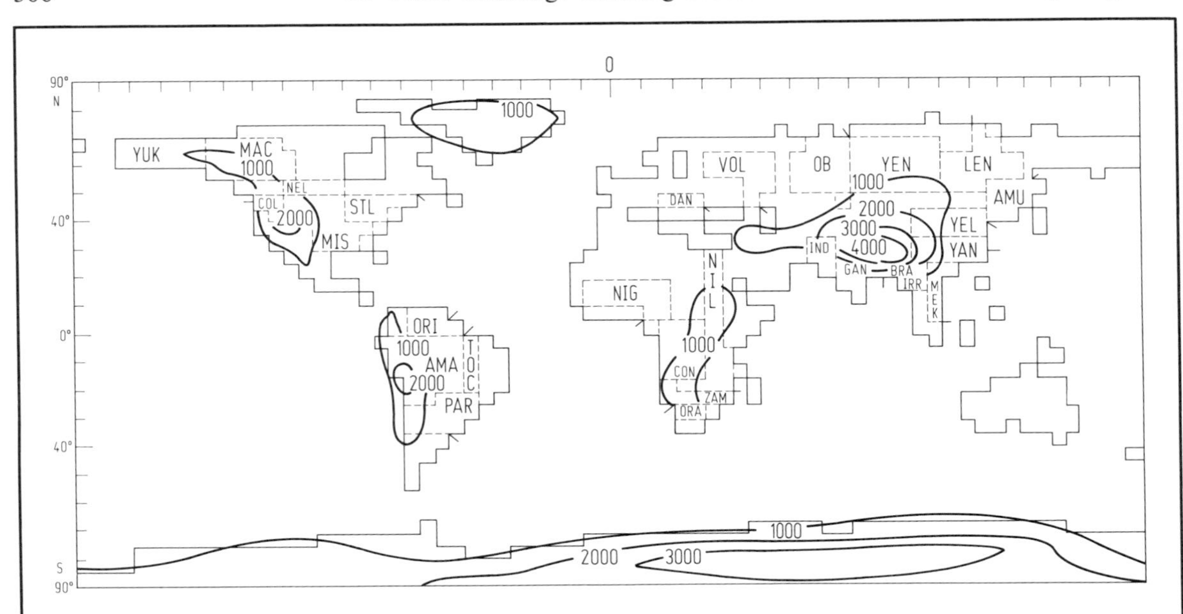

Fig. 75a. The NCAR 5° average elevation map showing the drainage basins and discharge sites of 29 important rivers, AMA: Amazon, AMU: Amur, BRA: Brahmaputra, COL: Columbia, CON: Congo, DAN: Danube, GAN: Ganges, IND: Indus, IRR: Irrawaddy, LEN: Lena, MAC: Mackenzie, MEK: Mekong, MIS: Mississippi, NEL: Nelson, NIG: Niger, NIL: Nile, ORA: Orange, ORI: Orinoco, PAR: Parana, STL: Saint Lawrence, TOC: Tocantins, VOL: Volga, YN: Yangtze, YEL: Yellow River, YUK: Yukon, ZAM: Zambezi. From [83 H].

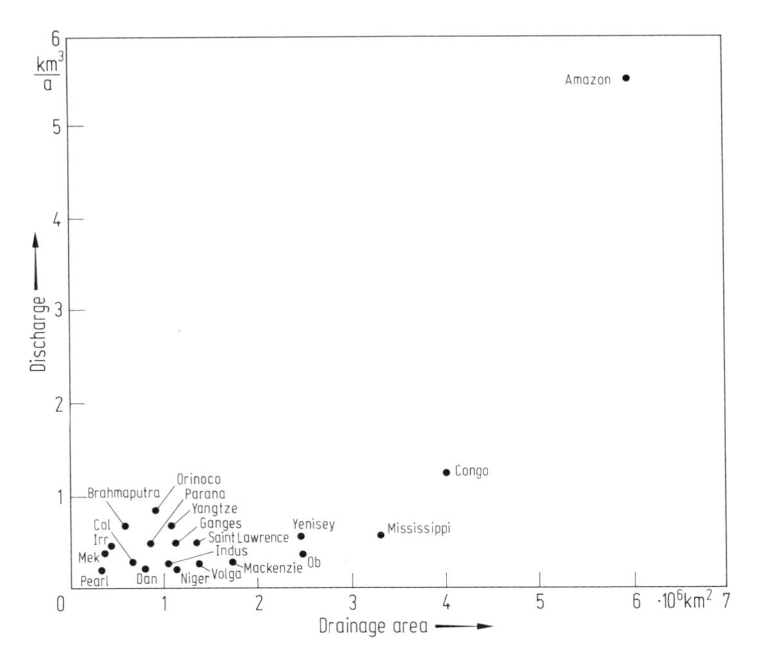

Fig. 75b. Annual discharge as a function of drainage area for the 20 rivers having the largest discharge. Col: Columbia, Dan: Danube, Irr: Irrawaddy, Mek: Mekong. From [83 H]. Fig. 75b is a plot of the annual discharge versus drainage area for the 20 rivers having the largest discharge (data from Holland 1978). The flow of the 10 largest rivers is $1.13 \cdot 10^9$ cm^3/a, and represents 25% of the total input of fresh water into the ocean. The 20 largest rivers supply $1.41 \cdot 10^9$ cm^3/a, about 31% of the world total. One of the 20 largest rivers, the Volga (14th in rank) drains to the Caspian Sea and is not included in the total discharge to the ocean. Most of the largest rivers have drainage basins spanning 10° or more in latitude. After [83 H].

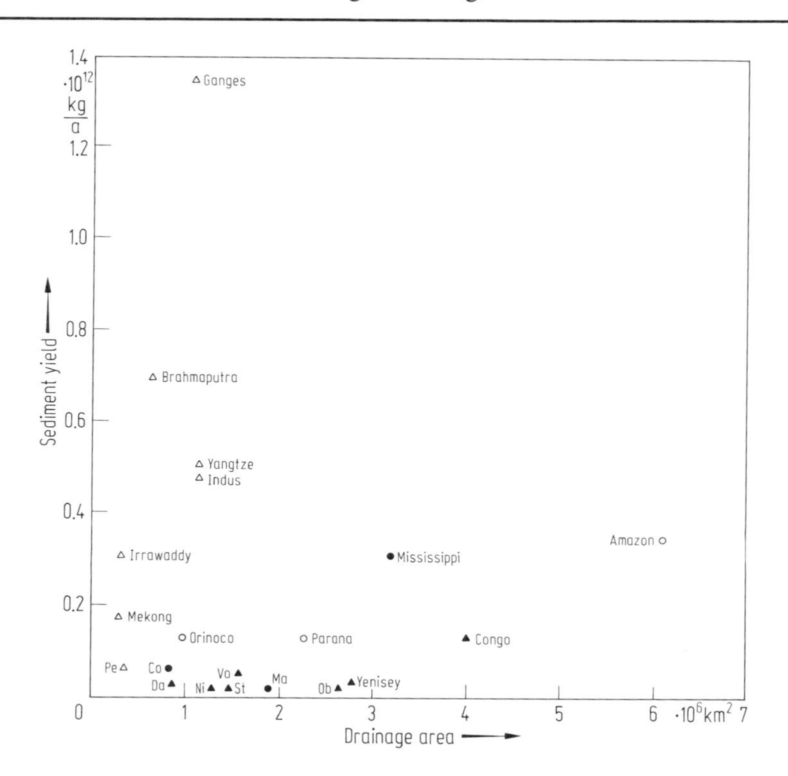

Fig. 75c. Sediment load of the 20 largest rivers plotted against drainage area. Co: Columbia, Dan: Danube, Mac: Mackenzie, Nig: Niger, Pe: Pearl, Stl: Saint Lawrence, Vol: Volga. From [83 H]. Fig. 75c shows that the sediment load brought to the ocean by large rivers is dominated by that of rivers that drain high uplifts. The six major rivers with headwaters in the Himalayas or on the Tibetan plateau carry 18% of the total detrital sediment delivered to the ocean margin. The Amazon and Mississippi, with headwaters in the Andes and Rockies, also carry significant loads. After [83 H]. Sediment load of 1st 8 rivers $= 3.9 \cdot 10^{12}$ kg a^{-1}, of 1st 20 rivers $= 4.2 \cdot 10^{12}$ kg a^{-1}, and of all rivers $= 18.3 \cdot 10^{12}$ kg a^{-1}. Open circles: Andes uplift drainage, solid circles: Rocky Mt. uplift drainage, open triangles: Himalayan uplift drainage, solid triangles: other drainage.

9.9.2 River discharge in polar regions

Table 48. Discharge and drainage basins in the USSR [75 G 4].

Basin	Rivers	Estuary (mouth)	Discharge km^3	Groundwater %
North-west	Dwina Petchora	White Sea Barents Sea	530	22
Ural-Westsibiria	Irtysch-Ob	Kara Sea	447	28
Ostsibiria	Yenisei	Kara Sea	1085	23
Far East	Lena a. Indigirka Kolyma	Laptew Sea East Sibirean Sea	1467	16
USSR/total			4350	22

For river discharge and influence on coastal waters after ice break-up, see Figs. 75 and 76.

The Colville river delta area on the Alaskan north coast is characterized by permafrost. Each spring, the sudden breakup of the frozen rivers causes overice flooding of the sea ice as well as extended subice floodwater penetration into the Beaufort Sea.

The discharged waters consist mainly of fluvial meltwater, a certain quantity is derived from the melting of snow and ice on the sea. Quality and quantity of this process were analyzed by H. J. Walker and presented in numerous publications: [70 W, 73 W 2, 84 W 4, 82 W 4, 74 W 1, 66 W 1, 73 W 3, 73 W 4, 77 W 1, 84 W 1].

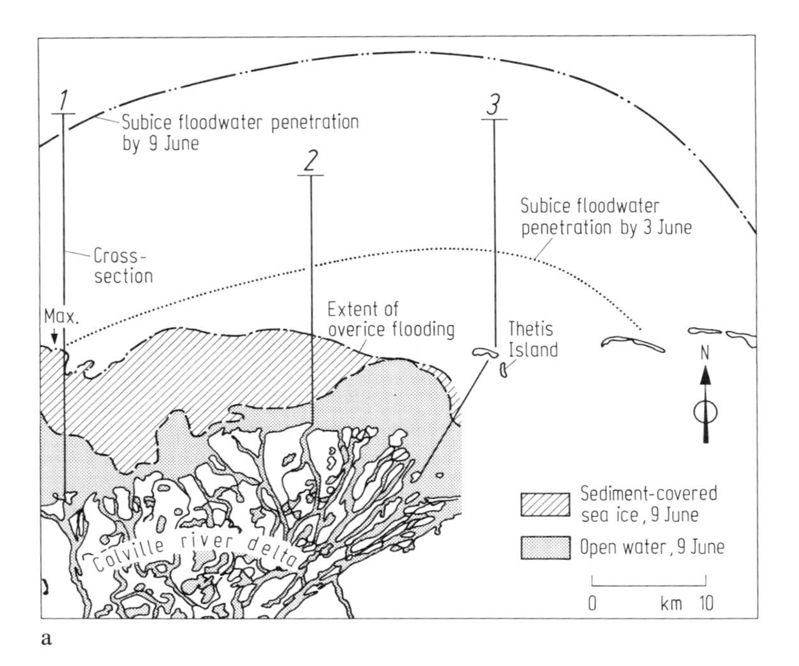

a

Cross section 1

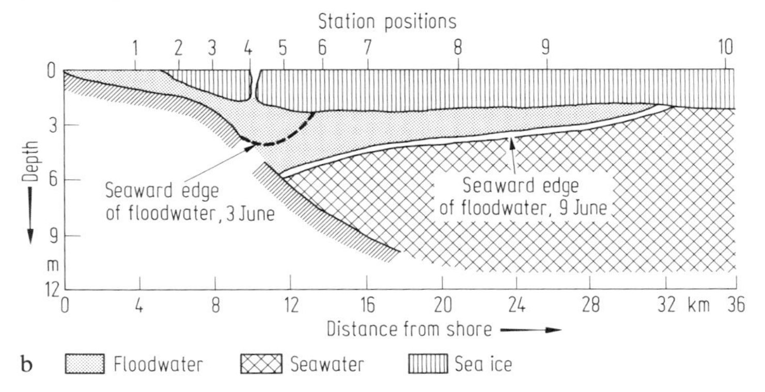

b

Fig. 76a, b. Floodwater discharge after the spring break-up (end of May – beginning of June, 1971) of the Colville River delta, Alaskan north coast, Beaufort Sea. [70 W].

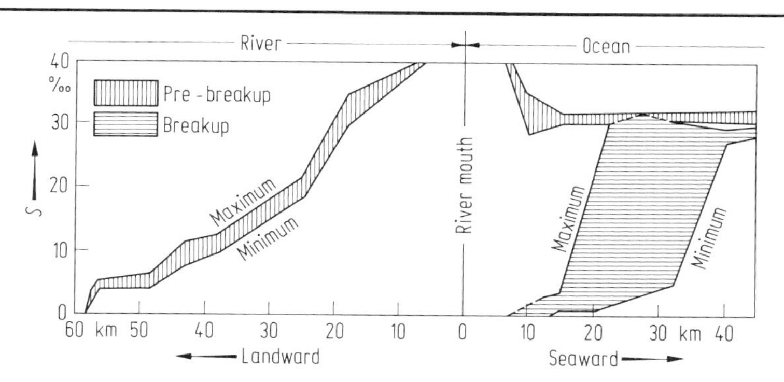

Fig. 76c. Change of ocean water salinity S in front of the Colville River delta, Alaska [70 W].

9.9.3 Deltas

Occurrence and distribution of deltas.

Deltas are subaerial and subaqueous accumulations of river-derived sediments deposited at the coast when a stream decelerates by entering and interacting with a larger body of water. Deltas occur wherever a stream debouches into a receiving basin as an ocean gulf, inland sea, bay, estuary or lake. See Figs. 77, 78, Tables 49a, 50a, b, c. In addition to these major deltas literally thousands of minor deltas are distributed over all of the world's coasts.

The first requirement for their occurrence is the existence of a major river system that carries substantial quantities of clastic sediment. The interactions of delta processing are: fluvial regime, wave regime, tide regime, ocean current regime, shelf configuration, receiving-basin geometry and tectonics.

The delta shoreline is the temporary land-sea interface and seaward boundary of the subaerial delta. Compared with the somewhat more stable shorelines of coastal barriers or beach systems, delta shorelines are ephemeral and unstable with respect to their position. In the case of rapidly prograding deltas experiencing minimal wave energy, the delta shoreline may be crenulate and highly irregular, consisting of mudflats or marsh. In higher energy situations, where powerful waves have slowed the advance of the delta and re-sorted the deltaic sediments, the delta shoreline commonly tends to be straight and to take the form of a sandy beach. For delta variability, hydraulic processes, features and delta models see [78 W 1 and 85 W in 85 D]. [75 C, 85 W].

Deltaic classification and process.

For Fig. 77, see page 312.

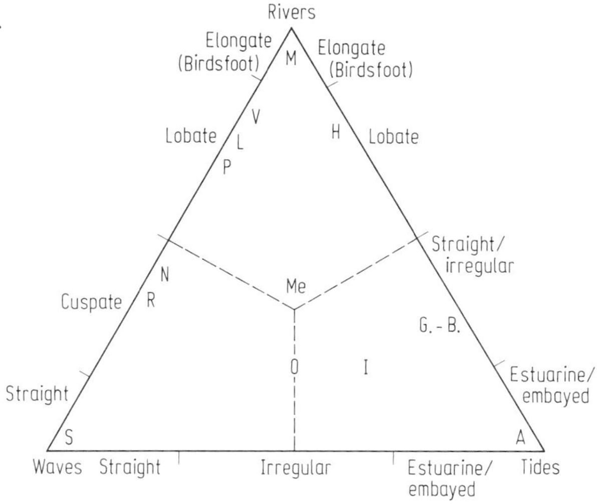

Fig. 78. Classification of deltas [84 K 1] according to dominating processes: rivers (cf. Table 49a), waves (cf. Table 51), tides (cf. Table 50). A: Amazonas; G.–B.: Ganges–Brahmaputra; M: Mississippi; Me: Mekong; I: Indus; L: Lena; N: Nile; O: Orinoco; R: Rhone; P: Po; S: Senegal; V: Volga; H: Hwangho; triangle frame: deltaic depositional type (prevailing) [85 K, 85 W]. The three dominating processes, rivers, waves, tides, are located at the corners of the triangle.

Table 49a. Deltas and rivers: topographic and hydrologic parameters [84 K 1 replenished]. Oceans: A: Atlantic O., I: Indian O., P: Pacific O., Arc: Arctic O., M: Mediterranean Sea, B: Black Sea.

Delta river	Ocean	Country	Delta area km²	River length km (rank)	Drainage area km² (rank)	Av. sediment discharge 10^6 t/a	Tidal range *) m
1. Amazon	A	Brazil	467 000	6 518 (2.)	5 778 290 (1.)	363···1 000	6.0
2. Ganges-Brahmaputra	I	India/ Bangladesch	105 600	2 901 (18.)	1 825 000 (10.)	700···1 800	6.0
3. Mekong	I	Vietnam	93 700 (52 000)	4 350 (9.)	906 500 (20.)	80	4.0
4. Yangtze Kiang	P	China	66 600 (124 000)	5 827 (4.)	1 808 000 (11.)	499···600	4.2
5. Lena	Arc	USSR	43 500 (28 500)	4 318 (10.)	2 421 000 (7.)	12···15.4	0.3
6. Hwangho	P	China	36 272 (127 000)	4 850 (5.)	761 500 (22.)		3.4
7. Indus	I	Pakistan	29 500	3 186 (17.)	963 480 (17.)	400···435	4.2
8. Mississippi	A	USA	28 500	6 420 (3.)	3 211 600 (3.)	516	0.5
9. Volga	–	USSR	27 200 (11 100)	3 700 (15.)	1 358 000 (13.)	26	0.0
10. Orinoco	A	Venezuela	20 600 (56 900)	2 500 (22.)	1 086 000 (16.)	45	2.2
11. Irrawaddy	I	Burma	20 500 (31 000)	2 150 (25.)	429 940 (25.)	299···350	4.3
12. Yukon	P	Alaska	20 000	3 700 (14.)	932 000 (18.)	88	1.2
13. Niger	A	Nigeria	19 100	4 200 (11.)	2 100 000 (9.)	40···67	1.7
14. Shatt-al-Arab	I	Irak	18 500	2 735 (20.)	1 113 700 (15.)		2.6
15. Po	M	Italy	13 400	760 (34.)	143 900 (34.)	18	0.6
16. Nile	M	Egypt	12 500	6 693 (1.)	2 878 500 (5.)	54···111	0.3
17. Chao Phraya	P	Thailand	11 400 (24 500)	866 (32.)	922 000 (19.)	12	3.5
18. Indigirka	Arc	USSR	9 200	1 791 (27.)	360 000 (27.)		0.1
19. Mackenzie	Arc	Canada	8 500	4 600 (7.)	1 660 000 (12.)	15	0.4
20. Zambezi	A	Moçambique	7 200	3 450 (16.)	1 328 600 (14.)	100	4.0
21. Godovari	I	India	6 300	1 448 (30.)	297 800 (30.)		2.0
22. Paraná	A	Brazil	5 440	3 718 (13.)	2 305 100 (8.)	96	1.0
23. Senegal	A	Senegal	4 250	1 190 (31.)	196 400 (33.)		1.2
24. Ord	P	Australia	3 900	405 (38.)	46 570 (38.)		6.0
25. Amu-Darya	–	USSR	3 500	2 441 (23.)	297 850 (29.)	94···100	0.7
26. Ob	Arc	USSR	2 850	4 505 (8.)	2 913 750 (4.)	15.8	0.6
27. Danube	B	Romania	2 740 (4 400)	2 832 (19.)	733 000 (21.)	100	0.1
28. Yenisei	Arc	USSR	2 460	3 797 (12.)	2 698 780 (6.)	13.2	0.4
29. Burdekin	P	Australia	2 100	613 (36.)	266 700 (31.)		2.5

Table 49a (continued).

Delta river	Ocean	Country	Delta area km²	River length km (rank)	Drainage area km² (rank)	Av. sediment discharge 10^6 t/a	Tidal range *) m
30. Kongo	A	Zaire	2 100	4 666 (6.)	4 014 500 (2.)	70	1.8
31. Colville	Arc	Alaska	1 700	567 (37.)	59 400 (37.)		0.2
32. Rhône	M	France	1 700 (2 586)	805 (33.)	95 830 (35.)		0.3
33. Magdalena	A	Columbia	1 680	1 529 (29.)	251 000 (32.)	100⋯150	1.0
34. São Francisco	A	Brazil	734	2 574 (21.)	630 000 (23.)		2.5
35. Don	B	USSR	647	1 963 (26.)	414 400 (26.)	5.7⋯6.7	0.0
36. Dnepr	B	USSR	639	2 253 (24.)	502 460 (24.)	11	0.0
37. Ebro	M	Spain	624	623 (35.)	89 800 (36.)	5	0.2
38. Rhine	A	Netherland	2 200	1 300	145 000		3.0
39. Colorado	P	USA–Mexico	1 980		640 000	135⋯160	8.0
40. Fraser	P	Canada	3 000	1 200	270 000		3.0
41. Red River	P	North-Vietnam	7 700		119 000	130	4.0

*) Spring tide, at mouth.

Table 49b. Surface-water suspended sediment concentrations in river and estuarine waters along muddy coasts [77 W3].

Location	Concentration [100 mg/l] maximum	Concentration [100 mg/l] minimum	Source
Haringvliet estuary (Netherlands)	1.00	0.38	Terwindt (1967)
Po River plume	1.10	0.07	Nelson (1970)
Ems estuary	1.80	0.10	Postma (1960)
Thames estuary	2.00	0.01	Inglis and Allen (1957)
Mississippi River (South Pass)	3.10	0.04	Wright (1970)
Chao Phraya River	6.90	0.14	NEDECO (1965)
Surinam River	9.20	0.06	NEDECO (1968)
Bristol Channel	14.00	0.30	Bassindale (1943)

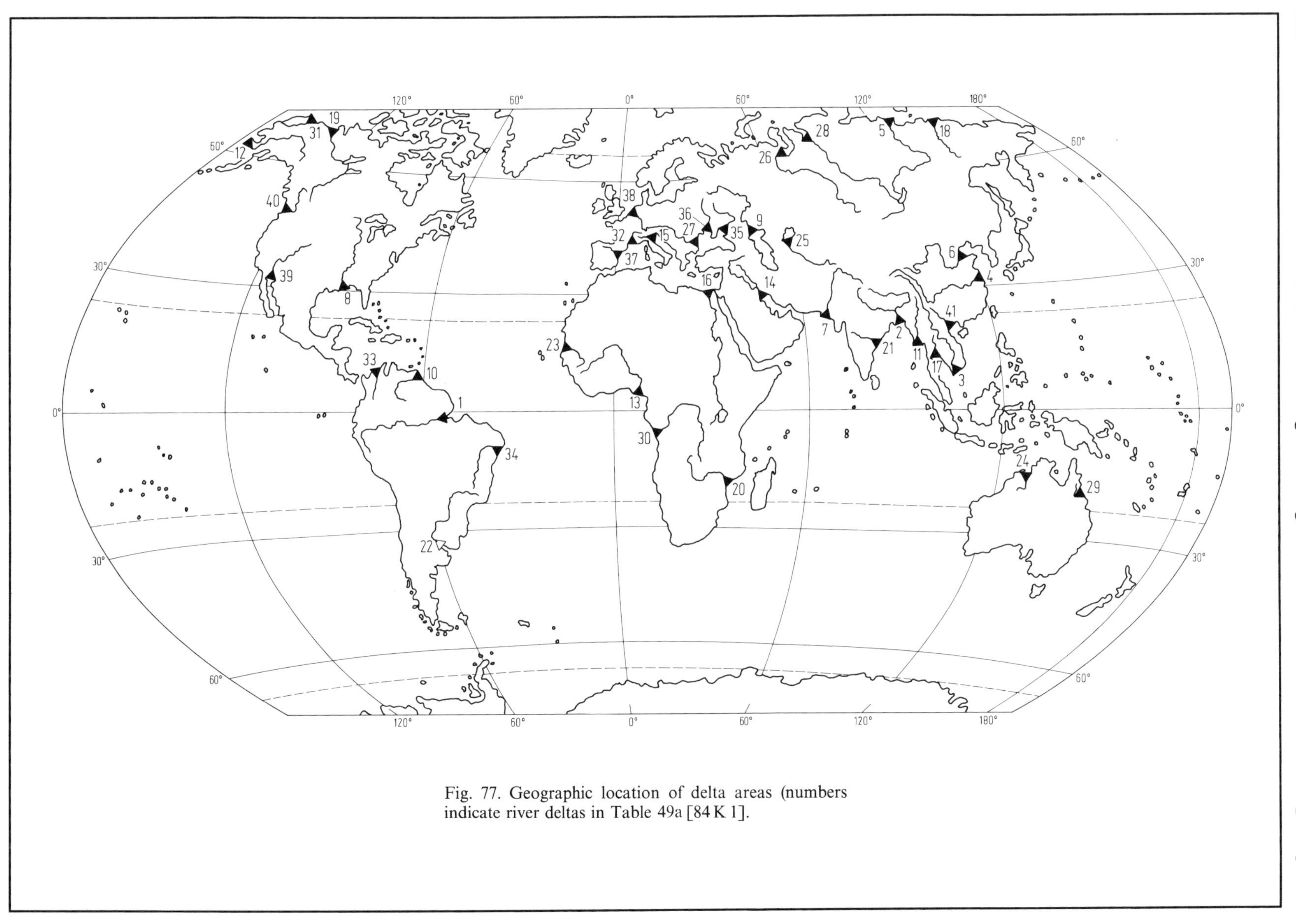

Fig. 77. Geographic location of delta areas (numbers indicate river deltas in Table 49a [84 K 1].

Table 50a. Tidal range and submarine slope inclination of selected deltas [56 S].

Delta	Tidal range (spring) m	Submarine slope inclination %
Ebro	0	0.6500
Danube	0.10	0.3500
Nile	0.33	0.2200
Mississippi	0.46	0.1600
Magdalena	1.00	1.0000
Niger	1.70	0.1350
Senegal	1.22	1.1000
Burdekin	2.53	0.1000
São Francisco	2.50	0.4000
Shatt-al-Arab	2.60	0.2100
Chao Phraya	2.60	0.0085
Hwangho	3.39	0.0550
Indus	4.00 (2.3)	0.2000
Irrawaddy	4.00	0.0450
Yangtze Kiang	4.21	0.0225
Ganges-Brahmaputra	5.65	0.0225
Amazon	5.70	0.0085
Ord	5.85	0.1000

Table 50b. Mean annual wave energy at deltas at the 10-m-isobath off the delta coast, computed according to absolute data [74 W 2, 84 K 1].

Energy dissipation of long waves is starting offshore, depending on submarine deltaic slope inclination.

Delta	Relative [1]) mean annual wave energy at the 10-m-isobath
Shatt-al-Arab	1.0
Ord	3.7
Danube	9.2
Yangtze Kiang	10.2
Hwangho	15.6
Burdekin	18.5
Nile	24.0
Niger	32.6
Ebro	33.7
Mississippi	34.0
Amazon	38.2
Chao Phraya	41.2
Irrawaddy	45.9
Senegal	53.6
São Francisco	111.2
Ganges-Brahmaputra	136.9
Indus	170.8
Magdalena	171.3

[1]) Delta with least wave energy placed as 1.0.

Table 50c. Wave energy at deltas at the delta coast (the shore) in relation to wave energy at the 10-m-isobath of these deltas, computed according absolute data [74 W 2, 84 K 1].
(The dynamic range of the spectrum of this ratio is very large.)

Delta	Wave energy at the coast of the delta	Wave energy at the 10-m-isobath
Senegal	1	2.5
Magdalena	1	4.4
Nile	1	13.0
Burdekin	1	15.4
Ord	1	19.6
São Francisco	1	20.0
Niger	1	53.0
Indus	1	64.6
Chao Phraya	1	300.0
Shatt-al-Arab	1	382.0
Hwangho	1	383.0
Yangtze Kiang	1	431.0
Amazon	1	1059.0
Ebro	1	1115.0
Ganges-Brahmaputra	1	1252.0
Irrawaddy	1	1272.0
Danube	1	1348.0
Mississippi	1	4687.0

9.10 Chemical and biological effects on the coastal zone

9.10.1 Corrosion, dissolution, evaporation, biochemistry, organic features

The dissolving of limestone is represented by the equation:

$$CaCO_3 + H_2O + CO_2 \rightarrow Ca(HCO_3)_2,$$

the rock passing into solution as calcium bicarbonate $Ca(HCO_3)_2$.

Beach rock is formed where a layer of beach sand becomes consolidated by secondary deposition of calcium carbonate at about the level of the water table [80 R 2, 62 R 2, 65 R 2]. The cementing of an aragonite material is precipitated from groundwater in the zone between HW and LW Tide level. The activity of microorganisms can be included.

The beach rock development process can happen in few decades, and is predominant at semiarid subtropical coast. See Table 51.

Eolianites: Under certain climatic conditions (as tropical or subtropical) a sand composed of calcium carbonate rather than quartz is common. Calcium carbonate sand, acted on by the wind, will produce a distinct type of coastal dune, referred to as eolianite or eolian calcarinate, commonly known as dune limestone. This type of dune forms where eolian accumulations of calcareous sand have become lithified.

Vertical zonation of steep coasts caused by chemical and biological processes:

A vertical zonation of steep coasts is caused by the effects of the sea water. Above the surf- and the spray-zones (supra-littoral), as well as below the sea level and low water level (sub-littoral), there are distinct morphologic and biologic zones caused by bio-constructional and bio-destructional processes. (Fig. 79) [85 K 2].

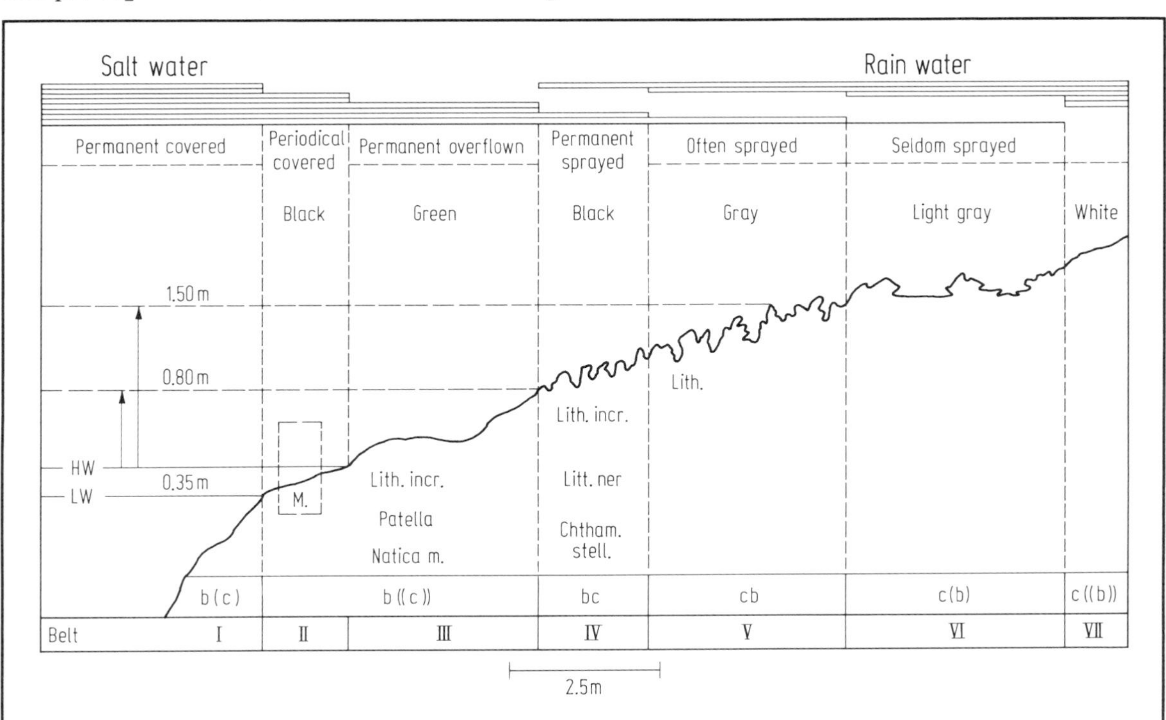

Fig. 79. Vertical zonation of the boundary layer of tidal levels and of the coastal surf- and spray-zones [74 K]. Stories ("étagement"), limestone coast, moderately steep. Zonation of salt water karst features on Eocene limestone near Cape S. Lorenzo, Southern Italy (b = biological, c = chemical erosion; M = Mytilus minimus, Lith-(ophyllum) incr-(ustans), Patella, Natica m, Litt-(orina) ner-(itoides), Chtham-(alus) stell-(atus) [74 K]. Belts I–III: green zone of prevailingly biological and chemical-biological effects, leading to smooth pits and pools, and with plenty of lime-absorbing organisms.

Belts IV–V: black to dark-gray spray-zone with living and dead algae and sea-pox, and sharp lapies and honeycomb-like karst features.

Belt VI: light-gray zone of large flat, shallow pools and honeycomb-like karst features, small lateral notches and strong chemical solution, without macroscopic organisms [74 K, 81 B 1, 76 H 2, 82 V].

Table 51. Climatic parameters in relation to beach rock at mediterranean coasts. [85 K 1].

Station	t_{air} yearly mean	t_{air} coldest month	t_{air} warmest month	Number of months <20 °C	Precipitation	Number of months arid	t_{water} minimum
	°C				mm/a		°C
Murcia	18.0	10.0	26.4	8	304	9	ca. 11
Athen	17.8	9.3	27.6	7	402	6	ca. 13
Iraklion	18.6	12.3	25.6	7	453	6	ca. 13.5
Izmir	17.4	8.3	27.0	8	652	4.5	ca. 11
Antalya	18.9	10.6	28.3	6	1052	4.5	ca. 13
Beirut	20.5	13.6	27.5	6	893	5	ca. 14.5
Haifa	21.4	13.9	28.3	5	668	6.5	ca. 15

9.10.2 Zonation of flat coasts caused by chemical and biological processes

Super-haline lagoons with sebhkas (i.e. evaporitic lagoonal sedimentation): sebhkas are salt encrusted mudflats, marshy only after rare rains, or temporary brackish lakes or saline marshs. See Fig. 80.

Biochemical features at coasts: algal mats and stromatolites. See Fig. 81. [76 W 1].

Application of stromatolite studies:

Holocene sea-level changes on the coast of Mauritania have been determined to be $\pm 0.2 \cdots 1.0$ m from mean sea level by noting algal mat formations during the change from lagoonal to sebhka sequences. Indeed the intertidal algal belt facies may be viewed as the final stage of infilling of marine lagoons that have salinities greater than that of ocean water [74 E 1, 76 W 1].

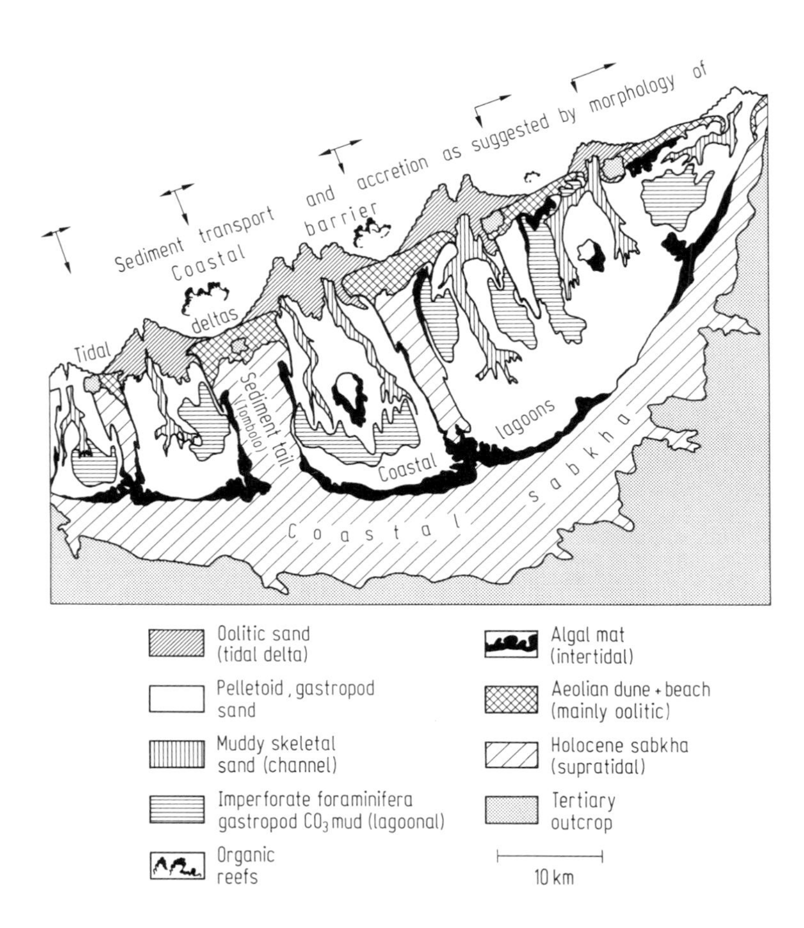

Fig. 80. Schematic sketch-map of the east-half of the central region (E. Abu Dhabi) showing the geometry of the principal sedimentary units [73 P].

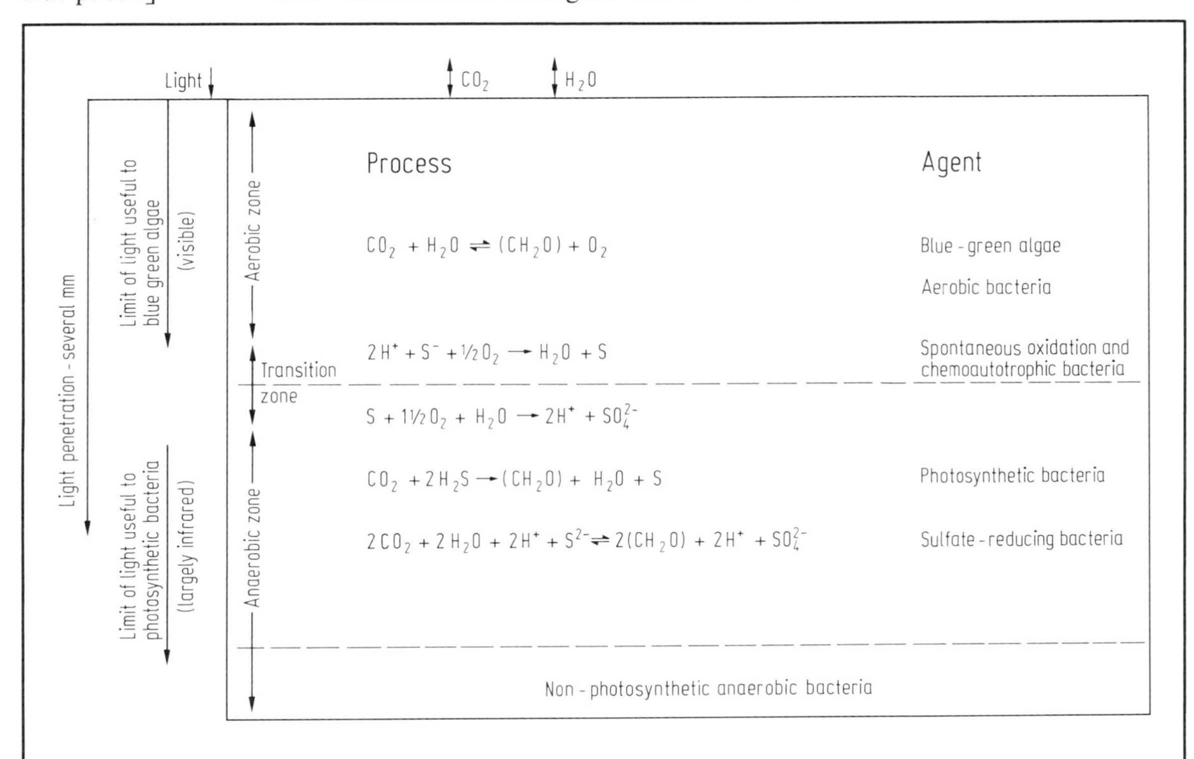

Fig. 81. Top several mm of an idealized laminated algal mat [74 E 1]. For the development of algal mats and stromatolites see [76 W 1].

9.10.3 Coral reefs and atolls: organic created coasts

Compare subsect. 9.8.4, Figs. 73 and 74.

In areas with negative temperature anomalies (areas of upwelling waters), the coral boundary is shifted towards the equator (SW-America, SW-Africa, SW-Australia). Corals and *coral islands* exist in oceanic areas, where doldrums and tropical cyclones determine the atmospheric wind belts. The wide distribution of coral islands in the Pacific Ocean indicates the occurrence of islands, and thus geological-structural features of the sea floor. See Fig. 82.

For terminology and taxonometry of coral reefs, see [56 G, 78 S, 1842 D, 28 D, 65 G, 69 G, 82 H, 77 O, 70 S 3, 68 V 2, 83 B 1]. For distribution of coral islands compare ch. 1.4.3.2, Fig. 57.

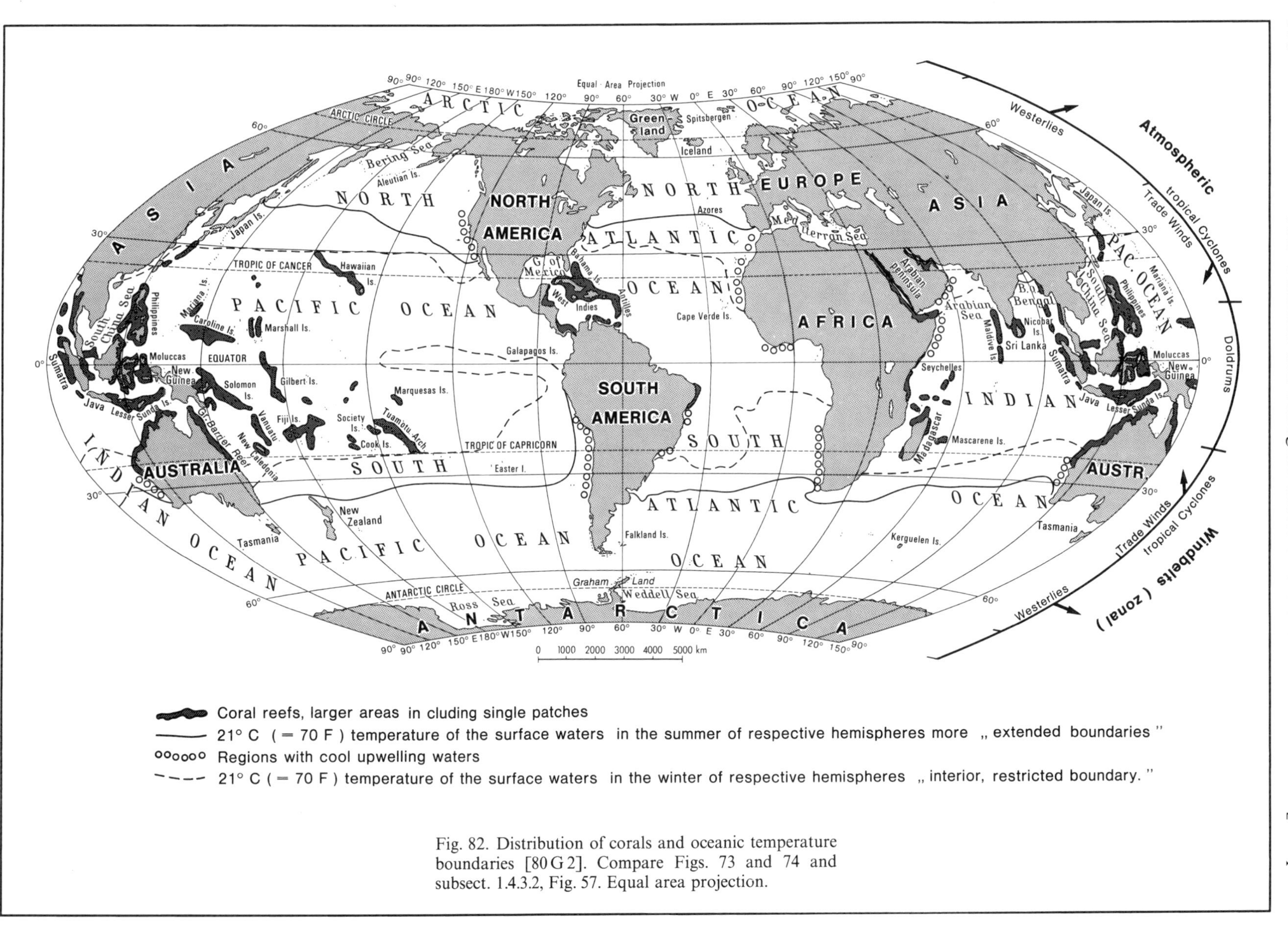

Fig. 82. Distribution of corals and oceanic temperature boundaries [80G2]. Compare Figs. 73 and 74 and subsect. 1.4.3.2, Fig. 57. Equal area projection.

9.11 Climatologic – meteorologic effects on coasts Oceanic – atmospheric coupled systems

9.11.1 Scales of boundary layers

The atmospheric boundary layer is the region of the atmosphere that is directly affected by friction caused by interaction with the earth's surface. Transport in this layer in nearshore and estuarine environments: The wind stress or momentum flux is one of the most essential driving forces in shallow-water circulation. Heat and convection are the origins of some localized coastal weather systems. Sensible heat and water vapor fluxes are necessary elements in radiation and heat budget considerations, including the computation of salt flux for a given estuarine system [70 H]. See Fig. 83. Large scale processes and events cause catastrophical or seasonal impacts on coasts, as there are storm floods. [62 R 1, 77 W 2, 78 I 2, 81 P, 82 R 2, 84 L 2].

The character of particle aerosols in the coastal zone, causing haze and the transport of dust and sand are treated in "Eolian transport to the world ocean" [81 P] in "The ocean lithosphere" [81 E]. [54 B, 75 J].

Time scales of thermal effects depend on radiation periodical processes as day–night and saisonal as summer–winter. The exposure interval of wadden areas is ruled by the ebb-flood period. Compare Fig. 60, Table 39.

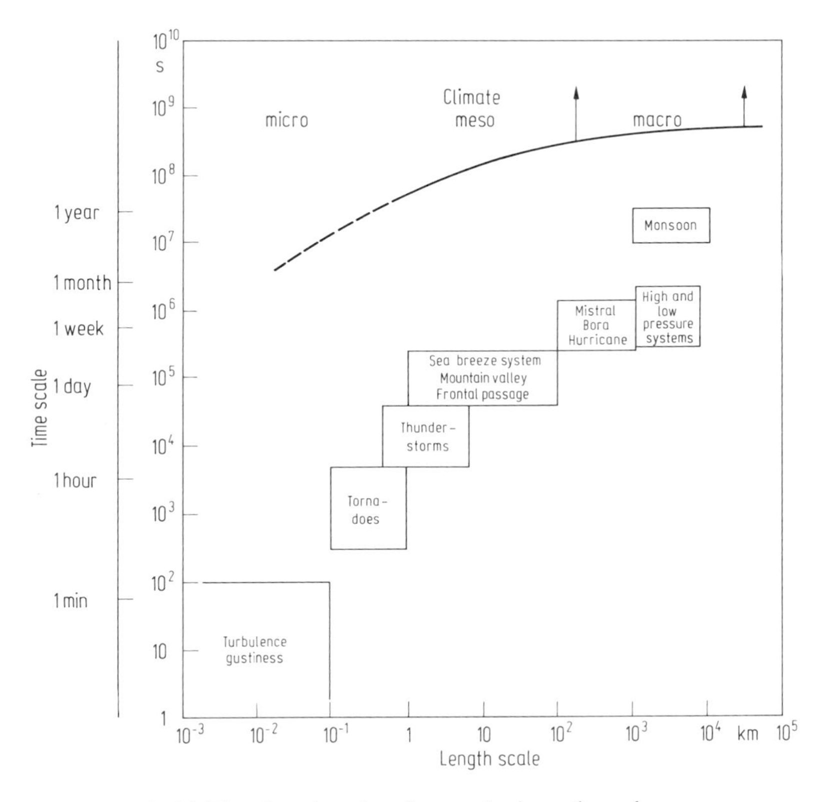

Fig. 83. Time-length scales of atmospheric motions along the coast. After [76 H 2].

For large time-scale of meteorologic effects on coasts see Tables 52, 54.
For medium time-scale of meteorologic effects on coasts see Tables 52, 53, Figs. 85, 86.
For small time-scale of meteorologic effects on coasts see Figs. 84, 85.

9.11.2 Wind stress of small space and time scales on coasts

The land-sea breeze:

The development of a land-sea wind circulation depends temporally and spatially on the direction of the coastline. Along the north-south-running continental coasts, the wind direction changes simultaneously over larger areas, because of the synchroneous sunrise and sunset. On east-west-running coasts, there are time lags from east to west, depending on the position of the sun in the initial phase, the land-sea breeze boundary shows a wedged pattern; the tip of the wedges points to the west. See Fig. 84a.

Coastal atmospheric circulation, meso space scale.

Coastal atmospheric circulation, small space scale.

Importance of meteorologic driving forces:

Varying horizontal pressure gradients because of relative heating and cooling of the land and sea surface generate onshore winds in the early morning hours. Currents near the coast can in fact be influenced strongly by this sea breeze wind system. The sea breeze, a coastal local wind that blows from sea to land, occurs when the temperature of the sea surface is lower than that of the adjacent land (Defant [61 D], Hsu [70 H]). It usually blows on relatively calm, sunny, summer days, and alternates with an oppositely directed, usually weaker, night-time breeze. As a sea breeze regime progresses, Coriolis deflection causes the development of a wind component parallel to the coast. [82 G 2, 84 H].

Sand transport at coasts is processed by diurnal or seasonal winds mainly. See Fig. 84b.

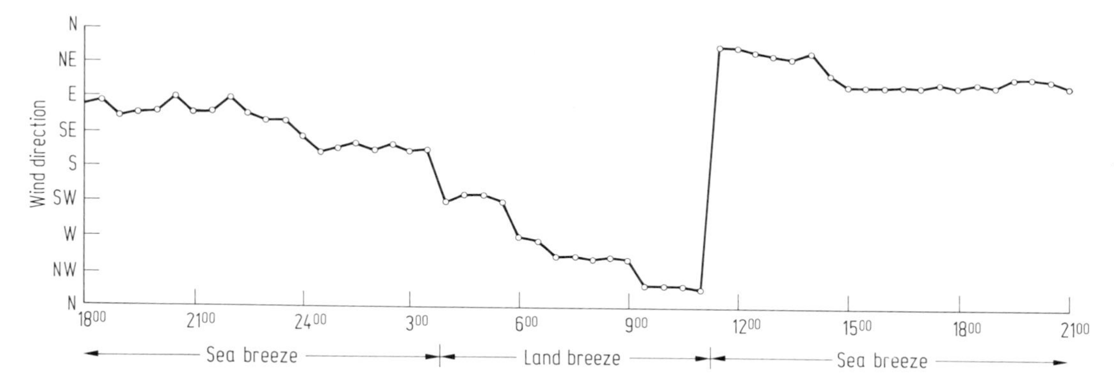

Fig. 84a. Three-hourly wind field in the region of the beach, ridge, and leeward side of a dune on Padre Island, Texas [77 H].

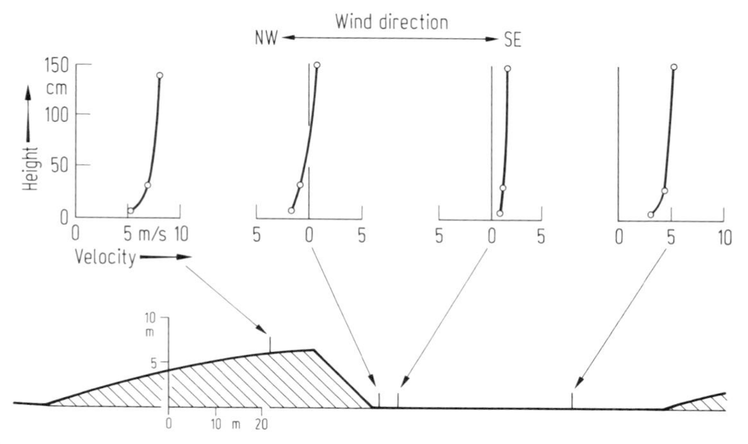

Fig. 84b. Wind velocity profile over a dune. Note the reversal of wind direction at the toe of the slip face. After [82 G 2].

Eolian ripples tend to have large (> 17) ripple indexes (ripple wavelength/height) and symmetry indexes between 2.0 and 4.0, although there are many exceptions. Dune travel distance under such conditions is observed at the Pacific coast with travel scale of 18 m/a.

Cold air outbreaks from continents over the coast to the ocean:

Cold air outbreaks are processes of the time scale of 1 day···5 days and of the space scale of some 100 km···1000 km. They occur, for instance, during the winter season at the coast of the Gulf of Mexico. Cold air outbreaks induce strong latent, sensible, and radiative heat fluxes from the warm sea to the cold atmosphere. Negative oceanic heat flux episodes so induced have an important influence on the physical properties of coastal and continental shelf waters. These processes are extremely efficient, and the total heat flux densities are known to exceed 485 W m^{-2} (1000 cal cm^{-2} day^{-1}). The sea floor or strong density stratification limits the available heat, salt, and mixing volume of the shelf, thus intensifying the resultant water mass transformations.

Cold air outbreaks also result in ocean thermal frontogenesis, high seas, and strong alongshelf currents. Because shallowest waters cool most rapidly, oceanic thermal fronts are formed between shallow and adjacent deepwater areas [84 H].

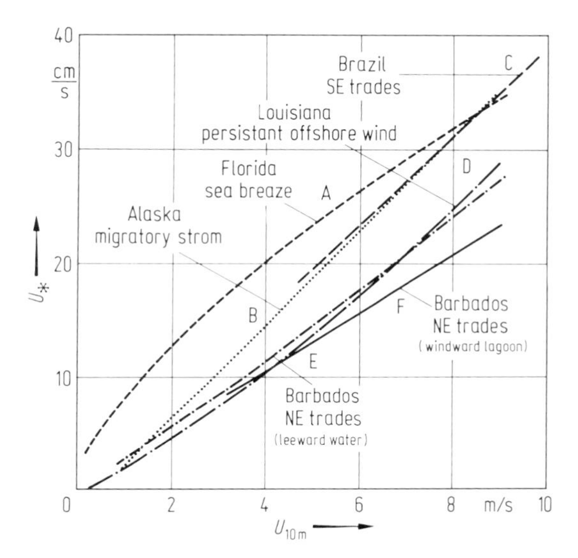

Fig. 85a. Examples of the variations of U_* as a function of U_{10m} over coastal waters. [80 H 4, quoted].

Shear velocity, U_*, plays a very important role in air – sea interaction. U_* has been correlated with the wind speed at 10 m height (U_{10m}), to obtain the drag coefficient. U_* is correlated not only with U_{10m} but also with wave conditions, particularly in coastal waters (Hsu, 1978). [80 Hsu]. Curve A in Fig. 85a was based on experiments performed under diurnal sea breeze conditions (limited duration and fetch) near Fort Walton, Florida; B, under migratory storm conditions (variable duration and fetch) near Point Lay, on the Alaskan Arctic coast; C, under southeast trade wind effects in the shallow water near Aracaju, northeastern Brazil (nearly unlimited duration and fetch); D, under relatively persistent onshore winds caused by the Bermuda high-pressure system in Caminada Bay, Louisiana (long duration and short fetch); E, under northeast trade wind effects (unlimited duration and somewhat limited fetch) in leeward coastal waters in Barbados, West Indies; and F, under trade wind effects in a windward lagoonal environment (unlimited duration but limited fetch) in Barbados. Note also that all experiments were performed by using the same type of instrument (C. W. Thornthwaite Wind Profile Register System) and instrumentation setups. Compare [80 H 4].

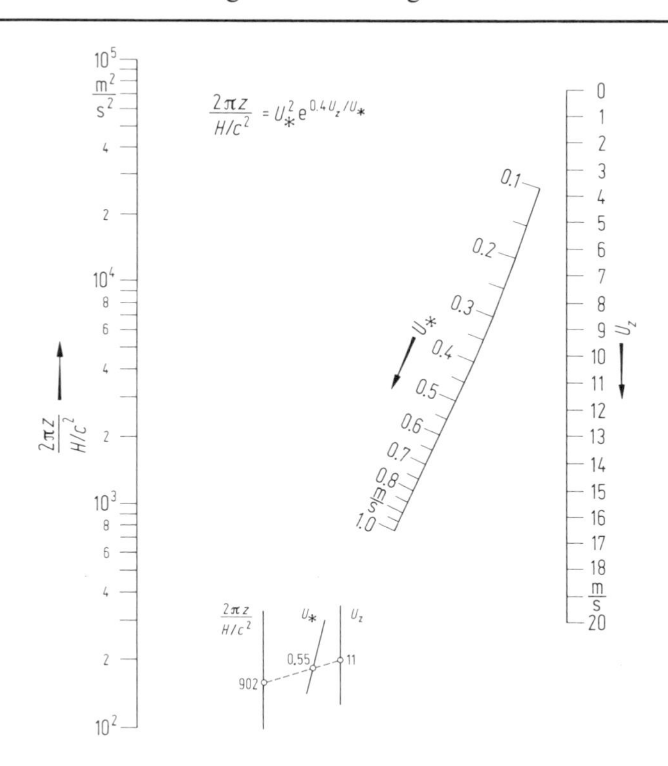

Fig. 85b. A nomogram to obtain U_* using commonly available wind speed at height z, U_z and average wave height H and phase velocity c parameters. [80 H 4].

9.11.3 Wind stress of medium and large space and time scales on coasts

Table 52. Examples of strong local winds [74 D 2].

Type	Name	Location	Direction	Max. speed kn	Season
I. Katabatic					
(a) Fall winds	Bora	Adriatic	NE	40···60	winter
	Mistral	Gulf of Lions	N	50···70	winter
	Papagayo	W. Costa Rica	NE	20···40	winter
	Williwaw	Magellan Street	Any	50···70	any
(b) Föhn winds	Southeaster	Cape Town	SE	50···60	summer
	Zonda	Argentina	W	60···80	winter
	Santa Ana	S. California	NE	30···50	fall
II. Anabatic	Sirocco	N. Mediterranean	SE	30···40	spring
	Virazon	Chile	W	20···30	summer
	Khamsin	N. Africa	SE	30···50	spring

Table 53. Mean annual frequencies of tropical cyclones [74 D 2]. Further tables and statistics, see [82 R 2] and "Marine Weather Log", USA, Department of the Interior.

Month	North Atlantic	North Pacific 100E···170E	North Pacific 80W···120W	Bay of Bengal	Arabian Sea	South Indian	South Pacific 140E···140W
January	0	0.4	<0.1	0.1	0.1	1.3	8.1
February	0	0.2	<0.1	0	0	1.7	4.9
March	0	0.3	<0.1	0.2	0	1.2	7.5
April	0	0.4	<0.1	0.2	0.1	0.6	1.6
May	0.1	0.7	0.2	0.5	0.2	0.2	0.3
June	2.4	1.0	1.0	0.6	0.3	0	0
July	2.6	3.2	0.9	0.8	0.1	0	0
August	2.6	4.2	1.5	0.6	0	0	0
September	3.7	4.6	2.6	0.7	0.1	0	0.3
October	2.5	3.2	1.5	0.9	0.2	0.1	0.3
November	1.5	1.7	0.1	1.0	0.3	0.2	0.8
December	0.1	1.2	0	0.4	0.1	0.8	3.2
Per year	10.9	21.1	7.9	6.0	1.5	6.1	27.0

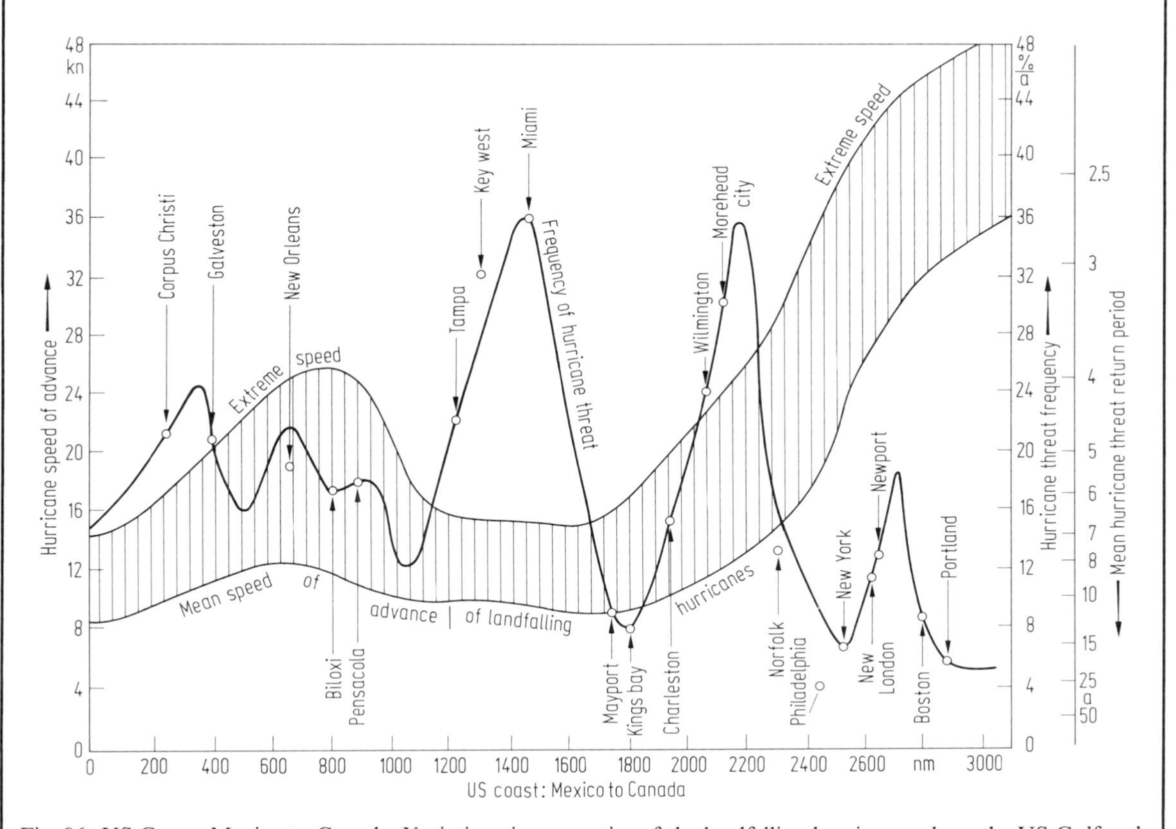

Fig. 86. US Coast: Mexico to Canada. Variations in properties of the landfalling hurricanes along the US Gulf and Atlantic coasts from Mexico to Canada. Variable quantities: the frequency per year of hurricane landfall along the coast, the likely error in the real-time forecast at 48 hours ahead of possible landfall, and the mean and extreme speeds of advance of hurricanes in that region of the coast. From: [82 Hurricane Havens Handbook and Marine Weather Log, USA, Dept. of the Interior]. The mean hurricane threat period is given by the inverse of the hurricane threat frequency. Compare [83 N].

9.11.4 Climatic characteristics related to coastal landscapes

Table 54. Climatic characteristics related to coastal landscapes [60 B1].

This classification contains 14 types, which are in coincidence with 14 types of vegetation.

1 Each pair of numerals in the table refers to data from climatic stations at the 25th and 75th percentiles of the frequency distribution appropriate to the climatic type and element. As only long-period means have been entered into the frequency distributions, the data in the table above show the spread in average conditions of climate in the most representative parts of the several climatic regions. Because approximately equal spacing was employed in the station network, it is also true that the data illustrate, for a given climatic type, conditions in about 50% of the aggregate length of coastline affected by that climatic type.

2 The winter concentration of precipitation is defined as the percentage of the mean annual total that falls in the winter half-year, October through March in the Northern Hemisphere, April through September in the Southern Hemisphere. The computation was not carried out for those places where the difference between the mean monthly temperatures of the warmest and coldest month was less than 3 °C.

Climatic type and percent of world coastline	Typical locality	Climatic conditions[1])				
		Temperature		Precipitation		
		Mean maximum warmest month	Mean minimum coldest month	Mean annual depth	Mean annual no. of days	Winter concentration of precipitation[2])
		°C	°C	cm	>0.004 cm	%
Rainy tropical (20%)	Inner tropics	30···32	19···23	198···310	134···185	14···49
Subhumid tropical (10%)	Border tropics	30···33	15···21	104···140	61···114	17···38
Warm semiarid (2%)	Tamaulipas, Venezuela	32···34	13···19	53···71	42···60	9···40
Warm arid (5%)	Horn of Africa, Sonora	33···37	11···20	13···25	10···32	36···94
Hyperarid (4%)	Cool-water coasts of subtropics	24···34	9···14	< 5	1···4	42···100
Rainy subtropical (6%)	East coasts, lat. 20···35°	29···32	6···9	114···147	93···142	29···49
Summer-dry subtropical (7%)	Mediterranean	27···31	6···9	43···69	54···103	74···87
Rainy marine (1%)	W. coasts, Tasmania, New Zealand	17···20	4···7	109···206	166···187	51···69
Wet-winter temperate (2%)	Oregon, Washington	17···22	0···6	99···170	120···198	67···78
Rainy temperate (9%)	NE United States, W. Europe	20···27	− 7···2	66···112	127···188	41···54
Cool semiarid (1%)	Bahia Blanca	19···31	− 3···9	30···53	45···87	37···52
Cool arid (2%)	Patagonia	21···26	1···7	10···15	24···41	54···88
Subpolar (6%)	Gulfs of Alaska, Bothnia	15···23	−23···−13	46···104	106···184	32···50
Polar (25%)	Arctic Sea border	9···13	−34···−13	18···66	91···131	30···49

9.12 Sea level change and the coastal zone

9.12.1 Eustatic changes of the sea level

Changes of the sea level occur in geologic time intervals because of tectonic or sedimentologic changes, or because of changes in the water-ice budget [1882 P, 59 Z, 65 R 1, 70 G, 79 P 1, 61 F 1, 76 B 2, 82 K 1, 83 F 1, 71 M 2, 82 M 2].

Processes which can effect changes in sea level through geologic time:

1. *Tectono-eustatic changes*, which result from tectonic modifications in the shape and volume of ocean basins [66 F, 61 F 1, 69 M].

Global tectonics and eustasy for the past $2 \cdot 10^9$ years are discussed and allow to be interpreted in time scales as small as $440 \cdot 10^6$ years [84 W 1].

2. *Sedimento-eustatic changes*, which are due to accumulation of sediments in ocean basins.

3. *Addition of juvenile water* due to submarine volcanism. The assumption of accumulation rates constant in time since the formation of the earth leads to a sea level rise of less than 1 m/10^6 years. Thus sea level in the middle Cretaceous ($100 \cdot 10^6$ years) could have been up to 100 m lower than present. Small amounts of water may be removed due to hydrothermal alteration of the oceanic crust.

4. *Glacial isostasy*, which results from ice loading and unloading. Glacial isostatic subsidence helps counteract a glacio-eustatic drop in sea level. The reverse is true when the ice sheets melt.

5. *Hydro-isostatic deformation*, which results from the loading effect of sea water over the continental shelf and nearshore regions. As sea level rises, subsidence occurs, thus amplifying a transgression.

6. *Tectonic erosion*, which results from deep-sea sediments being removed from the ocean basins to the earth's interior in the course of subduction. Little is known about the efficiency of this process compared with accretion at the active margins. [84 W 3, 80 S 1, 82 K 1].

7. *Geodetic sea-level changes* are a process of potential geological importance, causing undulations of several meters amplitude due to regional variations in the geoid [79 C]. The history of geodetic change is still unknown. Eustatic curves are valid only regionally, not globally [83 M 4, 81 W 1].

8. *Astronomic-geophysical parameters* are recently discussed by R. F. Fairbridge [83 F 1], [38 M].

Interacting of ocean, atmosphere and ice caps evolved in a complex way during the past.

The scheme of 4 ice ages as developed by Penck and Brückner (1909) appeared not satisfactory since Eberl (1930) found some 15 ice advances, which were adapted only as "stadials."

In 1947, Urey found that the oxygen-isotopic composition of the carbonate would depend significantly on the temperature of water at the time of carbonate deposition. The new geological thermometer, the ^{14}C dating method, from foraminiferal fauna in deep sea cores, the alternating representing warm and cold epoches could be detected. This way 9 to 10 warm and cold layers were identified, which gave a picture of "ice ages" very different from those as postulated in 4 [84 B].

A reliable time scale back to 900000 years by an isotopic curve from cores was presented in 1973, which showed a total of 8 ice ages during this very time of the geomagnetic Brunhes Epoche of 700000 years [81 E].

Glacial marine sedimentation [82 K 1] is of importance for coastal oceanographic sediment research to compare present day sedimentation with the sedimentation of former climatic ages as reference conditions.

9.12.2 Definitions

Glacial eustasy: changes of the sea level caused by the shifting of the water-ice-balance described in subsect. 9.12.1 [61 F 1, 71 F].

Ice-ages: larger time intervals (order of 1 million years), which can be subdivided by glacial-cycles, i.e. time intervals of 100000···10000 years, which caused sea level changes of the order of 100 m (135 m).

Stadials: low stands of sea level associated with ice advances.

Interstadials: higher stands associated with retreats of the ice sheets.

Transgressions and regressions: repeated sea level advances and retreats, which have been instrumental in shaping the continental shelf, especially during low sea level stands, when the shelf was exposed to subaerial processes.

Marine limit: the maximum elevation which the late- or post-glacial sea reached at a coastal site, measured with respect to present sea level.

Eustatic sea level: world-wide sea level measured with respect to present sea level.

Post-glacial uplift: the glacio-isostatic recovery of the earth's crust between the instant of deglaciation and the present. Post-glacial uplift Up is defined as the difference of the marine-limit elevation ML and the change in eustatic sea level E between the date of deglaciation and the present: Up = ML − E. (In Arctic Canada, the equation in general is Up = ML − (−E)).

Post-glacial emergence: the upward movement of land relative to sea level.

Isobase: a line joining points of equal post-glacial emergence (coastal sites) or uplift (glacial lake shorelines) operating over the same length of time.

Equidistant diagram: a diagram drawn in a plane orthogonal to the local system of isobases. On this graph, the y axis is elevation above present sea level and the x axis, distance. The projection of a strandline on the plane shows the extent and geometry of crustal deformation.

Strandline: the trace of a former sea level or lake level as indicated by a variety of raised (or submerged) morphological/sedimentological features, such as beach terraces, deltas and shingle ridges.

Shoreline relation (SR) diagram: a graph on which the x axis is the elevation above present sea level of a particular strandline and the y axis the elevation of other strandlines above this "reference level." Other strandlines are related to the reference strandline by an equation of the form: $y = a + bx$ [70 A 2].

9.12.3 Eustatic sea level parameters

Eustatic variation is the ultimate barometer of climatic change. The absolute range is about 200 m, corresponding in volume to $0.36 \cdot 10^{12}\,\text{m}^3$ of water per mm range, or $72 \cdot 10^6\,\text{km}^3$. The total volume of the world's land ice today is about $30 \cdot 10^6\,\text{km}^3$, and was about $(40 \cdots 45) \cdot 16\,\text{km}^3$ greater during glacial phases. See Table 55. [78 C].

Details of relative sea level curves for all areas known at the time are shown in an "Atlas of sea level curves" compiled by Bloom [77 B 1].

The research on eustatic sea level change was up to 1960 restricted on methods concerning coastal morphology and paleontology, which lead to relative sea level variation curves. The classic work is that of R. W. Fairbridge "Eustatic change in sea level" 1961 in "Physics and chemistry of the earth", Vol. 4, and [83 F 1]. Since two decades radioactive age testing methods lead to the present scene of sea level change [61 F 1, 65 F, 66 F, 57 E, 82 M 2, 83 M 1, 76 R 3, 81 R 2, 81 S 1, 82 S 2].

Changes of mean sea level:

A pioneer study by Gutenberg (1941) of data from worldwide tide gauge records suggested that the world mean sea level was rising at about 1.1 mm/a during the first half of the twentieth century. Last figure being 1.2 mm/a (1900⋯1950); and there are stages of very rapid rise (e.g. 1946⋯1956 at an average rate of 5.5 mm/a). [77 R 2].

Rates and amplitudes:

The rate of sea level changes depends upon the factor causing the change; tectono-eustasy is a slow process (less than 1 mm/a), glacial eustasy is only relevant during periods of glacial volume changes and may reach rates of up to 10 mm/a, and geoidal eustasy may reach rates of up to 30 mm/a and often amounts to 10 mm/a. [85 F 2].

Table 55. Causes of sea level fluctuations. [75 K 1].

Cause	Effect, [m]		Duration, [a]	Rate of change, cm/1000 a
Irreversible trends				
Differentiation of continents and mantle	Ocean creation		$4 \cdot 10^9$	0·1
Accumulation of Pacific submarine volcanics	Raise sea level		$20 \cdot 10^8$	0·02
Accumulation of unconsolidated sediments	Raise sea level		$160 \cdot 10^9$	0·02
Accumulation of 2nd seismic layer	Raise sea level		$180 \cdot 10^9$	0·02
Fluctuations				
Melting of existing ice	Raise sea level	46	–	–
Formation of max. Pleistocene glaciers	Lower sea level	200	$(1 \cdots 2) \cdot 10^5$	>100
Melting of most recent Pleistocene glaciers	Raise sea level	120	$2 \cdot 10^4$	860
Elevation of existing ocean rises	Raise sea level	300	10^8	0·3
Sinking of Darwin Rise	Lower sea level	100	10^8	0·1

Time units and amplitudes of sea level changes range from $1 \cdots 10^6$ time units equivalent to 0.01 m⋯>200 m amplitudes. [83 M 3].

Calculation of eustatic sea level [66 F]. See Fig. 87:

Actual elevation of sea level at any stage of the quaternary can be calculated by use of the formula:

$$T \cdot G - (Wt - Wi) = 0\,,$$

where

T = time (expressed as a percentage from the beginning)

G = negative eustatic effect of geotectonic processes (such as downwarp of ocean basins, not locally compensated, between 200 m and 50 m, suggested figure 135 m)

Wt = total water removed from ocean during glacial maxima expressed as a eustatic fall (165 m ± 20 m)

Wi = water returned to ocean in interglacials, expressed as eustatic rise (75⋯110 m the higher figure during warmer interglacials).

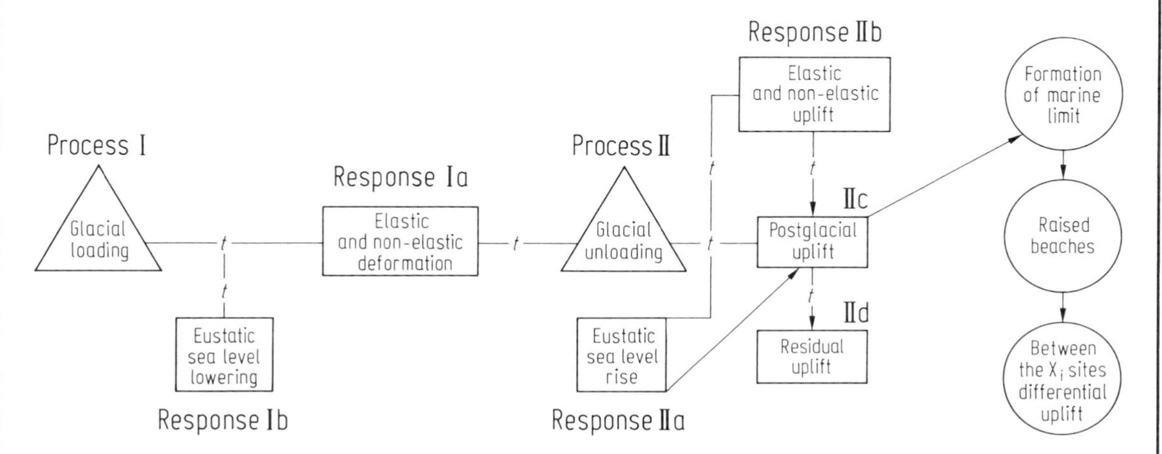

Fig. 87. A process-response model for glacio-isostatic studies [70 A 2].

9.12.4 Quaternary sea level history

The Flandrian Transgression, time-scale 1⋯12000 years. See Fig. 88.

It is now accepted that sea level has risen 100⋯130 m between 18000 and 6000 years ago after which time sea level has remained close to that of the present day. During the most rapid phase of deglaciation, from about 10000⋯7000 years ago, sea level probably rose at a rate of 10 mm/a. This eustatic sea level rise is called the "Flandrian Transgression" in Europe and other parts of the world [62 S 2, 74 E 2, 61 F 1, 80 H 1, 66 H 1, 67 J, 77 K, 77 P, 58 S, 76 Z 2, 83 F 1, 52 P 1, 57 C, 69 C, 71 E, 71 F 1, 79 J 2, 79 J 3, 79 O 1, 83 R 1, 84 M, 85 K 2].

For Quaternary coastlines and marine archeology compare [83 M 4] and Fig. 89.

The pleistocene-holocene boundary is discussed by R. F. Fairbridge at the INQUA Congress, the most recent review 1983 (83 F, p. 215⋯244].

Time scale $10 \cdots 12 \cdot 10^3$ years.

Late Quaternary sea level history: time scale $10 \cdots 40 \cdot 10^3$ years.

The history of the most recent sea level oscillation, during the last 30000⋯40000 years, has been studied using radiocarbon dates of materials known to have close relations with sea level, including salt-marsh and fresh-water peat deposits, very shallow-water marine or brackish molluska, coralline algae, and beach rock. See Fig. 89. [65 C, 68 M, 79 M 2, 72 S 1].

Earlier Quaternary sea level history: time scale $10^3 \cdots 140 \cdot 10^3$ years. Because radiocarbon dating is limited to materials younger than 40000 years, an understanding of older sea level history is obtained using ^{230}Th dating of fossil corals in uplifted terraces and oxygen isotopic studies. These studies indicate that a sequence of glacio-eustatic sea-level oscillations must have extended throughout the earlier Quaternary.

Gierloff-Emden

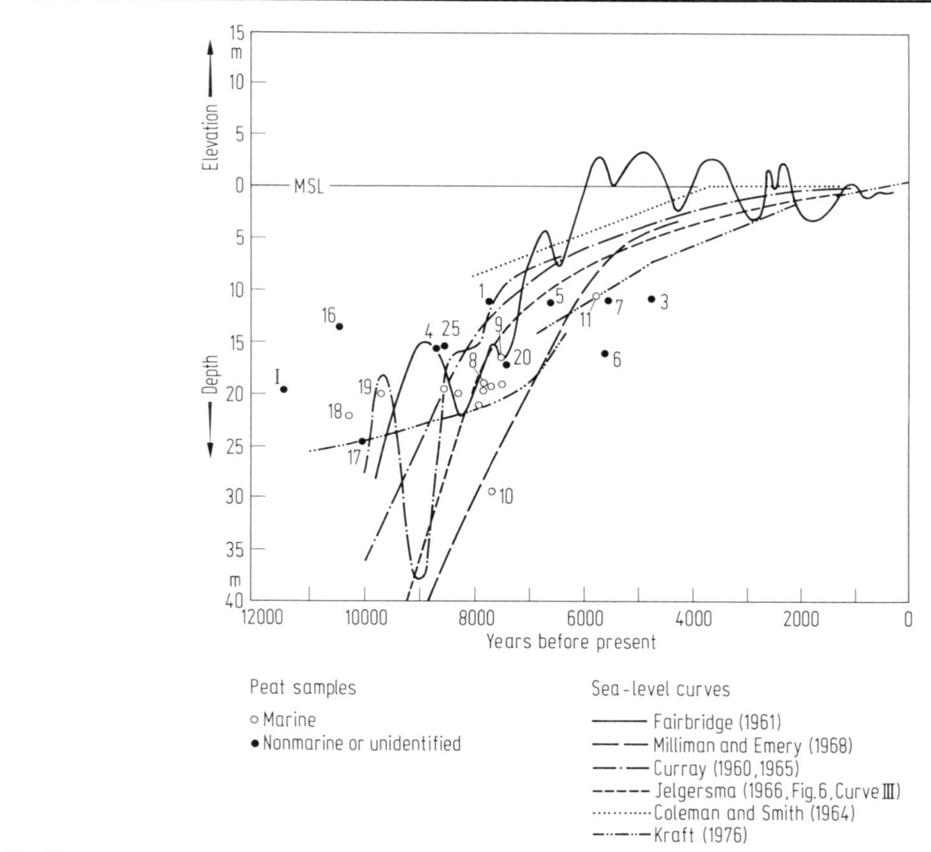

Fig. 88. Age-depth relations of Holocene peat samples from east coast of the United States and family of sea level curves for the last 10 000 years (modified by Field et al., 1979 after Curray 1969). (From M. E. Field et al., Geol. Soc. Amer. Bull. Part I, vol. 90, p. 626, 1979, courtesy: The Geological Society of America) [79 F].

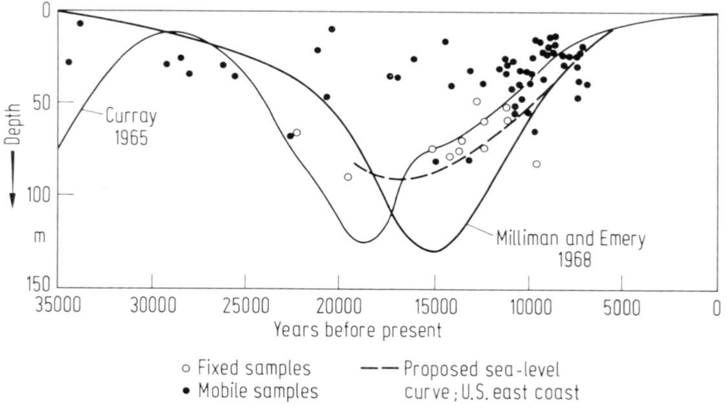

Fig. 89. Sea level curves for the late Quaternary inferred from radiocarbon-dated samples along the coast of the USA. The most recent curve (dashed in figure) suggests a late (Wisconsin low stand of less than 100 m below present sea level. After [78 D 2]. [68 E, 72 E].

The Barbados date.

A detailed curve representing sea level oscillations has been compiled by Bloom and others (1974) for the last 140 000 years, based on the dated terraces of New Guinea and Barbados. It shows the sea level 6 m above the present level 125 000 years ago, succeeded by pulses of sea level fall, with the decrease interrupted by brief intervals of high sea levels, i.e. levels still well below the present sea level [82 K 1, 73 M]. See Fig. 90.

Emiliani and Rona analyzed Caribbean cores to set up a new absolute chronology [69 E, 66 E].

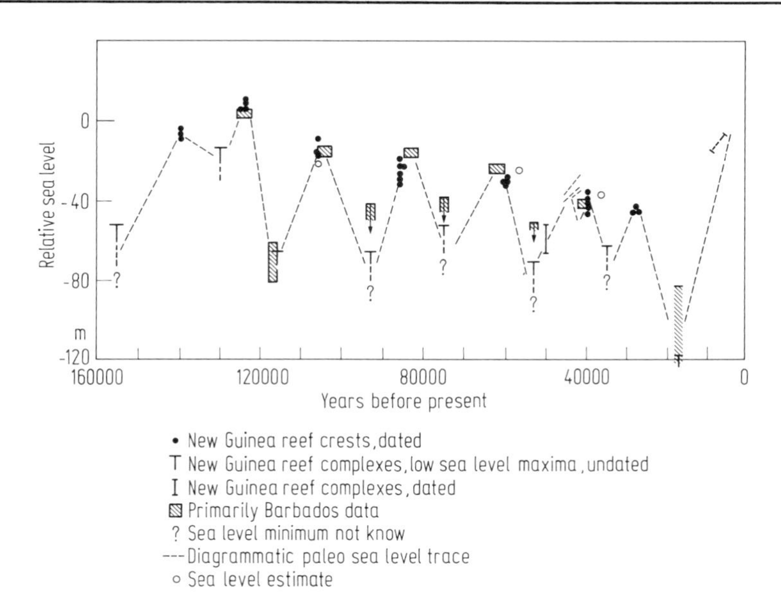

Fig. 90. Sea level fluctuations during the last 140 000 years determined from ages and altitudes of coral-reef terraces on tectonic coasts. The amplitudes of the oscillations were calculated using the assumption of constant rate of tectonic uplift. (From A. J. Bloom et al., Quaternary Research, vol. 4, no. 2, p. 203, 1974) [74 B].

Eustatic sea level and geologic classification of Quaternary time tables: Large time scale $10^4 \cdots 2 \cdot 10^6$ years. See Table 57.

Marine archaeology became an important science for the investigation of the continental shelves to gather facts and information on the relationship of early human settlement, environment, and the eustatic sea level changing, i.e. a discipline to be encompassed in coastal oceanography. Table 56. [83 M 1].

The area of the continental shelf is 5% of the entire area of the earth, equivalent to a continent the size of North America. This continental area was once available for exploitation by Stone Age peoples. [67 H, 85 F, 84 S 2].

Table 56. Archaeological periods. Approximate beginning and ending dates for the archaelogical periods. The terms refer to different styles and technologies for making flint tools. The transitions from one style to the next took place at different dates in different parts of the world. These dates are generally accepted for Europe and the Middle East. From [85 F 1].

Geological time	Geological sub-division	Archaeological period	Stone tool type	Years before present
Holocene				3 000
		Bronze age		5 000
		Neolithic		8 000
		Mesolithic	Microliths	10 000
Pleistocene		Upper	Magdalenian	25 000
		paleolithic	Aurignacian	35 000
	Upper	Middle	Levolloiso-	
	pleistocene	paleolithic	Moussterian	50 000
			Acheullian	
			Flake tools	100 000
	Middle	Lower	Handaxe	250 000
	pleistocene	paleolithic		500 000
	Lower		Chopper tools	3 million
	pleistocene		first tools	

Table 57. The marine Quaternary [77 B 2].
Partly based on Rep. Deep Sea Drilling Project 69–74.

Classification	Geologic time years BP	Magnetic epoch (years BP)	Europe: steps (Berggren, v. Couvering)	Europe: mediterranean (K. H. Kaiser, P. Woldstedt, M. Schwarzbach)	North-America Californian steps	Europe glacial stratigraphy (in comparison)
Holocene	10 300			Flandrian (Nizza)-transgression		Holocene
Late-	100 000	*Laschamp event* (20···30 000)		Post Monastir regr.		Weichsel-(Würm) glacial stage
				Monastir-transgr. (+20 m)		Eem interglacial stage
Middle-	200 000			(Ougartien) (Sahara)		Saale (Riss) glacial stage
				Posttyrrhenian regression		
	300 000				Hallian	
		Brunhes (normal)		Tyrrhenian transgression (+30 m)		Holstein interglacial stage
	400 000					
	500 000			(Taouritien) (Sahara)		Elster (Mindel) glacial stage
Early-				Roman regression		
	600 000			(−200 m)		
	700 000	690 000	Sicilian	Milazzo transgression (+60 m)		Cromer interglacial stage
	800 000					
	900 000			Syrian regression		Menap (Günz?) glacial stage
Pleistocene		*Jaramillo event*	Emilian			
				Sicilian transgression (+100 m)		Waal interglacial stage
	1 000 000					
	1 100 000			Post-Calabrian regression		Eburon glacial stage
	1 200 000					

Gierloff-Emden

Table 57 (continued).

Classi-fication	Geologic time years BP	Magnetic epoch (years BP)	Europe		North-America Californian steps	Europe glacial stratigraphy (in comparison)
			steps (Berggren, v. Couvering)	mediterranean (K. H. Kaiser, P. Woldstedt, M. Schwarzbach)		
Earliest-	1 300 000	Matuyama (revers)				Tegelen interglacial stage
	1 400 000 1 500 000 1 600 000		Calabrian	Calabrian transgression (> +150 m) ("Emiliano" sensu K. H. Kayser)	Wheelerian	
	1 700 000	*Gilsa event*				Prätegelen glacial stage
	1 800 000					
Bottom layer Pliocene		Olduvai (1 950 000 ··· 2 130 000)	(Astium/ Piacenzium)	("End/Post-Pliocene regression")	(Venturian)	

Table 58. Sea state codes corresponding to the Beaufort wind scale (after Defense Mapping Agency Hydrographic Center, 1977).

Beaufort wind scale	Wind speed		Seaman's term	U.S. Weather Bureau term	Effects of wind observed at sea (Petersen scale)	U.S. Navy Hydrographic Office (Douglas scale)		International (W.M.O.) code	
	mi/h	m/s				Term and height of waves ft	Code	Term and height of waves ft	Code
0	under 1	0.0…0.2	Calm		Sea like mirror.	Calm, 0	0	Calm, glassy, 0	0
1	1…3	0.3…1.5	Light air	Light	Ripples with appearance of scales; no foam crests.	Smooth, less than 1	1		
2	4…7	1.6…3.3	Light breeze		Small wavelets; crests of glassy appearance, not breaking.	Slight, 1…3	2	Rippled, 0…1	1
3	8…12	3.4…5.4	Gentle breeze	Gentle	Large wavelets; crests begin to break; scattered whitecaps.	Moderate, 3…5	3	Smooth, 1…2	2
4	13…18	5.5…7.9	Moderate breeze	Moderate	Small waves, becoming longer; numerous whitecaps.			Slight, 2…4	3
5	19…24	8.0…10.7	Fresh breeze	Fresh	Moderate waves, taking longer form; many whitecaps; some spray.	Rough, 5…8	4	Moderate, 4…8	4
6	25…31	10.8…13.8	Strong breeze	Strong	Larger waves forming; whitecaps everywhere; more spray.			Rough, 8…13	5
7	32…38	13.9…17.1	Moderate gale		Sea heaps up; white foam from breaking waves begins to be blown in streaks.				
						Very rough, 8…12	5		

Table 58 (continued).

Beaufort wind scale	Wind speed mi/h	Wind speed m/s	Seaman's term	U.S. Weather Bureau term	Effects of wind observed at sea (Petersen scale)	U.S. Navy Hydrographic Office (Douglas scale) Term and height of waves ft	U.S. Navy Code	International (W.M.O.) code Term and height of waves ft	International Code
8	39···46	17.2···20.7	Fresh gale		Moderately high waves of greater length; edges of crests begin to break into spindrift; foam is blown in well-marked streaks.			Very rough, 13···20	6
9	47···54	20.8···24.4	Strong gale	Gale	High waves; sea begins to roll; dense streaks of foam; spray may reduce visibility.	High, 12···20	6		
10	55···63	24.5···28.4	Whole gale		Very high waves with overhanging crests; sea takes white appearance as foam is blown in very dense streaks; rolling is heavy and visibility reduced.	Very high, 20···40	7	High 20···30	7
11	64···72	28.5···32.6	Storm	Whole gale	Exceptionally high waves; sea covered with white foam patches; visibility still more reduced.	Mountainous, 40 and higher	8	Very high, 30···45	8
12	73···82	32.7···36.9	Hurricane	Hurricane	Air filled with foam; sea completely white with driving spray; visibility greatly reduced.	Confused	9	Phenomenal, over 45	9
13	83···92	37.0···41.4							
14	93···103	41.5···46.1							
15	104···115	46.2···50.9							
16	115···125	51.0···56.0							
17	126···136	56.1···61.2							

9.12.5 Change of sea level during the Tertiary

Time scale $10^6 \cdots 65 \cdot 10^6$ years

For comparison and cause of sea level change due to sea floor spreading, see sect. 3.1 [82 K 1].

For Fig. 91, see next page.

Fig. 91. Global cycles of relative change of sea level during Jurassic-Tertiary time. Cretaceous cycles (hachured area) not yet published. (From P. R. Vail, R. M. Mitchum, jr. and S. Thompson III, Seismis Stratigraphy-Applications to Hydrocarbon Exploration, Memoir **26** (1977) pp. 66, 70, 72, 78, reproduced with permission of the American Association of Petroleum Geologists). Compare "Sea-floor spreading", subsect. 1.6.4.
The relative change of sea level encompasses a scale of approximately 650 m.

9.12.6 Sea level history of geologic ages in relation to coal and oil source rocks

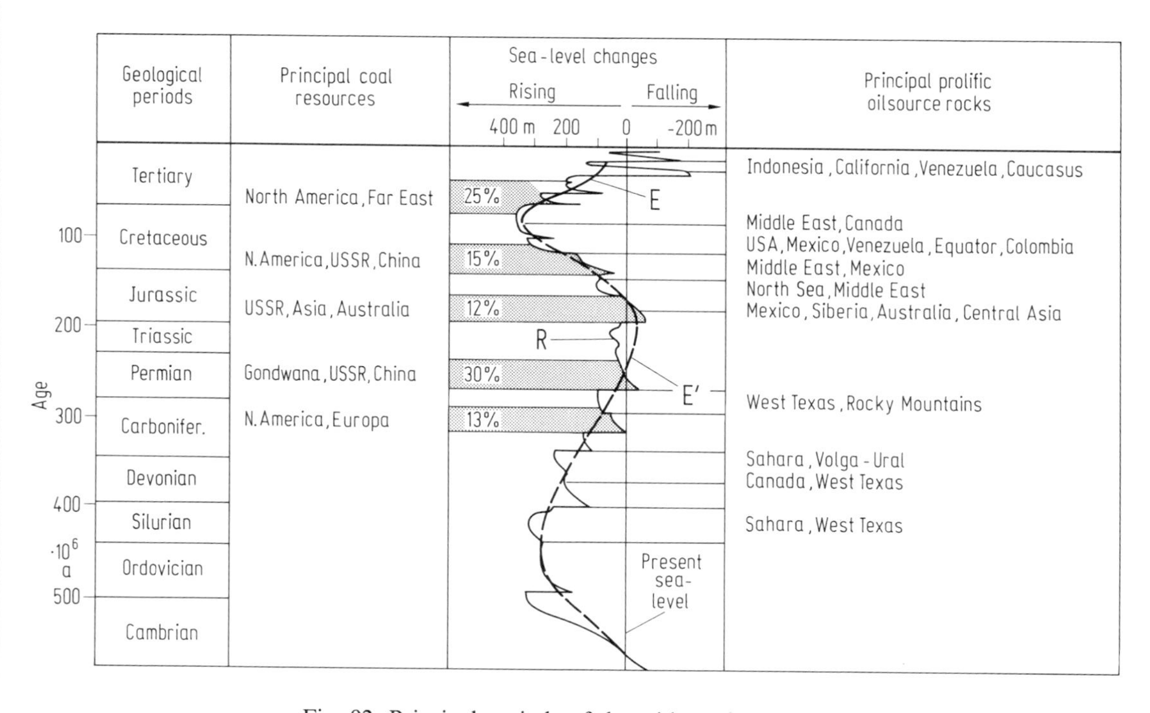

Fig. 92. Principal periods of deposition of prolific oil source rocks and major coal deposits are compared with worldwide transgressions and regressions as determined by Vail et al. (1979) = R, and Pitman (1978) = E. The periods considered for petroleum source rocks are responsible for more than 10^8 t of oil Myr^{-1} during the first cycle (570···200 Myr) and more than 10^9 t Myr^{-1} during the second cycle (200···0 Myr). From [82 K 1]. For the geologic time-scale compare subsect. 1.7.1. Compare [77 V 1].

Fig. 91

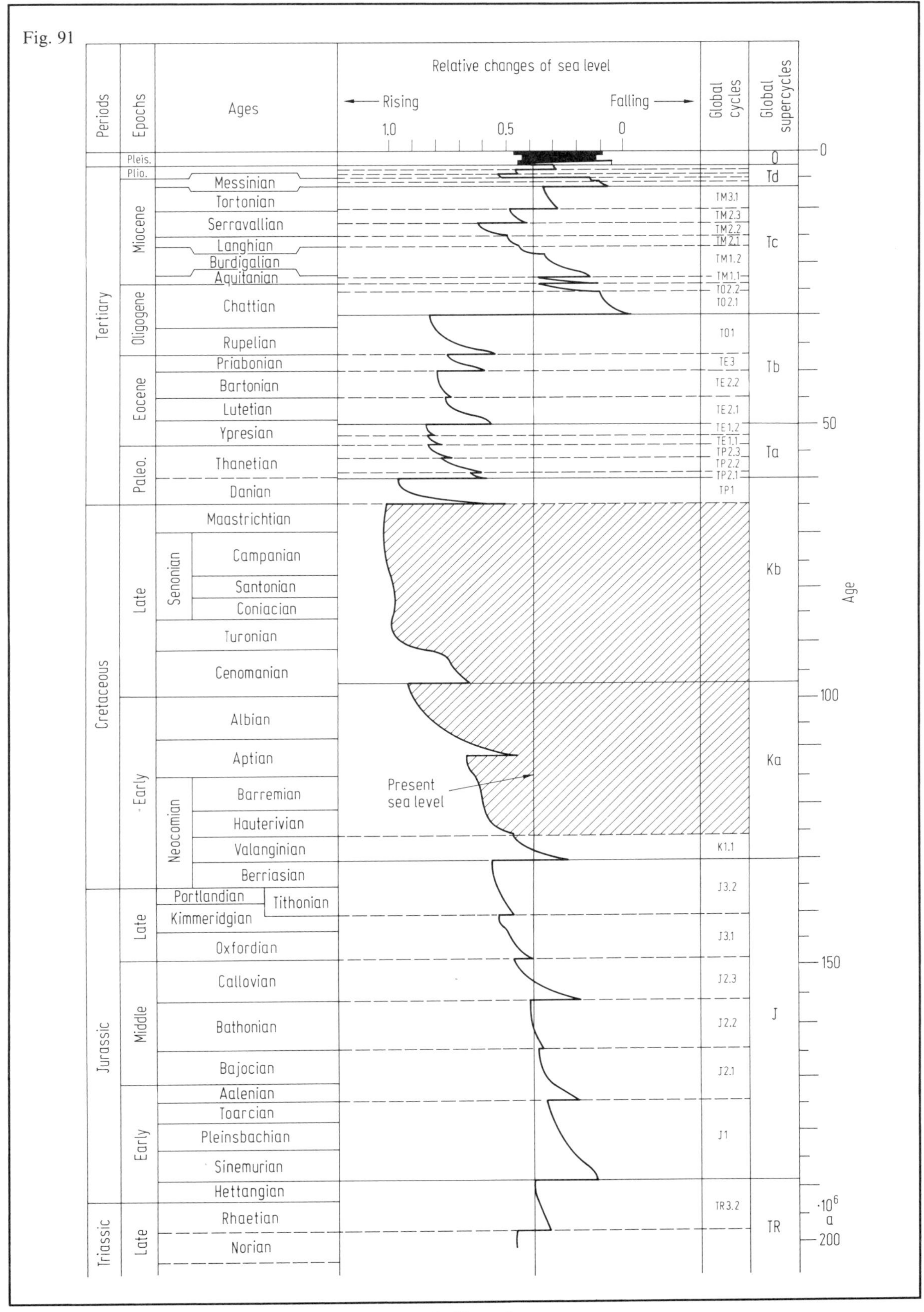

Periods
Epochs
Ages
Relative changes of sea level
Rising
Falling
1.0
0.5
0
Global cycles
Global supercycles
Pleis.
Plio.
Messinian
Tortonian
Serravallian
Langhian
Burdigalian
Aquitanian
Chattian
Rupelian
Priabonian
Bartonian
Lutetian
Ypresian
Thanetian
Danian
Maastrichtian
Campanian
Santonian
Coniacian
Turonian
Cenomanian
Albian
Aptian
Barremian
Hauterivian
Valanginian
Berriasian
Portlandian
Tithonian
Kimmeridgian
Oxfordian
Callovian
Bathonian
Bajocian
Aalenian
Toarcian
Pleinsbachian
Sinemurian
Hettangian
Rhaetian
Norian
Tertiary
Cretaceous
Jurassic
Triassic
Miocene
Oligogene
Eocene
Paleo.
Late
Senonian
Early
Neocomian
Middle
Present sea level
TM3.1
TM2.3
TM2.2
TM2.1
TM1.2
TM1.1
TO2.2
TO2.1
TO1
TE3
TE2.2
TE2.1
TE1.2
TE1.1
TP2.3
TP2.2
TP2.1
TP1
K1.1
J3.2
J3.1
J2.3
J2.2
J2.1
J1
TR3.2
Q
Td
Tc
Tb
Ta
Kb
Ka
J
TR
0
50
100
150
200
$\cdot 10^6$ a
Age

9.13 References for 9

1842 D Darwin, Ch.: Structure and distribution of coral reefs. Berkeley: **1962**, Univ. of Calif. Press, p. 241 (1. ed. 1842).

1882 P Penck, A.: Jb. Geogr. Ges. München **7** (1882).

1888 S Suess, E.: The faces of the earth, Vol. II (Engl. transl. H. B. Sollas, 1906), London: Oxford University Press **1888**, 5 vols.

11 K Krümmel, O.: Handbuch der Ozeanographie, Vol. II: Die Bewegungsformen des Meeres (Wellen, Gezeiten, Strömungen) 2. ed., Stuttgart: J. Engelhorns Nchf. **1911**.

15 D Daly, R.A.: Proc. Am. Acad. Arts Sci. **51** (1915) 155.

19 J Johnson, D.W.: Shore processes and shoreline development, London: **1919**.

27 S Sverdrup, H.V.: Geofysiske Publ. Oslo **4**, 5 (1927) 75.

28 D Davis, W.M.: The coral reef problem. Am. Geogr. Soc., Spec. Pub. No. **9** (1928) p. 596.

33 B Bauer, H.A.: Geogr. Rev. **23** (1933) 259.

38 M Milankovitch, M.: Bull. Acad. Sci. Math. Nat. Belgrade **4** (1938) 49.

39 H Hjulström, F.: Am. Ass. Petr. Geol. (1939).

44 H Horn, W., Stocks, Th.: Mitt. d. Hydrogr. Dienstes Berlin **1944**.

45 S Schou, A.: Det marine forland. Folia Geogr. Danica, IV. Geografiske studier over Danske fladkystlandskabers Dannelse og formudvikling, Copenhagen: **1945**.

47 B Bigelow, H.B., Edmondson, W.T.: Wind waves at sea, breakers and surf., H.O. Pub. 602, Washington, D.C.: US Navy Dept. **1947**, p. 177.

47 S Sverdrup, H.U., Munk, W.H.: Wind, sea and swell-theory of relationships in forecasting. H.O. Pub. 601, Washington, D.C.: US Navy Dept **1947**.

50 A Academy of Sciences, Ministry of the Navy, UdSSR: Atlas of the oceans "Morskoi"; Vol. I: Topography, Moscow **1950**; Vol. II: Physics, Moscow **1952–1953**; Vol. III: History, Moscow **1958**.

50 F Fairbridge, R.W., Teichert, C.: J. Geol. **58** (1950) 330.

50 K Kuenen, P.H.: Marine Geology, New York: Wiley-Interscience **1950**, p. 568.

50 S Shepard, F.P., Inman, D.L.: Trans. Am. Geophys. Union **31**, 2 (1950) 196.

50 T Tricart, J., Schaeffer, R.: Revue de Géomorphologie Dynamique **4** (1950) 151.

51 M Marmer, H.A.: Tidal datum planes. Spec. Publ. No. 135, US Coast and Geodetic Survey **1951**.

52 P 1 Pfannenstiel, M.: Abh. Akad. Wiss., Math.-Naturwiss. Kl. **7** (1952/53) 373.

52 P 2 Pritchard, D.W., in: Adv. in Geophys. (Landsberg, H.E., ed.), Vol. 1, New York: Academic Press **1952**, p. 243–280.

52 S Stommel, H.M., Farmer, H.G.: On the nature of estuarine circulation. Woods Hole Oceanogr. Inst. Ref. no. 52–88, 52–51, 52–63, Woods Hole, Mass. **1952**.

52 U Ursel, F.: Proc. R. Soc. London, Ser. A, **214** (1952) 79.

52 V Valentin, H.: Peterm. Geogr. Mitt., Ergänzungsheft **246** (1952) 118.

53 B Bruun, P.: Univ. Calif. Inst. Eng. Res. Ser. 3 **347** (1953) 1.

53 I I.H.O.: Limits of oceans and seas. Int. Hydrographic Organization, Spec. Publ. **28** (1953) 38.

53 K King, C.A.M.: Trans. and Papers Inst. Brit. Geogr. **19** (1953) 13.

53 W Wiegel, R.L.: Waves, tides, currents and beaches: glossary of terms and list of standard symbols, Berkeley: Council on Wave Res., Eng. Foundation **1953**, p. 113.

54 A Arx, W.S., in: Bikini and nearby atolls, Pt. II: Circulation systems of Bikini and Rongelab lagoons, US Geol. Surv. Profess. Paper **260**-B (1954) 265.

54 B Bagnold, R.A.: The physics of blown sand and desert dunes, London: Methuen & Co. **1954**, p. 265.

54 E Emery, K.O., Tracey, J.I., jr., Ladd, H.S.: Geology of Bikini and nearby atolls. Pt. 1, US Geol. Surv. Profess. Paper **260**-A (1954).

54 G Gierloff-Emden, H.G.: Hamburger Geogr. Stud. **4** (1954) 40.

54 V Valentin, H: Die Küsten der Erde. Erg. H. Pet. Geogr. Mitt. 246, Gotha: VEB Haack **1954**, p. 118.

55 L Lisitzin, E.: Les variationes annuelles du niveau des océans, B.C.O.E.C. vol. **7** (1955) 235.

55 P Pattullo, J.G., Munk, W., Revelle, R., Strong, E.: J. Mar. Res. **14** (1955) 88.

56 G Guilcher, A.: Ann. Inst. Océanogr., Monaco **33** (1956) 64.

56 R Robinson, A.H.W.: J. Inst. of Navigation **9** (1956) 20.

56 S Samojlov, J.V.: Die Flußmündungen, Gotha: Haack **1956**, p. 647.

57 C Charlesworth, J.K.: The Quaternary era, 2 Vols, London: Edward Arnold **1957**, p. 1700.

57 E Emery, K.O., Stevenson, R.: Geol. Soc. Am. Mem. **67**, 1 (1957) 673.

57 G Gierloff-Emden, H.G.: Die Naturwissenschaften **44**, 15 (1957) 418.

57 W Wells, J.W.: Geol. Soc. Am. Mem. **67** (1957) 609.
58 G Guilcher, A.: Coastal and submarine morphology, London: Methuen & Co. **1958**, p. 274.
58 M 1 Maull, O.: Handbuch der Geomorphologie, Wien: Deuticke **1958**, p. 500.
58 M 2 McGill, J.T.: Geogr. Rev. **48** (1958) 402.
58 P Pritchard, D.W.: The John Hopkins Univ. Chesapeake Bay Inst. Techn. Rep. **4** (1952).
58 S Sauramo, M.: Die Geschichte der Ostsee. Acad. Sci. Fennicae, III, Helsinki: **1958**, p. 522.
58 Z Zenkovitch, A.: The morphology and dynamics of the Soviet Black Sea shores, Vol. 1, 2, Moskau: Academy of Science **1958**, **1960**.
59 A Ahnert, F.: Geogr. Rev. **49** (1959) 340.
59 G Gierloff-Emden, H.G.: Die Küste von El Salvador. Eine morphologisch-ozeanographische Monographie. Acta Humboldtiana, Ser. geografica et ethnographica, 2, Wiesbaden: Franz Steiner **1959**, p. 183.
59 K King, C.A.M.: Beaches and coasts, London: Arnold **1959**, p. 403.
59 M McGill, J.T.: Coastal classification maps, a review. 2nd Coastal Geogr. Conf. **1959**, p. 1–21.
59 S 1 Sager, G.: Ebbe und Flut, Gotha: Haack **1959**, p. 251, 133 Figs., 19 Tables.
59 S 2 Stocks, Th.: Kartograph. Nachrichten **1** (1959).
59 T Tricart, J., Cailleux, A.: Initiations à l'étude des sables et des galets, Vols. 1, 2, 3, Paris: C.D.V., **1959**, p. 376, p. 194, p. 202.
59 Z Zeuner, F.E.: The Pleistocene period; its climate, chronology and faunal successions. 2. ed., London: Hutchinson **1959**, p. 447.
60 A Axelrod, D.I.: Coastal vegetation of the world, in: Putnam, W.C., et al.: Natural coastal environments of the world, Los Angeles: Univ. Calif. Press **1960**, 43–58, 1 map.
60 B 1 Bailey, H.P.: Climates of coastal regions, in: Putnam, W.C., et al.: Natural coastal environments of the world, Los Angeles: Univ. Calif. Press **1960**, 59–77, 1 map.
60 B 2 Bruun, P., Gerritsen, F.: Stability of coastal inlets, Amsterdam: North Holland Publishing Comp. **1960**, p. 123.
60 G Granö, O.: Fennia **83**, 3 (1960) 5–36, Fig. 35. See also: Coast and archipelage, Int. Geogr. Congr., Guide Book to Exc. E.F. 1.
60 H Holtedahl, H.: Mountain, fjord, strandflat: geomorphology and general geology of parts of western Norway. Guide to excursions 21st session of the Int. Geol. Congress, Norden **1960**.
60 P Putnam, W.C., Axelrod, D.I., Bailey, H.P., McGill, J.T.: Natural coastal environments of the world, Los Angeles: Univ. Calif. Press and: Washington, D.C.: Geography Branch, Office of Naval Research **1960**, p. 140.
60 R Roden, G.: J. Geophys. Res. **65** (1960) 2809.
60 S Schou, A.: Geogr. Tidsskr. **59** (1960) 10.
61 D Defant, A.: Physical oceanography, Vol. 1, 2, New York: Pergamon Press **1961**, p. 729, **1963**, p. 598.
61 F 1 Fairbridge, R.W.: Eustatic changes in sea level, in: Physics and chemistry of the earth, Vol. 4, London: Pergamon Press **1961**, p. 99–185.
61 F 2 Fischer, A.G.: Am. Assoc. Pet. Geol. Bull. **45** (1961) 1656.
61 G 1 Gierloff-Emden, H.G.: Peterm. Geogr. Mitt. **105**, 2 (1961) 81–92, 3 (1961) 161–176.
61 G 2 Gierloff-Emden, H.G.: Luftbild und Küstengeographie am Beispiel der deutschen Nordseeküste, in: Landskundliche Luftbildauswertung im mitteleuropäischen Raum. Schriftenfolge d. Inst. f. Landeskde. i. d. Bundesanstalt f. Landeskde., 4, Bad Godesberg: Selbstverlag d. Bundesanstalt f. Landeskde. u. Raumforsch. **1961**, p. 118.
61 J Jelgersma, S.: Med. Geol. Stichting, Ser. C, **6** (1961) p. 100.
61 M Murray, G.L.: Geology of the Atlantic and Gulf coastal provinces of North America, New York: Harpers Brothers **1961**, p. 692.
61 R Rex, R.W., in: Geology of the Arctic, Proc. First Intern. Sympos. on Arctic Geol. (Raasch, G.O., ed.), Toronto: University of Toronto Press **1961**, p. 1021–1043.
62 B Bruun, P.: Am. Soc. Civ. Eng. Proc., J. Waterways Harbors Div. **88** (1962) 117.
62 C Carson, C.E., Hussey, K.M.: J. Geol. **70** (1962) 417.
62 H Hansen, W.: Mitt. Inst. f. Meereskde. d. Univers. Hamburg **1** (1962).
62 R 1 Reineck, H.E.: Sturmfluten. Eine Aufstellung der historischen Sturmfluten in der Nordsee seit dem 9. Jhdt. **1962**.
62 R 2 Russell, R.J.: Geomorph. N. F. **6**, 1 (1962) 1.
62 S 1 Shalowitz, A.L.: Shore and sea boundaries. Vol. 1, 2, Washington D.C.: US Government Printing Office, Publ. 10–1, **1962**, p. 420, p. 749.
62 S 2 Sindowski, K.H.: Geogr. Rdsch. **8**, 14 (1962) 322.

62 W Wiens, H.J.: Atoll environment and ecology, New Haven: Yale University Press **1962**, p. 532.

63 A Ahnert, F.: Distribution of estuarine meanders. Final Report, Project No. 388–069 Contract No. (G)-00048-62, Office of Naval Research, Dept. of Geography, Univ. of Maryland **1963**.

63 B Bagnold, R.A.: Beach and nearshore processes. I. Mechanics of marine sedimentation, in: The Sea (Hill, M.N., ed.), New York: Wiley-Interscience **1963**, 507–528.

63 D 1 Deutsches Hydrographisches Institut: Atlas der Gezeitenströme No. 245, Hamburg: DHI **1963**.

63 D 2 Dietz, R.S.: Bull. Geol. Soc. Am. **74** (1963) 971.

63 G Guilcher, A.: Estuaries, deltas, shelf, slope, in: The Sea (Hill, M.N., ed.), New York: Wiley-Interscience **1963**.

63 I Inman, D.L., Bagnold, R.A.: Beach and nearshore processes, II: Littoral processes, in: The Sea (Hill, M.N., ed.), New York: Wiley-Interscience **1963**, 529–553.

63 R 1 Reineck, E.: Abh. Senckenberg. Naturf. Ges. **505** (1963) 1.

63 R 2 Rossiter, J.R.: Long-term variations in sea-level, in: The Sea (Hill, M.N., ed.), New York: Wiley-Interscience **1963**, 590–610.

63 S 1 Seibold, E.: Geological investigation of nearshore sand-transport-examples of methods and problems from the Baltic and North Seas, in: Progress in oceanography 1 (Sears, M., ed.), New York: Pergamon Press **1963**, 1–70.

63 S 2 Shepard, F.P.: Submarine geology. 2. ed., (with chapters by Inman, D.L., and Goldberg, E.D.), New York: Harper & Row **1963**, p. 557.

64 D Davies, J.L.: Morphogenetic approach to world shorelines. Geomorph., N. F. **8**, 5 (1964) 127.

64 S Steers, J.A.: The coastline of England and Wales. 3. ed., Cambridge: Cambridge Univ. Press **1964**, p. 750.

64 V Van Straaten, L.M.J.U. (ed.): Deltaic and shallow marine deposits, Amsterdam: Elsevier Scientific Publ. Comp. **1964**.

64 W Wiegel, R.L.: Oceanographical engineering, New Jersey: Prentice Hall **1964**, p. 522.

64 Y Yasso, W.E.: J. Geol. **73** (1964) 702.

65 C Curray, J.R.: Late quaternary history. Continental shelves of the USA, in: The Quaternary of USA (Wright, L.D., jr., Frey, D.G., eds.), Princeton: Princeton Univ. Press **1965**, p. 723–735.

65 F Fairbridge, R.W.: Ancient shorelines and absolute dating. Rep. 6th INQUA Congr., Warsaw 1961, 1, **1965**, p. 653–675.

65 G Guilcher, A.: Grand récif sud. Récifs et lagon de Tuo, in: Exped. Franc. sur les récifs coralliens de la nouv. Calédonie, 2, Paris: **1965**, p. 135–240.

65 K Kinsman, B.: Wind waves, their generation and propagation on the ocean surface, Englewood Cliffs: Prentice Hall **1965**, p. 676.

65 R 1 Richards, H.G., Fairbridge, R.W.: Annoted bibliography of Quaternary shorelines (1945–1964). Acad. Sci., Spec. Publ., Philadelphia **1965**.

65 R 2 Russel, R.J., McIntire, W.G.: Geogr. Rev. **55** (1965) 17.

65 R 3 Russel, R.J., McIntire, W.G.: Geol. Soc. Am. Bull. **76** (1965) 307.

65 S Straaten, L.M.J.U. van: Med. Geol. Stichting **5**, 17 (1965) 41.

65 Y Yasso, W.: J. Geol. **73** (1965) 702.

66 B 1 Baker, B.B., jr., Deebel, W.R., Geisenderfer, R.D.: Glossary of oceanographic terms. 2. ed. Spec. Publ. 35, US Naval Oceanographic Office, Washington D.C. **1966**.

66 B 2 Bowen, A.J., Inman, D.L.: Coastal Eng. Res. Cent. Techn. Mem. 19, US Army Corps of Eng. Washington D.C. **1966**.

66 C Caldwell, J.M.: J. Soc. Civil Eng. **53**, 2 (1966) 142.

66 E Emiliani, C.: J. Geol. **74** (1966) 109.

66 F Fairbridge, R.W. (ed.): The encyclopedia of oceanography. Encyclopedia of earth sciences series, Vol. 1, New York: Reinhold **1966**, p. 1021.

66 G Gjessing, J.: Norsk Geogr. Tidsskr. **20** (1966) 273.

66 H 1 Hafemann, D.: Abh. Akad. Wiss., Math.-Naturwiss. Kl. **12** (1966) 609.

66 H 2 Hansen, D., Rattray, jr.: Limnol. Oceanogr. **11** (1966) 319.

66 P Pattullo, J.G.: Mean sea level, in: The encyclopedia of oceanography (Fairbridge, R.W., ed.), New York: Reinhold **1960**, p. 475–478.

66 S Schmidt-Falkenberg: Kartographie, in: Westermann Lexikon der Geographie, Braunschweig: G. Westermann Verlag **1969**.

66 W 1 Walker, H.J., Arnborg, L.: Permafrost and ice-wedge effect on riverbank erosion. Proc. Permafrost Inter. Conf., NAS-NRC Publ. 1287. **1966**, p. 164–171.

66 W 2 Wilson, W., in: Encyclopedia of earth science series, Vol. 1 (Fairbridge, R.W., ed.), New York: Reinhold **1966**, 804–817.

67 D Delaney, P.J.V.: Geomorphology and Quaternary coastal geology of Uruguay, Porto Alegre, Brazil: Univ. do Rio Grando do Sul **1967**, p. 39.

67 E Emery, K.O.: Estuaries and lagoons in relation to continental shelves, in: Estuaries (Lauff, G.H., ed.), Am. Ass. Adv. Sci. Publ. 83, Washington D.C. **1967**, p. 9–11.

67 G 1 Griffin, W.L., Jones, B.G., McAlinden, J.M.: Establishing tidal datum lines for sea boundaries. – Paper presented at the Joint American Society of Photogrammetry-American Congress on Surveying and Mapping Convention, Washington, D.C. **1967**.

67 G 2 Gunter, G.: Some relationship of estuaries to the fisheries of the Gulf of Mexico, in: Estuaries (Lauff, G.H., ed.), Am. Ass. Adv. Sci. Publ. 83, Washington D.C. **1967**, p. 621–638.

67 H Hopkins, D.M. (ed.): The Bering land bridge, Stanford, Cal: Stanford University Press **1967**

67 I Iida, K., Cox, D.C., Pararas-Carayannis, G.: Preliminary catalog of tsunamis occurring in the Pacific Ocean, Hawaii Inst. Geophys., Report No. HIG-67-10, Honolulu: University of Hawaii **1967**, p. 277.

67 J Jelgersma, S.: Sea-level changes during the last 10000 years. World climate from 8000 to 0 B.C. Proc. of the Int. Symp. on World Climatol., London: R. Meteorological Soc. **1967**, p. 54–71.

67 L Lauff, G.H. (ed.): Estuaries. Am. Ass. Adv. Sci., Publ. 83, Washington D.C. **1967**, p. 757.

67 P Phleger, F.B.: Some general features of coastal lagoons. Lagunas Costeras Symposium, UNAM, Unesco, Mexico City **1967**, p. 5–20.

67 R 1 Rossiter, J.R.: Geophys. J. R. Astron. Soc. **12** (1967) 259.

67 R 2 Russel, R.J.: River and delta morphology. Coastal Stud. Inst., Tech. Rep., Coastal Stud. Ser., 20 Baton Rouge: Louisiana State Univ. Press **1967**, p. 63.

67 S 1 Schroeder-Lanz, H., Wieneke, F., Schmidt, W.: Mitt. Geogr. Ges. München **52** (1967) 267.

67 S 2 Schwartz, M.L.: J. Geol. **75** (1967) 76.

67 S 3 Shepard, F.P.: Submarine geology, New York: Harper & Row **1967**, p. 557.

67 Z Zenkovich, V.P.: Process of coastal development, New York: Wiley-Interscience **1967**, 738, and London: Oliver & Boyd.

68 A Allen, J.R.L.: Current ripples, their relation to patterns of water and sediment motion, Amsterdam: North Holland Publ. **1968**, p. 433.

68 D 1 Dale, E.D.: Photogr. Engin. **34**, 5 (1968).

68 D 2 Dietrich, G., Ulrich, J.: Atlas zur Ozeanographie. Meyers Großer Physischer Weltatlas. BI-Hochschulatlanten, Vol. 7, Mannheim: Bibliogr. Inst. **1968**, p. 76.

68 D 3 Dolan, R., Ferm, J.C.: Science **159** (1968) 627.

68 E Emery, K.O.: Am. Assoc. Pet. Geol. Bull. **52** (1968) 445.

68 F Fairbridge, R.W. (ed.): The encyclopedia of geomorphology. Encyclopedia of earth sciences series, Vol. III, New York: Reinhold Book Corp. **1968**, p. 1000.

68 G Galas, D.: Vergleichende geomorphologische Untersuchungen an den Rias-Küsten SW-Englands und der Bretagne. Arb. Geogr. Inst. TH Hannover, Jb. Geogr. Ges. Hannover, Sonderheft 2, **1968**, p. 176.

68 M Milliman, J.D., Emery, K.O.: Science **162** (1968) 1121.

68 S 1 Schou, A.: Basse Bretagnes kyster. – Geogr. Tidsskr. **67**, 2 (1968) 200.

68 S 2 Schülke, H.: Morphologische Untersuchungen an bretonischen, vergleichsweise auch an korsischen Meeresbuchten. Ein Beitrag zum Riaproblem, Arb. Geogr. Inst. d. Univ. d. Saarlandes **11** (1968) 192.

68 S 3 Shepard, F.P.: Coastal classification, in: The encyclopedia of geomorphology (Fairbridge, R.W., ed.), New York: Reinhold Book Corp. **1968**, p. 131–133.

68 T Takenaga, K.: Geogr. Tokyo **77** (1968) 329.

68 V 1 Verger, F.: Marais et wadden du littoral français, étude de gémorphologie littorale, Bordeaux: Bisquaye frères, imprimeurs **1968**, p. 544.

69 B Bowen, A.J., Inman, D.L.: J. Geophys. Res. **23** (1969) 5479.

69 C Curray, J.R., Emmel, J.F., Crampton, P.J.S.: Holocene history of a strand plain, lagoonal coast, Nayarit, Mexico. Lagunas Costeros Simposio, UNAM, Unesco, Mexico D.F.: **1969**, p. 77.

69 E Emiliani, C., Rona, E.: Science **166** (1969) 1551.

69 G Guilcher, A., Berthois, L., Doumenge, F. et al.: Géomorph., Sédim., Fonctionnement Hydrologique. O.S.T.R.O.M. **38** (1969) p. 103.

69 H Herm, D.: Marines Pliozän in Nord- und Mittel-Chile unter besonderer Berücksichtigung der Entwicklung der Mollusken-Faunen. Zitteliana, 2, Altötting-München: Geiselberger **1969**, p. 159.

69 M Mörner, N.-A.: The late quaternary history of the Kattegatt Sea and the Swedish west coast. Deglaciation, shorelevel displacement, chronology and eustacy. Sver. Geol. Unders. C, 640 **1969**, p. 487.
69 O O'Brian: Dynamics of tidal inlets, in: Coastal Lagoons Sympos. UNAM (Castanares, A., Phleger, F.B., eds.), Unesco, Mem. Simp. Intern., Lagunas Costeras, Mexico DF **1969**, pp. 686.
69 P 1 Phleger, F.B.: Some general features of coastal lagoons, in: Coastal Lagoons Sympos. UNAM (Castanares, A., Phleger, F.B., eds.), Unesco, Mem. Simp. Intern., Lagunas Costeras, Mexico DF **1969**, pp. 686.
69 P 2 Postma, H.: Chemistry of coastal lagoons, in: Coastal Lagoons Sympos. UNAM (Castanares, A., Phleger, F.B., eds.), Unesco, Mem. Simp. Intern. Lagunas Costeras, Mexico DF **1969**, pp. 686.
69 V Visher, G.: J. Sediment. Petrol. **39** (1969) 1074.
70 A 1 Allen, J.R.L.: Physical processes of sedimentation, London: Allen & Unwin **1970**, p. 248.
70 A 2 Andrews, J.T.: A geomorphological study of post-glacial uplift, with particular reference to Arctic, Canada, London: Institute of British Geographers **1970**, p. 156.
70 C Curray, J.R., Shepard, F.P., Veeh, H.H.: Bull. Geol. Soc. Am. **81** (1970) 1865.
70 G Guilcher, A. (ed.): Quaternaria **12** (1970) 229.
70 H Hsu, S.A.: Monthly Weather Rev. **98** 7 (1970) 487.
70 J Jelgersma, S., De Jong, J., Zagwijn, H., Regteren Altena, J.F.: Medelingen Rijks Geol. Dienst, N.S. **21** (1970) 93.
70 L 1 Lang, A.W.: Hamburger Küstenforschung **12** (1970) 195.
70 L 2 Longuet-Higgins, M.S.: J. Geophys. Res. **75**, 33 (1980) 6778.
70 N Newell, N.D., Bloom, A.: Geol. Soc. Am. Bull. **81** (1970) 1881.
70 S 1 Sarntheim, M.: Geol. Rdsch. **59** (1970) 649.
70 S 2 Sigl, R.: Report on the symposium on coastal geodesy, München **1970**.
70 S 3 Stoddart, D.R. (ed.): Atoll Res. Bull. **136** (1970) 1.
70 S 4 Swift, P.J.P.: Mar. Geol. **8** (1970) 5.
70 T Tricart, J.: Geomorphology of cold environments, New York: Macmillan **1970**, p. 320.
70 W Walker, H.J.: Some aspects of erosion and sedimentation in an arctic delta during breakup. Proc. Symp. Hydrology of Deltas, Unesco **1970**, p. 209–219.
71 B Bowen, A.J., Inman, D.L.: J. Geophys. Res. **76**, 36 (1971) 8662.
71 D Dolan, R.: Coastal landforms: Crescentic and rhythmic. Geol. Soc. Am. Bull. **82**, 1 (1971) 177.
71 E Emiliani, C.: The amplitude of Pleistocene climatic cycles at low latitudes and the FSO topic composition of glacial ice in the late Genozoic glacial ages, in: Turekian, K.K. (ed.), New Haven, Connecticut: Yale University Press **1971**, p. 183–197.
71 F Flint, R.F.: Glacial and quaternary geology, New York: Wiley-Interscience **1971**, p. 892.
71 I 1 Inman, D.L.: Nearshore processes, in: Encyclopedia of science and technology, Vol. 9, **1971**, 26–33.
71 I 2 Inman, D.L., Nordstrom, C.E.: J. Geol. **79** (1971) 1.
71 K Komar, P.D.: Geol. Soc. Bull. **82** (1971) 2643.
71 M Mörner, N.A.: Palaeogeogr., Palaeoclimatol., Palaeoecol. **9** (1971) 153.
71 R Reineck, H.E., Singh, I.B., Wunderlich, F.: Senckenbergiana Maritima **3** (1971) 93.
71 S 1 Schwartz, M.: J. Geol. **79** (1971) 91.
71 S 2 Shepard, F.P., Wanless, H.R.: Our changing coastlines, New York: McGraw-Hill **1971**, p. 579.
71 S 3 Steers, J.A. (ed.): Introduction to coastline development, Cambridge, Mass.: MIT Press, **1971**, p. 229.
71 S 4 Steers, J.A. (ed.): Applied coastal geomorphology, London, Basingstoke: Macmillan **1971**, p. 227.
71 S 5 Stoddart, D.R., Yonge, C.M. (eds.): Regional variation in Indian Ocean coral reefs. Symp. Zool. Soc. of London, Acad. Press for Zool. Soc. of London, 28, **1971**, p. 614.
71 W 1 Wagner, K.H.: Atlas zur Physischen Geographie, Mannheim: Bibliograph. Inst. **1971**, p. 59.
71 W 2 Wieneke, F.: Münchener Geogr. Abh. **3** (1971) 1.
71 W 3 Wieneke, F., Kritikos, G.: Umschau **24** (1971) 903.
72 D 1 Davis, R.A., jr., Fox, W.T.: Sediment. Petrology **42** (1972) 401.
72 D 2 Dolan, R., et al.: Classification of the coastal environments of the world, part I: The Americas. Univ. Virginia Techn. Rep. 1, Geogr. Prog., Off. Nav. Res. Charlottesville **1972**.
72 E Emery, K.O., Uchupi, E.: Am. Ass. of Pet. Geol. Mem. **17** (1972) 532.
72 H Hayden, B., et al.: Classification of the coastal environments of the world, part I: The Americas, Part II: Africa. Univ. Virginia Techn. Rep. 3, Charlottesville **1972/73**, p. 1–46.
72 K King, C.A.M.: Beaches and coasts. 2. ed., London: Arnold **1972**, p. 570.
72 L Laucht, H.L.: Hamburger Küstenforschung **22** (1972).

72 N 1 Nelson, B.W. (ed.): Environmental framework of coastal plain estuaries. Symposium on estuaries 18th meeting of southeastern section. The Geol. Soc. of Am. South Carolina, 1969. Geol. Soc. Am., Mem. 133, **1972**.

72 N 2 Nichols, M., Kelly, M.: The sensing and analysis of coastal water. 8. Inter. Symp. on Remote Sensing and Environment, Oct. 72, Willow Run, Ann Arbor, Mich. **1972**.

72 S 1 Sarntheim, M.: Mar. Geol. **12**, 4 (1972) 245.

72 S 2 Schwartz, M.L. (ed.): Spits and bars. Benchmark papers in geology, Stroudsburg, Penns.: Dowden, Hutchinson & Ross **1972**, p. 452.

72 U Ulrich, J.: Die Küste **23** (1972) 112.

72 V Valentin, H.: Gött. Geogr. Abh. **60** (1972) 355.

73 A Armstrong, R.W.: Atlas of Hawaii, Honolulu: Dept. of Geophys. Univ. Press of Hawaii **1973**, p. 222.

73 B Biewald, D.: Berliner Geogr. Abh. **15** (1973).

73 C 1 Coastal Engineering Research Center: Shore protection manual, Vol. 1, Ft. Belvoir, Va.: US Army Corps of Engineers **1973**, p. 750.

73 C 2 Coates, D.R. (ed.): Coastal geomorphology, New York: State University Binghampton **1973**.

73 C 3 Crofts, R.S.: A method of determine shingle supply to the coast. Inst. British Geogr., Transact. 62, London: **1973**, 125–127.

73 D Davies, J.L.: Geographical variations in coastal development, Edinburgh: Oliver and Boyd, 1972, p. 240. Geomorphological Text 4, New York: Hafner **1973**.

73 G Gellert, J.F.: Wasserwirtschaft – Wassertechnik **23**, 2 (1973) 58.

73 H 1 Hansen, W.: Hamburger Küstenforschung **26** (1973).

73 H 2 Hasselmann, K.: Dtsch. Hydrogr. Z. Erg.-H. Reihe A **12** (1973) p. 95.

73 H 3 Hicks, S.D.: NOAA Techn. Mem. No. 12, Washington, D.C.: US Dept. of Commerce **1973**, p. 11.

73 I Inman, D.L., Brush, B.M.: Science **181** (1973) 20.

73 K 1 Kaplin, P.A.: Recent history of the world ocean coast, Moscow: Moscow State University **1973**, p. 264.

73 K 2 Komar, P.D.: Bull. Geol. Soc. Am. **84** (1973) 2217.

73 M Matthews, R.K.: Quatern. Res. **3**, 1 (1973) 147.

73 P Purser, B.H.: The Persian Gulf. Holocene carbonate sedimentation and diagenesis in a shallow epicontinental sea, Berlin, New York: Springer **1973**, p. 471.

73 S 1 Schwartz, M.L. (ed.): Barrier islands, Benchmark Pap. in Geol., **9**, Stroudsburg, Penns.: Dowden, Hutchinson & Ross **1973**, p. 451.

73 S 2 Short, A.D., Wiseman, W.J.: State Univ. Louisiana, Coastal Stud. Bull. **7** (1973) 23.

73 S 3 Sindowski, K.H.: Das ostfriesische Küstengebiet. Inseln, Watten und Marschen. Sammlung Geol. Führer, 57, Berlin, Stuttgart: Borntraeger **1973**, p. 192.

73 S 4 Steers, J.A.: The coastline of Scotland, Cambridge: Cambridge Univ. Press **1973**, p. 335.

73 S 5 Sündermann, J., Vollmers, H.: Die Küste **24** (1973).

73 U 1 Ulrich, J.: Dtsch. Hydrogr. Z., Erg.-H. B **14** (1973).

73 U 2 Ulrich, J., Pasenau, H.: Die Küste **24** (1973) 95.

73 V 1 Verger, F.: Rev. Géomorph. dynamique **22**, 4 (1973) 161.

73 V 2 Verger, M.P.: Bull. Assoc. Geogr. France **1973**, 411.

73 W 1 Walker, H.J.: The morphology of the North-Slope, in: Alaskan arctic tundra (Britton, M.E., ed.), AINA Techn. Pap. 25, **1973**, p. 49–92.

73 W 2 Walker, H.J.: The nature of seawater-freshwater interface during breakup in the Colville River Delta, Alaska (1), Permafrost. The North American Contrib. to the Intern. Confer., Nation. Acad. of Sci., Washington **1973**, p. 474.

73 W 3 Wright, L.D.: Am. Assoc. Petrol. Geol. Bull. **57** (1973) 370.

73 W 4 Wright, L.D.: Alaskan arctic coastal processes and morphology. Baton Rouge: Coastal Stud. Inst. Louisiana State Univ., Techn. Rep. 149, **1973**.

74 A Abraham, H.J.: Das Seerecht. 4. ed., Berlin, New York: Walter de Gruyter **1974**, p. 284.

74 B Burk, C.A., Drake, C.L. (eds.): The geology of continental margins, Berlin, Heidelberg, New York: Springer **1974**, p. 1069.

74 C Chapman, V.J.: Salt marshes and salt deserts of the world. 2. ed. **1974**, p. 392.

74 D 1 Dolan, R., Vincent, L., Hayden, B.: Z. Geomorph. **18**, 1 (1974) 1.

74 D 2 Dorh, W.G. van: Oceanography and seamanship, New York: Dodd, Mead **1974**, p. 481.

74 E 1 Einsele, G., Herm, D., Schwartz, H.U.: Quat. Res. **4**, 3 (1974) 282.

74 E 2 Emiliani, C., Shackleton, N.J.: Science **183** (1974) 511.

74 G Guza, R.T., Davis, R.: J. Geophys. Res. **79** (1974) 1285.

74 K Kelletat, D.: Zschr. Geom., Supp. Bd. **19** (1974).

74 L Lisitzin, E.: Sea-level changes. Elsevier Oceanography Series 8, Amsterdam: Elsevier **1974**, p. 292.

74 P Perkins, E.J.: The biology of estuaries and coastal waters, London, New York: Academic Press **1974**, p. 640.

74 R Reimnitz, E., Barnes, P.W.: Sea ice as a geologic agent on the Beaufort Sea shelf of Alaska, in: The coast and shelf of the Beaufort Sea (Reed, J.C., Sater, J.E., Gunn, W., eds.), Proc. of a Symp., Arctic Inst. of North Amer., Arlington, Va. **1974**, p. 301–353.

74 S 1 Seibold, E.: Das Meer. Die Küstenregionen, in: Lehrbuch der Allgemeinen Geologie (Brinkmann, R., ed.), Vol. I: Festland – Meer. 2. ed., Stuttgart: F. Enke **1974**, p. 291–489.

74 S 2 Suhayda, J.N.: Determining nearshore infragravity wave spectra, in: Intern. Symp. on Ocean Wave Measurements and Analysis, New York: Am. Soc. Civ. Eng. **1974**, p. 54–63.

74 W 1 Walker, H.J., in: The coast and shelf of the Beaufort Sea (Reed, J.C., Sater, J.E., eds.), Washington DC: Arctic Institute of North America **1974**, 513–540.

74 W 2 Wright, L.D., Coleman, J.M., Erickson, M.W.: Coastal Studies Inst. Techn. Rep., Louisiana State Univ. **156** (1974) 1.

75 B 1 Battistini, R.: Téthys **7**, 1 (1975) 1.

75 B 2 Baumgartner, A., Reichel, E.: The world water balance, Amsterdam: Elsevier **1975**, p. 182.

75 B 3 Bloom, A.L.: Geomorphology. A systematic analysis of late Cenocoic landforms, Englewood Cliffs, N.J.: Prentice-Hall **1975**, p. 510.

75 C Coleman, J.M., Wright, L.D., in: Deltas: Models for exploration (Broussard, M.L., ed.), Houston: Geol. Soc. **1975**, p. 99–149.

75 D 1 Dietrich, G., Kalle, K., Krauss, W., Siedler, G.: Allgemeine Meereskunde. Eine Einführung in die Ozeanographie, Berlin, Stuttgart: Borntraeger **1975**, p. 593.

75 D 2 Dolan, R., Hayden, B., Vincent, M.: Z. Geomorphol. N.F. Suppl. **22** (1975) 72.

75 E Embleton, C., King, C.A.M.: Glacial geomorphology, London: Edward Arnold **1975**, p. 573.

75 G 1 Ginsburg, R.N. (ed.): Tidal deposits. A casebook of recent examples and fossil counterparts, Berlin, Heidelberg, New York: Springer **1975**, p. 428.

75 G 2 Göhren, H.: Die Küste **27** (1975) 28.

75 G 3 Greenwood, B., Davidson-Arnott, R.G.D., in: Nearshore sediment dynamics and sedimentation (Hails, J., Carr, A., eds.), New York, London: Wiley-Interscience **1975**, p. 123–150.

75 G 4 Grin, A.M.: Discharge and drainage basins in the USSR, Abfluß und -regionen in der Sowjetunion **1975**.

75 G 5 Guza, R.T., Inman, D.L.: J. Geophys. Res. **80** (1975) 2997.

75 H 1 Holtedahl, H.: Nor. Geol. Unders. **323** (1975) 87.

75 H 2 Huntley, D.A., Bowen, A.J.: J. Geol. Soc. London **131** (1975) 69.

75 J Jennings, J.N.: Desert dunes and estuarine fill in the Fitzroy Estuary (North-Western Australia), Catena **2** (1975) 215.

75 K 1 King, C.A.M.: Introduction to physical and biological oceanography, London: Arnold **1975**.

75 K 2 Klemas, V., Bartlett, D., Rogers, R.: Photogrammetric Eng. Remote Sensing **41** (1975) 499, 509.

75 N NOAA: Chart No. 1, United States of America. Nautical chart symbols and abbreviations. US Department of Commerce, National Oceanic and Atmospheric Administration. 6. ed. July **1975**.

75 R 1 Reineck, H.E.: North Sea tidal flats, in: Tidal deposits, a casebook of recent examples and fossil counterparts (Ginsburg, R.M., ed.), New York: Springer **1975**.

75 R 2 Reineck, H.E., Singh, I.B.: Depositional sedimentary environments, Berlin, Heidelberg, New York: Springer **1975**, p. 579.

75 T Tunstall, E.B., Inman, D.L.: J. Geophys. Res. **80** (1975) 3475.

75 U Unesco, IAHS: Hydrology and marsh-ridden areas. Studies and Reports in Hydrology 19, **1975**, p. 562.

75 W Winant, C.D., Inman, D.L., Nordstrom, C.E.: J. Geophys. Res. **80** (1975) 1779.

76 B 1 Bird, E.C.F.: Coasts. An introduction to systematic geomorphology, Canberra: Australian Nat. Univ. Press **1976**, p. 282.

76 B 2 Bird, E.C.F., May, V.J.: Shoreline changes in the British Isles during the past century, Bournemouth: Div. Geogr. College of Technology **1976**, p. 44.

76 C 1 Chapman, V.J.: Coastal vegetation, Oxford: Pergamon Press **1976**, p. 292.

76 F Fairbridge, R.W.: Science **191** (1976) 353.

76 G Guilcher, A.: Ann. Geogr. **85**, 472 (1976) 641.

76 H 1 Hansen, U.A.: Mitt. des Leichtweiß-Instituts für Wasserbau der Technischen Universität Braunschweig **52** (1976).

76 H 2 Hsu, S.A., Whelan, T.: Environm. Sci. Techn. **10** (1976) 281.

76 I IHO: General bathymetric chart of the oceans (GEBCO), IHO: Monaco **1976** ownward.
76 K 1 Klein, G. de V., in: Geology 30, Stroudsburg, Pennsylvania: Dowden, Hutchinson & Ross **1976**.
76 K 2 Komar, P.D.: Beach process and sedimentation, Englewood Cliffs, N.J.: Prentice-Hall **1976**, p. 429.
76 L Lüders, K., Luck, G.: Kleines Küstenlexikon, Hildesheim: A. Lax **1976**, p. 242.
76 P Pfeil, H.P.v.: Oceans, coasts and law, Vol. 1, 2, New York: Oceana Publ. **1976**, p. 887.
76 R 1 Radok, R.: Australia's coast, an environmental atlas guide with baselines, Adelaide, Sidney, Melbourne: Rigby Ltd. **1976**.
76 R 2 Reineck, H.E.: Hamburger Küstenforschung **35** (1976).
76 R 3 Rust, U., Wieneke, F.: Münchener Geogr. Abh. **19** (1976) 74.
76 S 1 SCOR: Coastal lagoon survey-results of a world-wide inquiry made by the SCOR/Unesco ad hoc. Advisory panel on coastal lagoons (1976–1978). Coastal lagoon research: present and future. I: Guidelines-Unesco/IABO Seminar, Duke University, Sept. 78. **1976** onward.
76 S 2 Shepard, F.P.: Coastal research. Geoscience and Man **14** (1976) 53.
76 S 3 Stanley, P.J., Swift, D.J.P. (eds.): Marine sediment transport and environmental management, New York: Wiley-Interscience **1976**.
76 S 4 Swift, D.J.P.: Coastal sedimentation, in: Marine sediment transport and environmental management (Stanley, D.J., Swift, D.J.P., eds.), New York: Wiley-Interscience **1976**.
76 U Unesco: Geological world atlas. 1 : 10 000 000; in progress since 1976. Ocean sheets 1 : 36 000 000, 22 sheets, Washington: Unesco, **1976** onward. Commission for the Geological Map of the World. (CGMW), Int. Union of Geological Sciences.
76 W 1 Walter, M.R. (ed.): Stromatolites. Developments in sedimentology, Vol. 20, Amsterdam: Elsevier **1976**, p. 804.
76 W 2 Wiley, M. (ed.): Estuarine processes, Vol. I: Uses, stresses, and adaption to the estuary, Vol. II: Circulation, sediments, and transfer of material in the estuary, **1976**, p. 558, p. 444.
76 W 3 Williams, R.S., Carter, W.D.: ERTS-1. A new window on our planet. US Geological Survey Professional Paper 29, Washington D.C. **1976**.
76 Z 1 Zimmermann, J.T.F.: Neth. J. Sea Res. **10** (1976) 149, 397.
76 Z 2 Zunica, M.: Coastal changes in Italy during the past century. Italian Contrib. to the 23rd Int. Geogr. Congr., Rom **1976**, p. 275–281.
77 B 1 Bloom, A.J.: Atlas of sea level curves. Int. Geol. Correl. Program, Project 61, IUGS, Unesco, Ithaca, NY: Dept. Geol. Sci. Cornell Univ. **1977**, p. 100.
77 B 2 Brinkmann, R., Krömmelbein, K.: Abriß der Geologie, Vol. 1, 2. Historische Geologie, Stuttgart: F. Enke **1977**, p. 400.
77 C 1 Chapman, V.J.: Wet coastal ecosystems. Ecosystems of the world, Vol. 1, Amsterdam: Elsevier **1977**, p. 428.
77 C 2 Cox, D.C., Morgan, J.: Local Tsunamis and possible local Tsunamis in Hawaii, Honolulu: Hawaii Inst. Geophys., Univ. of Hawaii **1977**, p. 128.
77 E Engel, M., Zahel, W: Die Küste **31** (1977) 114.
77 G 1 Gierloff-Emden, H.G.: Orbital remote sensing of coastal and offshore environments, a manual of interpretation, Berlin: Walter de Gruyter **1977**, p. 176.
77 G 2 Gorsline, D.S., Swift, P.J.P. (eds.): Shelf sediment dynamics. A national overview. Report of a workshop held at Vail, Colorado, Nov. 2–6, 1976, **1977**.
77 H Hsu, S.A.: Research Techn. Coastal Environments, Geoscience and Man **18** (1977) 99.
77 K Kolp, O.: Z. Geol. Wiss., **5**, 7 (1977) 853.
77 L Leatherman, S.P.: Overwash hydraulics and sediment transport, coastal sediments, Proc. Am. Soc. Civ. Eng. **1977**, p. 135–148.
77 M 1 Mandelbrot, B.B.: Fractals, San Francisco: W.H. Freeman **1977**, p. 350.
77 M 2 McDowell, D.M., O'Connor, B.A.: Hydraulic behaviour of estuaries, London: Macmillan **1977**.
77 O Orme, A.R., Flood, P.G.: The geological history of the Great Barrier Reef. A reappraisal of some aspects in the light of new evidence. Proc. 3rd. Intern. Coral Reef Symp., Miami, 2, **1977**, p. 37–43.
77 P Pirazzoli, P.A.: Z. Geomorph. **21**, 3 (1977) 284.
77 R 1 Robinson, L.A.: Mar. Geol. **23** (1977) 237.
77 R 2 Robinson, L.A.: Mar. Geol. **23** (1977) 257.
77 R 3 Rohde, H.: Sturmfluthöhen und säkularer Wasseranstieg an der deutschen Nordseeküste, **1977**.
77 S 1 Suhayda, J.N., Hsu, S.A., Roberts, H.H., Short, A.D.: Documentation and analysis of coastal processes, north-east coast of Brazil, Baton Rouge: Coastal Stud. Inst. Louisiana State Univ. **1977**.
77 S 2 Sunamura, T.: J. Geol. **85** (1977) 613.

77 T Tanner, W.F. (ed.): Coastal sedimentology. Proc. Symp., Florida State Univ., 29 January 1977, Tallahassee: Coastal Res. Geol. Dept., Florida State Univ. **1977**, p. 315.

77 U US Army Coastal Engineering Research Center: Shore protection manual, Washington, D.C.: US Govt. Printing Office **1977**, p. 514.

77 V 1 Vail, P.R., Mitchum, R.M., jr., Thompson, S., in: Seismic stratigraphy application to hydrocarbon exploration (Paytom, C.E., ed.), Am. Ass. of Petrol. Geol. Mem. 26, **1977**, p. 83–97.

77 V 2 Vanney, J.R.: Geomorphologie des plates-formes continentales Paris: Doin-Editeurs **1977**, p. 302.

77 V 3 Verstappen, H.Th.: Remote sensing in geomorphology, Amsterdam: Elsevier Scientific Publ. Comp. **1977**.

77 W 1 Walker, H.J.: Depositional environments in the Colville River delta (Alaska). Louisiana State Univ., Baton Rouge, Coastal Studies Inst., Techn. Rep., TR-227 **1977**, p. 26.

77 W 2 Warncke, W.: Hamburger Küstenforschung **37** (1977).

77 W 3 Wells, J.T., Coleman, J.M.: Mar. Geol. **24** (1977) M47.

78 B Bird, E.C.F.: The geomorphology of the Gippsland lake region. Ministry for Conservation, Victoria, Environm. Studies Ser., Publ. no. 186 **1978**, p. 158.

78 C Clark, J.A., et al.: Global change in postglacial sea level, a numerical calculation **1978**.

78 D 1 Davis, R.A. (ed.): Coastal sedimentary environments, Berlin, Heidelberg, New York: Springer **1978**, p. 420.

78 D 2 Dillon, W.D., Oldale, R.N.: Geology **6**, 56 (1978).

78 E 1 Ellis, M.Y. (ed.): Coastal mapping handbook, Washington, D.C.: US Dept. of the Interior, USGS **1978**, p. 200.

78 E 2 El-Sabh, M.I., Dionne, J.C.: L'océanographie de l'estuaire du St. Laurent **1978**.

78 F Frey, R.W., Basan, P.B., in: Coastal sedimentary environments (Davis, R.A., ed.), Berlin, Heidelberg, New York: Springer **1978**.

78 H Higelke, B.: Regensburger Geograph. Schriften **1978**, p. 153.

78 I 1 International Hydrographic Organisation: IHB tidal data request, Monaco, International Tidal Data Bank **1978**.

78 I 2 IOC: Oceanographic components of the Global Atmospheric Research Programme (GARP). Intergovernmental Oceanographic Commission Techn. Ser., 17, Paris: Unesco **1978**, p. 35.

78 K Kliewe, H., Janke, W.: Stratigraphie und Entwicklung des nordöstlichen Küstenraumes der DDR. Peterm. Geogr. Mitt. **122**, 2 (1978) 81.

78 L 1 LeBlond, P.H., Mysak, L.A.: Waves in the ocean. Elsevier Oceanography Series, 20, Amsterdam: Elsevier Scientific Publ. Comp. **1978**, p. 616.

78 L 2 Lucchini, L., Voelckel, M.: Les états et la mer. Le nationalisme maritime, Paris **1978**, p. 460.

78 N 1 Nihoul, J.C.J.: Hydrodynamics of estuaries and fjords. Proc. 9th Int. Liège Coll. on Ocean Hydrodyn., Elsevier Oceanography Series, 23, Amsterdam: Elsevier Scientific Publ. Comp. **1978**, p. 546.

78 N 2 Nummedal, D., Fisher, I.A.: Process-response models for depositional shorelines: the German and the Georgia Bights. Proc. of the 16th Coastal Engen. Conf. ASCE Hamburg **1978**.

78 R 1 Reineck, H.E.: Das Watt – Ablagerungs- und Lebensraum. 2. ed., Frankfurt: Kramer **1978**, p. 185.

78 R 2 Ritchie, W., Smith, J.S., Rose, N.: The beaches of North-East-Scotland. Dept. Geogr., Univ. Aberdeen **1978**, p. 278.

78 R 3 Rosen, P.S.: Mar. Geol. **26** (1978) M7.

78 S Stoddart, D.R., Johannes, R.E. (eds.): Coral reefs. Research methods, Paris: Unesco **1978**, p. 581.

78 W 1 Wright, L.D., in: Coastal sedimentary environments (Davis, R.A., ed.), Berlin, Heidelberg, New York: Springer **1978**, p. 350.

78 W 2 Wright, P.J., in: Climatic change and variability, a southern perspective (Pittock, A.B., et al., eds.), Cambridge: Cambridge Univ. Press **1978**.

79 C Chappell, J., Eliot, I.G.: Mar. Geol. **32** (1979) 231.

79 D 1 Deutsche Forschungsgemeinschaft: Sandbewegungen im Küstenraum: Rückschau, Ergebnisse u. Ausblick. Ein Abschlußbericht, Boppard: Boldt **1979**, p. 416.

79 D 2 Deutsch, M., Wiesnet, D.R., Rango, A.: Satellite Hydrology. 5th Annual W.T. Pecora Memorial Symp. on Remote Sensing, Sioux Falls, S. Dakota, June 10–15, 1979, Minneapolis, Minnesota: Am. Water Res. Assoc. **1979**, p. 1000.

79 D 3 Dyer, K.R. (ed.): Estuarine hydrography and sedimentation. A handbook, Cambridge: Univ. Press **1979**.

79 E 1 Ellenberg, L.: Berliner Geogr. Stud. **5** (1979).

79 E 2 Elliott, T.: Geol. en Mijnbouw **58**, 4 (1979) 479.

79 E 3 El-Sabh, M.I., Dionne, G.C.: L'océanographie de l'estuaire du St. Laurent, Quebec: Presse d'Université Laval **1979**, p. 276.

79 F 1 Frakes, L.A.: Climates throughout geologic time, Amsterdam: Elsevier **1979**, p. 310.

79 F 2 Führböter, A., Bösching, F., Dette, H.H., Hausen, U.A.: Energieumwandlung in Brandungszonen. DFG: Sandbewegung im Küstenraum, Boppard: Boldt **1979**, p. 80–96.

79 G Guilcher, A.: Précis d'hydrologie, marine et continentale, Paris: Masson & Cie. **1979**, p. 344.

79 H Hayes, M.O., in: Barrier islands: from the Gulf of St. Lawrence to the Gulf of Mexico (Leatherman, S.P., ed.), London, New York: Academic Press **1979**, p. 1–27.

79 I IOC: Bruun memorial lectures, 1977. Presented at the 10th session of the IOC Assembly, Unesco, Paris, 8. Nov. 1977: The importance and application of satellite and remotely sensed data to oceanography. Intergovernmental Oceanographic Commission Techn. Ser. 19 **1979**, p. 64.

79 J 1 James, N.P., Ginsburg, R.N.: The seaward margin of Belize Barrier and atoll reefs. Spec. Publ. of the Int. Assoc. Sedimentol., 3, Oxford, London, Edinburgh: Blackwell Scientific Publ. **1979**, p. 191.

79 J 2 Jelgersma, S., in: The Quaternary history of the North Sea (Oele, E., Schüttenhelm, R.T.E., Wiggers, A.J., eds.), Acta Univ. Ups. Symp. Univ. Ups. Annum Quingentesimum Celebrantis 2, Uppsala **1979**, p. 233–248.

79 J 3 Jelgersma, S., Oele, E., Wiggers, A.J., in: The Quaternary history of the North Sea (Oele, E., Schüttenhelm, R.T.E., Wiggers, A.J., eds.), Acta Univ. Ups. Symp. Univ. Ups. Annum Quingentesimum Celebrantis 2, Uppsala **1979**, p. 115–142.

79 K Kelletat, D.: Abh. Akad. Wiss. Göttingen, Math.-Phys. Kl. **3**, 32 (1979) 105.

79 L 1 Lasserre, P.: Sanctuary ecosystems, cradles of culture, targets for economic growth. Coastal lagoons, Vol. **15**, 4 (1979) 2.

79 L 2 Leatherman, S.P. (ed.): Barrier islands from the Gulf of St. Lawrence to the Gulf of Mexico, New York: Academic Press **1979**, p. 325.

79 L 3 Leatherman, S.P., in: Environmental geologic guide to Cape Cod National Seashore, Field guide book, Univ. Amhearst, Mass. **1979**, p. 249.

79 M 1 McGreal, W.S.: Mar. Geol. **32** (1979) 89.

79 M 2 Mörner, N.A.: Geol. Fören. Förhandl. **100**, 4 (1979) 381.

79 M 3 Muir, R., in: L'oceanographie de l'estuaire du Saint-Lawrence (El-Sabh, M.L., ed.), Naturaliste Canadier **106** (1979) 27.

79 O 1 Oele, E., Schüttenhelm, R.T.E., Wiggers, A.J. (eds.): The Quaternary history of the North Sea. Symposia Universitatis Upsaliensis Annum Quingentesimum Celebrantis 2, Uppsala 1979, Societas Upsaliensis pro Geologia Quaternaria Uppsala, Stockholm: Almquist & Wikseil **1979**, p. 248.

79 O 2 Otvos, E.G., jr., Price, W.: Mar. Geol. **31**, 3–4 (1979) 251.

79 P 1 Paskoff, R., Bird, E.C.F.: Relationships between vertical changes of land and sea level and the advance and retreat of coastlines. Proc. 1978 Intern. Symp. Coastal Evolution Quaternary, São Paulo **1979**, p. 29–40.

79 P 2 Pyökäri, M.: Mixed sand and gravel shores in the South Western Finnish Archipelago, Fennicae, Ser. A III Geol-Geogr. 128, Helsinki: Suomutainen Tiedeakatemia **1979**, p. 126.

79 S 1 Sallenger, A.H., jr.: Mar. Geol. **29** (1979) 23–37.

79 S 2 Sharma, G.O.: The Alaskan shelf, Berlin, Heidelberg, New York: Springer **1979**, p. 500.

79 S 3 Short, A.D.: J. Geol. **87**, 5 (1979) 553.

79 U UN Office of the Law of the Sea Negotiations: A guide to the Law of the Sea. UN Dept. of Public Informat., Ref. Pap. 18, **1979**.

79 V Vollbrecht, K., Wünsche, B.: Dynamik der Sandbewegung vor Sylt. Suspensionstransport in der Brandungszone. Sandbewegung im Küstenraum. DFG Abschlußbericht, Boppard: Boldt **1979**, p. 351–371.

79 W Wright, L.D., Chappell, J., Thom, B., Bradshaw, M., Cowell, P.: Mar. Geol. **32** (1979) 105.

80 A ASCE: Proceedings of the 17th International Coastal Engineering Conference ASCE, Sidney, Australia **1980**.

80 B 1 Barnes, R.S.K.: Coastal lagoons: the natural history of a neglected habitat. Cambridge Stud. in Modern Biology, Cambridge: Cambridge Univ. Press **1980**, p. 106.

80 B 2 Barthel, V.: Die Küste **35** (1980) 57.

80 C Coleman, J.M., Prior, D.B.: Assoc. Petrol. Geol. **15** (1980) 171.

80 D Davies, J.L.: Geographical variation in coastal development, Edinburgh: Oliver & Boyd **1980**, p. 212.

80 F Freeland, H.J., Farmer, D.M., Levings, C.D. (eds.): Fjord oceanography, New York: Plenum Press **1980**, p. 715.

80 G 1 Gardner, L.R., Bohn, M.: Mar. Geol. **34**, 3/4 (1980) M91.
80 G 2 Gierloff-Emden, H.G., in: Geographie des Meeres, Ozeane und Küsten, Vol. 1, 2. Berlin: Walter de Gruyter **1980**, p. 847, 648.
80 G 3 Goldsmith, V., Golik, A.: Mar. Geol. **37** (1980) 147.
80 H 1 Haarnagel, W., in: Transgressies en occupatiegeschiedenis in de kustgebieden van Nederland en Belgie (Verhulst, A., ed.), Coll. Gent, Handelingen, Belgisch Centrum voor Landelijke Geschiedenis 66, **1980**, p. 209–239.
80 H 2 Hofmeister, B., Steinike, A. (eds.): Beiträge zur Geomorphologie der Küsten, Berliner Geogr. Stud., 7 Berlin **1980**, p. 350.
80 H 3 Huntley, D.A., in: Coastline of Canada: Littoral processes and shore morphology (McGann, S.B., ed.), Geol. Surv. of Canada, Pap. 80–10, Rep. 2 **1980**, p. 11–121.
80 H 4 Hsu, S.A.: Research in coastal meteorology, basic and applied. 2nd Conf. on Coastal Meteorol. Jan.–Febr. 1980, Los Angeles, Cal., Boston: Am. Meteorol. Soc. **1980**.
80 K Kaiser, K.: Berliner Geogr. Stud. **7** (1980).
80 N Norrman, J.O., in: Coasts under stress (Orme, A.R., Prior, D.B., Psuty, N.P., Walker, H.J., eds.), Z. Geomorph. Suppl. **34** (1980) 20.
80 O 1 Orme, A.R., Prior, D.B., Psuty, N.P., Walker, H.J.: Coasts under stress. Z. Geomorph. Suppl. **34** (1980) 261.
80 O 2 Olausen, E., Cato, I.: Chemistry and biochemistry of estuaries, New York: Wiley & Sons **1980**, p. 452.
80 P 1 Pickard, G.L., Stanton, B.R.: Pacific fjords – a review of their water characteristics, in: Fjord oceanography, Vol. 4 (Freeland, H.J., Farmer, D.M., Levings, C.D., eds.), New York: Plenum Press **1980**.
80 P 2 Prior, D.B., Renwick, W.H., in: Coasts under stress (Orme, A.R., Prior, D.B., Psuty, N.P., Walker, H.J., eds.), Z. Geomorph., Suppl. **34** (1980) 63.
80 R 1 Regrain, R.: Géographie physique et télédétection des marais charentais. U.E.R. Lettres, Géogr., Univ. de Picardie, Amiens, Thèse **1980**, p. 512.
80 R 2 Rust, U., Leontaris, S.N.: Valentin Festschrift, Berliner Geogr. Stud. **1980**, p. 115–135.
80 S 1 Schoefield, J.C., in: Earth rheology, isostasy and eustasy. (Mörner, N.A., ed.), Proc. Symp. Stockholm 1977, London, New York: Wiley-Interscience **1980**, p. 517–521.
80 S 2 Stäblein, G.: Berliner Geogr. Stud. **7** (1980) 217.
80 T Trenhaile, A.S.: Progr. Phys. Geogr. **4** (1980) 1.
80 U Ulrich, J.: Dt. Hydrogr. Z. **33** (1980) 236.
80 W West, R.G.: The pre-glacial Pleistocene of the Norfolk and Suffolk coasts, London, New York: Cambridge University Press **1980**, p. 203.
81 B 1 Battistini, R.: Rev. Géomorph. Dynam. **30**, 3 (1981) 81.
81 B 2 Belknap, D.F., Kraft, J.C.: Mar. Geol., **42**, Amsterdam (1981) 429–442.
81 B 3 Boon, J.D., III, Byrne, R.J.: Mar. Geol. **40** (1981) 27.
81 E Emiliani, C., in: The Sea, Vol. 7, The oceanic lithosphere (Emiliani, C., ed.), New York: Wiley-Interscience **1981**, p. 1738.
81 H 1 Harrison, C.G.A., Brass, G.W., Saltzman, E., Sloan, J., Southam, J., Whitman, J.M.: Earth Planet. Sci. Lett. **54** (1981) 135.
81 H 2 Holland, H.D., in: The Sea, Vol. 7. The oceanic lithosphere (Emiliani, C.D., ed.), New York: Wiley-Interscience **1981**, p. 763.
81 I 1 IABO: Coastal lagoon research, present and future. Report and guidelines, Paris: Unesco **1981**, p. 97.
81 I 2 IABO: Coastal lagoon research, present and future. Proceed. of Semin. Duke University, Aug. 1978, Paris: Unesco **1981**, p. 348.
81 K 1 Krämer, I., Zitschner, F.F. (eds.): Die Küste **1981**.
81 K 2 Die Küste, Ausschuß Küstenschutzwerke, Kuratorium für Forschung im Küsteningenieurwesen. Sonderheft **36** (1981) 364.
81 L Leatherman, S.P.: Overwash processes. Benchmark Papers in Geology 58, Stroudsburg, Pennsylvania: Hutchinson & Ross **1981**, p. 376.
81 N 1 Nichols, M., Allen, G.: Sedimentary processes in coastal lagoon research, present and future. Proceed. of Smin. Duke Univ. Beaufort. NC, USA, Unesco Techn. Pap. in Marine Science 33, Paris: IABO, Unesco **1981**.
81 N 2 Nittrover, C.A. (ed.): Sedimentary dynamics of continental shelves. Developments in sedimentology 32, Amsterdam: Elsevier Scientific Publ. Comp. **1981**, p. 449.
81 O Officer, C.B.: Mar. Geol. **40** (1981) 1.

81 P Prospero, J.M., in: The Sea, Vol. 7. The oceanic lithosphere (Emiliani, C.D., ed.), New York: Wiley-Interscience **1981**, p. 801–874.
81 R 1 Ruddle, K., Manshard, W.: Renewable natural resources and the environment. United Nation University, Dublin: Tycooly Int. Publ. **1981**, p. 326.
81 R 2 Rust, U., Schmidt, H.H.: Mitt. Geogr. Ges. München **66** (1981) 141.
81 S 1 Semeniuk, V.: Mar. Geol. **43** (1981) 21.
81 S 2 Southam, J.R., Hay, W.W., in: The Sea, Vol. 7 (Emiliani, C., ed.), New York: Wiley-Interscience **1981**, p. 1617–1684.
81 T TenCate, J.A.M.: Report of the working group from Sept. 13, 1981 till May 31, 1982. Working group on the geomorphology of river and coastal plains. IGU, Soil Survey Institute, Wageningen, The Netherlands **1981**.
81 W 1 Whitman, J.M.: Tectonic and bathymetric evolution of the Pacific Ocean basin since 74. Ma.M.Sc. thesis. Univ. Miami, Miami, Fla. **1981**, p. 176.
81 W 2 Wright, L.D.: Beach cut in relation to surfzone morphodynamics. Proc. 17. Int. Conf. Coastal Eng., Am. Soc. of Civil Eng. New York **1981**, p. 978.
82 A Aronow, S., in: The encyclopedia of beaches and coastal environments (Schwartz, M.L., ed.), in: Encyclopedia of earth sciences series, Vol. 15 (Fairbridge, R.W., ed.), Stroudsburg, Pennsylvania: Hutchinson & Ross **1982**, p. 754.
82 B Brenninkmayer, B.M., in: The encyclopedia of beaches and coastal environment (Schwartz, M., ed.), Stroudsburg, Pennsylvania: Hutchinson & Ross **1982**, 693–695.
82 C Cracknell, A.P., MacFarlane, N., McMillan, K.: Int. J. Remote Sensing **3** (1982) 113.
82 D Deutsches Hydrographisches Institut (DHI): Gezeitentafeln für das Jahr 1983, Hamburg: DHI, 2115 **1982**, p. 244.
82 G 1 Gierloff-Emden, H.G.: Interest of remote sensing in coastal lagoon research. Oceanologia Acta, Spec. Issue, Paris **1982**, p. 139–150.
82 G 2 Goldsmith, V., in: Encyclopedia of beaches and coastal environments (Schwartz, M., ed.), Stroudsburg, Pennsylvania: Hutchinson & Ross **1982**, p. 235–243.
82 G 3 Greenwood, B., in: Encyclopedia of beaches and coastal environments (Schwartz M., ed.), Stroudsburg, Pennsylvania: Hutchinson & Ross **1982**, p. 133–139.
82 G 4 Guilcher, A., Andrade, B., Dantec, M.H.: Norois **114**, Poitiers (1982) 205.
82 H 1 Hopley, D.: The geomorphology of the great barrier reef. Quaternary development of coral reefs, London, New York: Wiley-Interscience **1982**.
82 H 2 Howarth, M.J., in: Offshore tidal sands (Stride, A.H., ed.) London: Chapman & Hall **1982**.
82 K 1 Kennett, J.P.: Marine geology, Englewood Cliffs, N.J.: Prentice-Hall **1982**, p. 752.
82 K 2 Klemsdal, T.: Geogr. Tidskrs. **36**, 3 (1982).
82 L 1 Lasserre, P., Postma, H. (eds.): Coastal lagoons. IABO-SCOR-Unesco Sympos. Bordeaux 1981. Proc. of the Sympos. on Coastal Lagoons. Oceanologica Acta. Revue Européenne d'Océanologie. Special issue, Paris: Gauthier-Villers **1982**, p. 460.
82 L 2 Leeder, M.R.: Sedimentology, London: Allen & Unwin **1982**.
82 L 3 Lynch, D.K.: Spektrum der Wissenschaft **12** (1982) 100.
82 M 1 Mandelbrot, B.B.: The fractal geometry of nature, San Francisco: W. H. Freeman **1982**.
82 M 2 Mörner, N.-A. (ed.): Bull. of the INQUA Neotectonic Commission **5** (1982) 75.
82 M 3 Mosetti, F., in: The encyclopedia of beaches and coastal environments (Schwartz, M.L., ed.), in: Encyclopedia of earth sciences series, Vol. 15 (Fairbridge, R.W., ed.), Stroudsburg, Pennsylvania: Hutchinson & Ross **1982**, p. 276.
82 N NOAA: Nautical chart-catalog, Rockville Riverdale, Maryland, USA **1982.**
82 P Phleger, F.B., Ewing, E.: Soc. Am. Bull. **73** (1982) 145.
82 R 1 Richards, K.S.: Rivers, London: Methuen **1982**.
82 R 2 Rudloff, W.: World climatology **1982**.
82 S 1 Schwartz, M.L.: The encyclopedia of beaches and coastal environments, in: Fairbridge, R.W.: Encyclopedia of earth sciences series, Vol. 15, Stroudsburg, Pennsylvania: Hutchinson & Ross **1982**, p. 940.
82 S 2 Seibold, E., Berger, W.H.: The sea floor. An introduction to marine geology, Berlin, Heidelberg, New York: Springer **1982**, p. 288.
82 S 3 Snead, R.E.: Coastal landforms and surface features, a photographic atlas and glossary, New York: Van Nostrand Reinhold **1982**, p. 272.
82 S 4 Stride, A.H. (ed.): Offshore tidal sands, London: Chapman & Hall **1982.**

82 V Van der Haar, S., Javor, B., in: The encyclopedia of beaches and coastal environments (Schwartz, M.L., ed.), in: Encyclopedia of earth sciences series, Vol. 15 (Fairbridge, R.W., ed.), Stroudsburg, Pennsylvania: Hutchinson & Ross **1982**.

82 W 1 Walker, H.J., in: The encyclopedia of beaches and coastal environments (Schwartz, M., ed.), Stroudsburg, Pennsylvania: Hutchinson & Ross **1982**, 57–61.

82 W 2 Wieczorek, U.: Münchener Geogr. Abh. **27** (1982).

82 W 3 Wright, L.D., in: The encyclopedia of beaches and coastal environments (Schwartz, M., ed.), Stroudsburg, Pennsylvania: Hutchinson & Ross **1982**, p. 358–366.

82 W 4 Wright, L.D., Nielson, P.N., Short, A.D., Green, M.O.: Mar. Geol. **50**, 97 (1982).

82 W 5 Wright, L.D., Guza, R.T., Short, A.D.: Mar. Geol. **45** (1982) 41.

82 Z Zanke, U.: Grundlagen der Sedimentbewegung, Berlin, Heidelberg, New York: Springer **1982**, p. 350.

83 B 1 Barnes, D.J. (ed.): Perspectives on coral reefs. Australian Inst. of Mar. Sci., Manuka A.C.T., Australia: Brian Clouston Publ. **1983**, p. 277.

83 B 2 Bertram-Lyko, A.: Die Küste **39** (1983) 148.

83 C 1 Canadian National Committee: Fjord oceanography. NATO Conference Series IV: Marine Sciences, Victoria B.C., Canada 1979, **1983**.

83 C 2 CNC/SCOR: Proceedings of the Joint Oceanographic Assembly 1982. General Symposia, Ottawa: Canadian National Committee/Scientific Committee on Oceanic Research **1983**, p. 189.

83 C 3 Couper, A.D.: The Times atlas of the oceans. London: Times Books Limited **1983**, p. 272.

83 C 4 Czaya, E.: Rivers of the world **1983**, p. 248, Figs. 250.

83 D 1 Dean, R.R., Maurmeyer, E.M., in: CRC Handbook of coastal processes and erosion (Komar, P.D., ed.), Boca Raton, Florida: CRC Press Inc. **1983**, p. 151.

83 D 2 Dolan, R., Hayden, B., May, S., in: CRC Handbook of coastal processes and erosion (Komar, P.D., ed.), Boca Raton, Florida: CRC Press Inc. **1983**, p. 285.

83 F 1 Fairbridge, R.W.: Quaternary Science Rev. **1** (1983) 215.

83 F 2 Farmer, D.M., Freeland, H.J.: Progress in oceanography (Angel, M.V., O'Brien, J.J., eds.), **12**, 2, special issue (1983) 74.

83 H Hay, W.W.: Significance of runoff to paleoceanographic conditions during the Mesozoic and Clues to locate sites of ancient river inputs, in: Proc. of the Joint Oceanogr. Assembly 1982-General Symposia, Ottawa: Canadian National Committee/Scientific Committee on Oceanic Research (CNC/SCOR), **1983**, 9–17.

83 I 1 Iida, K., Iwasaki, T. (eds.): Tsunamis: their science and engineering. Proc. of the Intern. Tsunami Symp. 1981, IUGG Tsunami Comm., Sendai-Ofunato-Kamaishi, Japan. Advances in earth and planetary sciences **1983**, p. 563.

83 I 2 IOC: Operational sea level stations. IOC Techn. Ser. 23, Paris: Unesco **1983**, p. 40.

83 J Johns, B. (ed.): Physical oceanography of coastal and shelf seas. Elsevier oceanography series, 35, Amsterdam: Elsevier Scientific Publ. Comp. **1983**, p. 470, 484.

83 K 1 Kelletat, D.: Internationale Bibliographie zur regionalen und allgemeinen Küstenmorphologie (ab 1960). Essener Geogr. Arb., 7, Paderborn: Schöningh **1983**.

83 K 2 Ketchum, B.H. (ed.): Estuaries and enclosed seas. Ecosystems of the world, 26, Amsterdam: Elsevier Scientific Publ. Comp. **1983**, p. 500.

83 K 3 Komar, P.D. (ed.): CRC Handbook of coastal processes and erosion, Boca Raton, Florida: CRC Press Inc. **1983**, p. 305.

83 M 1 Masters, P.M., Flemming, N.C. (eds.): Quarternary coastlines and marine archaeology, London, New York: Academic Press, **1983**, p. 641.

83 M 2 Matthes, G., Ubell, K.: Allgemeine Hydrogeologie, Grundwasserhaushalt, in: Lehrbuch der Hydrogeologie, Vol. I (Matthes, G., ed.), Berlin: Borntraeger **1983**, p. 438.

83 M 3 Melchior, P.: The tides of the planet earth. 2. ed. **1983**, p. 648.

83 M 4 Mörner, N.A.: Sea level indicators and the role of sea level dynamics. Council of Europe, Strasbourg June 83, AS Science Ocean 35, **1983**.

83 N Nalivkin, D.V.: Hurricanes, stormes and tornadoes. Geographic characteristic and geologic activity. Russian translation series, Rotterdam: A.A. Balkema **1983**.

83 P 1 Park, D.W., in: Geomorphology of river and coastal plains, Proc. 3rd Meeting of the Working Group, Dec. 1983, Bangkok, Thailand (TenCate, J.A.M., ed.), UNO **1983**, p. 98.

83 P 2 Pugh, D.T., Faull, H.E.: Operational sea level stations. IOC Techn. Ser. Paris: Unesco **1983**, p. 30.

83 R 1 Rabassa, J.: Quaternary of South America and Antarctic Peninsula, Vol. I, Rotterdam: A.A. Balkema **1983**.

83 S 1 Stapor, F.W., jr.: Mar. Geol. **51**, 3, 4 (1983) 217.

83 S 2 Sunamura, T., in: CRC Handbook of coastal processes and erosion (Komar, P.D., ed.), Boca Raton, Florida: CRC Press Inc. **1983**, p. 233–265.
83 W Wright. L.D., Short, A.D., in: CRC Handbook of coastal processes and erosion (Komar, P.D., ed.), Boca Raton, Florida: CRC Press Inc. **1983**, p. 35–64.
84 B Barth, M.C., Titus, J.G. (eds.): Greenhouse effect and sea level rise, A Challenge for this generation, New York: Van Nostrand Reinhold **1984**, p. 325.
84 C Clarke, D.J., Eliot, I.G., Frew, J.R.: Mar. Geol. **58** 3/4 (1984) 319.
84 G Greenwood, B., Davis, R.A. (ed.): Hydrodynamics and sedimentation in wave-dominated coastal environments. Developments in sedimentology, 39. Repr. from: Mar. Geol. Vol. 60, Amsterdam: Elsevier **1984**, p. 474.
84 H 1 Herbich, J.B., Schiller, R.E., Watanabe, R.K., Dunlap, W.A.: Design guidelines for ocean-founded structures, New York: Marcel Dekker Inc. **1984**, p. 320.
84 H 2 Huh, O.K., Rouse, L.J., Walker, N.D.: J. Geophys. Res. **84** (1984) 717.
84 K 1 Kelletat, D.: Deltaforschung: Verbreitung, Morphologie, Entstehung und Ökologie von Deltas. Erträge der Forschung, Vol. 214, Darmstadt: Wissenschaftliche Buchgesellschaft **1984**, p. 158.
84 K 2 Kuhn, G.G., Shepard, F.P.: Sea cliffs, beaches and coastal valleys of San Diego County, Berkeley: Cal: University of California Press **1984**, p. 193.
84 L 1 Langenaar, W.: Surveying and charting of the seas. Elsevier oceanography series, 37, Amsterdam: Elsevier Scientific Publ. Comp. **1984**, p. 630.
84 L 2 Loon, H. van: Climates of the oceans. World survey of climatology, Vol. 15, Amsterdam: Elsevier Scientific Publ. Comp. **1984**, p. 716.
84 L 3 Luck, G. (ed.): Die Küste **41** Heide (1984).
84 M Mahaney, W.C. (ed.): Quaternary dating methods. Developments in paleontology and stratigraphy, 7, Amsterdam: Elsevier **1984**, p. 448.
84 P Pethick, J.: An introduction to coastal geomorphology, London: Arnold **1984**, p. 260.
84 R Reineck, H.E.: Aktuogeologie klastischer Sedimente. Senckenberg Buch 61, Frankfurt: W. Kramer **1984**, p. 348.
84 S 1 Shackleton, J.C., van Andel, T.H., Runnels, C.N.: J. Field Archaeol. **11** (1984) 307.
84 S 2 Sindern, J.: Die Küste **40** Heide (1984) 13.
84 S 3 Sleath, J.F.A.: Sea bed mechanics, New York: John Wiley & Sons **1984**, p. 335.
84 W 1 Westermeyer, W.E., Shusterich, K.M. (eds.): United States arctic interests, The 1980s and 1990s, New York: Springer **1984**, p. 369.
84 W 2 Wieland, P., Fladung, B., Bergheim, V.: Die Küste **40** Heide (1984) 139.
84 W 3 Worsley, T.R., Nance, D., Moody, J.B.: Mar. Geol. **58** 3/4 (1984) 373.
84 W 4 Wright, L.D., Short, A.D.: Mar. Geol. **56** 1/4 (1984) 93.
85 B Bird, E.C.F., Schwartz, M. (eds.): The world coastline, New York: Van Nostrand Reinhold **1985**, p. 1071.
85 D 1 Davis, R.A., jr.: Coastal sedimentary environments, Berlin, Heidelberg, New York: Springer **1985**, p. 713.
85 D 2 Delesalle, B., Gabrie, C., Galzin, R., Harmelin-Vivien, M., Toffart, J.L., Salvat, B. (eds.): Proc. 5th Intern. Coral Reef Congress Tahiti, French Polynesia, 27 May–1 June 1985, 6 Vols., Papetoai, Moorea: Antenne Museum EPHE **1985**, p. 3486.
85 F 1 Fleming, N.C.: Oceanus **28** (1985) 18.
85 F 2 Führböter, A., Jensen, J.: Die Küste **42** (1985) 78.
85 G 1 Gellert, J.F., Heyer, E., Neumeyer, G.: Acta Hydrophysica, Berlin **29**, 2/3 (1985) 93.
85 G 2 Gierloff-Emden, H.G., in: The world coastline (Bird, E.C.F., Schwartz, M., eds.), New York: Van Nostrand Reinhold **1985**, 335.
85 K 1 Kelletat, D.: Geoökodynamik **6** (1985) 1.
85 K 2 Kelletat, D., in: Geographie der Küsten und Meere, Symp. Mainz, Okt. 1984 (Hofmeister, B., Voss, F., eds.), Berliner Geogr. Stud. 16, Berlin: **1985**, p. 169.
85 N Nichols, M., Biggs, R., in: Coastal sedimentary environments (Davis, R.A., jr., ed.), Berlin, Heidelberg, New York: Springer **1985**, p. 77–186.
85 R Robinson, I.S.: Satellite oceanography, Ellis Horwood series in marine science, Chichester: Ellis Horwood Ltd **1985**, p. 455.
85 T Thurman, H.: Introductory Oceanography, 4th ed., Columbus, Ohio: Bell & Howell Comp. **1985**, 503 p.
85 W Wright, L.D., in: Coastal sedimentary environments (Davis, R.A., jr., ed.), Berlin, Heidelberg, New York: Springer **1985**, p. 1–76.